Methods of Experimental Physics

VOLUME 10
PHYSICAL PRINCIPLES OF FAR-INFRARED RADIATION

METHODS OF EXPERIMENTAL PHYSICS:

L. Marton, *Editor-in-Chief*

Claire Marton, *Assistant Editor*

1. Classical Methods
   Edited by Immanuel Estermann

2. Electronic Methods
   Edited by E. Bleuler and R. O. Haxby

3. Molecular Physics
   Edited by Dudley Williams

4. Atomic and Electron Physics—Part A: Atomic Sources and Detectors, Part B: Free Atoms
   Edited by Vernon W. Hughes and Howard L. Schultz

5. Nuclear Physics (in two parts)
   Edited by Luke C. L. Yuan and Chien-Shiung Wu

6. Solid State Physics (in two parts)
   Edited by K. Lark-Horovitz and Vivian A. Johnson

7. Atomic and Electron Physics—Atomic Interactions (in two parts)
   Edited by Benjamin Bederson and Wade L. Fite

8. Problems and Solutions for Students
   Edited by L. Marton and W. F. Hornyak

9. Plasma Physics (in two parts)
   Edited by Hans R. Griem and Ralph H. Lovberg

10. Physical Principles of Far-Infrared Radiation
    L. C. Robinson

Volume 10

Physical Principles of Far-Infrared Radiation

L. C. ROBINSON

School of Physics
University of Sydney
Sydney, Australia

1973

ACADEMIC PRESS • New York and London

COPYRIGHT © 1973, BY ACADEMIC PRESS, INC.
ALL RIGHTS RESERVED.
NO PART OF THIS PUBLICATION MAY BE REPRODUCED OR
TRANSMITTED IN ANY FORM OR BY ANY MEANS, ELECTRONIC
OR MECHANICAL, INCLUDING PHOTOCOPY, RECORDING, OR ANY
INFORMATION STORAGE AND RETRIEVAL SYSTEM, WITHOUT
PERMISSION IN WRITING FROM THE PUBLISHER.

ACADEMIC PRESS, INC.
111 Fifth Avenue, New York, New York 10003

United Kingdom Edition published by
ACADEMIC PRESS, INC. (LONDON) LTD.
24/28 Oval Road, London NW1

LIBRARY OF CONGRESS CATALOG CARD NUMBER: 68-56970

PRINTED IN THE UNITED STATES OF AMERICA

CONTENTS

FOREWORD ix

PREFACE xi

1. Introduction to the Far-Infrared

1.1. The Spectrum. Basic Equations 1
1.2. Historical Background 4

2. Far-Infrared Wave Generation

Introduction 10
2.1. Incoherent Sources 13
2.2. Harmonic Generators 18
2.3. Electron Tubes 30
2.4. Relativistic Electrons 42
2.5. Gas Lasers 45
2.6. Solid State Lasers and Junctions 64
2.7. The Electron Cyclotron Maser 77

3. Wave Transmission and Transmission Systems

Introduction 82
3.1. Preliminary Optical and Microwave Concepts 82
3.2. Waves in Material Media 87
3.3. Quasioptical Components 112
3.4. Waveguides and Quasimicrowave Components 113
3.5. Grating Monochromators 130
3.6. Multiplex Spectrometry 140
3.7. Extensions of the Techniques of Microwave Spectroscopy . 150
3.8. The Fabry–Perot Spectrometer 156

4. Detection of Far-Infrared Radiation

	Introduction	157
4.1.	Point-Contact Crystal Diodes	158
4.2.	Thermal Detectors	166
4.3.	Photoconductive Detectors	174
4.4.	An Electron Cyclotron Resonance Method of Radiation Detection and Spectral Analysis	188
4.5.	The Josephson Junction	201
4.6.	Limits of Detection Set by Random Fluctuations	210

5. Cyclotron Resonance with Free Electrons and Carriers in Solids

	Introduction	229
5.1.	Absorption by Nonrelativistic Free Charges	230
5.2.	Cyclotron Resonance and Magnetoplasma Effects in Semiconductors and Metals	238
5.3.	Faraday Rotation and Related Magnetooptic Effects in Semiconductors	257
5.4.	Positive and Negative Absorption by Weakly Relativistic Electrons	259

6. Wave Interactions in Plasmas

	Introduction	265
6.1.	Plasma Conductivity and Tensor Dielectric Constant	266
6.2.	Wave Propagation and Dispersion Relations	272
6.3.	Propagation across a Magnetic Field	273
6.4.	Propagation along a Magnetic Field	276
6.5.	Comments on Warm Plasma Effects	278
6.6.	Phase Shift and Attenuation Measurements	280
6.7.	Changes of Polarization	286
6.8.	Bremsstrahlung and Blackbody Radiation from Plasmas	287
6.9.	Cyclotron and Synchrotron Radiation	295
6.10.	Longitudinal Plasma Waves	300

7. Spectra of Gases, Liquids, and Solids in the Far-Infrared

	Introduction	302
7.1.	Rotational Spectra of Molecules	306
7.2.	Vibration–Rotation Spectra	321
7.3.	Molecules with Electronic Angular Momentum	328
7.4.	Intensities and Shapes of Spectral Lines	339
7.5.	Dispersion in Liquids and Solids	347
7.6.	Crystal Lattice Vibrations in Solids	358
7.7.	Ferroelectric Crystals	372
7.8.	Magnetic Resonances	385
7.9.	Raman Effect	420
7.10.	The Energy Gap in Superconductors	431

AUTHOR INDEX 442

SUBJECT INDEX 452

FOREWORD

All the previous volumes of "Methods of Experimental Physics" were multiauthor works. This policy is now somewhat changed as we avail ourselves of outstanding authors who can cover a broad field singlehanded. We were indeed fortunate to find Professor Robinson willing to write a monograph on far-infrared radiation.

In Volume 9 I announced our intention to expand the series with volumes dealing with optics and with the use and handling of data. To this list we can add two others, a multiauthor volume dealing with astrophysics, and a single-author volume, by Professor Jesse W. Beams, on high rotational speed methods.

I would like to express my pleasure in welcoming Professor Robinson to this Treatise as a representative of Australian physics.

L. Marton

PREFACE

This book is an account of the physics of radiation and its interaction with matter in a particular spectral region. It is concerned with physical principles as they relate to far-infrared electromagnetic waves. In studying physics from such a point of view one naturally cuts across many of the usual boundaries of specialist research, but, in a field that has been beset by a sea of experimental difficulties, it is particularly desirable to be familiar with many, or most, aspects of the field. Thus we treat the processes of far-infrared wave generation, transmission and detection —always with an emphasis on basic principles—and the interaction of these waves with matter in its various forms: free electrons, semiconductors, plasmas, magnetic and nonmagnetic solids, etc.

It is the author's hope that the graduate student, the specialist research worker, and the scholar interested in physics seen in another rather special context will find interest in the following pages.

The value of my wife's encouragement and help during the labor is immeasurable. I am also particularly indebted to Mrs. Elaine Rodgers for all the help she has given, to Dr. L. U. Hibbard and Dr. W. I. B. Smith for many background discussions of physical principles, and to Dr. J. Campbell, Mr. E. Wood, and Mr. L. B. Whitbourn for their critical comments.

L. C. ROBINSON

1. INTRODUCTION TO THE FAR-INFRARED

1.1. The Spectrum. Basic Equations

Far-infrared waves lie between light and microwaves. They are intimately connected with them, but there are differences. These differences have their origins not in fundamental laws but in circumstances that have, historically, been so consistently adverse to this part of the spectrum that one must wonder how this comes about. It will appear in the following chapters that the reasons are many and varied; they may be associated with the proximity of $h\nu$ to kT, with purely technical limits set by the short wavelength, with the restoring forces and masses determining vibrational absorption in solids, and so forth. But the very existence of physical effects that set the limits of applicability of experimental methods bears witness to the wealth of phenomena accessible to this part of the spectrum. This then sets our scene. We shall look at the physical processes, their limitations to experimentation, and their relation to the body of understanding about electromagnetic waves and their interaction with matter.

Far-infrared wavelengths range from, say, ~ 20 to ~ 2000 μm, but we shall not hesitate to take our discussion below 10 μm and above 1 cm on occasions. The position of these waves in the spectrum is shown in Fig. 1.1.

It is of value to think in terms of wavelengths measured in micrometers (μm), and frequencies measured in wavenumbers (cm^{-1}) and in GHz (10^9 Hz). Each of these is favored by one specialist group or another. It is useful, too, to remember the following relations:

$$100 \; \mu\text{m} \rightarrow 100 \; \text{cm}^{-1} \rightarrow 3000 \; \text{GHz},$$

$$10 \; \mu\text{m} \rightarrow 1000 \; \text{cm}^{-1} \rightarrow 3 \times 10^{13} \; \text{Hz},$$

$$1 \; \text{eV} \; (=1.60 \times 10^{-19} \; \text{joule}) \rightarrow \lambda = 1.24 \; \mu\text{m},$$

$$1 \; \text{meV} \rightarrow \lambda = 1.24 \; \text{mm} \rightarrow 1/\lambda = 8.065 \; \text{cm}^{-1},$$

where the arrows mean "correspond to." When photon energy $h\nu$ is

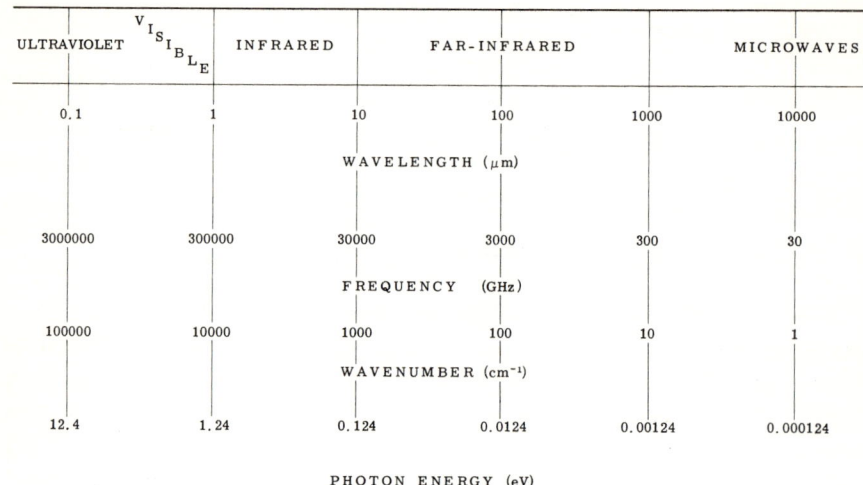

FIG. 1.1. The position of far-infrared waves in the spectrum.

equated to kT, we obtain the correspondence

$$1 \text{ cm}^{-1} \rightarrow 1.436°\text{K}.$$

Planck's constant h and Boltzmann's constant k have the values

$$h = 6.63 \times 10^{-34} \text{ joule sec},$$
$$k = 1.38 \times 10^{-23} \text{ joule}/°\text{K}.$$

The fundamental laws of electrodynamics underly the physical processes of interest in this book. Maxwell's equations have the form

$$\nabla \cdot \mathbf{B} = 0, \tag{1.1}$$

$$\nabla \cdot \mathbf{D} = \varrho, \tag{1.2}$$

$$\nabla \times \mathbf{E} = -\partial \mathbf{B}/\partial t, \tag{1.3}$$

$$\nabla \times \mathbf{H} = \mathbf{j} + \partial \mathbf{D}/\partial t, \tag{1.4}$$

and these are supported by Newton's laws of dynamical motion.

The quantum descriptions of the physical systems are based on Schrödinger's equation

$$H\Psi = W\Psi. \tag{1.5}$$

The classical Hamiltonian is the sum of the kinetic energy T and the

1.1. THE SPECTRUM. BASIC EQUATIONS

potential energy V. That is,

$$H = T + V = W. \tag{1.6}$$

V is a function of position and $T = p^2/2m$ is a function of the momentum p, and in the Schrödinger equation, H becomes a differential operator when we replace p_x, p_y, and p_z by

$$\frac{\hbar}{i}\frac{\partial}{\partial x}, \quad \frac{\hbar}{i}\frac{\partial}{\partial y}, \quad \text{and} \quad \frac{\hbar}{i}\frac{\partial}{\partial z},$$

respectively. Replacing W by $(\hbar/i)(\partial/\partial t)$ Schrödinger's equation becomes

$$-\frac{\hbar^2}{2m}\nabla^2\Psi + V\Psi = -\frac{\hbar}{i}\frac{\partial \Psi}{\partial t}, \tag{1.7}$$

and this, provided the form of V is prescribed, can be solved to give the wavefunction Ψ describing the physical system. Stationary states separated by discrete energy intervals emerge from the solutions of the equation, and transitions between these states are associated with quanta of energy.

In Schrödinger's picture of quantum mechanics, the operators are constant and the wave function is time dependent in accordance with

$$\Psi = e^{-iHt/\hbar}\psi, \tag{1.8}$$

where ψ is a function of position only. The alternative representation of quantum mechanics, the Heisenberg representation, makes the wave functions constant and the operators vary with time.[1] If time-dependent operators $O(t)$, and time-independent operators O, are related by

$$O(t) = e^{iHt/\hbar} O e^{-iHt/\hbar}, \tag{1.9}$$

operations by $O(t)$ on ψ and by O on Ψ give the same expectation values

$$\Psi^* O \Psi = \psi^* O(t) \psi \tag{1.10}$$

and the same physical consequences. In Heisenberg's picture the operators $O(t)$ vary according to

$$i\hbar \dot{O}(t) = [O(t), H], \tag{1.11}$$

---

[1] E. M. Henley and W. Thirring, "Elementary Quantum Field Theory." McGraw-Hill, New York, 1962.

as we can see from the differentiation of Eq. (1.9). The commutator bracket

$$[O(t), H] = O(t)H - HO(t) \qquad (1.12)$$

is the operator form of the Poisson bracket in classical mechanics.[2]

In the chapters that follow we shall make full use of results derived from these equations. In particular, the results of the electron theory of matter built up largely by Lorentz just before the turn of this century will be frequently required, as will be results calculated from Schrödinger's equation for the linear harmonic oscillator, and other simple systems. On occasion, it will be necessary to use such results without deriving them, but we will give references to appropriate sources of detailed information. This will be so particularly in early chapters where, in the context of the topic of central interest, a digression would be distracting.

1.2. Historical Background

Before we proceed to the detailed physical processes it is of interest to survey the historical development of the study of the far-infrared and its relation to the evolution of physics over the past century or so.

The observational science of light developed in early centuries because biological evolution had endowed man with a sensitive detector—the human eye—with which to observe the incoherent optical radiation from hot objects. Optics also derived great benefit from the abundance of silica on the earth's surface, for glass provided the essential building blocks of optical instruments. Thus advances in geometric optics with sophisticated telescopes, dispersive prisms, and so forth, came very early, and the beautiful theories of optical wave phenomena and interference due to such people as Young and Fresnel came in the very early 1800s. No prohibitive experimental restrictions inhibited the steady growth of this science from then on; on the contrary, it developed steadily until, at various times, it took great leaps forward when other developments in physics brought new techniques to the laboratory. Photoelectric detection and laser generation provided two such notable stimuli to optics.

The infrared developed too. In 1800 Sir William Herschel separated from the red end of the optical spectrum the invisible rays that showed a marked heating effect. Herschel's experimental methods were, of course,

---

[2] L. I. Schiff, "Quantum Mechanics." McGraw-Hill, New York, 1955.

extensions of the optical methods of the period, but he moved beyond the range of the human eye and into the region where glass tended to be opaque. The search for transparent solid materials was begun, but to this day a material for the long wavelength region with properties comparable with those of glass in the visible spectrum has not emerged. Under the stimulus of a controversy caused by Herschel's conclusion that the essence of infrared radiation was different from that of light,[3] transmission studies by Melloni led to the useful natural crystal rock salt (NaCl). Systematic studies of the transmission properties of solids, and the use of natural crystals for the separation of far-infrared waves by crystal reflection and transmission, advanced considerably in the 19th century. Solid state spectroscopy in the far-infrared thus acquired an early foundation.

Herschel's move beyond the range of direct visual detection was also of great consequence. In replacing the eye with a very insensitive and very slow mercury-in-glass thermometer detector, the need for improved detection methods was highlighted. Seebeck's discovery in 1821 of the thermoelectric effect enabled Nobili and Melloni to introduce the thermopile as a receiver of infrared in 1831. Langley introduced the bolometer in 1881. In 1947 the thermometer reemerged, but this time it was the gas-filled device now known as the Golay cell and used extensively in modern laboratories. Despite its recent origin it is very insensitive relative to the detectors of the neighboring optical and microwave regions. Improved materials from the era of solid state physics (since 1950), and the use of low temperatures and superconductivity have, however, resulted in improvements. Bolometers made with carbon, superconducting tin and germanium radiation sensing elements, and the extension of superheterodyne crystal detection, have yielded some improvement in the situation.

Further advances in detection have come in the past decade, notably from the free-carrier photoconductive effect and from the Josephson effect. The physical principles of these detectors are discussed in Chapter 4. Their merits are that they combine sensitivity comparable with, or (in the case of the Josephson effect) superior to, bolometers, and they are much faster. The improved speed of response is of great value in certain aspects of modern research, for example, in the study of short laser pulses.

From the time of Herschel until recently, infrared and far-infrared

[3] E. S. Barr, *Amer. J. Phys.* **28**, 42–54 (1954).

radiation was derived from the incoherent emission of hot bodies. The power levels were very low indeed and, with the relatively insensitive detection methods available, far-infrared physics made only slow progress. Techniques from optics, in particular, grating spectrometers and interference spectrometers, were the main experimental methods of measurement. Grating dispersion was, of course, natural to a region where the lack of transparent materials severely restricted the use of prisms. Interference spectrometry[4]—which Michelson used in the 1890s at optical wavelengths to determine the structure of the red Balmer line from his "interferometric visibility" curves—was employed by Rubens and his collaborators in the far-infrared from about 1910.

From the observation of the output of an interferometer as the path difference between the interfering beams is varied, one can derive the spectral composition of the radiation. Given that the detectors of radiation in this spectral region are severely limited by noise, interference spectroscopy has something important in its principle. Firstly, as Jacquinot pointed out in 1948, its light "throughput" or "*étendue*" is high, so that efficient use is made of the available radiation. Secondly, because it enables continuous observation of the entire radiation spectrum, rather than "piece by piece" observation as in the grating spectrometer, it has a valuable advantage with respect to signal to noise ratio. This so-called "multiplex advantage" was clearly enunciated by Fellgett in 1951.[5]

Experimentation with incoherent radiation progressed steadily from the time of Hertz. Lebedew in 1895, and later his student Glagolewa-Arkadiewa, improved the spark gap techniques of Hertz and extended their operation towards shorter wavelengths. By 1923 Nichols and Tear took them to wavelengths as short as 420 μm.[5a] Waves generated by this means were suitable for propagation and polarization experiments, etc., but it stands to the record of the entire spark gap era that they revealed not a single discrete spectral line.

Early in this century Coblentz, Rubens, Paschen, Randall, and others built up spectroscopy laboratories by working from the infrared towards longer wavelengths.[5b] In 1913 von Bahr observed fine structure on the infrared vibrational bands of HCl at about 3.5 μm.[6] This was at once

---

[4] E. V. Loewenstein, *Appl. Opt.* **5**, 845–853 (1966).

[5] See, for example, P. Fellgett, *J. Phys. (Paris)* **28**, 165 (1967).

[5a] E. F. Nicols and J. D. Tear, *Phys. Rev.* **21**, 587–610 (1923).

[5b] See, for example, J. Lecomte, "Le Rayonnement Infrarouge," Vol. 1. Gauthier-Villars, Paris, 1948.

[6] H. M. Randall, *J. Opt. Soc. Amer.* **44**, 97–103 (1954).

recognized as due to molecular rotation, and it proved to be a major advance, for it showed that the notions of quantized energy states applied to rotational motion. By 1908 Coblentz had pioneered spectroscopy in the infrared with thermal sources to wavelengths as long as 15 μm and had applied it as an analytic tool to a wide range of organic compounds.[5b] His work paved the way for the development of structural molecular spectroscopy. The wavelength range was extended beyond 100 μm by 1927, when Czerny succeeded in observing the pure rotational spectrum of HCl from 30 to 120 μm. In 1911 Randall returned from Paschen's spectroscopy laboratory in Germany and set the course of far-infrared spectroscopy in the U.S.A.[6] With researchers of the magnitude of Randall, Strong, and Dennison, far-infrared molecular spectroscopy made great progress in that country.

The advent of far-infrared experimental methods based on coherent radiation generation and detection may be dated from 1945 when the war-time techniques of microwave radar were extended in various directions in the physics laboratory. Cleeton and Williams had carried out microwave spectroscopy measurements in the early 1930s using experimental arrangements that were partly optical in concept. However, they used a split-anode magnetron and therewith generated coherent radiation. In 1939 the needs of centimeter radar required new concepts in wave generation. In response to this need, the klystron was invented by the Varian brothers in the U.S.A., while in England Boot and Randall built the multicavity magnetron, and Kompfner invented the traveling wave tube. Postwar extensions of these generators towards submillimeter wavelengths was essentially a task of sophisticated and difficult technology. It progressed slowly in the 1950s and 1960s and, ultimately, generators based on the klystron and traveling wave-tube principles reached wavelengths below a half millimeter. During this time it was apparent that new principles of wave generation were needed, and to this end some elaborate and expensive research programs were undertaken, while many scientists pondered new possibilities. Ultimately the maser came along.

A few groups, notably at universities in the U.S.A., carried the task of extending and applying microwave spectroscopic techniques to short millimeter and submillimeter wavelengths. Gordy's group at Duke University[7] based their work largely on harmonic generation of klystron radiation in nonlinear crystal diodes, and they extended the methods of

---

[7] W. Gordy, "Millimeter Waves," Vol. 9, pp. 1–23. Brooklyn Polytech. Press. Brooklyn, New York, 1960.

video detection with such diodes to the submillimeter band. The application of this work was largely to the spectroscopy of gases. Gaseous spectroscopy with microwave techniques was also underway at Columbia University after 1945, and here the magnetron harmonic generator found application.

It was in the Columbia microwave spectroscopy laboratory that Townes invented the maser, and it was the need for a source of submillimeter waves that provided the stimulus for this advance in physics. That such a development originated from a research program where the properties of molecules were studied is not surprising. Einstein had clarified the notions of spontaneous and stimulated transitions in 1916, and quantum state changes by these processes were the essence of research into the absorption and emission properties of molecules. However, although the maser idea arose from a pressing need for a source of far-infrared waves, it led first (in 1954) to microwave amplification and emission, and in 1960 to the generation of light in Maiman's laser. The far-infrared remained untouched. But in the 1960s water vapor and hydrogen cyanide were made to lase in the far-infrared, and a number of other molecules followed suit.

Lasers now provide a large number of emission lines throughout the far-infrared but, in general, they derive from quantum-mechanical systems with fixed energy separations and are not tunable. However, laser power levels are high, and this can be used to advantage. In particular, high-power laser scattering from systems of oscillators tunable in frequency (e.g., electron spin transitions in a solid immersed in a magnetic field) can lead to tunable stimulated Raman emission at levels quite suitable for spectroscopy.

In the far-infrared, the contact between the development of experimental techniques and their application to the study of physical processes has always been particularly close. Indeed, the limitations on generators and detectors have demanded the attention of the experimenter to every aspect of his instrumentation. The inherent experimental difficulties notwithstanding, the range of applications to the study of physical systems has been extensive and remarkably successful in its contributions to the understanding of their nature. In the last decade or two this has been particularly so. Far-infrared waves have been used in gases, liquids, solids, and plasmas. Crystal lattice dynamics, magnetic resonances, cyclotron resonance and magnetoplasma effects in semiconductors, the probing of gaseous laboratory plasmas, and the excitation of transitions across the energy gap in superconductors have been highlights of this development.

1.2. HISTORICAL BACKGROUND

Studies of resonances in the motion of crystal lattices have developed since those early measurements of crystal transmission and reststrahlen bands and have been based on extensions of the dynamical models of masses linked by springs. The problem of the specific heats of solids as generalized by Debye, Born, and others from the early Einstein model is closely connected with this work. More recently this field has been directed towards the study of lattice absorption which results from the presence of impurity ions within the crystal lattice.

In contrast, magnetic resonance studies in the far-infrared are of relatively recent origin. Antiferromagnetic resonance, in particular, was first treated as recently as 1951, and this, because it involves the precessional motion of electron spins in the high fields of interpenetrating crystal sublattices, leads naturally to far-infrared resonance frequencies in some materials.

Cyclotron resonance in semiconductors was seen by Shockley in 1953 as a means of measuring the effective mass of carriers. The suggestion was followed up with remarkable success and continues in this role as a major tool in the determination of band structure. With the progress of time these techniques have been applied to semiconductors with higher electron densities and the resonance phenomena have linked up with plasma effects in the solids. Similar magnetoplasma effects (e.g., Faraday rotation) occur in both solid state and gaseous plasmas, but in the latter case the parameter generally determined by the measurement is the electron density rather than particle mass. This linkup between solid state and plasma research is particularly interesting in view of the widely different origins and emphases of the two research fields: solid-state physics developed from a desire to understand and make use of solid materials, while plasma research was stimulated by the quest for controlled thermonuclear energy that started in the mid 1950s.

Between 1957 and 1959 Tinkham and Richards used far-infrared radiation to demonstrate the existence of an energy gap between the normal and paired-electron states of superconductors.[8] Although this application was somewhat isolated from the main stream of far-infrared physics, it was of great significance in the development of an understanding of superconductivity. It confirmed what microwave and optical data had previously suggested, for these two regions had shown quite different values of superconductor resistance. Subsequently, the theoretical descriptions of superconductivity incorporated the energy gap into their structure.

---

[8] M. Tinkham, *Science* **145**, 240–247 (1964).

2. FAR-INFRARED WAVE GENERATION

Introduction

Electromagnetic radiation is produced by changes of energy states of charged particles. Processes such as the deceleration of free electrons or the rearrangement of charge configurations or particle spin in an atom or molecule can result in photon emission. The emitted waves may be coherent or incoherent according to the particular generation mechanism. For example, incoherent waves may be emitted when electrons are decelerated during collisions in a discharge, or when they undergo phase-unrelated transitions between energy levels within an atom or molecule. Otherwise they can be induced to give up energy together, that is in a phase-related way as in an electron tube or laser, to emit coherent waves.

Far-infrared waves lie in the spectral region between wavelengths of 10 and 2000 μm, or thereabouts. Historically, these waves have proven particularly difficult to generate, and this has handicapped their development relative to the neighboring optical and microwave regions. Many of the mechanisms used in light and microwave generation experience limitations which are particularly marked in this part of the spectrum.

The earliest laboratory sources, particularly of infrared waves, were hot-body radiators. Herschel used red-hot pokers in his pioneering investigations of the infrared in 1800, however, they are quite feeble sources. While at temperatures of a few hundred degrees the peak of the spectrum of blackbody radiation is near 10 μm, at more elevated and useful temperatures it shifts in accordance with Wien's displacement law to the region of 1 μm. There is, of course, a power increase with temperature at all wavelengths, but, in accordance with the Rayleigh–Jeans law, the emitted power per unit frequency interval declines as λ^{-2} as the wavelength increases through the far-infrared.

The quest for more powerful far-infrared waves began soon after the historic wave generation experiment of Hertz in 1886, and for many decades it was focused on mechanisms of incoherent generation. Using the principle of the "mass radiator" first proposed by Oliver Lodge[1] in the year 1890, Lebedew[1a] produced millimeter wavelengths at Moscow University in 1895. In 1924 Madame Glagolewa-Arkadiewa,[2] also at Moscow University, produced 90 μm waves with a mass radiator. In these generators polarized metal spheres or dipoles radiate incoherently when the electric polarizing field is suddenly reduced to zero. This process is initiated in a spark discharge through an oil stream containing metal spheres or dipoles in suspension, the polarizing force suddenly collapsing with the onset of breakdown between the discharge electrodes. In 1923 Nicols and Tear[3] generated 220-μm waves with a mass radiator. They were used by spectroscopists through the 1930s, and were investigated by Daunt at Oxford University[4] early in World War II, when there was a pressing need for microwave generators.

The 1930s saw the emergence of coherent microwave electron tube oscillators. In the form of the split-anode magnetron[5] they produced centimeter waves early in that decade and were used by Cleeton and Williams in 1934 to launch the science of microwave spectroscopy.[6] At the close of the 1930s, the microwave klystron, the resonant cavity magnetron, and the traveling wave tube were invented, although the value of the latter generator was not appreciated until 1946[7] and its development was accordingly delayed. With extensive subsequent developments, these microwave generators and their many variants have produced centimeter and millimeter radiation with properties ideally suited to scientific experimentation: high, medium, or low power levels of pulsed or continuous radiation can be produced; it can be readily tuned in frequency, is amplitude and frequency modulated easily, and so forth. Unfortunately, the extension of electron tubes towards the far-infrared has proven a formidable and slow task. Nevertheless, the backward-traveling wave tube, or carcinotron, has reached as low as 345 μm and klystrons have

[1] O. Lodge, *Nature (London)* **41**, 462–463 (1890).

[1a] P. N. Lebedew, *Ann. Phys. Chem.* **56**, 1–17 (1895).

[2] A. Glagolewa-Arkadiewa, *Nature (London)* **113**, 640 (1924).

[3] E. F. Nichols and J. D. Tear, *Phys. Rev.* **21**, 587–610 (1923).

[4] R. Q. Twiss, *J. Electron.* **1**, 502–507 (1956).

[5] G. B. Collins, "Microwave Magnetrons," Chapter 1. McGraw-Hill, New York, 1948.

[6] C. E. Cleeton and N. H. Williams, *Phys. Rev.* **45**, 234–237 (1934); *Phys. Rev.* **50**, 1091 (1936).

[7] J. R. Pierce, "Travelling Wave Tubes." Van Nostrand-Reinhold, Princeton, New Jersey, 1950.

operated at 430 μm, while other electron tube oscillators are gradually being extended towards the 1-mm mark.

Ways of getting coherent radiation below millimeter wavelengths were devised by spectroscopists during the 1950s. They exploited devices with nonlinear current–voltage characteristics to generate higher harmonics of millimeter waves, and, although they obtained only low power, it proved sufficient for the needs of gaseous spectroscopy. The point-contact crystal diode acted as a particularly useful nonlinear multiplier and enabled extensions down to 500 μm.[8]

Although harmonic generators, together with hot-body radiators, enabled the extension of research in the far-infrared, the need for more powerful sources comparable with the generators of microwaves was very much in the minds of physicists. In 1952 the problem of generating these waves was so much in the mind of one spectroscopist, Townes at Columbia University, that he invented the maser in an attempt to solve the far-infrared generation problem.[9,10] At that stage the maser generated only centimeter waves. Several years later lasers did break through to shorter wavelengths,[11] but they went all the way to light waves, leaving the far-infrared untouched. Recently, however, progress has been made. Water vapor lasers have yielded a large number of emission lines between 16 and 120 μm,[12] while other molecular gas lasers have given emission up to and above the lower limits of present carcinotron operation.

At this time there are many sources of radiation suitable for laboratory experimentation, but the situation is not comparable with that at microwave frequencies. Gas lasers are essentially fixed-frequency oscillators and, therefore, not entirely suitable for many applications. In spectroscopy, for example, it is usually required to sweep the frequency in order to measure spectral line shapes. Harmonic generators are too weak for many applications. The carcinotron is in most ways an ideal source for experimentation, but presently it is very expensive and is still limited in its coverage of the far-infrared spectral region. Many other generation mechanisms have capabilities in the far-infrared; the merits and limitations of these will be discussed in the following sections.

[8] G. Jones and W. Gordy, *Phys. Rev. A* **136**, 1229–1232 (1964).
[9] J. P. Gordon, H. J. Zeiger and C. H. Townes, *Phys. Rev.* **95**, 282–284 (1954).
[10] N. G. Bassov and A. M. Prokhorov, *Sov. Phys. JETP* **3**, 426–429 (1956); *Dokl. Akad. Nauk SSSR* **101**, 47–48 (1955).
[11] T. H. Maiman, *Nature (London)* **187**, 493–494 (1960).
[12] L. E. S. Mathias and A. Crocker, *Phys. Lett.* **13**, 35–36 (1964).

2.1. Incoherent Sources

Contemporary hot-body radiators are generally gas discharge tubes. In this form they are relatively simple devices which generate an extensive spectrum, from which usable narrow bands of radiation can be selected with diffraction gratings or interferometer filters. High temperatures can be attained, but nevertheless, the power capabilities per unit bandwidth are low.

The power output can be calculated from Planck's equation for the radiation from a blackbody at temperature T. In the far-infrared, the Rayleigh–Jeans approximation is applicable; for the power radiated by unit area (1 cm²) in a narrow frequency interval dv about the frequency v, it gives

$$P(v)\, dv = 8.6 \times 10^{-15}\, T\, dv/\lambda^2. \tag{2.1}$$

In this expression, T is in degrees Kelvin, the wavelength λ is in μm, dv is in Hz, and the power $P(v)\, dv$ is in watts. The emissivity of the discharge is taken as unity. For any reasonably narrow band the output predicted by Eq. (2.1) is very small indeed, and falls drastically with increasing wavelength. The power emitted in a 1-MHz band by a blackbody at 5000°K is plotted in Fig. 2.1. This temperature is of the right order for high-pressure mercury discharge tubes, although in some laboratory arc discharges temperatures of 50,000°K are attainable, but generally with small emitting areas. Even at this elevated temperature only a little over 10^{-10} W per unit area is emitted in a 1-MHz band about 1000 μm. High-pressure mercury arc discharge lamps are frequently used as laboratory sources. Their total emission is some hundreds of watts, with about 10^{-7} W in a 10% bandwidth at 500 μm.

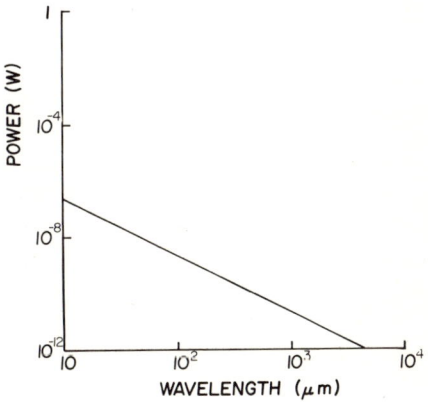

FIG. 2.1. Power radiated by 1 cm² of a blackbody at 5000°K in a bandwidth of 1 MHz.

The physics of bremsstrahlung emission from the collisional deceleration processes in the plasma of a discharge tube is discussed in Chapter 6. Radiation emission from an arc lamp is determined not only by the temperature and emissivity of the plasma, but by the transparency of the envelope enclosing the discharge. Thin fused quartz envelopes are usual, but these are certainly not completely transparent in the wavelength region up to 100 μm (see Chapter 3, Section 3.2.2), where the plasma emission is reasonably high.

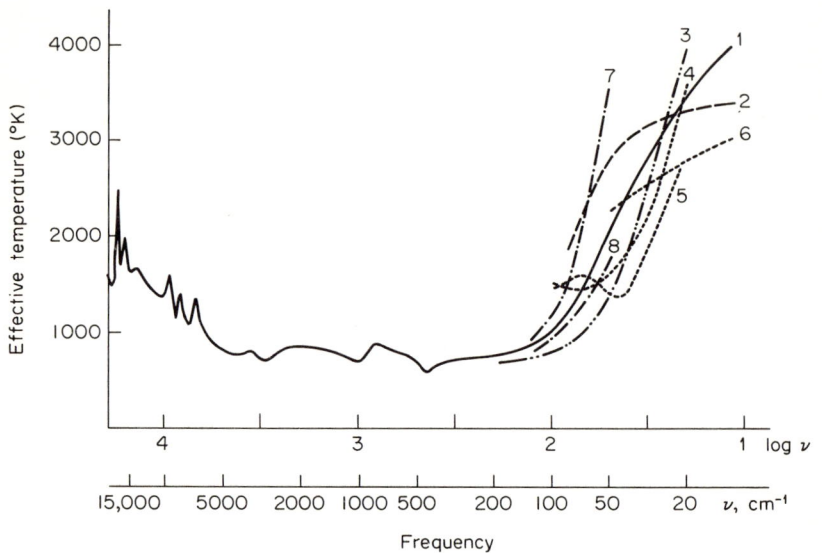

FIG. 2.2. The effective temperature of a number of commercial mercury lamps. Curves 1, 2, and 3 are for the PRK-4; curves 4 and 5 are for Philips model 93110; curve 6 is for the IA-2; curve 7 is for the HPK-125; and curve 8 is for the General Electric AM-4 (after Stanevich et al.[13]).

Measurements have shown that temperatures in excess of 12,000°K can be obtained on the axis of arc discharges in argon and xenon. However, for the high-pressure mercury arc lamp it has been found that, despite lower temperatures (6000–6500°K), the intensity of long wavelength emission is higher.[13] Stanevich et al., have published curves comparing several commercially available arc lamps. They are reproduced in Fig. 2.2, where the emission is given as an effective blackbody temper-

---

[13] A. E. Stanevich, I. A. Broadskiy and N. G. Yaroslavskiy, *Sov. J. Opt. Technol.* **36** (1969).

ature—the temperature of a hypothetical blackbody which would give the same radiation in the frequency band of interest.

Various solid body radiators may be used as radiation sources in the infrared. Historically, the Nernst glower, the globar, and the gas mantle have played important roles in spectroscopy. The Nernst glower is a filament composed of rare earth oxides which becomes conductive when heated, and which may then be kept hot by ohmic heating; gas mantles are mixtures such as 90% ThO_2 and 1% CeO_2; and the globar is an electrically heated rod of silicon carbide.

Performances of solid radiators and mercury lamps have been discussed recently by Stanevich et al., who show the superiority of the arc for wavelengths above about 100 μm. In the region from 10 to 100 μm good performance is found for the globar and for a platinum ribbon coated with yttrium oxide.

For the wavelength range 2–40 μm, Ramsey and Alishouse[14] have compared the spectral energy distributions of commercially supplied Nernst glowers, globars, and gas mantles with a commercial 900°C blackbody source. The operating conditions used for these sources are given in Table 2.1, and their emission spectra are compared in Fig. 2.3 and 2.4.

TABLE 2.1. Operating Conditions of Sources[a]

Source	Conditions	Color temperature (°K)	Dimensions (mm)
Globar	200 W, 6 A	1470	5.1 × 20.3
Nernst glower	45 W each, 0–6 A	1980	3.1 × 12.7
Mantle	Propane heated	1670	25.4 × 38.1
Blackbody		1173	27.9 × 19.6

[a] After W. Y. Ramsey and J. C. Alishouse, Infrared Phys. **8**, 143–152 (1968).

These workers find that the Nernst glower emits the most energy from 2 to 14 μm, but as the wavelength increases its emission drops below that of the globar and that of the gas mantle. From 14 to 25 μm, the mantle is superior, and at wavelengths beyond 25 μm both the globar and the mantle emit 40% more power than the glower.

[14] W. Y. Ramsey and J. C. Alishouse, Infrared Phys. **8**, 143–152 (1968).

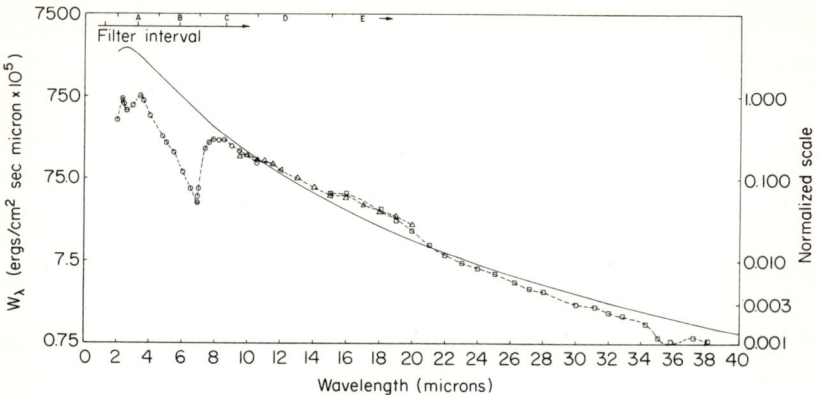

Fig. 2.3. Spectral emittance of a blackbody at 900°C. The solid curve is theoretical and the dashed curve experimental (after Ramsey and Alishouse[14]).

We have mentioned the mass radiator as a source of incoherent radiation. The spectrum generated by this device is not as broad as the hot-body spectrum, but the power is greatly enhanced in a region about the natural resonance frequency of the suspended metal particles. Perfectly conducting ball bearings, for example, have a fundamental mode of oscillation at a wavelength equal to 3.63 times the diameter D of the spheres. When the polarizing field drops to zero at the onset of the spark breakdown, the displaced charges execute heavily damped oscillatory

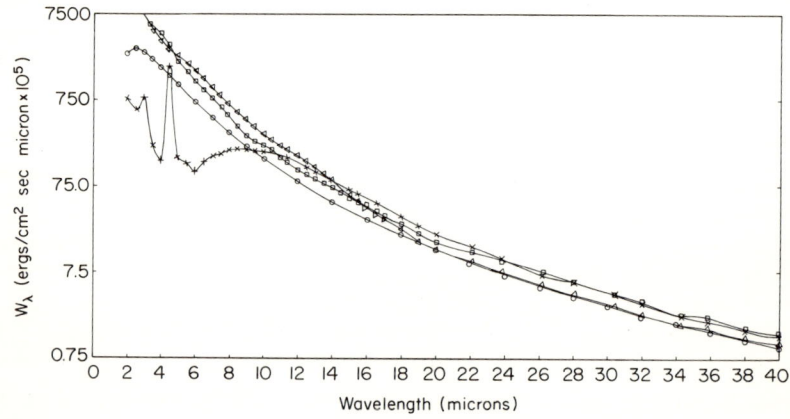

Fig. 2.4. Spectral emittance of a globar, a Nernst glower, a gas mantle, and a blackbody at 900°C. The conditions of operation of the sources are shown in Table 2.1 (after Ramsey and Alishouse[14]). Circle is blackbody, square is globar, triangle is Nernst glower, and cross is mantle.

2.1. INCOHERENT SOURCES

motion and emit damped waves of the form[15,16]

$$e^{-3.63\nu t} \cos 2\pi\nu t, \qquad (2.2)$$

where ν is the natural resonance frequency, equal to the velocity of light divided by $3.63D$. This expression for the field variation with time shows a decrease in amplitude by a factor e in the time taken by the wave to travel a distance from the sphere equal to its diameter. The extremely heavy damping results in short pulses only 10^{-11}–10^{-12} sec in duration,[17] with a resulting broad Fourier spectrum about the frequency ν.

Within the region of the discharge several spheres are set into oscillation. These are randomly distributed within the discharge, and the emitted waves are therefore randomly distributed in phase, that is, the emission is incoherent. Of course, if the spheres or dipole emitters have a variety of sizes the spectrum will be broader and the power generally below the peak produced with uniform emitters. The power performance of this generator can be expressed in terms of an effective blackbody temperature. A mass radiator described by Cooley and Rohrbaugh in 1945[18] was made with aluminium particles of assorted sizes suspended in a discharge through oil. The discharge was pulsed 1000 times per sec at a voltage exceeding 70 kV. Emitted waves, selected with a transmission grating spectrometer and measured with a bismuth–antimonide thermopile, were found to range from 200 to 2200 μm. In the experiments conducted by Daunt,[4] 6-mm waves were emitted by an oil discharge containing an average of 25 oscillating spheres, all of the same size. An applied field of 1.5×10^5 V cm^{-1} was collapsed 80 times per sec, producing an average power in the total spectrum of some 30 mW. The length of each pulse was 5×10^{-12} sec, and the peak power in this short pulse was some 70 MW. The average power generated near the peak of the spectrum in a 1-MHz bandwidth was 4×10^{-7} W, corresponding to an effective blackbody temperature near 2×10^9 °K.

The scaling of Daunt's mass radiator to a wavelength of 100 μm has been discussed by Twiss, who concludes that the performance at this

---

[15] W. R. Smythe, "Static and Dynamic Electricity," 1st ed., pp. 457–460. McGraw-Hill, New York, 1939.

[16] L. Page and N. I. Adams, "Electrodynamics." Van Nostrand-Reinhold, Princeton, New Jersey, 1940.

[17] M. H. N. Potak, *J. Brit. Inst. Radio Eng.* [N.S.], **13**, 490–497 (1953); *Proc. Inst. Elect. Eng. Part B* **103**, 781–786 (1956).

[18] J. P. Cooley and J. H. Rohrbaugh, *Phys. Rev.* **67**, 296–297 (1945).

short wavelength would be no better than that of a gas discharge.[4] Probably the most serious drawback of the mass radiator is the large variation of power from pulse to pulse. This is caused by the fluctuation $\pm N^{1/2}$ in the number N of spheres within the collapsing field of the spark discharge, and random variations in their relative positions.

2.2. Harmonic Generators

The primary requirement for the conversion of electromagnetic energy at one frequency to a wave at a higher frequency is that the converting material or device be nonlinear. The current can then be expressed, by Fourier's theorem, as a wave series with each of the higher terms multiple frequencies of the fundamental. The form of the nonlinear relation between current and voltage will determine the relative amplitudes of the harmonics, and may favor the production of some harmonics and the suppression of others.

The efficiency of energy transfer to a higher harmonic depends on the nonlinear law, on the magnitude of the fundamental frequency field, and also on the loss processes in the device. In principle, it is possible for all the fundamental energy to be converted to a particular set of harmonics or to an individual harmonic. One hundred per cent conversion efficiency requires a lossless reactive element,[19] while with a purely resistive element, as Page has shown,[20] the harmonic efficiency cannot exceed n^{-2}, n being the harmonic number. Although practical harmonic converters are neither purely resistive nor purely reactive, these idealized efficiencies are a useful guide to the possibilities of harmonic generating processes.

2.2.1. The Crystal Diode

The point-contact crystal diode is used in the far-infrared as both a frequency multiplier and as a detector. For this reason the description of physical processes in this section overlaps to some extent with that of Section 4.1.

To get an elementary view of the rectification process in a crystal diode, let us consider a metal n-type semiconductor junction. The equilibrium

---

[19] J. M. Manley and H. E. Rowe, *Proc. IRE* **44**, 904–913 (1956).
[20] C. H. Page, *J. Res. Nat. Bur. Stand.* **56**, 179–182 (1956); *Proc. IRE* **46**, 1738–1740 (1958).

voltage forward conduction knee. The hardness of these two materials results in pressure contacts of small area and low capacity.

Diodes with very small effective areas can be made from ion bombarded silicon.[23,24] A low resistivity surface layer is produced if the silicon is bombarded with phosphorous ions, while carbon ion bombardment produces a high resistivity layer. This bombardment process results in a p-n junction near the surface to which a near-ohmic high-pressure whisker contact is made. At high frequencies the very thin surface layer, and the depletion layer structure formed by it and the bulk semiconductor, act as a high resistance radial transmission line along the surface from the junction. The only low impedance path is perpendicular to the surface in a small area immediately beneath the point contact. Silicon diodes of this type have shown particularly useful harmonic generation capabilities.

Extensive studies of diode fabrication techniques and performance as millimeter and submillimeter generators have been carried out at the Bell Telephone Laboratories and at Duke University.[25,26] With carefully designed diodes made from ion-bombarded silicon crystals, Ohl and his coworkers have penetrated deep into the far-infrared. They have found that while the second and third harmonics are adequately generated by operation about the forward current knee of the characteristic, higher harmonics are more efficiently produced when the diode is operated near the reverse current break. This operation is accomplished by applying a reverse dc bias voltage from a low-impedance source. The conversion efficiencies obtained by Ohl *et al.*,[23] are shown in Table 2.2. With a 24-GHz fundamental the power decrease for the lower harmonics is -10 dB per harmonic, while for the higher harmonics it is only -2 dB. Half-millimeter waves can be produced at a power level of about 10^{-8} W. This wave is the 12th harmonic of a 50-GHz fundamental.

Radiation mixing with nonlinear point-contact diodes has, in recent years, been extended to a very high level of refinement which has enabled the mixing of infrared and far-infrared signals. These developments have taken place at Massachusetts Institute of Technology, where they have been applied principally to the measurement of absolute frequencies. They provide an extension of a frequency multiplier chain operating throughout the submillimeter band down to wavelengths as short as 5 μm.

[23] R. S. Ohl, P. P. Budenstein and C. A. Burrus, *Rev. Sci. Instrm.* **30**, 765–774 (1959).
[24] C. A. Burrus, *Proc. IEEE* **54**, 575–587 (1966).
[25] C. A. Burrus and W. Gordy, *Phys. Rev.* **93**, 897–898 (1954).
[26] W. C. King and W. Gordy, *Phys. Rev.* **93**, 407–412 (1954).

TABLE 2.2. Conversion Loss of Point-Contact Crystal Diode Frequency Multipliers

Harmonic number	Conversion efficiency (dB below the 200 mW, 24 GHz fundamental)
2	−11
3	−19
4	−26
5	−33
6	−36
7	−42
8	−46
9	−50
10	−54
12	−60
15	−66

The technique consists of mixing harmonics of a far-infrared laser whose absolute frequency has been previously determined with the fundamental of another laser and a microwave signal at the difference frequency to produce a zero beat frequency.

In one experiment by Hocker et al.[27] the fourth harmonic of a 337-μm HCN laser was mixed with the 84-μm line from a D_2O laser. The difference in frequency falls in the microwave C-band where it gives a zero beat with a 5.9-GHz signal. Previously the 337-μm line had been accurately determined by harmonic mixing with a microwave oscillator signal. Later experiments[28] successfully mixed a 28-μm water vapor laser line with a line at 9.3 μm from a CO_2 laser and radiation from a 20-GHz klystron, and, more recently, the technique has been extended to 5.2 μm. by the mixing of this CO laser line with second harmonic CO_2 laser radiation at 10.4 μm.

Mixing is done in special fast-response metal–silicon and metal–metal diodes. The metal–metal junction is used for the highest frequency mixing. It consists of a thin tungsten wire antenna about 2 μm in diameter etched to a pointed tip and pressed against a nickel or cold-rolled steel

[27] L. O. Hocker, J. G. Small and A. Javan, *Phys. Lett. A* **29**, 321 (1969).
[28] V. Daneu, D. Sokoloff, A. Sanchez and A. Javan, *Appl. Phys. Lett.* **15**, 398–401 (1969).

post. The tungsten whisker is electrochemically etched to about a 1000 Å point. The contact presumably involves a thin oxide layer sandwich between the metal surfaces, with tunneling of electrons through potential barriers in the oxide giving rise to the nonlinear voltage–current characteristic.

As we have mentioned earlier, more efficient harmonic generation is theoretically possible with a purely reactive multiplier. "Varactor" or variable capacity diode operation can be achieved in junctions with a barrier resistance much larger than the reactance of the nonlinear capacity. This occurs in certain diodes operating with reverse voltage bias.[24] Millimeter and submillimeter investigations of point-contact varactor diodes have been very limited, and the few published results are disappointing. For example, second harmonic production of 108 GHz from a 54-GHz fundamental has been achieved with a conversion loss of -11 dB.[24]

2.2.2. The Arc Discharge

A method of harmonic generation in an arc discharge has been developed by Froome[29] at the National Physical Laboratory, England. This device emerged from Froome's discovery of the high current densities at the cathode spot of a narrow electric arc, and the realization by Bleaney that the nonlinearity of such an arc might provide a mechanism of harmonic conversion. The generator has a resemblance to a solid-state point-contact diode but instead of a semiconductor–metal junction, a plasma–metal junction provides the nonlinear region for frequency multiplication. However, unlike the solid-state diode, the arc can cope with quite high fundamental frequency power.

The essential elements of the arc are shown in Fig. 2.7. A narrow dc arc is struck between electrodes separated by only 1.2×10^{-4} cm in argon at a pressure of 100 atm. Extending from near the anode, the greater part of the arc is composed of a plasma column where equality of electron and ion density prevail at a level of some 10^{19} cm^{-3}.

We can visualize the arc as a superposition of electron space-charge and positive-ion space-charge. Under the influence of the applied dc field, the electron cloud is displaced slightly towards the anode and the ion cloud towards the cathode. They overlap in the plasma region but in the near vicinity of the anode electrons predominate. Near the

---

[29] K. D. Froome, *Proc. Int. Conf. Quantum Electron.*, 3rd, *Paris*, 1963, **2**, pp. 1527–1539, Columbia Univ. Press, New York, 1964.

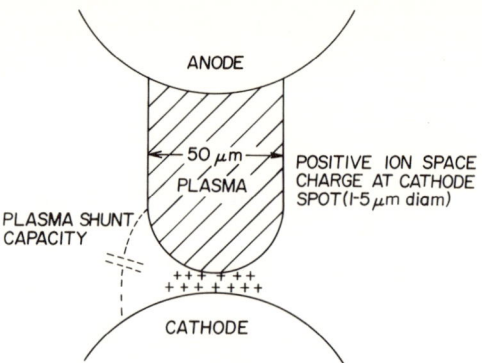

FIG. 2.7. Illustration of the plasma–metal junction of an arc discharge harmonic generator (after Robinson[22a]).

cathode surface there is an excess of positive ions, which results in an intense electric field at the cathode and the well-known cathode fall in potential. In the transition layer between the plasma and the cathode the intense electric field ($\sim 10^7$ V cm^{-1}) can cause field emission of electrons from the metal cathode.

The arc is located within interconnecting crossed waveguides after the fashion of the crystal converter previously described. When a microwave signal at the fundamental frequency is matched into the arc, microwave fields are superimposed on the dc field maintained by the arc. Throughout the period of the microwave oscillation the electrons in the transition layer, and to a lesser extent the positive ions, are density modulated. Consequently, the electric field at the cathode surface is modulated. According to Froome, the nonlinearity of the field emission process produces the observed harmonic conversion.

The arc carries a dc current of 0.5 A; the emission density at the cathode spot is between 2.5×10^6 and 10^8 A cm^{-2}. The fundamental microwave signal is propagated into the region of the cathode spot via the anode wire and plasma column. It is there superimposed on the dc voltage of 3–5 V—this voltage is chosen for optimum conversion efficiency.

After generation, the harmonic waves propagate through the plasma column in being launched into the harmonic waveguide. They are diminished by collision losses in the column and by capacitive shunting between the anode and cathode electrodes and across the junction itself, that is, from the end of the plasma column to the cathode. While the narrowness of the column and the small area of the cathode spot (as small as 10^{-9} cm^2) minimize this latter capacitance, the effect increasingly

2.2. HARMONIC GENERATORS

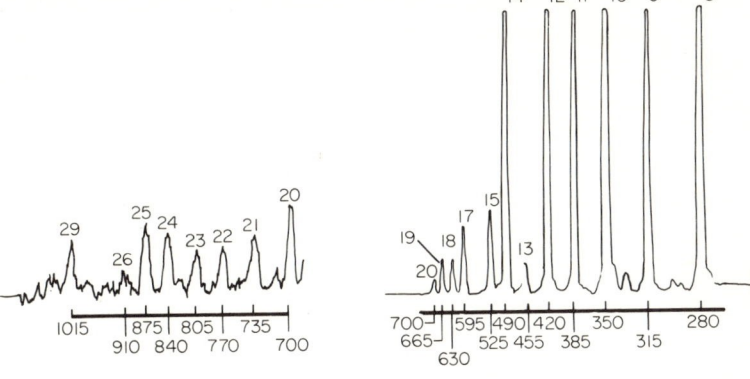

FIG. 2.8. The output of the plasma–metal junction microwave harmonic generator. The harmonics are numbered by multiples of the fundamental frequency, namely 35 GHz. Atmospheric water vapor absorbs the 16th harmonic at 0.54 mm. The output in the 13th harmonic appears low because of poor output coupling at this frequency. The deflexions between the 8th and 9th, and 9th and 10th harmonics are grating 2nd-order effects (after Froome[29]).

bypasses the higher harmonics. This shunt capacitance is in the region of 0.05 pF, about the same as that of the point-contact crystal diode.

Froome has used a 35-GHz fundamental frequency at a power level of 10 W, and has successfully generated harmonics up to the 29th, with a wavelength of 298 μm. In Fig. 2.8 the harmonic output recorded by Froome is displayed, the power level of some selected harmonics as published in 1963 being given in Table 2.3. For these measurements Froome separated the harmonics with a reflection grating spectrometer and detected them with a Golay pneumatic cell.

TABLE 2.3. Harmonic Generation in a Froome Arc Discharge, with a 10-W, 35-GHz Fundamental

Harmonic number	Wavelength (μm)	Power output (μW)	Conversion efficiency (dB below the 10 W fundamental power)
6	1440	100	−50
12	720	1	−70
20	432	0.01	−80
29	298	0.001	−100

It is interesting to note (by extrapolation) that in the vicinity of the 15th harmonic the conversion loss is of the order of −2.5 dB per harmonic, similar to the value achieved in Ohl's crystal converter. Between the 20th and 29th harmonic the loss is only −1 dB per harmonic. Of a number of materials tested for the anode and cathode (mercury, platinum, gold, copper, molybdenum, gallium), platinum has proven the most satisfactory. Froome has overcome the problem of arc erosion of the electrodes by reversing the dc voltage at a rate of 5000 times per sec with a switchover time of 0.3 μsec. This enables operation at steady harmonic power levels over several hours.

2.2.3. Nonlinear Interactions

A variety of nonlinear phenomena have possibilities for harmonic conversion. Effects in ionized gases, solids, and electron beam devices have been widely studied with a view to far-infrared generation, and some useful generators have emerged. Laser beams have been mixed in nonlinear crystals and have generated signals at far-infrared difference frequencies.

Sources of nonlinearities in ionized gases can be seen from the well-known Boltzmann equation for the distribution function of particle density in phase space.[30] This equation contains nonlinear terms associated with spatial gradients of the distribution function, through energy dependent collision frequencies, and through Lorentz forces given by vector cross products of particle velocity with the B-field of an electromagnetic wave. Although the frequency conversion properties of these nonlinearities have not been well studied at submillimeter wavelengths some of the results of work with microwaves is worth noting. In a gas discharge generator exploiting spatial gradient effects Swan[31] has found conversion efficiencies for the first three harmonics of a continuous 3000-MHz signal of −6.4, −13.2, and −16.6 dB, respectively. With a similar fundamental frequency, Hill and Tetenbaum[32] have investigated interactions where the nonlinear part of the Lorentz force is effective. With a frequency near the electron cyclotron frequency in a magnetized plasma these workers produced second, third, and fourth harmonics down on the fundamental power by −15, −35, and −50 dB, respectively.

---

[30] D. G. Montgomery and D. A. Tidman, "Plasma Kinetic Theory." McGraw-Hill, New York, 1964.
[31] C. B. Swan, *Proc. IRE* **49**, 1941–1942 (1961).
[32] R. M. Hill and S. J. Tetenbaum, *J. Appl. Phys.* **30**, 1610–1611 (1959).

2.2. HARMONIC GENERATORS

Frequency multiplication of pulsed millimeter radiation has been studied in magnetized ferrites. In these materials efficient harmonic conversion results when a magnetized ferrite is irradiated at high power levels. The nonlinearity results from interactions between the precessing magnetization vector and the wave field, producing a frequency doubled output proportional to the square of the input power.[33] Higher harmonics are also produced. The doubling action arises when the uniformly precessing magnetization vector is influenced by the synchronously rotating applied field to precess in an elliptical orbit. The projection in the direction of the magnetic field of the constant-length saturation magnetization then has a component of twice the precession frequency. This nonlinearity, at least for second-harmonic production, shows up from the classical equation of gyromagnetic motion.[34]

The effect has been used by Roberts et al.,[33] to generate 2-mm harmonics at a peak power level of 50 W, from a 4-mm fundamental about ten times more powerful. For large output a ferromagnetic material with a large saturation moment and a narrow resonance linewidth is desirable. Yttrium iron garnet has suitable properties. Small ferrite samples are located within a waveguide in such a way as to reduce serious perturbations of the waveguide modes. Their small size results in low average power capabilities, and presents a serious obstacle to their extension to shorter wavelengths.

Microwave electron beam devices are highly nonlinear, and can be productive sources of harmonics. When the electron beam in such a generator is density modulated through interaction with an electromagnetic field the electron current contains not only the fundamental frequency but also higher components.[35,36] These harmonics can induce wave growth in a suitable wave-supporting structure.

Harmonic beams can be produced in linear and other accelerators, in klystrons, and in magnetrons. In the linear accelerator, waves at frequencies corresponding to harmonic numbers up to 30–50 have been obtained for a 3000-MHz fundamental frequency.[34,37] These and other relativistic devices yield very tight, harmonic-rich electron bunches. In

---

[33] R. W. Roberts, W. P. Ayres and P. H. Vartanian, *Proc. Int. Conf. Quantum Electron.*, 1st, pp. 314–323. Columbia Univ. Press, New York, 1960.
[34] P. D. Coleman and R. C. Becker, *IRE Trans. Microwave Theory Tech.* **7**, 42–61 (1959).
[35] A. H. W. Beck, "Velocity-Modulated Thermionic Tubes." Cambridge Univ. Press, London and New York, 1948.
[36] D. L. Webster, *J. Appl. Phys.* **10**, 501–508 (1939).
[37] P. D. Coleman, *IRE Trans. Microwave Theory Tech.* **11**, 271–288 (1963).

one such linear generator developed at the University of Illinois,[38] an average current of 5–30 mA of modulated 1 MeV electrons has been used. Harmonic power extraction has been achieved in higher mode cavities, Fabry–Perot resonators[37] and by means of beam–dielectric interactions leading to Cerenkov radiation.[34,38] The microtron accelerator[39] has been studied as a means of forming tightly bunched beams of electrons. One such device with a 6-MeV electron beam bunched at 2800 MHz has recently excited waves between 0.5 and 1 mm in a Fabry–Perot resonator, and has produced power levels of tens of microwatts.[40] These experiments have confirmed the theoretical predictions of high harmonic content in microtron beams. At the 200th harmonic ($\lambda = 0.5$ mm) the calculated harmonic current amplitude is 10% of the dc beam current. High-energy accelerators are essentially pulsed devices with a high order of technical sophistication and are accordingly limited in their range of applications.

The linear accelerator multiplier has a close resemblance to the frequency multiplier klystron, a counterpart which is itself capable of submillimeter waves. The potentialities of klystron frequency multipliers were apparent from the theory of velocity modulation of electron beams developed by Webster in 1939.[36] Webster's small signal theory neglects space-charge forces and assumes boundaries at infinity. It predicts that the amplitude of the current harmonics in a bunched beam are in the ratio 100:83:75:64:52.

Recent calculations by van Iperen and Nunnink[41] for beams of circular cross section enclosed by cylindrical metal walls show that moderate space-charge densities may enhance the generation of harmonics. They find that the amplitude of the 10th harmonic can be half that of the fundamental. Based on these calculations van Iperen and Kuypers[42] have designed a klystron multiplier operating at 870 μm with a continuous power level of 35 mW. This is the 10th harmonic of a 15-W fundamental signal. It is generated in a beam of 28 kV carrying a current of about 20 mA. Beam modulation is imposed by the 8.7-mm fundamental field in a conventional reentrant resonant cavity, while harmonic extraction is accomplished by interaction with a TE_{012} mode in a resonant cavity with dimensions $1 \times 0.8 \times 0.15$ mm³. The cavity can be tuned by about

---

[38] I. Kaufman, *Proc. IRE* **47**, 381–396 (1959).

[39] E. Brannen, H. Froelich and T. W. W. Stewart, *J. Appl. Phys.* **31**, 1829 (1960).

[40] E. Brannen, V. Sells and H. R. Froelich, *Proc. IEEE* **55**, 717–718 (1967).

[41] B. B. van Iperen and H. J. C. A. Nunnink, *Philips Res. Rep.* **20**, 432–461 (1965).

[42] B. van Iperen and W. Kuypers, *Philips Res. Rep.* **20**, 462–468 (1965).

5%. A recent extension of this work has extracted the 20th harmonic to generate 0.1 mW of continuous power at a wavelength of 430 μm.[43]

The electron space-charge in a microwave magnetron has high harmonic content and can yield pulses of short wavelength radiation. Because the magnetron is a multicavity structure it supports a multitude of natural modes of resonance to some of which the harmonics of the bunched electrons may couple. Empirical studies of these processes have shown that higher mode oscillations are aided by operating the magnetron into a mismatched waveguide presumably through the effects of mismatching on mode tuning and coupling to the load.

At Columbia University's Radiation Laboratory, this technique enabled the generation of a few hundred microwatts at 1.1-mm wavelength, this frequency being the third harmonic of the fundamental mode of oscillation. Second harmonic generation with such a magnetron has given 0.5-W power pulses.

In the last few years developments in nonlinear optics have resulted in successful laser beam mixing in nonlinear crystals and the production of far-infrared radiation. Under conditions of high laser intensity, certain transparent crystals exhibit a sufficient level of nonlinearity to mix two optical beams and generate a useful far-infrared difference signal.

Zernicke and Berman[44] have used a pulsed neodymium–glass laser to produce radiation near 100 cm^{-1}, and Yajima and Inoue[45] have generated a fixed difference frequency of 29 cm^{-1} by mixing lines from a single-pulsed ruby laser in nonlinear media. Van Tran and Patel[46] have followed these workers with experiments that show the possibility of tunable far-infrared generation by nonlinear mixing of two CO_2 lasers in bulk InSb. For a given carrier concentration, and an appropriate applied magnetic field, magnetoplasma effects in the semiconductor can be used to give a phase-matched nonlinear interaction. Microwatt potentiality, and magnetic field-tunable generation throughout the far-infrared is suggested via the mixing of lines from the 9.1–9.8 μm and 10.2–10.8 μm bands of CO_2 lasers.

Experiments with two temperature-tuned Q-switched ruby lasers have enabled Faries *et al.*[47] to generate tunable radiation over the frequency

---

[43] W. Kuypers, Private communication, 1967.

[44] F. Zernicke and P. R. Berman, *Phys. Rev. Lett.* **15**, 999 (1965).

[45] T. Yajima and K. Inoue, *Phys. Lett. A* **26**, 281 (1968).

[46] N. Van Tran and C. K. N. Patel, *Phys. Rev. Lett.* **22**, 463–466 (1969).

[47] D. W. Faries, K. A. Gehring, P. L. Richards and Y. R. Shen, *Phys. Rev.* **180**, 363–365 (1969).

range 1.2–8.1 cm^{-1} (that is, a wavelength range 8.3–1.2 mm). With a LiNbO$_3$ crystal they have measured peak pulsed power of 20 mW, and with crystal quartz powers lower than this, because it has an electro-optic coefficient smaller by a factor of 3.7.

In concluding this section, the successful generation of microwave harmonics based on the Josephson effect should be mentioned. The Josephson junction, which is discussed in Sections 2.6.3 and 4.4, has a highly nonlinear V–I characteristic, and can respond to frequencies well into the far-infrared. It is capable only of ultralow power output. Shapiro[48] has tested its harmonic production capabilities by converting both 4- and 6-GHz fundamentals to harmonics at 12 GHz with power of 2×10^{-10} W.

2.3. Electron Tubes

The essential components of an electron tube are a moving stream of electrons and a wave-supporting structure. The structure may support intense fields through properties of resonance, or it may support weaker fields which travel in near synchronism with the electrons, extracting energy over a long interaction path. Interactions affect the velocity of each electron in a manner depending on the phase of the wave as seen by the electron. In a klystron, for example, an electron interacts with the strong field for only the short time of electron transit across the gap of a reentrant resonant cavity. Those electrons which pass through the interaction gap at a time of zero field are unaltered in velocity, while those a little ahead are retarded, and those a little behind accelerated. A similar phenomena occurs in traveling wave interactions, electrons near the crest and troughs of the traveling wave being forced towards one another. In this manner the initially homogeneous beam clusters into bunches carrying the characteristic frequency of the wave and its harmonics. Direct current is thereby translated into alternating current.

In passing through a second klystron cavity (the "catcher" or "extractor"), fields are induced in such a phase as to slow the velocity of the bunches and thereby extract energy. Through coupling between the fields of the buncher and catcher cavities regenerative oscillations build up provided the energy transfer from the electron beam can exceed the losses of the system. In the case of traveling wave interactions the pro-

---

[48] S. Shapiro, *J. Appl. Phys.* **38**, 1879–1884 (1967).

cesses of density modulation and energy extraction occur simultaneously, feedback enabling the build up of oscillations.

There is a large variety of particular generator types, and indeed the distinction between traveling and standing wave interactions is not always clear. Further, the energy extraction is not always from beam kinetic energy; in the case of electron streams in crossed electric and magnetic fields, it comes from the potential energy of the electrons. This process is operative in the magnetron.

2.3.1. The Klystron

As we have indicated, the klystron is based on the principle of velocity and, thence density modulation brought about when the electron beam passes through a resonant cavity.[35] This mechanism is depicted in Fig. 2.9.

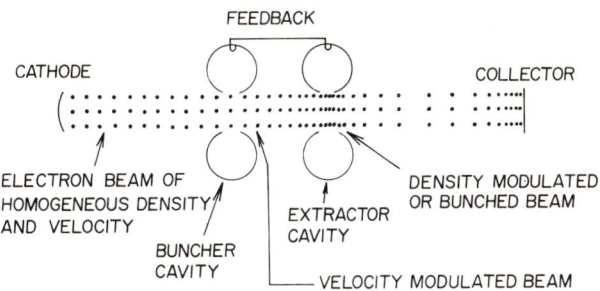

FIG. 2.9. Illustration of electron bunching in a drifting velocity-modulated beam (after Robinson[22a]).

The high-frequency limitations of the klystron, when extended from centimeter to millimeter wavelengths, are set by cavity losses, and by the maximum current that can be focused down to interact with the electric field stored in the cavity. The linear dimensions of a cavity supporting the lowest resonant mode are of the order of the wavelength λ, and the diameter of the interaction gap is somewhat smaller. Ohmic cavity losses, as predicted by skin-effect theory, increase as the square root of the frequency, but if there is surface strain or roughness the losses will rise more rapidly. Thus to overcome losses and provide useful output power, it is clear that the highest densities of electrons must be concentrated through these small cavities.

The task of developing short millimeter-wave klystrons is, then, one of fabricating very small cavities, and designing electron guns which multiply the cathode emission densities by large factors (present cathodes can emit up to 10–60 A cm^{-2} continuously, and up to 300–400 A cm^{-2} in short pulses). Thereafter the beam must be confined to this concentrated geometry while it is being density modulated and while its energy is converted to electromagnetic oscillations. Modern electron gun designs can produce current densities $\sim 10^3$ A cm^{-3}. This means volume densities of electrons exceeding 10^{12} cm^{-3}.

Klystron oscillators are commercially available down to about 2-mm wavelengths. As mentioned in Section 2.2.3, outstanding progress in the extension of frequency-multiplier klystrons into the far-infrared has been made at the Philips Research Laboratories, Eindhoven.[42,43] This work has been based on a theoretical analysis of harmonics in a velocity-modulated electron beam of circular cross section traveling within a

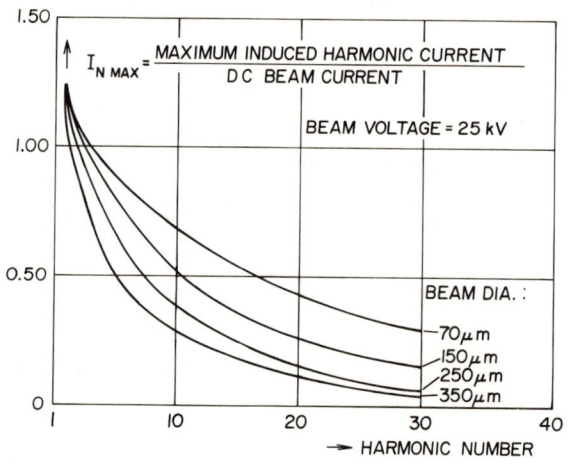

FIG. 2.10. Theoretical plot of the amplitudes of the current harmonics versus harmonic number for several values of the beam diameter and a beam voltage of 25 kV (after van Iperen and Kuypers[42]).

cylindrical conducting drift tube.[41] The effects of space-charge forces are included in the theory. Results of these calculations are shown in Fig. 2.10, where the ratio of the harmonic current amplitude to the dc beam current is plotted as a function of harmonic number. The beam is at 25 kV and is assumed to be modulated during passage through a narrow cavity interaction gap which is gridless and equal in diameter to the electron beam. There is a marked decrease in harmonic amplitude with increasing beam diameter. This is due to variations of the harmonic current over the beam cross section and to the decrease of coupling between the beam and the field which is especially marked for those electrons furthermost from the drift tube walls (this is related to the decline of wave-particle interactions discussed in Section 2.3.2 for the traveling wave tube, and in Section 2.4.1 for relativistic beam interactions). The figure shows that, for beam diameters which are not large, the 10th harmonic has about half the amplitude of the fundamental, and the 20th harmonic is about one quarter of the fundamental.

The Philips workers have used 25-kV, 20-mA beams 0.2 mm in diameter. The current is emitted from a cathode 0.6 mm in diameter in a Pierce gun. The beam is focused down to the drift-tube diameter and confined by a magnetic field of 3700 gauss. In the 430-μm multiplier, velocity modulation is not achieved in a single cavity but during transit through a system of four identical cavities. As explained in the last paragraph of this section, this has the advantage of reducing the power input necessary to give optimum bunching. With a four-cavity buncher 4 W of 8.7-mm radiation is required. The extraction of 20th harmonic power at a level up to 0.1 mW is carried out during beam transit through a waveguide whose narrow dimension is tapered to a very small size. Output power is transmitted through a mica window, and a variable mismatch in the waveguide on the output side of this window is employed to give a low-Q resonance to the catcher. The klystron is shown in cross section in Fig. 2.11. In contrast to the buncher system, the catcher in this early version of the far-infrared klystron appears to be far from optimum. One can expect progress to higher powers and frequencies to result from improved catchers, and from the exploitation of the still higher harmonics indicated in Fig. 2.10.

In addition to the technical klystron design problems mentioned, the single-cavity bunchers and energy extractors are not as efficient as the wave-supporting structures used in traveling wave tubes. The reason for this can be seen from elementary microwave electronic principles, by considering the buildup of an extended interaction structure by the successive addition of cavities. Across the interaction gap of a single cavity with shunt resistance R_{sh}, power P will give rise to a voltage $(PR_{sh})^{1/2}$. For two coupled cavities, each dissipating power $P/2$, the voltage across each will be $(PR_{sh}/2)^{1/2}$, and the total interaction voltage $2^{1/2}(PR_{sh})^{1/2}$. Similarly for an n-cavity system, the effective interaction voltage will increase by $n^{1/2}$ over that of a single cavity. This multicavity situation is the limiting case of a traveling wave tube, namely, the limit of weak coupling between the separate sections of a periodic wave-supporting structure.

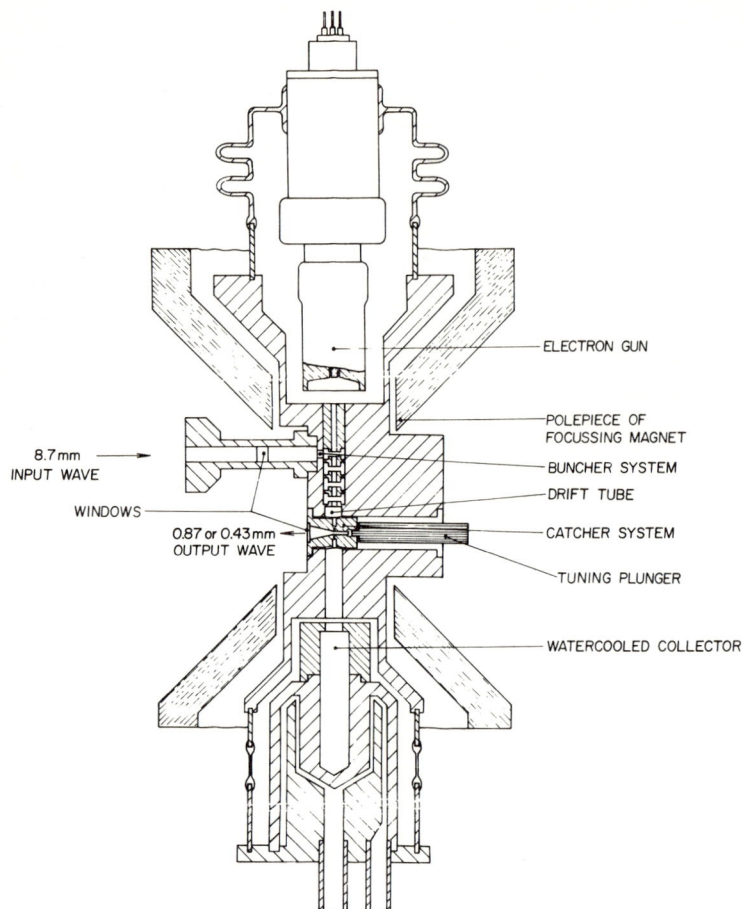

Fig. 2.11. Cross section of a klystron frequency multiplier used to generate 870- and 430-μm waves. The buncher system consists of four coupled cavities, and the output circuit is a tapered rectangular waveguide made resonant by a variable mismatch in the guide (after van Iperen and Nunnink[41]).

2.3.2. The Traveling-Wave Oscillator

In a traveling-wave tube, wavegrowth occurs on a periodic structure capable of supporting field components which travel in near synchronism with the electron beam.[7] As we have suggested, the periodic structure may be thought of as a system of coupled resonators with strong coupling. As the electrons travel through the structure the two effects of beam density modulation and excitation of growing waves by the modulated beam occur simultaneously. We can picture the process in the following synthetic way: Electrons traveling at the same speed as a (primary) wave become density modulated. The electrons cluster at points of zero field strength of the wave, ahead of which positions the field is retarding and behind which it is accelerating. This bunching is produced with no net energy transfer, since as many electrons are retarded as are accelerated. As the bunches form in the electron beam, they induce a secondary wave on the structure which retards the bunches and so gains energy from the decelerating electrons. The secondary wave has the same frequency but lags the initial wave by $\pi/2$. The process continues, resulting in the excitation of a tertiary wave, and so on. The action of each of these induced waves is to force the bunches from the positions of maximum retarding field to the positions of zero field. Thus the bunches are gradually slowed down and the induced waves grow with distance along the slow-wave structure. If regenerative effects are present, self-induced oscillations can build up in the tube.

The type of field distribution supported by the periodic structures is illustrated in Fig. 2.12. This spatial field pattern can be Fourier analysed into a series of spatially sinusoidal components[49] with periods S, $S/2$,

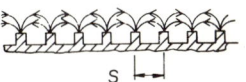

FIG. 2.12. Diagram of a periodic wave-supporting structure used in traveling-wave oscillators. The arrows represent the electric field (after Robinson[22a]).

$S/3, \ldots, S/m, \ldots$. Each of these components will vary (as does the entire field pattern) with time as $e^{i\omega t}$, ω being the angular frequency. Thus each component is a wave of the form

$$\exp[i(\omega t - \beta_m z)] \qquad (2.3)$$

traveling along the structure (in the z-direction), with its own particular propagation constant, and, of course, with an appropriate Fourier coefficient or amplitude. If θ is the phase shift over the periodic length S, the propagation constants are easily seen to be

$$\beta_m = (\theta + 2\pi m)/S, \qquad (2.4)$$

[49] J. C. Slater, "Microwave Electronics." Van Nostrand-Reinhold, Princeton, New Jersey, 1954.

where $m = 0, \pm 1, \pm 2, \ldots$. These component waves are called spatial or Hartree harmonics.†

In the traveling-wave tube, the electron beam can build up the field pattern by interactions via any one of the spatial harmonics, provided the beam velocity is substantially equal to the phase velocity of that harmonic. Interaction with the mth harmonic occurs when the beam velocity [given by Eq. (2.4)] is equal to

$$v_m = \omega/\beta_m = S\omega/(\theta + 2\pi m). \tag{2.5}$$

For the particular and important case when there is π phase shift between successive sections of the structure (this so-called "π-mode" is the case shown in Fig. 2.9),

$$v_m = S\omega/(\pi + 2\pi m). \tag{2.6}$$

This interaction via spatial harmonics can be seen from another point of view: If a bunch of electrons experiences maximum opposing field at the middle of a section in the periodic structure of Fig. 2.12 (and thus loses energy to the field), it will again be opposed in the next section if it takes $(m + \tfrac{1}{2})$ oscillation periods $(2\pi/\omega)$ to travel the distance S. Thus the bunch velocity for cumulative interactions is

$$v_m = \frac{S}{(2\pi/\omega)(m + \tfrac{1}{2})},$$

in agreement with Eq. (2.6). Interactions via the fundamental spatial harmonic corresponds to one half-period transit time between sections; the first spatial harmonic interaction is that corresponding to $m = 1$, that is, $1\tfrac{1}{2}$ periods transit time between sections. From this point of view, it is also apparent that velocities in the reverse direction can lead to cumulative interactions. They correspond to $m = -1, -2$, etc.

Associated with the field in the traveling-wave tube, there is a group velocity which is the velocity of power flow. It may be in the same direction or in the opposite direction to the phase velocity of the interacting spatial harmonic.[49] If the group and phase velocities are oppositely directed, the energy exchange is said to be via a backward wave. In such an interaction, electrons moving in one direction impart energy to a field

---

† It should be appreciated that these harmonics are part of a theoretical synthesis; they neither individually satisfy the boundary conditions nor do they have separate existence.

which grows in the opposite direction. This wave-growth mechanism is inherently regenerative. If the beam current is sufficient to overcome losses, the system will oscillate at a frequency determined partly by the electron velocity, for this is part of the regenerative loop. Such devices are called backward-wave oscillators or carcinotrons.

The problems of extending backward-wave oscillator performance from millimeter wavelengths to the far-infrared are quite like those which confront the scaling of klystrons. However, traveling-wave devices have the advantage, as we indicated at the end of Section 2.3.1, of providing better beam–field energy exchange to overcome losses. The size of the periodic structure can be seen from Eq. (2.6). For the fundamental spatial harmonic, this equation gives the phase velocity as $2CS/\lambda$, so that interaction with 10 kV electrons (velocity $\sim 0.2C$) requires $S \sim \lambda/10$. Thus short wavelengths require the finest of periodic structures. The dimensional requirement can be relaxed somewhat for higher space harmonic tubes, for in this case $S \sim (1 + 2m)\lambda/10$.

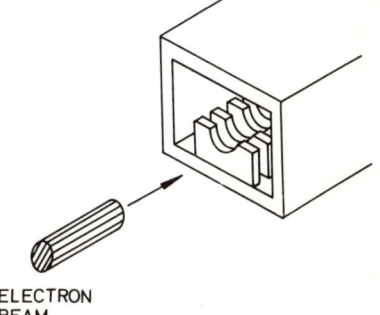

FIG. 2.13. Slow-wave structure of the type used in millimeter and submillimeter carcinotrons (after Robinson[22a]).

ELECTRON BEAM

In Fig. 2.13 a periodic structure of the type used for short millimeter and submillimeter carcinotron oscillators is shown. This structure will support slow waves with longitudinally directed **E**-vectors. The field amplitude is strongest near the periodic surface but it falls off in the transverse direction as $\exp(-\beta_m y)$. Thus there will be a "skin depth," $\delta = 1/\beta_m$, for effective wave-particle interaction.[50,51] For the 10-kV electron beam example used above, $\delta \sim \lambda/30$. It is clear that within this small distance of the periodic surface the largest currents are called for, and hence high-density electron beams are required. The initiation of

---

[50] A. Karp, *Proc. IRE* **43**, 41–46 (1955).
[51] A. Karp, *Proc. IRE* **45**, 496–503 (1957).

buildup of oscillations demands a forward gain in excess of the losses in the feedback system. If the interaction current is insufficient to overcome losses oscillations will not build up.

In the present range of carcinotrons operating below 1000 μm, the beam current densities are as high as 10^3–5×10^3 A cm^{-2} and the beam voltages are around 10 kV. Electrons are emitted from a cathode at a density of about 10 A cm^{-2} or a little higher, and the beam area is then compressed by a factor of 120. A range of carcinotrons developed by C.S.F., France, have generated continuous waves at 2000, 1000, 700, 500, and 345 μm, respectively. At the shorter wavelengths some hundreds of microwatts have been generated, and at 700 μm, 10 mW. Power levels ~1 W at 1000 μm are attainable.

2.3.3. The Magnetron

Within the broad classifications of microwave electron tubes there is a third class which is characterized by the role of a magnetic field in the generation process. Within this class there are many generators with differing detailed mechanisms, just as there are many variants within the klystron and traveling-wave tube families. The magnetron is prominent among these generators. Although the short wavelength achievements of the magnetron are in general above 2 mm, its interaction mechanisms and its role as a source of high-power pulsed radiation fringing on the far-infrared are of importance.

The magnetron is a cylindrical structure with a cathode on the axis and an enclosing multicavity resonator or anode.[5] The cathode and anode are separated by an annular interaction space. Electrons are produced by a radial dc electric field, and influenced by an axial magnetic field to move in curved trajectories about the cathode. The motion is a cyclotron orbit of small radius superimposed on an azimuthal drift velocity equal to the ratio of the electric and magnetic fields.[52] That is, the azimuthal drift velocity is

$$v_\theta = E_r/B_z, \qquad (2.7)$$

where r, θ, and z are the cylindrical coordinates. *En masse*, the electrons form a drifting cloud of space-charge which can impart energy to the field stored in the resonant structure. The transfer can be through fundamental or space harmonic interactions as described in the previous section. However, the electron dynamics are quite different.

[52] W. W. Harman, "Electronic Motion." McGraw-Hill, New York, 1953.

2.3. ELECTRON TUBES

Because the motion of an electron in crossed electric and magnetic fields is at right angles to the force causing the motion, azimuthal bunching is produced by radial components of the oscillating electric field. Also, azimuthal fields cause radial displacements. Bunching mechanisms can be understood by reference to the motions of electrons at positions C and C' in the field pattern shown in Fig. 2.14(a). At these two positions the radial field components are oppositely directed. Electrons at point C have an azimuthal velocity

$$\frac{E_r - E_{r,\,\text{osc}}}{B_z}, \tag{2.8}$$

and those at C' have velocity

$$\frac{E_r + E_{r,\,\text{osc}}}{B_z}, \tag{2.9}$$

where $E_{r,\,\text{osc}}$ is the radial component of the oscillating electric field. There is a difference in velocities, which means that electrons overtake one another and cluster in bunches or "spokes" of rotating space-charge, as shown in Fig. 2.14(b). The effect is called field focusing. As the electrons rotate they interact with the azimuthal component of the oscillating electric field. This does not alter the azimuthal drift velocity, but it does change the radial motions of individual electrons. Those electrons which

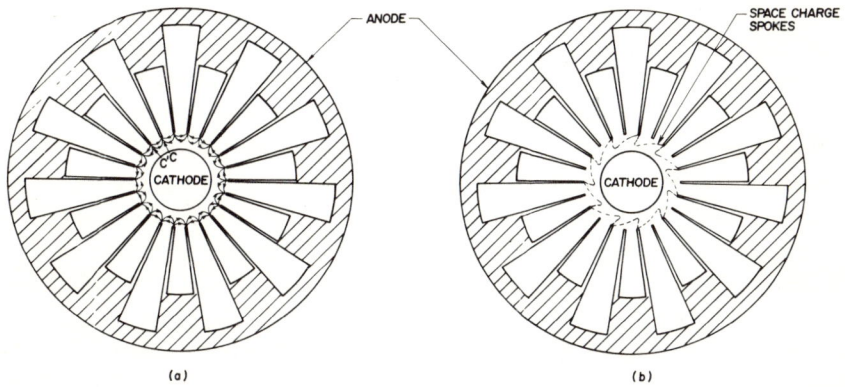

FIG. 2.14. Cross section of a 22-vane rising sun magnetron of the type used at short millimeter wavelengths. (a) The field lines of the π-mode. The position C and C' are those where the radial components of the oscillating field respectively oppose and aid the dc field. At C the azimuthal electron velocity is decreased by the effect of the radial field, and at C' it is increased. (b) The electron space-charge formed into rotating spokes (after Robinson[22a]).

are so placed that they extract energy from the field have the radii of their cyclotron orbits increased, and this causes them to be quickly returned to the cathode. On the other hand, those electrons which impart energy to the field negotiate tighter cyclotron orbits and consequently, after a relatively long distance of interaction, reach the anode. This selection process enables the build up of oscillations. Energy is gained from electron displacements in the radial direction, and is accordingly electron potential energy.

The magnetron resonator illustrated in Fig. 2.15 is of the type usually employed at short millimeter wavelengths. The anode is a "rising-sun" resonator.[5] This geometry has fabrication advantages over designs employing strapping.[5] As it is a multicavity structure, it can support a number of distinct standing-wave modes. Of these, the π-mode shown

FIG. 2.15. Photograph of the section of a rising-sun magnetron used in the Philips 2.5-mm magnetron (after Robinson[22a]).

in Fig. 2.14(a) is generally preferred, for this has the merit that it is a nondegenerate mode, that is, there is only one field distribution at its excitation frequency.[5] Because it is very difficult to eliminate oscillations in other modes, it is standard practice to develop short millimeter magnetrons by scaling well-understood centimeter wavelength designs. The consequences of frequency scaling are similar to those experienced with the previously described electron tubes. The dimensions, like those of the klystron and traveling wave tube, scale linearly with wavelength. Resistive losses increase. However, from the point of view of heat dissipation the magnetron is in a much worse position. The fragile radial vanes of the resonator must collect the electrons, while in the other tubes the functions of beam collection and the support of waves are separated. At high pulsed power, arcing can occur between the closely spaced vanes. The cathode is also in a bad position. Not only is there no possibility of beam density multiplication to relax the current emission demands, but the tiny cathode is subjected to heavy back-bombardment by returning electrons.

The most valued role of the magnetron is as a generator of microsecond-length pulses. At centimeter wavelengths it is a compact source of multimegawatt power, but at short millimeter wavelengths kilowatt powers are achieved. A peak output of 8 kW has been generated at 3 mm by a Columbia University magnetron. This is a 22-vane rising-sun magnetron, operated at 11 kV, 13 A, with a field of 23,000 G.[53] Wavelengths as low as 2.5 mm have been generated at a peak power level of 2.5 kW with 0.5 W average power and a pulse length of 0.1 μsec.[54] The resonator of a 2.5-mm, 22-vane magnetron designed at the Philips Research Laboratories, Eindhoven, is shown in cross section in Fig. 2.15.

2.3.4. Other Electron Tubes

The subject of radiation generation lends itself to invention and to the introduction of novel techniques and approaches. Consequently, between the three major electron tube types described there are many hybrid combinations, each with its particular advantages and disadvantages and prospects for far-infrared generation. In recent experimentation there has been interest in processes involving electron beams constrained to particular motions by magnetic forces. One such process is the production of cyclotron radiation from orbiting electrons. In another, undulating beams interact with fast waves in a smooth-wall waveguide, rather than with a slow wave. In the former,[55] electrons rotating at microwave Larmor frequencies are suddenly subjected to a pulsed increase in the magnetic field. As the orbits tighten, the accelerating charges emit radiation at the characteristic angular frequency $\omega = qB/m$. This principle does not demand the usual slow-wave structure, but submillimeter emission does require fields exceeding 100 kG.

In the ubitron,[56] the beam has some transverse energy and is able to support fast space-charge waves[57] which can interact with waveguide fields traveling at the velocity of light or faster. A periodic magnetic field induces beam undulations which in turn allow periodic extraction of energy by the wave. The low-loss TE_{01} mode in circular waveguide and the TE_{10} mode in rectangular waveguide can be used. The waveguide

---

[53] Columbia Radiation Lab. Quart. Reps. 1956–1958. Columbia Univ., New York.

[54] G. H. Plantinga, *Philips Tech. Rev.* **25**, 217–226 (1964).

[55] H. O. Dressel, S. M. Stone and G. E. Weibel, *Int. Congr. Microwave Tubes*, 4th, Scheveningen, Holland, September 1962.

[56] R. M. Phillips, *IRE Trans. Electron Devices* **7**, 231–241 (1960).

[57] A. H. W. Beck, "Space-Charge Waves and Slow Electromagnetic Waves." Pergamon, Oxford, 1958.

can be relatively large because the electric field intensities of fast waves increase with distance from the walls, instead of falling exponentially as do slow waves in periodic waveguides. Problems of design are really transferred to the periodic magnet, for the field of this falls exponentially with distance from the magnetic structure.

With the structures used to date, satisfactory operation is obtained only for voltages of 80 kV or more. The ubitron has not yet been extended into the far-infrared, but at a wavelength of 6 mm it has generated pulsed power levels of 150 kW. Small periodic magnets need not be essential obstacles to far-infrared fast-wave interaction devices. The well-known tendency of magnetically confined electron beams to exhibit rippled beam behavior offers a means of achieving fast-wave interactions with the use of strong uniform magnetic fields.[58]

2.4. Relativistic Electrons

Wave interactions with electrons traveling with energies in the megavolt range have particular merit in that they favor the maintenance of very tight electron bunches which can interact efficiently with waves supported by metal structures. They also offer the possibility of giving Doppler frequency multiplication of radiation.

The advantages of relativistic electrons were realized in the early 1950s,[59,60] and the subsequent decade saw efforts to exploit them in the generation of far-infrared waves. These generators are essentially pulsed sources, and their rather formidable structure is sufficient to curtail their widespread use. While they have considerable interest from a physical point of view, for many applications they are surpassed by carcinotrons, klystrons, and far-infrared lasers.

In this section we give a brief outline of some of their special physical features. Some discussion has already been given in Section 2.2.3.

2.4.1. Electron Bunching and Energy Extraction

We mentioned in Section 2.2.3 the capabilities of microtrons and linear accelerators for the production of bunched beams with high harmonic content. These devices are, of course, quite closely related to electron

---

[58] *New Sci.* **6**, 19 (1966).
[59] K. Landecker, *Phys. Rev.* **86**, 852–855 (1952).
[60] H. Motz, *J. Appl. Phys.* **22**, 527–535 (1951).

tubes, the differences arising from the different electron velocity ranges used. Electrons moving with velocities approaching the velocity of light in vacuum, c, have a number of advantages over slower electrons for the production of short wavelengths. Of particular value are the very dense electron bunches which can be formed and which can excite fields at a greater distance from the surface of a periodic structure. These properties stem from relativistic modifications of Coulomb repulsion forces and the interaction fields.

To examine the Coulomb forces between electrons moving with velocity $v \sim c$, we can imagine ourselves moving in a frame of reference in which the charges are at rest.[52] In this frame the laws of electrostatics are valid and may be applied in their usual form. Effects manifest in the laboratory frame of reference can then be found from the well-known transformation laws of relativistic electrodynamics. Let us denote quantities measured in the frame of reference moving with the electrons by primes and those in the laboratory frame without primes. The Coulomb force between two electrons moving in the z-direction and separated by a distance $\Delta z'$ in that direction is

$$F_z' = q^2/4\pi\varepsilon_0(\Delta z')^2. \qquad (2.10)$$

In a transformation to the laboratory frame, the force is unaltered but the separation undergoes a Lorentz contraction $\Delta z = \Delta z'(1 - \beta^2)^{1/2}$. The Coulomb force in the laboratory is then

$$F_z = F_z' = q^2(1 - \beta^2)/4\pi\varepsilon_0(\Delta z)^2, \qquad (2.11)$$

where $\beta = v/c$. Thus longitudinal space-charge debunching forces are reduced by a factor $(1 - \beta^2)$. Similar considerations of electrons moving in a direction at right angles to their separation show that transverse debunching forces are also reduced[52] when $v \sim c$.

Because of the large mass of a relativistic electron, and the tendency of an energy change to cause a mass change rather than a velocity change, it is difficult to bunch an electron stream after the particles have reached full energy. In the linear accelerator the beam is prebunched before injection into the periodic structure, and only the final compression is carried out during acceleration. Once the relativistic bunches are formed, however, the tendency to spread is small and they can interact with fields over long distances.

The electron interaction with the field is enhanced by relativistic effects, both during bunching and energy extraction. Whereas in a nonrelativistic structure the field falls off away from the periodic surface as $e^{-2\pi y/\lambda\beta}$ (see Section 2.3.2), in a relativistic structure the field decreases more gradually, as $\exp[-2\pi(1-\beta^2)^{1/2}y/\lambda\beta]$.† In contrast to the 10-kV case of Section 2.3.2, where electron flow must be within $\lambda/30$ of the periodic surface, for a 2-MeV beam the electrons need only be within approximately $\lambda/2$ of the surface for significant energy extraction.

Some attainments of relativistic generators have been given in Section 2.3.2. In most of the linear accelerator work, 3000-MHz waves have been used for beam bunching because, at the time the research was initiated, this frequency was the highest at which high-power microwave sources were readily available. In the work carried out by the Illinois group, 2-mm, 50th harmonics have been reached with a 1-MeV electron beam. Similar wavelengths have been generated by the Russian group[61] with a microtron buncher, but Brannen et al.,[40] have extended their 2800-MHz microtron to produce harmonic output signals in the range 500–1000 μm. Their 4-mA, 6-MeV bunched beam can produce between 10 and 50 μW in each harmonic, but with full beam current (40 mA) a pulsed output of 5 mW is anticipated at 1000 μm. There is little doubt that, with electron bunching at frequencies of 35 GHz or higher, relativistic electron generators can operate effectively throughout the far-infrared.

2.4.2. Doppler Frequency Multiplication

In schemes somewhat different from the harmonic multipliers, relativistic electrons have been studied as Doppler frequency multipliers.[59,60] In the "undulator," a relativistic bunched beam zigzags through a transverse periodic magnetic field. If l_0 is the period of the magnet structure, the Doppler shifted radiation has the frequency[60]

$$\nu = \frac{c\beta}{l_0(1-\beta\cos\theta)}, \qquad (2.12)$$

where θ is the angle between the motion and the direction of observation. As with other beam devices, the electron bunches must be tight to give reasonable power levels. If the bunches are longer than one-half wave-

---

[61] S. P. Kapitza, Lecture at Sydney Univ., 1967.

† For the usual cylindrical structures of linear accelerators, the field varies with radius r as the zero-order modified Bessel function $J_0[i2\pi(1-\beta^2)^{1/2}r/\lambda\beta]$.[49]

length of the observed radiation, the individual electrons will radiate waves so spread in phase that they tend to cancel one another. If the bunches of N electrons are short enough, the emission will be coherent, and the total emitted power will approach N^2 times the emission from an individual electron.

The undulator experiments described by Motz et al.[62] in 1953, yielded 1.9-mm radiation from a 3-MeV linear accelerator beam with a peak power of about 1 W. Grishaev et al.,[63] have generated 0.5–8-mm waves with a 2-MeV beam, and with 100-MeV electrons have observed light output. One might expect a reasonable degree of coherence in millimeter radiation emitted by the undulating bunches, but at infrared and optical wavelengths the size of bunches would be too large for coherence in the Doppler shifted radiation.

2.5. Gas Lasers

Thus far in this chapter we have considered classical oscillators, the principles of which were understood prior to the advent of quantum oscillators in the 1950s. We now turn to the processes by which coherent far-infrared radiation can be derived from stimulated photon emission. These processes rely on stimulated energy level transitions within quantized systems that have been understood for a long time. The application of the principles to lasers and masers[†] are now well described in the literature[64-66] and we can, accordingly, restrict our preliminary discussion to a few basic elements of quantum oscillator principles. We follow this with some detailed descriptions of gas lasers, for these have proven particularly significant with respect to far-infrared operation. In Sections 2.6 and 2.7 we treat solid-state and other quantum oscillators.

The first requirement for far-infrared transitions is the existence of energy eigen states (W_1 and W_2) with separations between approximately

---

[62] H. Motz, W. Thon and R. N. Whitehurst, *J. Appl. Phys.* **24**, 826–833 (1953).

[63] I. A. Grishaev, V. I. Kolosov, V. I. Myakota, V. I. Beloglasov and B. V. Yakinova, *Dokl. Akad. Nauk SSSR* **131**, 61–63 (1960).

[64] J. P. Wittke, *Proc. IRE* **45**, 291–316 (1957).

[65] B. A. Lengyel, "Lasers." Wiley, New York, 1962.

[66] A. Yariv and J. P. Gordon, *Proc. IEEE* **51**, 4–29 (1963).

---

† We will use the term "laser" rather than "maser" because the far-infrared techniques used in their construction have closer resemblance to those of light rather than microwaves.

0.1–0.001 eV. The Bohr frequency rule

$$\nu = (W_2 - W_1)/h, \qquad (2.13)$$

where $h = 6.6 \times 10^{-34}$ joule sec, then gives transition frequencies in the range 3×10^{11}–3×10^{13} Hz. Further, it is necessary to change the thermal equilibrium state populations in such a way that the higher energy state W_2 is more populated than the lower state W_1. It is well known that in thermal equilibrium at temperature T the state population ratio, given by the Boltzmann distribution law, is

$$N_1/N_2 = e^{h\nu/kT}. \qquad (2.14)$$

In thermal equilibrium the lower energy state is more populated than the higher state, but this distribution can be changed by any one of a number of means.[64,65] When the system is in an inverted state (i.e., when $N_2 > N_1$) it can be induced by a photon flux to radiate coherent waves of frequency ν.

The processes of photon absorption and emission were clarified by the treatment of Einstein in 1916.[67] From consideration of equilibrium of the quantized system with blackbody radiation Einstein showed that transitions may result from absorption, stimulated (or induced) emission, and spontaneous emission. The probability of absorption is equal to that of stimulated emission, so the ratio of photons absorbed to photons emitted under stimulation will be equal to the population ratio. An inverted system will then give net emission. Both absorption and stimulated emission are proportional to the energy density in the radiation field, but spontaneous emission is independent of the radiation. Thus if we place the inverted quantized system into a resonant wave-supporting structure, the radiation from spontaneous transitions can initiate stimulated transitions. The succession of emitted waves will then add in phase to build up strong oscillations. As the field grows the losses increase, and steady-state operation is established when all emitted power is dissipated in losses and useful power output.

Let us now consider the particular case of gas lasers. We start by generally reviewing a number of sources of lines throughout the far-infrared, and follow this with a more detailed description of some selected lasers.

Of the many energy state transitions that have been studied in laser and maser research those between vibrational and rotational states of gas

---

[67] A. Einstein, *Verh. Deut. Phys. Ges.* **18**, 318 (1916); *Phys. Z.* **18**, 121 (1917).

molecules have contributed most effectively to the far-infrared. The electronic state transitions of atoms and molecules tend to be in the visible spectrum with only higher quantum number states separated by far-infrared transition energies. Vibrational energy level transitions tend to be in the near infrared extending towards the far-infrared. Rotational level transitions are generally in the far-infrared to microwave range. In the case of polyatomic molecules, which can possess many modes of vibration, mutual interactions between various types of motion can arise leading to complicated vibrational–rotational transitions.

Population inversion in gases can be established by passing an ionizing current through the gas. The processes by which the thermal equilibrium state is changed in a gas discharge are complicated, and, in general, are incompletely understood. In complex gas mixtures there can be many dissociation products interacting with one another through collisions as well as through photon exchanges, recombining into neutral atoms and molecules, diffusing to the walls, and so forth. For most gas lasers throughout the far-infrared, these detailed processes have not yet been adequately investigated. The onset of laser action after the initiation of the discharge varies in time from one emission line to the next. In some cases emission occurs as the ionizing current builds up, and in some cases it occurs in the afterglow of the discharge.[68]

In the carbon dioxide laser vibrational–rotational transitions give emission at the long infrared wavelengths around 10 μm.[69,70] Most of the photon energy is derived from vibrational energy. The vibrational energy levels of CO_2 have associated with them many closely spaced rotational levels. Emission arises from a simultaneous vibrational–rotational transition wherein the mode of vibration changes, and at the same time the rotational quantum number changes by unity. Population inversion of CO_2 molecules is accomplished in a gas discharge via collisional resonance transfer from vibrationally excited nitrogen molecules. This is an extremely efficient "pumping" (i.e., inverting) mechanism, because a large percentage of the nitrogen in the discharge can be in an excited state from which radiative transitions are forbidden, and because of the closeness of the CO_2 and N_2 levels involved. The addition of helium to the mixture enhances the operation further. Consequently the CO_2 laser can

---

[68] W. Q. Jeffers and P. D. Coleman, *Appl. Phys. Lett.* **10**, 7–9 (1967).

[69] C. K. N. Patel, W. L. Faust and R. A. McFarlane, *Bull. Amer. Phys. Soc.* [2], **9**, 500 (1964).

[70] C. K. N. Patel, P. K. Tien and J. H. McFee, *Appl. Phys. Lett.* **7**, 290–292 (1965).

be a very powerful generator, capable of kilowatts of continuous power and very high levels of pulsed power.

Since 1963 many gas lasers have been developed with emission lines throughout the far-infrared. Atomic neon and xenon lines have given wavelengths between 9 and 25 μm, from well-understood atomic transitions.[71,72] Atomic helium has given 3-W, 1-μsec pulses at 95.788 μm.[73] Deeper in the far-infrared, transitions in molecular systems have generated many lines. About 50 lines have been reported from water vapor and deuterium dioxide discharges extending from 16 to 120 μm.[74] With a discharge several meters long many of these lines have been generated in microsecond pulses with peak power levels ranging from a fraction of a watt up to some tens of watts in certain lines. For the water vapor line at 27.9 μm, 40-W pulses have been reported.[75] Continuous laser action can also be obtained. Some of the water vapor lines have been attributed by Witteman and Bleekrode to rotational levels of the OH radical.[76]

Gas discharges through compounds of carbon, nitrogen, and hydrogen have been investigated for laser action, notably at the National Physical Laboratory and Services Electronics Research Laboratory, England. Breakdown through dimethylamine vapor $(CH_3)_2NH$ and through acetonitrile vapour (CH_3CN) has resulted in rotational transitions leading to the generation of wavelengths between 126 and 372 μm.[72,77,78] Similar lasers using iodine cyanide have produced lines at 537.7 and 538.2 μm.[79] A summary of emission lines obtained with gas lasers is given in Table 2.4.

Normally there is a continuous flow of gas through the discharge tube which replenishes the emissive molecules and removes the decomposition products. Mirrors at the ends of the discharge vessel form a Fabry-

---

[71] W. L. Faust, R. A. McFarlane, C. K. N. Patel and G. C. B. Garrett, *Appl. Phys. Lett.* **1**, 85–88 (1962).

[72] L. E. S. Mathias, A. Crocker and M. S. Wills, *Serv. Electron. Res. Lab. Tech. Rep.* No. M.237. Undated.

[73] L. E. S. Mathias, A. Crocker and M. S. Wills, *Serv. Electron. Res. Lab. Tech. J.* **17**, 6.1–6.2 (1967).

[74] L. E. S. Mathias and A. Crocker, *Phys. Lett.* **13**, 35–36 (1964).

[75] A. Crocker, H. A. Gebbie, M. F. Kimmitt and L. E. S. Mathias, *Nature (London)* **201**, 250–251 (1964).

[76] W. J. Witteman and R. Bleekrode, *Phys. Lett.* **13**, 126–127 (1964).

[77] H. A. Gebbie, N. W. B. Stone and F. D. Findlay, *Nature (London)* **202**, 685 (1964).

[78] H. A. Gebbie, N. W. B. Stone, W. Slough and J. E. Chamberlain, *Nature (London)* **211**, 62 (1966).

[79] H. Steffen, J. Steffen, J. F. Moser and F. K. Kneubühl, *Phys. Lett.* **20**, 20–21 (1966).

2.5. GAS LASERS

TABLE 2.4. Some Gas Laser Emission Lines in the Far-Infrared

Wavelength (μm)	Emissive material	Wavelength (μm)	Emissive material
~9.6 and ~10.6 with many nearby	CO_2	39.698	Water vapor
		40.627	Water vapor
		40.994	Deuterium dioxide
10.908 and many nearby	N_2O	45.523	Water vapor
		47.251	Water vapor
11.299	Xenon	47.469	Water vapor
12.917	Xenon	47.693	Water vapor
16.893	Neon	48.677	Water vapor
16.931	Water vapor	53.906	Water vapor
16.947	Neon	55.077	Water vapor
17.158	Neon	56.845	Deuterium dioxide
17.888	Neon	57.660	Water vapor
18.506	Xenon	67.177	Water vapor
21.471	Ammonia	70.6	Methyl alcohol
21.752	Neon	71.965	Deuterium dioxide
22.542	Ammonia	72.429	Deuterium dioxide
22.563	Ammonia	72.747	Deuterium dioxide
22.836	Neon	73.337	Deuterium dioxide
23.365	Water vapor	73.402	Water vapor
23.675	Ammonia	74.545	Deuterium dioxide
24.918	Ammonia	76.305	Deuterium dioxide
25.423	Neon	78.455	Water vapor
26.282	Ammonia	79.106	Water vapor
26.666	Water vapor	81.5	Ammonia
27.974	Water vapor	84.111	Deuterium dioxide
28.054	Water vapor	84.291	Deuterium dioxide
28.273	Water vapor	85.50	Water vapor
28.356	Water vapor	89.775	Water vapor
31.951	Ammonia	95.788	Helium
32.929	Water vapor	107.71	Deuterium dioxide
33.033	Water vapor	115.42	Water vapor
33.896	Deuterium dioxide	118.65	Water vapor
35.000	Water vapor	118.8	Methyl alcohol
35.090	Deuterium dioxide	120.08	Water vapor
35.841	Water vapor	126.24	Dimethylamine
36.319	Deuterium dioxide	128.74	Dimethylamine
36.524	Deuterium dioxide	130.95	Dimethylamine
36.619	Water vapor	135.03	Dimethylamine
37.791	Deuterium dioxide	140.76	Sulphur dioxide[80]
37.859	Water vapor	151.19	Sulphur dioxide[80]
38.094	Water vapor	164.3	Methyl alcohol

TABLE 2.4 (continued)

Wavelength (μm)	Emissive material	Wavelength (μm)	Emissive material
170.6	Methyl alcohol	264.6	Methyl alcohol
171.6	Deuterium dioxide	278.8	Methyl alcohol
181.90	Deuterium + bromine cyanide	292.5	Methyl alcohol
		309.94	Dimethylamine
185.5	Methyl alcohol	310.8	Cyanide compounds
190.08	Deuterium + bromine cyanide	311.08	Dimethylamine
		336.4	Cyanide compounds
190.8	Methyl alcohol	336.5	Cyanide compounds
192.72	Sulphur dioxide[80]	336.85	Dimethylamine
193.2	Methyl alcohol	369.1	Methyl alcohol
194.83	Bromine cyanide	372.80	Methyl fluoride
198.8	Methyl alcohol	386.0	Vinyl chloride
201.19	Dimethylamine	392.3	Methyl alcohol
202.4	Methyl alcohol	417.8	Methyl alcohol
204.53	Deuterium + bromine cyanide	451.9	Methyl fluoride
		496.1	Methyl fluoride
211.14	Dimethylamine	507.7	Vinyl chloride
215.35	Sulphur dioxide[80]	537.7	Iodine cyanide[79]
216.3	Helium	538.2	Iodine cyanide
220.34	Water vapor	541.1	Methyl fluoride
223.25	Dimethylamine	570.5	Methyl alcohol
223.5	Methyl alcohol	634.4	Vinyl chloride
237.6	Methyl alcohol	699.5	Methyl alcohol
253.6	Methyl alcohol	3.4 mm	HCN molecular beam gas maser[81] (power $\sim 10^{-9}$ W)
254.1	Methyl alcohol		
263.4	Ammonia		
263.7	Methyl alcohol		

Perot resonator, the length of which determines the resonant frequency. Provided the cavity resonance is within the line-width of the molecular transition, and provided there is sufficient emissive material to overcome losses, laser action will ensue. For excitation of the various resonances throughout the far-infrared, the Fabry–Perot resonator can be tuned by altering the mirror separation or by tilting the mirrors. For a single setting of the cavity several lines may be emitted.

[80] G. Hubner, J. C. Hassler, and P. D. Coleman, *Proc. Symp. Submillimeter Waves*, Brooklyn, New York, March–April, *1970*, pp. 69–78. Polytechnic Press, Brooklyn, New York, 1971.

[81] D. Marcuse, *J. Appl. Phys.* **32**, 743 (1961).

The Fabry–Perot mirrors give about 99% reflection. They can be silver, gold, or aluminium deposited on glass or quartz, and they should be silicon coated for protection from the discharge. They may be plane or curved, or one may be plane and the other curved. Plane-curved and confocal combinations are less subject to diffraction losses out of the edges of the system.[82,83] Useful output can be taken from a hole in the center of one mirror, from radiation diffracted around the edge of one mirror, or it may be reflected out by a beam splitter (see Chapter 3) located between the mirrors but preferably in a position where it is separated from the discharge. This latter scheme has the disadvantage that the thin film used for beam splitting adds to the resonator losses, but it has the advantage of being adjustable in angle so that the power coupled out of the laser can be optimized.

Pulsed-power levels ranging from tens of milliwatts up to hundreds of watts can be generated with HCN lasers. With a length of a meter or two, and a diameter of 5–10 cm, there is sufficient emissive material to overcome losses in the system and give useful power output. The level of operation grows rapidly as the length and diameter of the discharge tube are increased.

Continuous wave emission in the HCN line at 336.8 μm can readily give power levels of 0.1–1 W according to the magnitude of the discharge current. For the continuous generation of 100 mW, a laser 1-m long, and a 2-kV, 1-A power supply are adequate.

Continuous laser action has also been obtained by pumping various gases with CO_2 and N_2O laser beams. Many molecular gases have vibrational–rotational transitions in close coincidence with lines from these lasers and can be pumped into higher vibrational states. Within a single vibrational state, rotational inversion can be established and, provided the relaxation times are favorable and the molecule has a permanent dipole moment, radiative rotational transitions can build up to give laser action.

The N_2O laser line at 10.78 μm coincides with levels of the ammonia molecule separated by a quantum of vibrational energy, and this has enabled Chang et al.[83a,83b] to pump a system of NH_3 molecules to an excited state. Subsequent stimulation of pure rotational transitions has given

[82] A. G. Fox and T. Li, *Proc. IEEE* **51**, 80–89 (1963).
[83] A. G. Fox and T. Li, *Bell. Syst. Tech. J.* **40**, 453–488 (1961).
[83a] T. Y. Chang, T. J. Bridges and E. G. Burkhardt, *Appl. Phys. Lett.* **17**, 357–358 (1970).
[83b] T. Y. Chang, T. J. Bridges and E. G. Burkhardt, *Appl. Phys. Lett.* **17**, 249–251 (1970).

emission at 82.5 μm, and transitions between two inversion states of the vibrating NH_3 molecules have resulted in laser emission at 263.4 μm. When the N_2O pump is at a level of 1.5 W, the submillimeter laser power is about 0.1 mW.

In the case of CO_2, laser radiation at 32 submillimeter wavelengths has been generated in gaseous methyl fluoride, methyl alcohol, and vinyl chloride.[83] Many of these lines are included in Table 2.3. They range from 70.6–699.5 μm, and the majority of them have oscillated continuously.

Chang et al. have pumped these gases with single lines from the 9.6- and 10.6-μm band of the CO_2 laser. Single lines can be generated at a level of about 2 W continuous wave (CW) when a replica diffraction grating (see Chapter 3) is used as one end mirror of the laser. The CO_2 line is focused into a cavity 77 cm long and 4.7 cm in diameter which is enclosed by Fabry–Perot mirrors, and where it executes a large number of round trips and excites the gas. Efficient absorption at the pump frequency leads to laser action from 0.1 mW to a few milliwatts and provides a useful source covering a wide range of wavelengths throughout the far-infrared.

2.5.1. Hydrogen Cyanide Lasers

We have described in a general way the capabilities of HCN lasers, and have listed in Table 2.4 the many emission lines they give. The line at 337 μm is particularly strong and easy to excite. Because it is located at a position in the far-infrared where there are very few competing sources, it is worthy of detailed description. In turning to a detailed discussion of the HCN laser we set out to describe some of the essential developments leading to the present level of understanding of this laser, and we give also some detailed design and performance data for a 337-μm laser based on a discharge through a mixture of methane and nitrogen. This mixture of gases is particularly easy to handle, and it makes a useful laboratory laser.

The work that has been done with the cyanide laser can be divided into four categories: (a) spectroscopic studies of the line emissions and the consequent identification of the HCN molecule as the source of the 337-μm line and other lines; (b) studies of modes of excitation of the resonator; (c) observations of the output power, pulse shapes, etc., under a variety of discharge conditions; and (d) studies of chemical and physical properties of the active emissive medium itself. Category (a) has been investigated in considerable detail by Hocker and Javan,[84,85] Lide and Maki,[86,87] and others, while the question of cavity modes has been treated by Kogelnik and Li,[88] Kneubühl and Steffen,[89,90] and others. Categories (c) and (d) have been less thoroughly investigated, but some of the essential features will be described in what follows.

Dimethylamine [$(CH_3)_2NH$], methyl cyanide (CH_3CN), ethyl cyanide (C_2H_5CN), ethylenediamine [$C_2H_2(NH_2)_2$], propylamine ($C_3H_7NH_2$), and mixtures of methane and ammonia, nitrogen and acetone [$(CH_3)_2O$], nitrogen and methane, as well as pure HCN gas, have been used to produce 337-μm laser radiation. Lide and Maki[86] have shown that the line originates from transitions involving the 11^10 and 04^00 vibrational states of the linear molecule HCN. The rotational levels associated with these vibrational states are nearly degenerate at $J = 10$, and coriolis forces result in mixing of wavefunctions and significant matrix elements for transitions between the $J = 10(11^10)$ and $J = 9(04^00)$ states. The energy level notation used here is described in Chapter 7.

The processes of the electrical discharge dissociate the molecules, partially ionize, and provide conditions for the formation of HCN with an excess population of 11^10 vibrons relative to 04^00 vibrons. It is known that a discharge through methylamine produces about twenty different species including about 4.5% CN radicals and 2.7% HCN molecules.[91] While the detailed chemical processes have not yet been unravelled, experiments by Murai[92] and Frayne[93] point to a prominent role being played by reactions at the walls of the discharge vessel. During operation of the laser, a brown polymer film forms on the walls of the vessel. Murai has observed spectroscopically that a discharge through nitrogen in such a contaminated vessel results in the formation of CN radicals, and that this production is enhanced by the addition of bromine to the discharge. He suggests the CN thus produced may react with hydrogen in a lasing mixture to produce HCN. In support of this it is found that a given static

---

[84] L. O. Hocker, A. Javan, D. Ramachandra Rao, L. Frenkel and T. Sullivan, *Appl. Phys. Lett.* **10**, 147–149 (1967).

[85] L. O. Hocker and A. Javan, *Phys. Lett. A* **25**, 489–490 (1967).

[86] D. R. Lide, Jr. and A. G. Maki, *Appl. Phys. Lett.* **11**, 62–64 (1967).

[87] A. G. Maki, *Appl. Phys. Lett.* **12**, 122–124 (1968).

[88] H. Kogelnik and T. Li, *Proc. IRE* **54**, 1312–1329 (1966).

[89] F. K. Kneubühl and H. Steffen, *Phys. Lett. A* **24**, 639–640 (1967).

[90] H. Steffen and F. K. Kneubühl, *IEEE J. Quantum Electron.* **4**, 992–1008 (1968).

[91] C. A. McDowell and J. W. Warren, *Trans. Faraday Soc.* **48**, 1084–1093 (1952).

[92] A. Murai, *Jap. J. Appl. Phys.* **8**, 250–254 (1969).

[93] P. G. Frayne, *Proc. Phys. Soc. London (At. Mol. Phys.)* **2**, 247–259 (1969).

filling of gas will lase for hours if bromine is included in the gas mixture, but only for minutes if it is not. Frayne has observed that the wall temperature influences the laser performance.

Thus it appears that wall effects play an important role in determining the amount of HCN in the laser. In the case of a sealed vessel containing a static filling of gas, several discharge pulses are required before lasing starts, the pulse power builds up for a few hundred pulses reaching a maximum beyond which it declines until lasing ceases after a few hundred further pulses.

That the function of the applied electric field is to cause breakdown of the gas molecules and, rather than directly excite HCN molecules to the necessary state of inverted population, to create conditions which influence the later formation of excited vibrons is suggested by the observations of time delays between the cessation of the current pulse and the onset of lasing. With short (microsecond) current pulses lasing occurs only in the afterglow.[94–96] In the experiments performed by Frayne with a methane–ammonia laser 7.3 m long, carrying current pulses of 50-μsec duration, it was necessary to delay the Q-switching mirror by at least 6 msec to ensure a state of negative absorption, and it was found that population inversion sufficient for laser action persisted for a further 18 msec. Notwithstanding, the effect on laser action of the voltage used to produce breakdown is quite pronounced and, as we shall see, there is an optimum value for maximum 337-μm radiation output.

Generally, the radiation is generated in the afterglow in short (tens of microseconds) pulses but their lengths have been increased to 500 μsec with more sustained discharge excitation,[94,97] and multiple pulse trains as long as 450 μsec have been generated with very large current pulses (kiloamperes) 1 μsec long.[95] With long current pulses of sufficient amplitude, lasing can occur during the period of active discharge excitation but under these conditions the output radiation occurs in a succession of short "spikes" followed by the usual pulse in the afterglow which is both larger in amplitude and longer in duration. Experimental records of the detailed features of HCN laser pulses showing spiking have been

[94] H. Yoshinaga, S. Kon, M. Yamanaka and J. Yamamoto, *Sci. Light* (*Tokyo*) **16**, 50–63 (1967).

[95] R. Turner, A. K. Hochberg and T. O. Poehler, *Appl. Phys. Lett.* **12**, 104–106 (1968).

[96] L. E. S. Mathias, A. Crocker and M. S. Wills, *IEEE J. Quantum. Electron.* **4**, 205–208 (1968).

[97] S. Kon, M. Yamanaka and J. Yamamoto, *Jap. J. Appl. Phys.* **6**, 612–619 (1967).

2.5. GAS LASERS

published by Kon *et al.*, and an explanation in terms of plasma tuning of the cavity has been put forward by Robinson and Whitbourn.[98]

A Laboratory Laser

The layout of a hydrogen cyanide laser and its pulsed power supply is illustrated in Fig. 2.16. Nitrogen and methane are mixed and flow through the glass vessel, where the pressure and gas flow rate are controlled by needle valves at the inlet and a baffle valve at the pump end. The vessel is 7.6 cm in diameter and 2.2 m long, and the gain is such that only the 337 μm line is excited to any marked extent. Breakdown is produced between annular electrodes 1.5 m apart when the firing of the ignitron discharges the capacitor C_1.

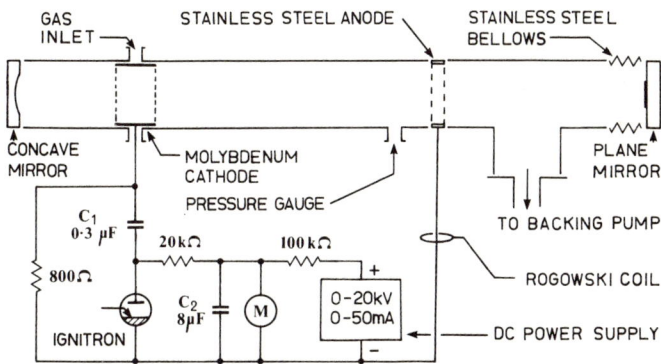

FIG. 2.16. Diagram of the HCN gas laser. The laser vessel has a length of 2.2 m between mirrors, a diameter of 7.6 cm, and discharge electrodes 1.5 m apart. The voltages recorded in this paper are measured with the meter M and currents with the Rogowski coil (after Robinson and Whitbourn[98]).

The ends of the discharge vessel are closed with the laser mirrors. One mirror is silvered to a diameter of 5.1 cm and is concave with radius of curvature 8 m, while the other is plane and silvered to 3.8 cm on a quartz plate. Such a plane–concave geometry is not oversensitive to mirror alignment, and the cavity can always be tuned into resonance by tilting the plane mirror by means of a tripod micrometer mirror mount. Power is taken from the laser by diffraction around the edge of the plane mirror, transmitted through the quartz, and then directed through a tapered light pipe to the detector. Alternatively, a mirror with a 1-cm diameter hole in the center can be used for the extraction of power.

[98] L. C. Robinson and L. B. Whitbourn, *Proc. IRE* (*Aust.*) **32**, 355–360 (1971).

The diffraction loss at the output mirror is about 20% per pass. Thus, one photon in five will escape from the resonator per transit, and it follows that the Q of the resonator must be $2\pi \times 5 \times 2L/\lambda = 3 \times 10^5$, L being the length of the resonator and λ the free-space wavelength.

The discharge is operated, typically, with voltages of 4–10 kV, currents of some hundreds of amperes peak amplitude, and durations of about 3–4 μsec. In addition to the capacitor C_1, the discharge circuit has an inductance of about 4 μH and a resistance that varies over a wide range during a pulse, and also changes with gas mixture, pressure, and the energy supplied to the discharge. Figure 2.17 shows the variation of the plasma impedance for several gas mixtures and voltages. As the energy ($\frac{1}{2}C_1V^2$) to the gas is increased, the discharge is more highly ionized and the plasma impedance decreases. The circuit is critically damped or overdamped if the discharge resistance is about 7 Ω or more. When it decreases to below 1 Ω or so, the circuit has an increasing tendency to ring, and reverse current may or may not be conducted by the ignitron.

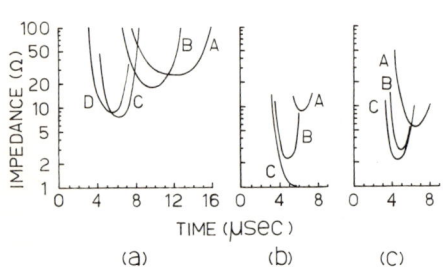

FIG. 2.17. Impedance of the HCN discharge as a function of time. (a) Ratio of nitrogen to methane, $R = 1$, total pressure 0.65 torr. Curves A, B, C, and D are for 5.0, 5.5, 6.5, and 7.0 kV, respectively. (b) $R = \frac{1}{2}$, total pressure 0.73 torr. Curves A, B, and C are for 5.1, 6.6, and 8.0 kV, respectively. (c) $R = \frac{2}{3}$, total pressure 0.63 torr. Curves A, B, and C are for 5.2, 6.9, and 7.9 kV, respectively (after Robinson and Whitbourn[98]).

When substantial reverse currents do flow, the laser output drops significantly, and, as the ignitron is erratic in its conduction of reverse current, wide fluctuations in power occur from pulse to pulse. The problem is overcome by operating near critical damping or in an overdamped regime. When care is taken to remove this and other electronic causes of pulse energy variation, one can achieve operation where the pulse-to-pulse variation is set by the reproducibility of the discharge itself. Figure 2.18 shows superpositions of pulses recorded when the laser operates at various voltages. Figure 2.18(c) is at the voltage for maximum laser output and shows better pulse-to-pulse reproducibility than is obtained with higher and lower voltages.

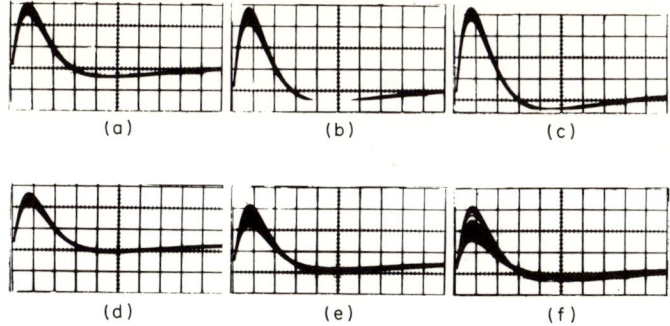

FIG. 2.18. Superpositions of 50 pulses recorded by the Golay cell. The oscilloscope time scale is 5 msec/large division. The gas discharge conditions correspond to those of curve 2, Fig. 2.19a, with: (a) 4.8 kV, (b) 5.0 kV, (c) 5.3 kV, (d) 6.3 kV, (e) 6.5 kV, and (f) 6.7 kV. Minimum pulse-to-pulse fluctuation is seen to occur at the voltage which produces maximum output, and the fluctuations increase for higher voltages (after Robinson and Whitbourn[98]).

Dependence of Laser Output on Various Parameters

I. CONTINUOUS GAS FLOW. The power generated by the laser is a function of the voltage, gas mixture, pressure, volumetric flow rate, and pulse repetition frequency. When the voltage is changed for a set of fixed pressures, one observes the marked variation of power in Fig. 2.19. Maximum power levels on these graphs are approximately 0.1 W. At any

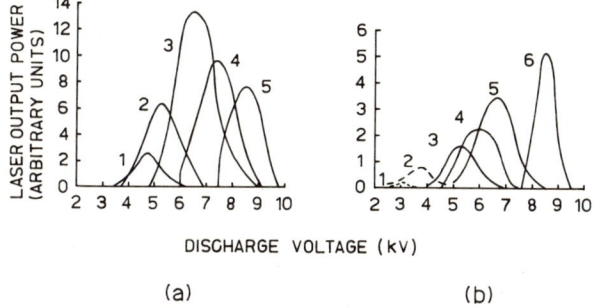

FIG. 2.19. Laser output power as a function of voltage at a repetition rate of 10 pulses/sec. (a) A mixture of nitrogen to methane in the ratio $R = \frac{1}{2}$; curves 1, 2, 3, 4, and 5 are for total pressures of 0.37, 0.55, 0.73, 0.80, and 0.96 torr, respectively, and flow rates of about 0.5 l/sec. (b) A mixture with $R = 1$; curves 1, 2, 3, 4, 5, and 6 are for total pressures of 0.24, 0.32, 0.45, 0.60, 0.65, and 1.1 torr, respectively. Flow rates are 0.5 l/sec for curves 1, 2, and 6, and 0.8, 0.7, and 0.9 l/sec for curves 3, 4, and 5, respectively. The dashed curves, 1 and 2, are five times actual size (after Robinson and Whitbourn[98]).

particular pressure there is an optimum voltage either side of which the power falls off. The reason for the falloff with increasing voltage is not known. In Fig. 2.19(b) the ratio of nitrogen to methane (1:2) found to give maximum pulsed output has been used. As shown in Fig. 2.19(b) for a mixture richer in nitrogen, the maximum power is less and the dependence on pressure is of the same form. For gas mixtures with reduced nitrogen content, the form of the voltage variation is the same but the maximum power is much lower. With a mixture of two parts nitrogen to nine methane the peak power is about 10% of that generated in the 1:2 mixture.

For constant voltage, pressure, and flow rate, there is an optimum pulse repetition frequency which produces maximum energy per laser pulse. Figure 2.20(a) shows this effect for two different flow rates. The existence of an optimum pulse repetition frequency which increases with increasing flow rate indicates that there is an optimum exposure of a volume element of the gas mixture to the succession of discharge pulses at this pressure. Under the conditions in Fig. 2.20(a) maximum power is generated when a molecule experiences very approximately 110 discharge pulses in negotiating the laser vessel, and it is natural to interpret this as due to a balance of the hydrogen cyanide production and deactivation chemistry, presumably involving processes at the walls of the vessel. The optimum number of discharge pulses increases approximately

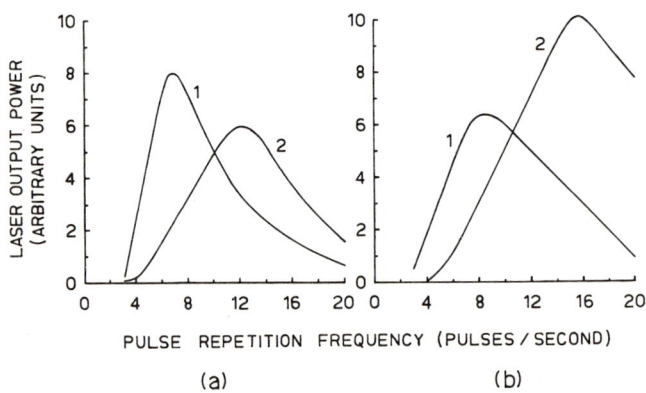

FIG. 2.20. Laser power output as a function of pulse repetition frequency. (a) A mixture with equal quantities of nitrogen and methane at a voltage of 6.0 kV, total pressure 0.58 torr. Curve 1 is for a flow rate of 0.5 l/sec, and curve 2 is for 0.75 l/sec. (b) A mixture with 1 part nitrogen to 2 parts methane, and a flow rate of 0.68 l/sec. Curve 1 is for a discharge voltage of 5.0 kV and a total pressure of 0.37 torr, and curve 2 is for 6.5 kV and 0.74 torr (after Robinson and Whitbourn[98]).

linearly with pressure. This is shown in Fig. 2.20(b) for a gas mixture with 1 part nitrogen to 2 parts methane.

For constant voltage, pressure, and pulse repetition frequency the power is sensitive to flow rate, and there is an optimum value of this parameter also. On the basis of the previous paragraph one might expect the flow rate and the pulse repetition frequency to vary in approximate proportion for the maintenance of optimum power conditions. This is found to apply only very roughly.

II. STATIC FILLING. With a static filling of gas, lasing starts only after several discharge pulses; the pulse power then rises to reach a maximum after a definite number of discharge pulses and falls gradually to zero by the time approximately twice this number of pulses has elapsed. This behavior is consistent with the observation of the previous section that maximum power output occurs under continuous flow conditions for particular combinations of flow rate and pulse repetition frequency which subject an element of the gas to an optimum number of discharge pulses.

Results of a static filling experiment plotted in Fig. 2.21 show not only that the power rises to a maximum and subsequently decays but that the pressure in the discharge vessel behaves similarly. It increases rapidly

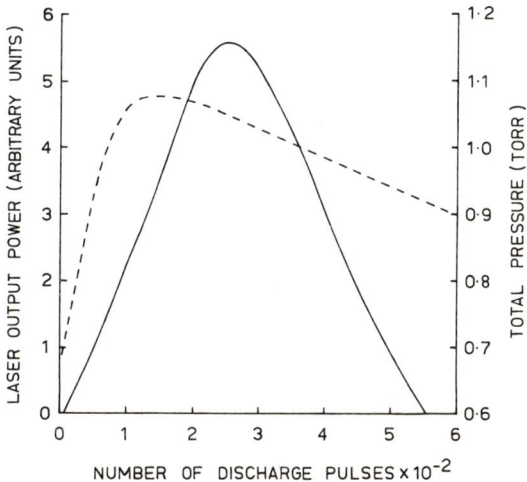

FIG. 2.21. The dependence of laser output power (solid curve) and gas pressure (dashed curve) on the number of discharge pulses in a static filling of gas. Both power and pressure show an initial increase and subsequent decline. The gas mixture is 1 part nitrogen to 2 parts methane, the voltage is 7.0 kV, and the repetition frequency 1 pulse/sec (after Robinson and Whitbourn[98]).

with the early discharge pulses and rises to a maximum somewhat before maximum laser output is reached. Beyond this there is a steady fall in pressure. The initial pressure rise is a natural consequence of the dissociation of the molecules in the gas mixture and the formation of new molecules such as HCN. The pressure drop is probably due to the formation of the brown wall deposit mentioned earlier and the associated removal of HCN from the gas mixture.

Plasma Effects

Electron densities in the discharge have been measured with an 8.8-mm Mach–Zehnder microwave interferometer with the wave propagating across the diameter of the laser vessel.[98] A wavelength of 8.8 mm sets an upper limit for electron density measurements of $n_c = 1.44 \times 10^{13}$ cm^{-3}, above which the effective refractive index of the plasma becomes imaginary and propagation is cut off, and it readily gives accurate measurements in a range one order of magnitude below this. The reader will find a discussion of the physics of microwave plasma interferometry in Chapter 6.

In Fig. 2.22 the variation with time of electron density produced for a number of voltage settings on the maximum power curve 3 of Fig. 2.19(b) is shown. With the higher voltages the density can rise above 1.44×10^{13} cm^{-3}, fall in a few microseconds to approximately 10^{12} cm^{-3}, and thereafter decay more slowly.

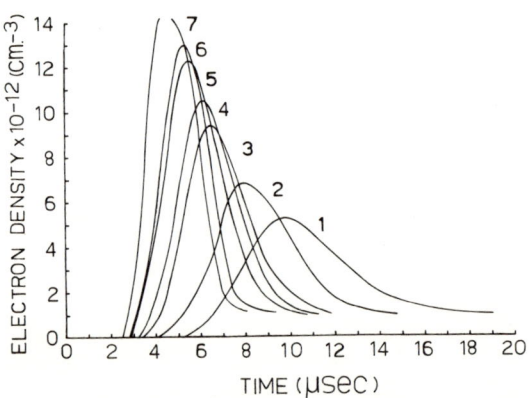

FIG. 2.22. Time-dependent electron density in the laser discharge as a function of time. The density curves correspond to curve 3 of Fig. 2.19, that is, 1 part nitrogen to 2 parts methane at a pressure of 0.73 torr, a repetition frequency of 10 pulses/sec, and a flow rate of 0.5 l/sec. The voltages for curves 1–7 are 4.3, 4.9, 5.4, 5.8, 6.5, 6.8, and 7.7 kV, respectively (after Robinson and Whitbourn[98]).

The most obvious effect of a changing electron density is that it will tune the resonances of the Fabry–Perot. The Doppler width of the HCN line is approximately 10 MHz, corresponding to a 20-μm range in cavity length, and the fundamental modes $\text{TEM}_{q,0,0}$ and $\text{TEM}_{q+1,0,0}$ of the Fabry–Perot are 168 μm apart. Higher modes $\text{TM}_{q,m,n}$ may be excited, and this will increase the density of modes, but, generally, the net mode spacing is greater than the gain width of the emission.[99] Furthermore, we have seen that the resonator Q is about 3×10^5, so the width of a Fabry–Perot resonance is about 3 MHz. Thus only one resonance will be within the laser gain curve.

The effective length of the resonator at 337 μm is (see Chapter 6)[98]

$$L' + L[1 - (n/N_c)]^{1/2} \simeq L' + L[1 - (n/2N_c)],$$

where $N_c = 9.8 \times 10^{15}$ cm^{-3} is the cutoff density corresponding to this far-infrared wavelength, L is the length of the plasma column, and $L' + L$ is the total length of the resonator. The rate of change of effective length is then $(L/2N_c)(dn/dt)$, and it is clear that the falling density in Fig. 2.22 rapidly sweeps successive modes through the Doppler line.

The work of Kon et al.[97] has shown that spiking pulses in the laser output occur when the current pulses are long enough for lasing to start while current is still flowing. Spiking pulses have been clearly recorded for 8 μsec current pulses. This spiking probably has its origin in the plasma tuning effects described here.[98,99a]

2.5.2. Carbon Dioxide Lasers

The CO_2 laser operates near 10 μm—a little outside the wavelength range of central interest in this text. However, because the H_2O laser and the HCN laser generate lines from 26 μm up to 774 μm, a brief description of the 10-μm source may serve to balance the spectral coverage.

The laser system illustrated in Fig. 2.16, composed of a gas discharge vessel with Fabry–Perot resonator mirrors at its ends, can be operated with a variety of gases. It is equally suitable for CO_2, water vapor and HCN, provided the transmitting window materials, and the output coupling, are chosen to suit the wavelength. With a single apparatus the experimenter then has access to a wide range of fixed laser lines.

In the case of CO_2, very high powers, by both continuous and pulsed waves, can be generated.

[99] H. Steffen, J. F. Moser and F. K. Kneubühl, *J. Appl. Phys.* **38**, 3410–3411 (1967).
[99a] L. B. Whitbourn, L. C. Robinson, and G. D. Tait, *Phys. Lett.* **38A**, 315–317 (1972).

Mixtures of carbon dioxide, nitrogen, and helium provide particularly efficient mechanisms of population inversion when they are ionized in a gas discharge. Primarily this has its origin in the near coincidence of a nonradiative excited vibrational state of N_2 at 2329.66 cm^{-1} and a state of the CO_2 molecule at 2349.3 cm^{-1}. Because N_2 is homonuclear, it has zero dipole moment, and its radiative transitions are forbidden. During a collision with a CO_2 molecule, the nitrogen readily transfers its vibrational energy thus populating the CO_2 energy state at 2349.3 cm^{-1} relative to lower states. Transitions to vibrational states at 1285.5 and 1388.3 cm^{-1} then give laser emission at 10.4 and 9.4 cm^{-1}, or rather, transitions between rotational states associated with the vibrational states can give emission in a number of lines near these frequencies.

Transitions involving several levels of the rotational branches of the vibrational levels give a multifrequency output in the region of 10.6 μm. More precisely, a line at 10.61 μm originates from a transition involving the rotational quantum number $J = 22$ of a P branch, and there are five equally spaced lines, one either side of this with separations of 0.02 μm and corresponding to values of J from 12–32. Details of the transitions, and the question of mode designation and notation, are discussed in a recent review by Tyte.[100]

Gas mixtures are usually in the ratio $CO_2:N_2:He = 1:1:8$ or $1:2:8$, or thereabouts. Nitrogen has the function, as we have said, of pumping the CO_2 molecules into their upper excited state. The main influence of the He is to depopulate the lower laser level. Thus with both mechanisms operative the CO_2 molecules are in a particularly favored position to lase with high efficiency.

As shown in Fig. 2.23 the optimum total pressure and discharge current for maximum CW oscillator output are functions of the diameter of the discharge tube. Efficient operation can be achieved in a sealed-off tube, but in most lasers a gas flow system which changes the gas some 400 times per sec is used. Power levels of 50 W cm^{-1} and higher can be obtained, and with long discharge tubes the total power can be many kilowatts.

Most of the interest and activity with CO_2 lasers has centered around high power CW operation. But pulsed operation has been studied to some extent. It is found that after breakdown with a short excitation pulse (current $\sim$0.5 A) there is a delay before lasing commences, and then the emission of pulses from a few tens to a few hundreds of micro-

---

[100] D. C. Tyte, *Advan. Quantum Electron.* **1**, 129–198 (1970).

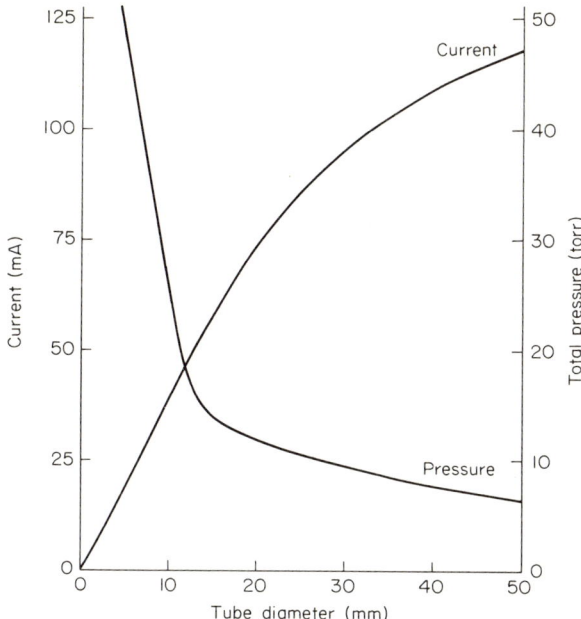

FIG. 2.23. Optimum current and total pressure for maximum output power from a CW CO_2 laser, as a function of tube diameter. The curves represent the general region of the optima rather than precise values, for there is some dependence on gas flow rate, gas mixture, etc. (after Tyte[100]).

seconds in length. The delay and the pulse length are determined by the gas pressure and other factors. Peak output power levels can be hundreds of times larger than that generated in CW operation. Hill[101] has generated peak power levels of 200 kW in pulses ∼25 μsec long by exciting the discharge with voltages of 200 kV to 1 MV. He used a mixture of carbon dioxide, nitrogen, and helium in a 15-cm diameter vessel at a total pressure of 500 Torr. However, improved methods of gas excitation have been devised which have generated peak powers of 100 MW in a laser 10 ft long operating at atmospheric pressure. In a laser only 1 m long, but with a high gas flow, 150-W mean power has been generated at an efficiency of 17%.[101]

The CO_2 laser is well suited to Q-switching. If one mirror of the resonator is rotated, the upper laser level can be densely populated when the cavity is off-tuned, while at the same time the lower level remains nearly empty. When the mirrors rotate into alignment, large pulses of

[101] A. E. Hill, *Appl. Phys. Lett.* **12**, 324–327 (1968).

radiation can be emitted. In practice, peak power levels of 10 kW m^{-1} in 100 nsec pulses have been obtained in this way.

The technique of mode locking[102] that is so effective at optical frequencies for the production of trains of very short high-power pulses, can also be used with the CO_2 laser. But it is much less effective in this case. Because the CO_2 linewidth is narrow, it is difficult to fit even three longitudinal modes of the resonator into the gain-width. Thus the short pulses associated with the momentary in-phase emission of these modes will not be particularly large in amplitude. Nevertheless, the technique has been used successfully for the production of 100-μsec long trains with pulses of over 10 kW peak power, 20 nsec long, and separated by about 100-nsec intervals.

2.6. Solid State Lasers and Junctions

2.6.1. p-n Junction Lasers

The p-n junction diode or "injection" laser was invented in 1962 as a source of near monochromatic radiation at a wavelength of 0.85 μm. It has since been extended to longer infrared wavelengths, and can operate at least in the lower part of the far-infrared. p-n junction lasers differ from the lasers of the previous section in that the energy level transitions are not between discrete energy levels but rather are transitions between energy bands of semiconductors. Population inversion can be produced by conduction currents through the junction, which inject minority carriers into the materials. These minority carriers (electrons in the p-material and holes in the n-material) in recombining with carriers of the opposite sign (that is, majority carriers: holes in the p-material and electrons in the n-material) may undergo radiative transitions. If the boundaries of the diode are cut perpendicular to the plane of the junction to form a plane-parallel Fabry–Perot resonator, the emitted waves can add coherently to build up laser action in the region of the depletion layer of the junction.

Electron injection into the p-material is illustrated in the energy-band diagram of Fig. 2.24. The probability of occupation of an energy state E is given by the Fermi–Dirac statistical distribution:

$$f = \frac{1}{1 + \exp[(w - \phi)/kT]}, \qquad (2.15)$$

---

[102] A. J. De Maria, D. A. Stetser and W. H. Glenn, Jr., *Science* **156**, 1557–1568 (1967).

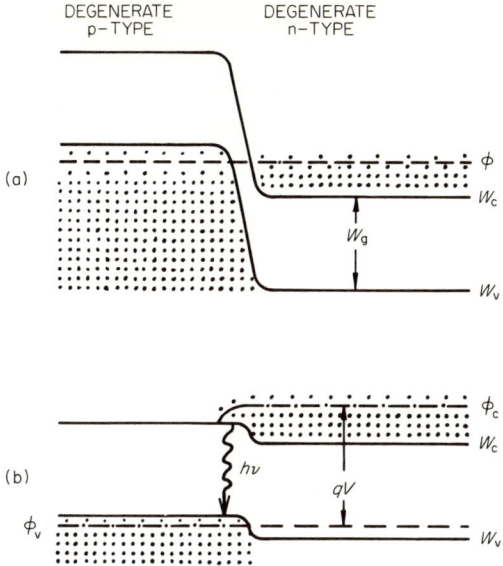

Fig. 2.24. Energy-band diagram of a p–n junction: (a) zero applied voltage, (b) applied voltage $V > W_g/q$ (after Rediker[104]).

where ϕ is the Fermi energy, defined as the energy of the state for which $f = \frac{1}{2}$. For the case of zero bias voltage [Fig. 2.24(a)] between heavily doped degenerate semiconductors, the Fermi level lies below the top of the valence band in the p-material and above the bottom of the conduction band in the n-material. With the application of a forward bias voltage across the junction (that is, the p-type crystal is made more positive) the flow of electrons from n to p is increased. For a bias voltage of the order of E_g/e there is a large spillover of electrons to the p-side.

The conditions for population inversion can be obtained by comparing the number of quanta emitted and absorbed, to find the circumstances in which stimulated emission exceeds absorption. If, in the biased diode, the electrons in the conduction and valence bands come to equilibrium among themselves in a time short compared with their lifetime, one can express the population distribution in the two bands in terms of quasi-Fermi levels:

$$f_{c,v} = \frac{1}{1 + \exp[(w_{c,v} - \phi_{c,v})/kT]}, \quad (2.16)$$

where the subscript c refers to the conduction band, and v to the valence band. The quasi-Fermi levels for the electrons in the conduction and valence bands, ϕ_c and ϕ_v, respectively, are shown in Fig. 2.24(b). The number of quanta emitted per unit time is proportional to the probability f_c of an electron being in the conduction band, and to the probability of there being an empty state in the valence band. That is,

$$N_{\text{emitted}} \propto f_c(1-f_v). \tag{2.17}$$

Conversely, the number of quanta absorbed per unit time is

$$N_{\text{absorbed}} \propto f_v(1-f_c). \tag{2.18}$$

Equations (2.17) and (2.18) have the same proportionality constant. This constant is determined by the Einstein coefficient for absorption (which is equal to that for emission), the energy density of the radiation field, and the densities of states in the valence and conduction bands.

Stimulated emission will exceed absorption when

$$N_{\text{emitted}}/N_{\text{absorbed}} = f_c(1-f_v)/f_v(1-f_c) > 1.$$

Substitution from Eq. (2.16) into this inequality gives the condition

$$\phi_c - \phi_v > h\nu, \tag{2.19}$$

where

$$w_c - w_v = h\nu. \tag{2.20}$$

Thus the condition for population inversion is that the forward voltage applied across the junction, which is equal to $\phi_c - \phi_v$, must exceed the energy of the emitted photon.[103,104] This was first pointed out by Bernard and Duraffourg.[105] The inequality, in turn, puts conditions on the concentration of impurities necessary to give laser action.[103,104].

We have assumed the photon emission arises from direct radiative recombination across the energy gap. However, there are other possible transitions. In compensated semiconductors where there are both donor and acceptor levels (or bands in the case of high concentration semiconductors) in the energy gap, transitions to the valence band can proceed via capture in these levels.[104] Such indirect radiative transitions can give rise to longer wavelength emission. Provided there are sufficient radiative transitions between any of the levels to overcome losses, laser action can occur.

[103] G. Burns and M. I. Nathan, *Proc. IEEE* **52**, 770–794 (1964).
[104] R. H. Rediker, *Phys. Today*, **18**, No. 2, 42–50 (1965).
[105] M. G. A. Bernard and G. Duraffourg, *Phys. Status Solidi* **1**, 699 (1961).

2.6. SOLID STATE LASERS AND JUNCTIONS

TABLE 2.5. Operating Wavelengths of p-n Junction Lasers

Semiconductor	Wavelength (μm)	Temperature (°K)	Reference
Ga(As$_{1-x}$P$_x$) alloy	0.71	77	105
GaAs	0.84	77	106
	0.90	300	106
In(P$_{0.49}$As$_{0.51}$)	1.6	77	107
InAs	3.11	4.2	108
	3.15	77	108
InSb	5.2	1.7	109
PbTe	6.5	12	110
Pb$_x$Sn$_{1-x}$Se	15	—	111
PbSe	7.3–22	77	112, 113, 111
Pb$_{1-x}$Sn$_x$Te	32	—	113a

Table 2.5 lists some of the wavelengths generated by p-n junction lasers. These lasers have commonly been operated in pulses, but continuous operation is possible provided the power dissipation in the junction is not too great. It has been indicated by Rediker[104] that the GaAs diode has emitted up to 6 W of continuous radiation at approximately 50% efficiency, when operated between 4 and 20°K. In pulsed operation at room temperature, they can produce 20 W of peak power in pulses 50 nsec long. The threshold injection current densities required to produce population inversion are generally some thousands of amperes per square centimeter.

[106] R. J. Keyes and T. M. Quist, *Proc. IRE* **50**, 822–823 (1962).
[107] F. B. Alexander *et al.*, *Appl. Phys. Lett.* **4**, 13–15 (1964).
[108] I. Melngailis, *Appl. Phys. Lett.* **2**, 176–178 (1963).
[109] R. J. Phelan, A. R. Calawa, R. H. Rediker, R. J. Keyes and B. Lax, *Appl. Phys. Lett.* **3**, 143–145 (1963).
[110] J. F. Butler, A. R. Calawa, R. J. Phelan, T. C. Harman, A. J. Strauss and R. H. Rediker, *Appl. Phys. Lett.* **5**, 75–77 (1964).
[111] J. F. Butler, A. R. Calawa, R. J. Phelan, A. J. Strauss and R. H. Rediker, *Solid State Commun.* **2**, 303–304 (1964).
[112] R. H. Rediker, Private communication, 1967.
[113] J. M. Besson, J. F. Butler, A. R. Calawa, W. Paul and R. H. Rediker, *Appl. Phys. Lett.* **7**, 206–208 (1965).
[113a] E. D. Hinkley and P. L. Kelley, *Science* **171**, 635–639 (1965).

A very significant feature of injection lasers is that they can be tuned in frequency by temperature or pressure changes and by the application of a magnetic field. They affect the transition frequency by changing the bandgap, and they also tune the cavity largely through altering the refractive index of the semiconductor, but also through temperature and pressure alterations of the Fabry–Perot cavity dimensions. As shown in Table 2.5 a temperature increase from 4.2–77°K changes the emission wavelength of InAs diodes from 3.11–3.15 μm. Tuning of about 4% has been demonstrated for InSb for magnetic field variations between 20–70 kG. This is shown in Fig. 2.25. Very substantial frequency tuning of a PbSe laser has been obtained by using high-pressure techniques.[113]

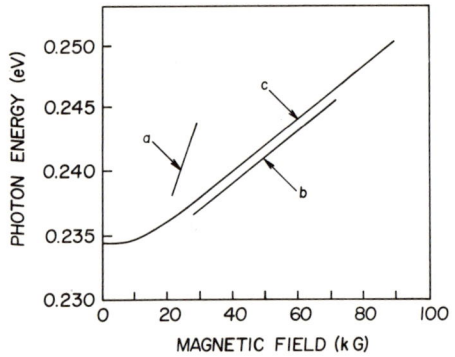

FIG. 2.25. Variation with magnetic field of the energy of photons emitted by InSb at a temperature of 1.7°K; a and b are stimulated and c is spontaneous emission (after Besson et al.[113]).

As indicated in Fig. 2.26, the wavelength of the radiation emitted by these diodes has been tuned in the range 7.3–11 μm with hydrostatic pressures up to 7 kbars. The energy gap decreases almost linearly with a slope of -8.50×10^{-6} eV/bar. Recently the M.I.T.–Harvard group has extended PbSe laser operation to 22 μm, by pressurizing to 14.1 kbars at 77°K.[111] Tuning to still longer wavelengths will require improvements in the techniques of producing high pressures.[113] At 77°K the energy gap in PbSe is expected to close at about 20 kbars. This group has also made semiconductor diode lasers of $Pb_xSn_{1-x}Se$ which have operated at about 15 μm.[111]

The Fabry–Perot resonators in junction diode lasers are often formed by cleaving two planes perpendicular to the plane of the junction. A typical cavity may have dimensions $0.4 \times 0.2 \times 0.2$ mm³, the longer di-

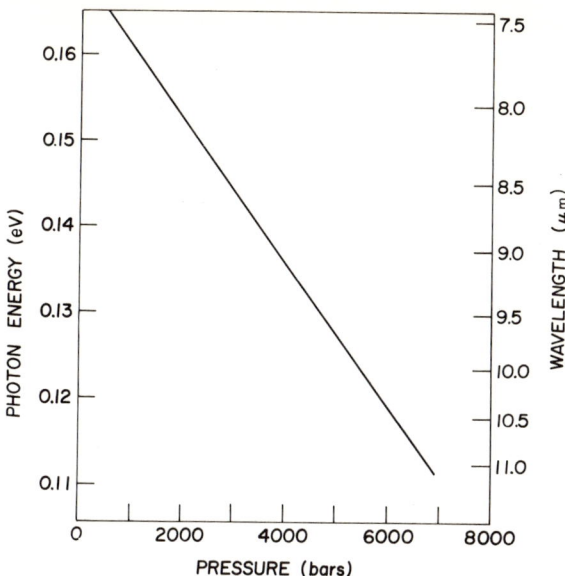

FIG. 2.26. Variation with pressure of the dominant emission modes from a PbSe diode laser at 77°K. Current density = 4800 A cm^{-2} (after Besson et al.[113]).

mension being the distance between the reflecting surfaces. Laser emission may occur along the entire length of the junction but it is often restricted to filaments some tens of microns wide. In the direction of current flow its depth is usually only a few microns within the depletion layer. The smallness of the active region aggravates the power dissipation problem, and it also determines the diffraction width of the emitted beam.

The reason for the small penetration of the region of population inversion is because the lifetime of carriers is generally quite short. In experiments carried out by Melngailis et al.,[114] minority carrier injection in n$^+$pp$^+$ InSb structures has had the effect of saturating traps and so increasing the carrier lifetime from about 10^{-10} to 10^{-7} or 10^{-6} sec. They have observed emission of 5.2-μm radiation over a distance of 50 μm in the direction of current flow. This is one order of magnitude larger than that in previous lasers. This step towards achieving emission from the bulk of the semiconductor not only reduces the spreading of the radiation beam, but the larger volumes are better suited to the amplification of infrared or far-infrared waves.

[114] I. Melngailis, R. J. Phelan and R. H. Rediker, Appl. Phys. Lett. **5**, 99–100 (1964).

2.6.2. Generation by Stimulated Raman Scattering

When light is scattered from a system of quantized oscillators, the photons may lose or gain discrete amounts of energy according to whether it excites or deexcites the oscillators. This is the Raman effect which is discussed in more detail in Chapter 7. If the light is scattered at a lower frequency because of absorption by the scatterer, we call this Stokes scattering; if the frequency is increased by the receipt of energy from the scatterer we speak of anti-Stokes scattering. The frequency shifts between the incident and the Stokes and anti-Stokes light may correspond to far-infrared photon energies for, as we shall see in later chapters, many of the energy eigen states of molecules, vibrations in crystals, etc., are separated by these energies.

Laser light may be used to excite or pump the oscillators to higher states, and from these states they may decay, by spontaneous or stimulated transitions. The spontaneous process is the relatively feeble scattering process that has been studied in Raman spectroscopy since 1928, but the stimulated process is of more recent origin, having arisen from experimentation with high power lasers. Spontaneous Raman photons are emitted randomly in all directions and at a level of 10^{-6}–10^{-7} per incident photon which makes them of little interest from the point of view of efficient wave generation. Stimulated emission, on the other hand, is of considerable interest in this respect because not only is the radiation emitted in a well-defined direction but it is produced with high intensity. Strong laser beams can pump the oscillators abundantly into excited states, and the consequent intensity of stimulated photon emission can result in the buildup of sustained oscillations in a surrounding cavity. Laser action at the Stokes or anti-Stokes frequencies, or the transition frequency of the quantum oscillator, can then be obtained. The laser gain mechanism is a sensitive function of frequency which leads to stimulated output in a narrow band about the frequency where the gain is highest.

Efficient generation by Raman laser action has been achieved by groups at Stanford University and Bell Telephone Laboratories. These groups have Raman scattered from oscillatory motions in solids that can be adjusted in frequency to yield radiation tunable in frequency. Patel *et al.*[115] have scattered CO_2 laser radiation from the electron spins of carriers in indium antimonide. By varying the applied magnetic induction field B, the frequency of transition of an electron spin between its parallel and

---

[115] C. K. N. Patel, E. D. Shaw and R. J. Kerl, *Phys. Rev. Lett.* **25**, 8–11 (1970).

2.6. SOLID STATE LASERS AND JUNCTIONS

antiparallel directions of alignment changed, and tunable Stokes lines at the difference frequency of the CO_2 laser line and the spin-flip frequency $g\beta B/\hbar$ were generated. Here β is the Bohr magneton, and g is the effective g-values of the conduction electrons in InSb. The wavelength of the observed Stokes radiation as a function of the B-field is shown in Fig. 2.27. Peak powers of 30–100 W in pulses $\sim$30 nsec long were generated for an input of 1.5 kW obtained from a Q-switched laser triggered at a repetition rate of 120 pulses per sec, and 3-nsec pulses were obtained with mode-locked laser operation.

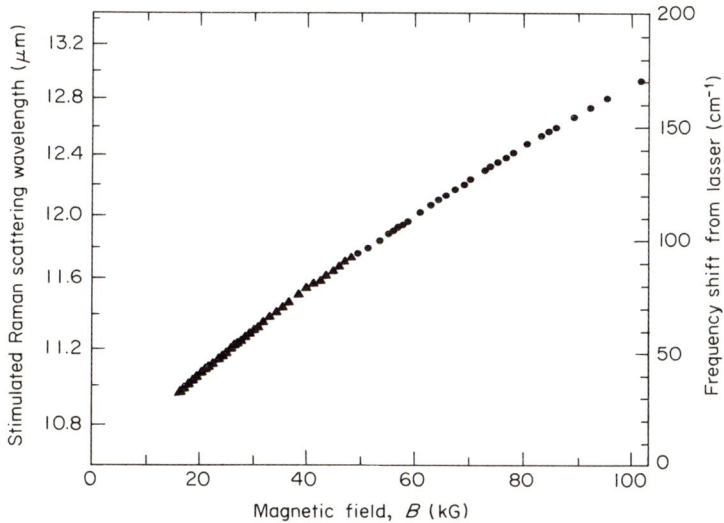

FIG. 2.27. The tuning characteristic of a spin-flip Raman laser pumped with a CO_2 laser at 10.6 μm. The frequency of the Raman (Stokes) line is $\omega_s = \omega_1 - g\beta B$, where ω_1 is the pump-laser frequency, g is the effective g-value of the electrons in InSb responsible for the scattering, β is the Bohr magneton and B is the magnetic induction field (after Patel et al.[115]).

Surface reflections from the boundaries of the InSb slab (whether coated or uncoated) are sufficient to enable the sample itself to be the laser resonator. Although the lengths of the sides of the slab are each only a few millimeters, at the Stokes frequency this represents a considerable number of wavelengths, and hence a sufficient density of overlapping modes to enable continuous tuning as the magnetic field is changed. The line width of the Stokes emission is less than 0.03 cm^{-1}, and the linearity of tunability and frequency resettability exceeds one part in 3×10^4.

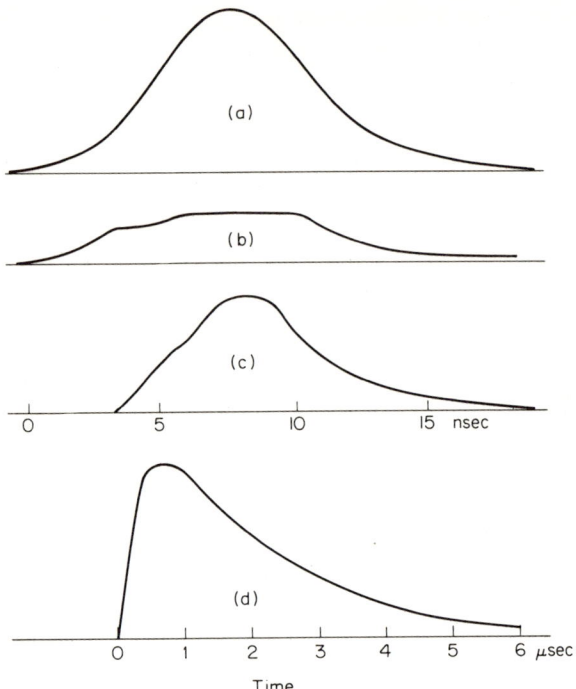

FIG. 2.28. Curves representing: (a) the 1.3-MW laser pulse input to a $LiNbO_3$ crystal at angle $\phi = 1°$ to the surface normal; (b) the pump power transmitted through the crystal; (c) the signal pulse at the Stokes frequency; (d) the far-infrared pulse at the crystal vibration frequency (after Yarborough et al.[117]).

With these capabilities the tuned spin-flip Raman source is well suited for high resolution spectroscopy, but it is rather restricted in its spectral coverage. However, as Patel et al. point out, this disadvantage can be overcome by using many different pump lasers and higher magnetic fields. Possibly by scattering in semiconductors such as $Pb_{1-x}Sn_xTe$, or $Hg_{1-x}Cd_xTe$, where the g-values of the conduction electrons are larger than that in InSb, further spectral coverage may be obtained.

Although these spin-flip experiments have been pulsed experiments, there is no reason why stimulated CW radiation may not be so generated. That this may be readily done has been demonstrated by Mooradian et al.[116] With 5-μm CW CO laser radiation scattered from a single crystal of n-type InSb, $0.5 \times 1 \times 3$ mm³, with an electron concentration of 10^{16}

---

[116] A. Mooradian, S. R. J. Brueck and F. A. Blum, Appl. Phys. Lett. **17**, 481–483 (1970).

cm^{-3}, the stimulated Raman effect has been observed with a threshold at only 200 mW. Continuous Raman oscillation tunable over a range of 30–100 cm^{-1} about the laser frequency has been obtained with fields between 17–50 kG.

Tuned Raman laser action by quite a different process has been demonstrated by the Stanford group. This device employs an infrared active lattice vibrational mode in the crystal lithium niobate, which is in the far-infrared, and which can be tuned by changing the angle between the laser and Stokes beams.

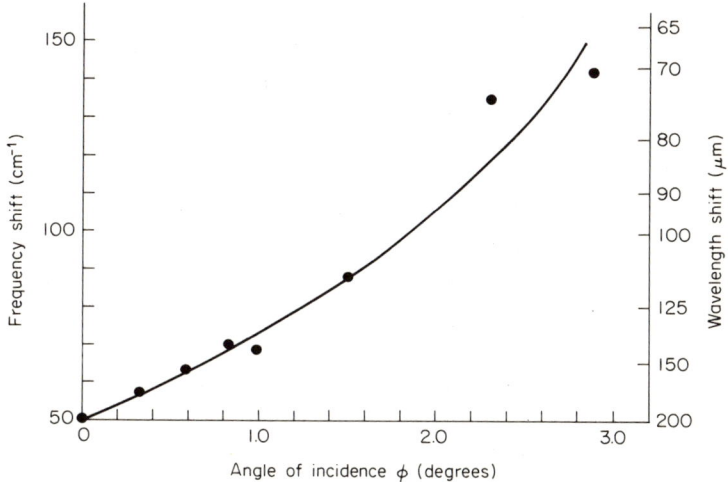

FIG. 2.29. Observed tuning curve for stimulated Raman scattering in LiNbO$_3$. The frequency shift is from the pump-laser frequency, and ϕ is the angle of incidence of the pump-laser beam with respect to the normal to the crystal surface (after Yarborough et al.[117]).

In an experiment where the pump was a Q-switched ruby laser beam of 1.3 MW entering the crystal at an angle $\phi = 1°$ to the surface normal, Yarborough et al.,[117] obtained the pulsed performance displayed by the curves of Fig. 2.28. Efficiencies up to 70% were achieved for conversion from the laser to the Stokes radiation. The observed tuning curve frequency shift versus angle of incidence ϕ is plotted in Fig. 2.29. It shows a shift of 50 cm^{-1} for normal incidence, increasing to 150 cm^{-1} as the laser beam angle is increased towards 3°.

[117] J. M. Yarborough, S. S. Sussman, H. E. Puthoff, R. H. Pantell and B. C. Johnson, Appl. Phys. Lett. **15**, 102–105 (1969).

Of significance to the far-infrared is the fact that not only does laser action occur at the Stokes frequency, but also at the crystal vibration frequency. Pantell has reported coherent emission continuously tunable from 50–200 μm at pulsed-power levels exceeding 10 W.

2.6.3. Radiation from Superconducting Junctions

Coherent radiation can be emitted by the supercurrents that oscillate across an insulating barrier separating two bulk superconductors. This is one of the manifestations of the Josephson quantum-mechanical tunneling effect which will be described in Chapter 4. Junction tunnel currents have extreme sensitivity to electric and magnetic fields and show remarkable associations between ac and dc physical phenomena. New applications to voltage measurement, magnetometry, and the precise determination of physical constants have followed Josephson's discovery. The far-infrared has gained, too. It has gained a new sensitive detector which, on the scale of performance of other detection schemes, is of primary importance. It has provided a source of monochromatic radiation of feeble intensity relative to other sources, but, nevertheless, deserving of a brief description.

The coupled electron pairs (Cooper pairs) which carry supercurrents in a bulk superconductor are associated with a macroscopic wavefunction which is coherent throughout the superconductor. When two superconductors are separated by a thin barrier and a potential difference V_0 is maintained between them, an electron pair can tunnel from one superconductor to the other in the direction aided by the potential if in so doing it conserves energy by the emission of a photon. Conservation of energy requires that every photon have a frequency

$$\omega_0 = 2qV_0/\hbar, \tag{2.21}$$

where $2q$ is the charge of an electron pair. Because every photon comes from an identical process the radiation emission is coherent.

Conversely, in the presence of radiation supplied to the junction the process can be reversed to give rise to photon-induced supercurrents which enable radiation detection.

The frequency-voltage ratio, $2q/\hbar$, is 483.6 MHz/μV, meaning that voltages of a millivolt or more can result in the generation of submillimeter waves. A voltage of 1 mV corresponds to a frequency of 16.12 cm^{-1} or 483.6 GHz.

2.6. SOLID STATE LASERS AND JUNCTIONS

The alternating supercurrent of frequency ω_0 can be written[118]

$$I = I_0 \sin(\delta_0 + \omega_0 t), \tag{2.22}$$

where δ_0 is a constant. The maximum junction current I_0 depends on the temperature and on the thickness and area of the junction,[119] and may be less than a microampere or larger than a milliampere. For a current flow of a few milliamperes, some 10^{16} electrons/sec tunnel across the junction, and the largest possible emission power could be about $10^{16} h\nu_0$ or $\sim 10^{-17} \nu_0$ W. Thus, at very best, a single Josephson junction is a source of microwatt level radiation. However, in practice, much less power than this is transferred out of the junction. The junction is highly nonlinear, and the radiation consists of photons at harmonic and sub-harmonic frequencies.[119] Thus while the radiation is coherent it is not monochromatic, and the energy output at the frequency $\omega_0 = 2qV_0/\hbar$ will be very much below the microwatt level. Inefficiency is also contributed to in no small part by the difficulty of matching the radiation from the low impedance ($\sim$milliohms[120]) junction source to an impedance of the order of that of free-space. In experiments at microwave frequencies powers of 10^{-10} W and lower have been measured.[120-123]

The dc voltage across the junction may be maintained by passing a current I' through a resistance R is series with the junction circuit as in Fig. 2.30. If I' is much larger than the current through the junction, the voltage V_0 is nearly $I'R$. Voltage fluctuations due to Johnson noise in the resistance R at the appropriate cryogenic temperature T will, by Eq. (2.21), lead to frequency modulation and hence broadening of the emission line. Silver[123] shows the line width to be

$$\delta \nu_0 = 2.57 \times 10^7 RT. \tag{2.23}$$

Broadening of this magnitude has been confirmed experimentally and

---

[118] B. D. Josephson, *Rev. Mod. Phys.* **36**, 216–220 (1964).

[119] V. Ambegaokar and A. Baratoff, *Phys. Rev. Lett.* **10**, 486–489 (1963); Erratum *Phys. Rev. Lett.* **11**, 104 (1963).

[120] D. N. Langenberg, D. J. Scalapino and B. N. Taylor, *Proc. IEEE* **54**, 560–575 (1966).

[121] I. K. Yanson, V. M. Svistunov and I. M. Dimitrenko, *Zh. Eksp. Teor. Fiz.* **48**, 976 (1965).

[122] D. N. Langenberg, D. J. Scalapino, B. N. Taylor and R. E. Eck, *Phys. Rev. Lett.* **15**, 294–297 (1965).

[123] A. Silver, *IEEE J. Quantum Electron.* **4**, 738–744 (1968).

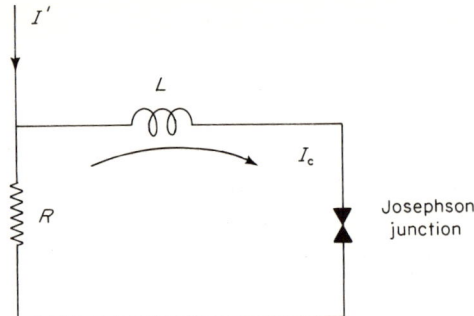

Fig. 2.30. Equivalent of a loop of superconducting wire with a Josephson junction. The current I' through the resistance R provides the dc voltage across the junction which produces an oscillating supercurrent.

is shown in Fig. 2.31 for a resistance $R = 25\ \mu\Omega$ and temperatures of 2, 4, and 8°K, respectively. The line widths are only a few kilohertz, so the level of coherence is good.

One may expect an upper frequency limit determined by the onset of internal absorption when the photon energy is sufficient to break the Cooper pairs, converting them to normal electrons. The binding energy of pairs is the energy gap of the superconductor and this, as we shall see in Chapter 7, is of the order of $3.5\ kT_c$, where T_c is the critical temperature below which the metal becomes superconducting. Values of T_c are given

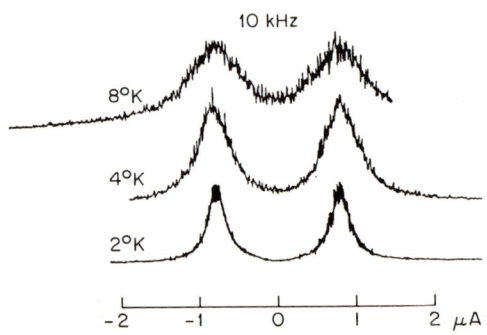

Fig. 2.31. The line width of a superconducting quantum oscillator at 10 kHz at temperatures of 2, 4 and 8°K. The Josephson junction is in the circuit shown in Fig. 2.30, with $R = 25\ \mu\Omega$, and the line width is a result of the Johnson noise voltage across R (after Silver et al.[124]).

[124] A. H. Silver, J. E. Zimmerman and R. A. Kamper, *Appl. Phys. Lett.* **11**, 209–211 (1967).

for some well-known superconducting metals in Table 7.13, Chapter 7. For tin, $T_c = 3.73°K$, and $h\nu \simeq 3.5\, kT_c$ at about 350 GHz. For Pb, T_c is 7.15°K and the upper frequency is about 650 GHz, while for Pb–Bi alloy it may be as high as 1000 GHz. In practice, however, it appears that junctions may operate above the frequency corresponding to the energy gap.

2.7. The Electron Cyclotron Maser

Electrons in orbital motion in high magnetic fields radiate energy at the cyclotron frequency. If the field exceeds $B = 100$ kG the cyclotron frequency, $\omega_c = qB/m$, is in the submillimeter wave band and is magnetic field tunable. If the electrons are mildly relativistic the radiation can be coherent. Thus with magnetic fields in excess of 150 kG now available with superconducting magnets, and with facilities up to 200–300 kG in various national laboratories, coherent tunable radiation throughout the millimeter band down to a submillimeter wavelength of 300 μm is possible. Through anharmonic content of the cyclotron orbital motion emission of harmonics may well extend the range to below 100 μm.

The mechanism leading to cyclotron resonance maser action can be described classically or quantum mechanically. It was first described in a paper by Twiss[125] in an analysis of radiation transfer and the conditions necessary for negative absorption in astrophysical systems of charges published in the Australian Journal of Physics in 1958. Schneider[126] in a Physical Review letter in 1959 explicitly proposed a free electron cyclotron maser on the basis of strictly quantum-mechanical considerations. The mechanism was realized experimentally by Hirshfield and Wachtel,[127] and by Bott,[128] at centimeter and millimeter wavelengths, and subsequently extended to submillimeter wavelengths.

In this text, the physics of positive and negative cyclotron resonance absorption is discussed in Chapter 5 in the general context of cyclotron resonance, and in Chapter 4 in connection with a proposed positive absorption cyclotron resonance detector. In the present section we describe qualitatively the physical mechanism of negative absorption as it applies to the electron cyclotron maser, and review the experimental work on

---

[125] R. Q. Twiss, *Aust. J. Phys.* **11**, 564–579 (1958).
[126] J. Schneider, *Phys. Rev. Lett.* **2**, 504–505 (1959).
[127] J. L. Hirshfield and J. M. Wachtel, *Phys. Rev. Lett.* **12**, 533–536 (1964).
[128] I. B. Bott, *Proc. IEEE* **52**, 330–332 (1964).

this means of wave generation. The description in Chapter 5 is both quantum mechanical and classical, but let us here consider the interaction classically.

Consider a beam of electrons that have been wound into circular motion coming into cyclotron resonance interaction in a region of uniform axial magnetic field with a transverse standing-wave electric field in a resonator. The electron motion is helical with a small pitch. That is, most of the kinetic energy of the electrons is in the x-y plane, this being the plane of the electric field vector, but there is a slow drift velocity in the z-direction which determines the transit time τ of the electrons through the cavity. Winding of electrons into tight helices may be achieved in magnetic mirror fields in the form in common use in plasma confinement experimentation. Some initial transverse motion, to start the mirror action on its way, may be imparted either by directing the beam at an angle to the z-axis, or by deflecting it into long-pitch helical motion with a helical magnetic field.[129,130] When the winding is completed, the electrons drift in time τ through a resonant cavity supporting a transverse electric field which we can consider as constituted of two circularly polarized components. Only the component rotating in the same sense as the electrons interacts significantly with the motion, and this influence is exerted for the interaction time τ.

The electrons are to be slightly relativistic so that the electron mass is a function of the particle energy. A few kilovolts is sufficient energy. Consider now the arrangement of electrons in the x-y plane shown in Fig. 2.32(a), where the uniform distribution of particles around the circle represents homogeneity of concentration with phase angle at the beginning of the interaction. Let us synthesize the interaction into, first, an electron bunching process, and, second, a process by which the electron bunches transfer energy to the radiation field. To appreciate the process of electron bunching under the influence of the field, consider the four representative particles A, B, C, and D rotating in the clockwise sense at the cyclotron frequency ω_c in near synchronism with the rotating electric field $E_0 e^{-i\omega t}$. Particle B is acted upon by a force of magnitude qE_0 in the direction of its motion and it acquires energy from the field. This applies to other electrons on the semicircle ABC—they gain energy and spiral outwards. On the other hand, the motion of D is opposed by the field. It, and other particles on the arc CDA, lose energy to the field. But we have assumed

[129] R. C. Wingerson, *Phys. Rev. Lett.* **6**, 446–448 (1961).
[130] L. C. Robinson, *Ann. Phys.* (*New York*) **60**, 265–276 (1970).

2.7. THE ELECTRON CYCLOTRON MASER

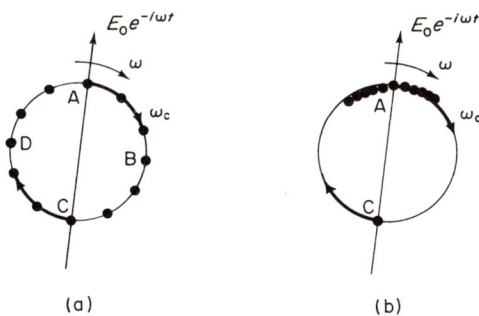

FIG. 2.32. Diagram used for the classical description of the bunching of electrons in the electron cyclotron maser, and the transfer of energy from the bunch to the rotating field $E_0 e^{-i\omega t}$. Bunching is caused by relativistic mass changes, and energy transfer by phase slippage of the bunch when $\omega_c < \omega$.

the electrons to have velocity-dependent mass, so we must expect those on arc ABC to increase in mass and therefore decrease in cyclotron frequency. Conversely, electrons on arc CDA lose mass and increase in cyclotron frequency. Thus the angular motions are modulated by the interaction, with the electrons on the semicircle ahead of A dropping back in phase position, and those on the semicircle behind A gaining in phase position, with the result that there is clustering in phase about the position A, and a rarefaction in the locality of C. With bunches thus formed we now have the essential ingredient of maser action—coherence.

We note that the argument thus far bears a resemblance to the bunching processes in electron tubes described earlier. Now let us consider the second stage in the synthetic argument, namely the net transfer of energy from the bunches to the field.

We suppose that ω_c is less than ω, so that the electron bunch about A slips in phase with respect to the rotating field. It slips towards D and in so doing acquires phase positions such that the bunches transfer energy to the field; and they do this coherently provided the bunches are tight in angular spread. It is clear from the diagram that net energy transfer is a transit time effect, for once the bunch slips past position C it begins to extract energy from the field. If the interaction time τ is such that $(\omega - \omega_c)\tau = 2\pi$, the bunch slips in phase by a complete cycle, returning to A with zero net energy transfer. Slippage beyond this angle can result again in transfer to the field, depending in a periodic way on $(\omega - \omega_c)\tau$.

If the electrons are nonrelativistic, no bunching occurs. In this case the cyclotron resonance absorption is positive and arises because the electrons that lose energy spiral inwards and travel a shorter distance in unit time than the electrons that gain energy and spiral outwards.

The alternative approach, due to Schneider,[126] brings out the maser description of the interaction. The salient features we have discussed here are, perhaps, brought out more clearly in the quantum-mechanical description given in Chapter 5; certainly, comparison of the two descriptions gives an interesting illustration of the correspondence principle. In essence, the quantum picture is based on stimulated transitions between energy eigen states of the free electron in a uniform magnetic field. The cyclotron oscillator has discrete allowed energy levels just as does the harmonic oscillator in quantum theory. They are called Landau levels, and are amenable to calculations of matrix elements describing transition probabilities, selection rules, etc. When the electrons are non-relativistic, the matrix elements for upward and downward transitions show net positive absorption. However, slightly relativistic motion modifies the separation of the Landau levels and gives rise to regions within the line shape of each level where the downward transitions exceed the upward transitions. As we shall see in Chapter 5, the line shape is a function of the transit time τ, and hence the maser action is transit time dependent.

Experimental electron cyclotron masers were developed in 1964 by Hirshfield and Wachtel[127] at microwave wavelengths.[128] Bott generated pulsed radiation at the cyclotron frequency tuned by the magnetic field from 2.2–0.95 mm with peak power of about 1 mW. In later experiments with a second maser[131] he generated pulsed powers up to 1 W with an efficiency of the order of 2%. He recorded the continuous magnetic-field tuning from 3.7–2.1 mm shown in Fig. 2.33, and obtained the minimum coherence $\Delta\lambda/\lambda < 10^{-3}$ by measuring the line width $\Delta\lambda$ of the line centred at wavelength λ. In addition to radiation at the cyclotron frequency, Bott measured second harmonic radiation at wavelengths between 1.06 and 1.44 mm with peak power in excess of 10 mW.

Electron cyclotron maser operation was extended into the far-infrared by the writer in 1968, with a maser built at Yale University by Hirshfield, and transported to the high magnetic field facility made available at the Francis Bitter National Magnet Laboratory, M.I.T., Boston. An overmoded resonator was used with the aim of providing a continuum of overlapping modes so that continuous magnetic-field tuning could be achieved. This feature is similar to that noted for Patel's Raman spin-

---

[131] I. B. Bott, *Phys. Lett.* **14**, 293–294 (1965).

2.7. THE ELECTRON CYCLOTRON MASER

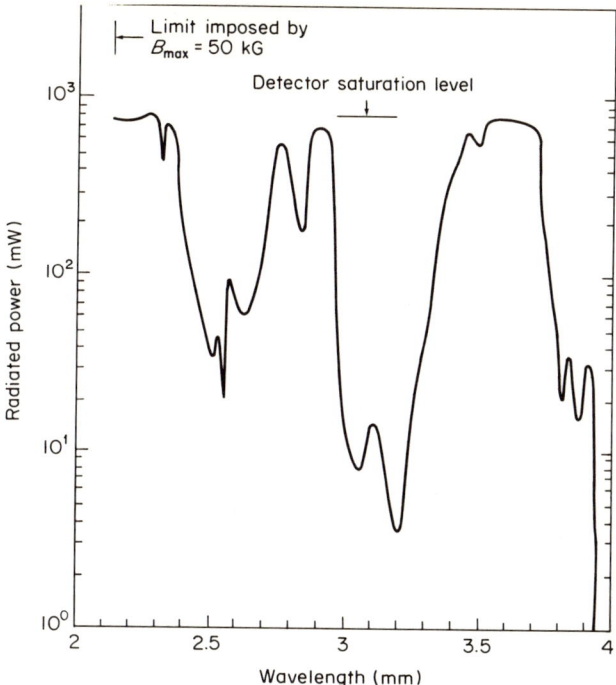

FIG. 2.33. Radiation Power generated in an electron cyclotron maser as a function of wavelength. The radiation is coherent, and tunable up to a limit set by the maximum available magnetic induction field, in this case 50 kG (after Bott.[131] Crown copyright. Reproduced by permission of the Controller of HM Stationery Office).

flip laser in InSb in Section 2.6.2. The outcome was that coherent radiation at the fundamental frequency was emitted from 5.82 mm–725 μm, at power levels up to 100 mW. Continuous wave operation, and continuous magnetic-field tuning over most of the range was recorded. Some preliminary observations of harmonics separated from the fundamental with cut-off waveguide filters, showed about 2 μW of radiation at the wavelength 488 μm.

3. WAVE TRANSMISSION AND TRANSMISSION SYSTEMS

Introduction

Between the source of radiation and the detector there is a transmission system which may be as simple as a light pipe or as complicated as a spectrometer with sophisticated optical or microwave components. We now consider the various forms and properties of these transmission systems.

What materials are transparent? What are the problems and capabilities of grating, interference, and microwave spectrometers when taken to the far-infrared? How does one filter radiation in this spectral region? These are some of the questions which must be answered: Practical matters such as these are, of course, a necessary prerequisite for investigations in the far-infrared.

3.1. Preliminary Optical and Microwave Concepts

In the far-infrared where optical and microwave experimental techniques meet, concepts and terminology derived from both fields come into use and principles commonly used in the two fields overlap. As a preliminary to the discussion of instruments it is desirable to review briefly such concepts and definitions as F-numbers and étendue[1] derived from geometric optics, and the quality factors finesse and Q used in connection with resonators.

3.1.1. Étendue and F-number

Consider paraxial rays in the simple lens system shown in Fig. 3.1, where media with refractive indices n and n' are on either side of the lens. Let α and α' be the angles to the optic axis of the extreme ray passed through the beam-defining aperture from the object to the image, and h and h' the object and image sizes. From simple geometric consid-

---

[1] W. H. Steel, "Interferometry." Cambridge Univ. Press, London and New York, 1967.

erations, it is readily seen that

$$\alpha(x+f) = \alpha'(x'+f'),$$

and

$$ni = n'i',$$

or

$$n\frac{h}{x+f} = n'\frac{h'}{x'+f'}.$$

Hence,

$$n\alpha h = n'\alpha'h' = \text{invariant}. \tag{3.1}$$

In two-dimensional form, the invariance can be written, for systems in vacuum,

$$A\Omega = \text{invariant}, \tag{3.2}$$

where A is the area of the object and Ω the solid angle that the beam-defining aperture subtends at its center. This product of the area and the solid angle of a beam is called the *étendue, optical extent,* or *light-gathering*

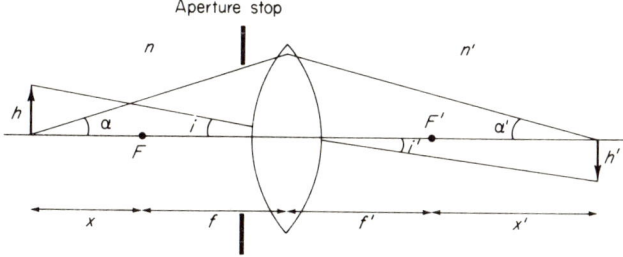

FIG. 3.1. Formation of an image h' of an object h by a simple lens in dielectric media with refractive indices n and n'. F and F' are the focal points of the lens, f and f' the focal lengths, and x and x' the positions of the object and image from the respective focal points.

power of an optical system. Neglecting losses by reflections and absorption, it remains constant along the path of a beam from the source, through an instrument, to the detector. While we have illustrated this property with ray propagation through a lens it is, of course, also valid for reflection optics (and, indeed, for particle beams in accelerators and other charged particle optical systems). As we shall see, in the far-infrared we are generally concerned with reflection optics because of the poor transmission properties of most refracting materials.

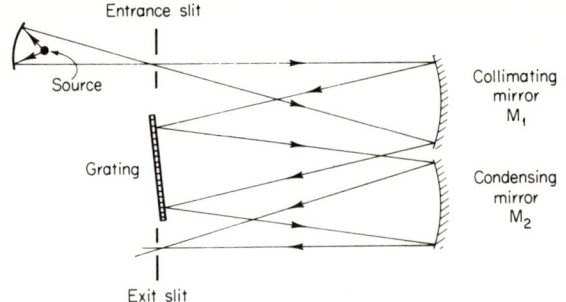

FIG. 3.2. Diagram of a grating spectrometer of the Czerny–Turner type. Radiation from the source is concentrated through the entrance slit, and thereafter it is collimated by mirror M_1, dispersed by the grating, and then focused by mirror M_2 through the exit slit.

To review the meaning of F-number, consider the grating spectrometer diagram in Fig. 3.2. Radiation from a lamp is condensed onto the entrance slit located at the center of focus of the collimating mirror M_1. The system collects and collimates radiation from the entrance slit, then, after dispersion by the grating, focuses an image of the entrance slit onto the exit slit by means of a similar mirror M_2. The solid angle of rays Ω collected by the collimating mirror is proportional to the square of the ratio of the mirror aperture diameter to its focal length, that is,

$$\Omega \propto (D/f)^2. \tag{3.3}$$

The F-number of the system is defined as

$$F = \frac{f}{D} = \frac{\text{focal length}}{\text{diameter of aperture}}. \tag{3.4}$$

It is a measure of the amount of radiation collected by the imaging system. For efficient use of a weak source the solid angle of accepted radiation should be large so that small F-number optics is required.

3.1.2. Q of a Resonator

Energy stored in a resonant cavity oscillates periodically between magnetic field energy associated with wall currents and electric field energy associated with surface charges. When excitation is removed, resistive wall dissipation results in exponential damping of the stored energy. The situation is similar to the LCR circuit where energy oscillates between inductive (magnetic field) storage and capacitive (electric field) storage with dissipation as the current passes through the resistance R, and the cavity can be analyzed in terms of an equivalent circuit.

In a series circuit with inductance L and capacitance C, the current I is given in terms of the applied voltage V by

$$L\frac{dI}{dt} + RI + \frac{1}{C}\int I\,dt = V. \tag{3.5}$$

Taking the case when the applied voltage is zero and assuming

$$I = I_0 e^{ipt}, \tag{3.6}$$

substitution readily gives

$$p = \pm \omega_0\left(1 - \frac{1}{2Q}\right)^{1/2} + i\frac{\omega_0}{2Q}, \tag{3.7}$$

where, in the usual notation,

$$\omega_0^2 = 1/LC, \qquad Q = L\omega_0/R. \tag{3.8}$$

p is a complex frequency with the imaginary part signifying amplitude decay according to the exponential $\exp(-\omega_0 t/2Q)$, and decrease of stored energy W as

$$W = W_0 \exp(-\omega_0 t/Q). \tag{3.9}$$

Differentiating, we obtain

$$\frac{dW}{dt} = -\frac{\omega_0}{Q}W,$$

or

$$Q = \frac{\omega_0 W}{dW/dt} = \frac{2\pi W}{T_0(dW/dt)}. \tag{3.10}$$

A similar analysis of a parallel circuit shows that Eq. (3.7) is the same for both cases, but for the parallel circuit

$$Q = RC\omega_0. \tag{3.11}$$

Thus while Eqs. (3.8) and (3.11) give Q-values only for particular circuits, Eq. (3.7), and hence Eqs. (3.9) and (3.10), allow a general definition suitable for any circuit or cavity resonator. From Eq. (3.10),

$$Q = \frac{2\pi \text{ energy stored}}{\text{energy lost per cycle}}. \tag{3.12}$$

In terms of concepts of stored photons, $1/Q$ is the fraction of photons lost, or the probability that a photon will be lost (by wall absorption or escape through holes in the cavity walls) in one radian of the oscillation period. Alternatively, $Q/2\pi$ is the number of oscillation periods for which a photon is stored before escaping.

Q is also a measure of the sharpness of the resonance line. By considering a sinusoidal voltage across the impedance of the equivalent circuit, one finds that the width in frequency $\delta\nu_0$ of the resonance curve (power input to the resonator versus frequency) between points at which the response is one-half that at the resonance frequency, ν_0 is related to Q by

$$Q = \nu_0/\delta\nu_0. \tag{3.13}$$

3.1.3. Finesse and Its Relation to Q

The multiple reflections of plane waves between parallel plates (mirrors), such as occur in the Fabry–Perot interferometer, are described by the Airy formula for the transmission $\tau(\phi)$, a function of the increment of phase ϕ between successive beams.[1] The transmitted intensity, or Airy function, is

$$\tau(\phi) = \frac{1}{[1 + A/(1-\mathscr{R})]^2} \frac{1}{1 + [4\mathscr{R}/(1-\mathscr{R})^2]\sin^2\phi/2}, \tag{3.14}$$

where A and $\mathscr{R}$ are the intensity absorptance and reflectance, respectively. The transmitted intensity is shown as a function of ϕ in Fig. 3.3. It is plotted to show two resonances corresponding to the lth and $(l+1)$th orders of interference. The finesse $\mathscr{F}$ is defined as the interorder spacing $\Delta\phi = 2\pi$ over the width $\delta\phi$ at the half power (i.e., half intensity) transmission points. In other words, it is the ratio of the interval between successive resonances to the width in an individual resonance expressed in either phase, wavelength, or frequency intervals:

$$\mathscr{F} = 2\pi/\delta\phi = \Delta\lambda/\delta\lambda = \Delta\nu/\delta\nu. \tag{3.15}$$

By Eq. (3.14)

$$\mathscr{F} = \pi\mathscr{R}^{1/2}/(1-\mathscr{R}). \tag{3.16}$$

The coefficient $\mathscr{F}$ does not depend on the mirror separation; it is sometimes called the "reflecting finesse" to indicate that it depends only the reflecting power of the plates.

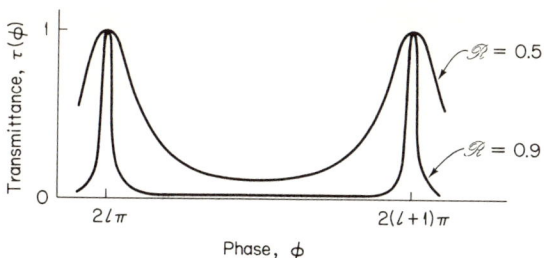

FIG. 3.3. Power transmission, $\tau(\phi)$, through a Fabry–Perot interferometer whose parallel plates have power reflectance $\mathscr{R} = 0.5$ and 0.9, as a function of the incremental phase ϕ between successive beams. With no power absorption the peak transmission at the resonances is unity.

The wavelength and frequency of the lth mode are $\lambda = l\,\Delta\lambda$, $\nu = l\,\Delta\nu$, and, since $Q = \lambda/\delta\lambda = \nu/\delta\nu$, the relation between finesse and Q is clearly

$$Q = l\mathscr{F}. \tag{3.17}$$

Thus the finesse is the Q of the first-order Fabry–Perot resonance. In terms of energy storage it is 2π times the ratio of the energy stored in the first-order resonance to the losses occuring in one cycle of oscillation; or it is 2π times the inverse of the probability of a photon escaping in two mirror reflections (two reflections take place in one period of oscillation of a first-order resonance).

3.2. Waves in Material Media

When electromagnetic radiation propagates in material media the field interactions with the constituent charges change the velocity of the wave and cause loss by scattering and absorption. According to classical physics the field drives the charges in oscillatory motion and as a consequence of this motion they reradiate (that is, scatter or reflect). The resulting wave is the superposition of the incident field and the many reradiated fields which add with such phases as to alter the phase velocity of the wave in the medium. Wave energy can be absorbed in collisional interactions which cause a transfer of energy from the oscillatory charge motion to randomized thermal motion.

Collections of free charges have a natural oscillation frequency, called the "plasma frequency," in the region of which there are marked changes in the wave propagation characteristics of the medium. For the case of

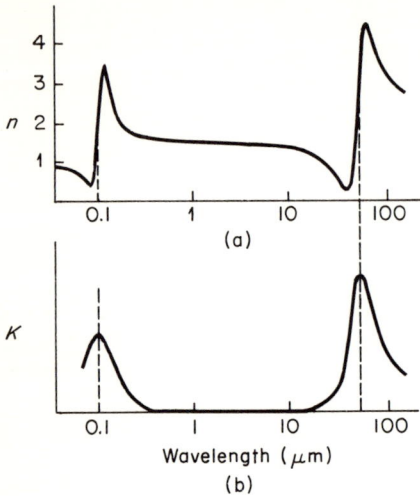

Fig. 3.4. Curves illustrating the behavior of (a) refractive index and (b) absorption coefficient at regions of anomalous dispersion. The particular resonances shown occur in the ionic crystal NaCl.

bound charges, such as occur in dielectric media, there are natural resonances at frequencies determined by the binding forces and the masses of the charged particles. Near these resonances there is considerable scattering and absorption, and the refractive index varies rapidly with frequency. These are regions of so-called anomalous dispersion where the refractive index and absorption behave in the characteristic way shown in Fig. 3.4. Single resonances such as this are normally observed in relatively dilute dielectric media, particularly gases. In many solids, where the oscillators are more densely arranged, the oscillators may couple to give bands of absorption.

The natural resonances in the infrared and far-infrared are largely associated with vibrational and (in the cases of gases and liquids) rotational motions. The fields excite mechanical motions through forces on the dipole moments of the molecules. The binding forces in many dielectric materials are such that there are many resonances in the far-infrared, with the consequence that transparent materials are hard to obtain in this region. A guide to the natural resonance frequencies can be seen by analyzing the rotational and vibrational motions separately (an accurate treatment would require account to be taken of the mutual interaction of the two motions). Consider, as a simple example, a diatomic molecule composed of two atoms with masses m_1 and m_2 separated by a constant distance r_0, rotating about the center of mass. According to classical mechanics, the rotational energy is

$$W_{\text{rot}} = M^2/2I, \tag{3.18}$$

where M is the angular momentum, and I the moment of inertia given by $I = mr_0^2$, with m the reduced mass $m_1 m_2/(m_1 + m_2)$. Now, according to quantum mechanics, M^2 has discrete allowed values given by[2]

$$M^2 = \hbar^2 J(J+1), \qquad J = 0, 1, 2, \ldots . \tag{3.19}$$

Hence the eigenvalues of the energy of rotational motion are

$$W_{\text{rot}} = \hbar^2 J(J+1)/2I \tag{3.20}$$

For the HCl molecule, for example, for which $m = 1.6 \times 10^{-27}$ kg and the separation $r_0 = 1.3$ Å, the series of rotational energy levels is

$$W_{\text{rot}} = 2.06 \times 10^{-22} J(J+1) \quad \text{joule}. \tag{3.21}$$

The selection rule governing transitions due to absorption or emission is that the quantum number J can change by amounts

$$\Delta J = \pm 1 \quad \text{or} \quad 0. \tag{3.22}$$

Transition frequencies for HCl are then seen to extend through the far-infrared. The frequency separation of the rotational states $J = 1$ and $J = 2$ is 240 μm, and that between $J = 2$ and $J = 3$ is 160 μm, and so forth. The longest wavelength is 479 μm, corresponding to the $J = 0$ to $J = 1$ transition.

Vibrational motion of a diatomic molecule can be treated as a harmonic oscillator with a binding energy taken as $\frac{1}{2} fr^2$, where r is the displacement from the equilibrium separation and f is the binding force constant. The well-known solution for the energy levels is

$$W_{\text{vib}} = (n + \tfrac{1}{2})\hbar\omega, \tag{3.23}$$

where

$$\omega = (f/\mu)^{1/2}. \tag{3.24}$$

Transitions between vibrational energy levels, consistent with the selection rule $\Delta n = \pm 1$, give lines typically in the infrared. For some molecules (e.g., the triatomic molecule CO_2) vibrational transitions occur at wavelengths of 10 μm and beyond.

Vibrational frequencies of ions bound in the lattices of crystalline materials extend well into the far-infrared. Whereas bands of absorption

---

[2] W. Heitler, "Elementary Wave Mechanics," 2nd ed. Oxford Univ. Press (Clarendon), London and New York, 1956.

are observed in some solid dielectrics, the absorption in ionic crystals tends to have a line spectrum.[3] In NaCl, for example, the lattice of Cl atoms vibrates at a particular frequency with respect to the lattice of Na atoms. This motion produces a variable electric moment in the crystal and hence absorption. Near the lattice resonance the absorption coefficient is large and, in consequence, there is large reflectivity. This region of enhanced reflection is referred to as the *"reststrahlen"* or the region where a band of "residual rays" is preferentially reflected.

Reststrahlen characteristics are also shown by some semiconductors (those such as indium antimonide and gallium arsenide where the binding between the two kinds of atoms is a covalent bond with a small ionic component[4]), but in these materials there are, in addition to dielectric properties, pronounced electronic effects. For frequencies sufficiently high to raise carriers to the conduction band from the valence band, or from impurity levels located within the energy bandgap, there is substantial absorption. While these effects prevent the transmission of short wavelengths, absorption by free carriers in the conduction band increasingly attenuates longer wavelengths (see Section 4.3.2). In the region between, the semiconductor can have reasonable transmission characteristics.

In metals and plasmas high-frequency waves interact largely with free electrons, and the high-frequency transmission and reflection properties are determined largely by the number density of these electrons and the collision frequency with background atoms and ions. Metals are generally good reflectors from low frequencies right through to the visible spectrum. In the case of plasmas transmission is determined largely by the relation of the frequency to the plasma frequency. Waves with frequencies below the plasma frequency are reflected (or "cut off" from transmission) at the plasma surface, while frequencies above can be transmitted. The plasma frequency is a function of electron density. For an electron density of 10^{15} cm^{-3} the critical cutoff wavelength is ~ 1000 μm; it is ~ 10 μm when the density has the very high value 10^{19} cm^{-3}. During transmission through a plasma the high-frequency wave sets the electrons into oscillatory motion and collisions with the atoms and ions of the plasma randomize this ordered motion. Dissipation is thus determined by the collision frequency. When the plasma is immersed in a magnetic field the wave propagation characteristics are considerably more complicated.

[3] J. T. Houghton and S. D. Smith, "Infra-Red Physics." Oxford Univ. Press (Clarendon), London and New York, 1966.
[4] D. A. Wright, "Semiconductors." Methuen, London, 1966.

3.2.1. Absorption by Air. Atmospheric Windows

The earth's atmosphere has a marked attenuating effect on radiation with wavelengths between about 14 and 1000 μm. On either side of this region, at light and radio frequencies, the atmosphere is relatively clear and it is because of this that astronomical observations have been largely confined to these two regions.

In the infrared region from 1 to 14 μm there are many bands of absorption, but there are also "windows" where the atmosphere is relatively transparent. Windows occur, as we see in Fig. 3.5, in the bands 2.0–2.4 μm, 3.4–4.1 μm, 8.0–12.5 μm, and there are somewhat less transparent windows between 12.6 and 13.4 μm and near 4.8 μm.[3,5] The bands between these infrared windows have been studied extensively. They arise from absorption by water vapor, carbon dioxide, ozone (O_3), and other minor atmospheric constituents such as carbon monoxide, nitrous oxide, and methane. Above 13 μm the absorption of carbon dioxide rises markedly as does the absorption due to water vapor. On the long wavelength side of the far-infrared there are bands of absorption corresponding to transitions in molecular oxygen and water vapor. The atmosphere is somewhat more transparent to millimeter than to submillimeter waves, but there are some absorption bands. Water vapor gives a weak absorption at 22.5 GHz, and a stronger absorption at 187 GHz. Oxygen has major absorption bands at 60 and 120 GHz.[6] Between the atmospheric absorption bands there are transmission windows, notably below 20 GHz, between 30 and 50 GHz, 75 and 100 GHz, etc. Above this the attenuation rises on the tail of the far-infrared absorption. The theory of the rotational transitions responsible for this absorption has been treated by Van Vleck and others.[7,8]

Atmospheric transmission from 18 μm throughout the far-infrared to 2500 μm is chiefly defined by the absorption of water vapor which has a multitude of purely rotational transitions in this region. Yaroslavsky and Stanevich[9] have measured the transmission characteristics of a 10-m length of atmospheric (room) air with absolute humidity 10.5 gm m^{-3}

---

[5] R. A. Smith, F. E. Jones and R. P. Chasmar, "The Detection and Measurement of Infrared Radiation." Oxford Univ. Press (Clarendon), London and New York, 1957.

[6] A. F. Harvey, "Microwave Engineering." Academic Press, New York, 1963.

[7] J. H. Van Vleck, *Phys. Rev.* **71**, 413–424, 425–433 (1947).

[8] H. H. Thiessing and P. J. Caplan, *J. Appl. Phys.* **27**, 538–543 (1956).

[9] N. G. Yaroslavsky and A. E. Stanevich, *Appl. Opt.* **7**, 380–382 (1959).

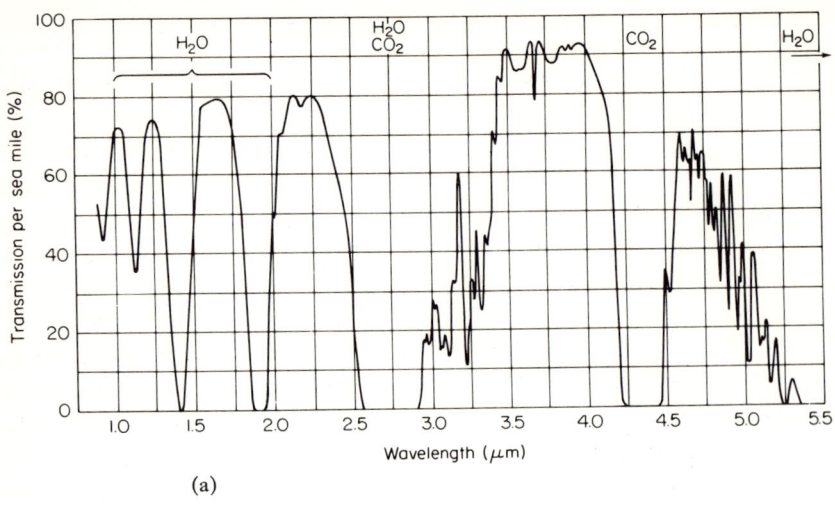

(a)

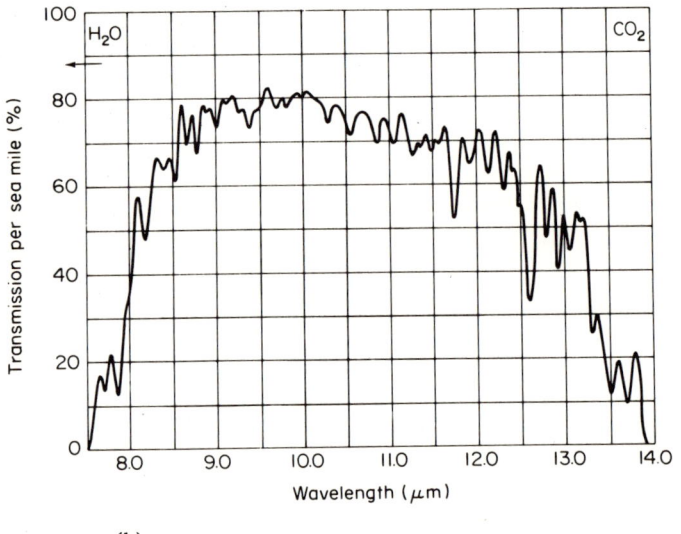

(b)

Fig. 3.5. Atmospheric transmission over 2000 yards at sea level; visual transmission (0.61 μm) was 60% over this distance; the water content was 17 mm. (a) Wavelength range 1–5.5 μm; and (b) Wavelength range 7.5–14 μm. Between 5.3 and 7.4 μm, water vapor absorbs the radiation. (After Gebbie et al.[10] Crown copyright. Reproduced by permission of the Controller of HM Stationery Office.)

---

[10] H. A. Gebbie, W. R. Harding, C. Hilsum, A. W. Pryce and V. Roberts, *Proc. Roy. Soc. Ser. A* **206**, 87–107 (1951).

and obtained the results shown in Fig. 3.6. Minimum loss is apparent in the longer wavelength regions 1200–1500 μm, and 1700–2500 μm. There are windows near 345 μm and 44 μm, and at many wavelengths below this.

Transmission of solar radiation through the earth's atmosphere has been measured in the bands between 50 and 1000 μm at the Kitt Peak Observatory, U.S.A.[11] The altitude of this observatory is 2150 m. Observations have shown a number of windows through which solar radiation

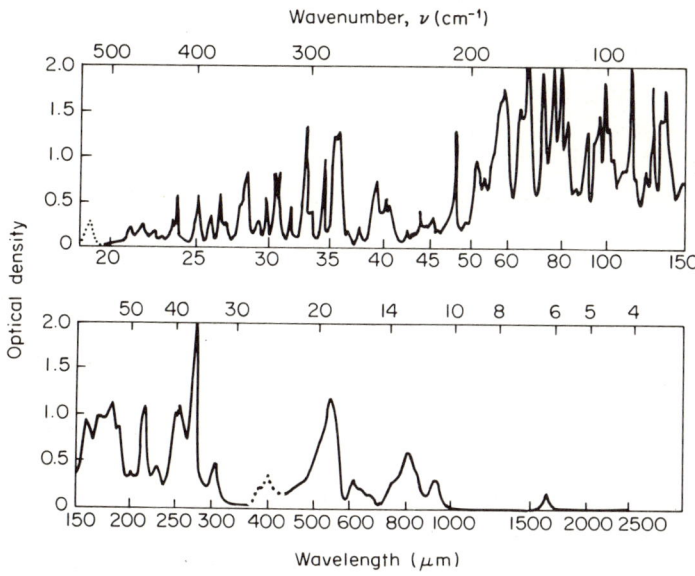

FIG. 3.6. Optical density (defined as the logarithm to the base ten of the ratio of the incident to transmitted intensity) of atmospheric (room) air in the region 20–2500 μm, for a layer of thickness 10 m at a temperature of 20°C, pressure 760 torr, and absolute humidity of 10.5 gm m^{-3}.

can penetrate on days of good visibility. Maximum transmission occurs at 345 μm, where 2% of the radiation from the sun, at approximately 30° from the zenith, is received when the total precipitation water vapor column is about 0.17 cm. Other less-clear windows are situated at 440, 590, 710, and 810 μm, respectively. At wavelengths shorter than 300 μm there are several partial windows. On days when the precipitable water vapor exceeds 5 mm the atmosphere is opaque to solar radiation.

[11] R. A. Williams and W. S. C. Chang, *Proc. IEEE* **54**, 462–470 (1966).

3.2.2. Transmission through Solid Dielectrics and Semiconductors. Windows, Filters, and Polarizers

Transparent solid materials are needed in far-infrared transmission systems for windows (for example, at the entrance to an evacuated spectrometer and at the aperture of a Golay cell or a high pressure p-n junction laser), for filtering wanted spectral bands from unwanted radiation, for beam splitting in the Michelson interferometer, and for lenses or dispersive prisms. As we have mentioned, for this part of the spectrum there is no "universal" material capable of playing a role equivalent to that of glass at optical frequencies. Glass is practically opaque to the far-infrared.[12] The limit of transmission of crown glass is 2.8 μm, although special infrared glasses extend to 5.5 μm and the toxic arsenic trisulfide glasses transmit to 12 μm. Crystalline quartz is transparent to a wavelength of a few microns but there are reststrahlen bands at 9 and 21 μm which result in poor transmission properties from about 4–40 μm. For wavelengths longer than 40 μm, quartz in the crystalline form becomes more transparent.

There are many materials with good transmission properties throughout the infrared up to some tens of microns. Rock salt (NaCl), for example, is transparent up to 16 μm, but beyond this absorption arises from the reststrahlen band centred a little higher, at 79 μm.

The essential features of the reststrahlen regions can be approximately described by the results of classical Drude–Lorentz dispersion theory.[3,13] According to this theory, wave propagation through material can be described by a complex refractive index $n - iK$, where the real part n is the ordinary refractive index and K is the absorption coefficient. For a single resonance n and K are given by the well-known relations[13]

$$n^2 - K^2 = \varkappa_\infty + \frac{Nq^2}{\varepsilon_0 m} \frac{\omega_0^2 - \omega^2}{(\omega_0^2 - \omega^2)^2 + \gamma^2 \omega^2}, \qquad (3.25)$$

$$2nK = \frac{Nq^2}{\varepsilon_0 m} \frac{\gamma \omega}{(\omega_0^2 - \omega^2)^2 + \gamma^2 \omega^2}, \qquad (3.26)$$

where N is the number of ion pairs per unit volume, m is the reduced mass, and $\varkappa_\infty$ the optical dielectric constant. The resonance frequency ω_0 is not predicted by the classical theory; it is given by quantum theory.

---

[12] E. M. Dianov, N. A. Irisova, and V. N. Timofeev, *Sov. Phys. Solid State* **8**, 2113–2116 (1967).

[13] P. Drude, "The Theory of Optics." Longmans, Green, New York, 1907.

The width of the absorption band is set by γ, the damping constant. Reflection from the material is determined from the boundary conditions obeyed by electromagnetic fields, namely the continuity of the tangential components of electric field. For normal incidence the familiar result for the magnitude of the power reflectivity is

$$\mathscr{R} = \frac{(n-1)^2 + K^2}{(n+1)^2 + K^2}. \tag{3.27}$$

Transmission characteristics of a number of materials with transparent regions ranging up to 50–60 μm are plotted in Fig. 3.7. The material irtran-2 is a sintered mixture of zinc sulphide. Irtran-4 and irtran-6 are also available commercially. They are mixtures of zinc selenium and cadmium telluride (respectively) with good mechanical strength. The former transmits up to 21.8 μm and the latter has high transmission between 1 and 30 μm.

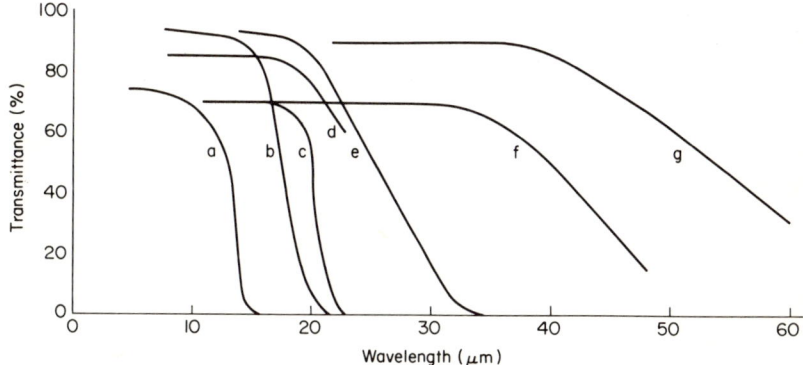

FIG. 3.7. Transmittance of materials suitable for the wavelength region 10–50 μm. The materials and thicknesses are as follows: (a) Irtran-2, 0.4 cm; (b) NaCl, 1 cm; (c) Irtran-4, 0.203 cm; (d) AgCl, 0.5 cm; (e) KBr, 1 cm; (f) KRS, 0.5 cm; (g) CsI, 0.5 cm.

In the far-infrared wavelength region up to 60 μm or so, pure semiconductors are quite transparent. Silicon has a limit of transmission at approximately 40 μm, germanium at 50 μm. The presence of impurities reduces the transparency of these materials. The short wavelength cutoff in transmission through semiconductors is particularly abrupt. Within a fraction of an electron volt near the gap energy the crystal suddenly becomes opaque. The long wavelength cutoff is much more gradual. Absorption is this region is due to "free" electrons in the conduction

TABLE 3.1. Far-Infrared Transmitting Materials

Material	Refractive index	Transmission regions (approx.) (μm)
Irtran-4	2.42 (at 11 μm)	$\lesssim 21.8$
Irtran-6	2.67 (at 10 μm)	$\lesssim 30$
Silicon	3.42 (at 10 μm)	$\lesssim 40$
Germanium	4.0 (at 16 μm)	$\lesssim 50$
Crystalline quartz	2.34 (at $\lambda > 150\,\mu$m)	$\lesssim 4;\ \gtrsim 40$
Sapphire	1.78 (at 2 μm)	$\lesssim 6\,\mu$m; $\gtrsim 200$
Diamond (type II)	2.38	$\gtrsim 11$
Polyethylene	1.46–(throughout far-infrared)	$\gtrsim 100$
TPX	1.43 (visible to far-infrared)	> 25

band and, in some semiconductors, to weak reststrahlen effects. As shown in Table 3.1, the refractive indices of semiconductors are high, and so the transparency is reduced by the reflection loss indicated by Eq. (3.27).

To obtain high transmittance it is necessary to coat semiconductors with appropriate antireflection films. In the case of a single-layer coating on a material with refractive index n, it is necessary to deposit a film of

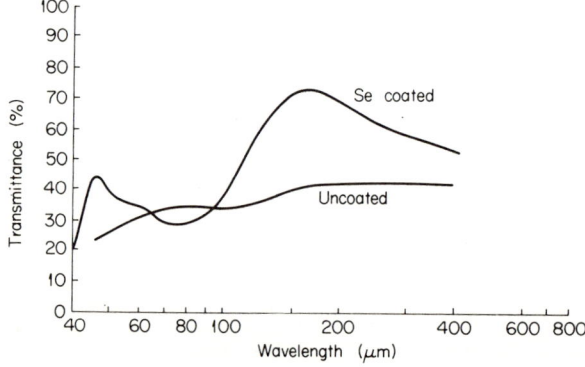

FIG. 3.8. Transmittance of germanium plate 2 mm thick, without coating and with a selenium coating 35.8 μm in thickness on one surface and 34.6 μm on the other (from Mitsuishi et al.[13a]).

[13a] A. Mitsuishi, H. Yoshinaga, K. Yata and A. Manabe, Jap. J. Appl. Phys. **4**, 581–587 (1965).

TABLE 3.2. A List of Polymers and Their Static Dielectric Constants

Polymer	Static dielectric constant
Polypropylene	2.1
Teflon	2.0
Ethylene–propylene copolymer	2.35
Polystyrene	2.4–2.65
Polycarbonate	3.0
Polyethylene terephthalate (synonymous with Mylar and Melinex)	3.1
Polyvinylidene chloride	3.0–4.0
Polyvinyle chloride	3.5–4.5
Di-acetyl cellulose	3.6
Tri-acetyl cellulose	4.5

refractive index $n^{1/2}$ to a thickness

$$\alpha = (2p + 1)\lambda/4n, \qquad (3.28)$$

where $p = 0, 1, 2, \ldots$. Thus coating materials with refractive indices about 2 are required. These, of course, will reduce reflections only at the far-infrared wavelength consistent with Eq. (3.28). At other wavelengths the reflection can be enhanced. Measurements of the properties of thin antireflection films have been described by Mitsuishi *et al.* With vacuum evaporated selenium on a 2-mm slab of germanium they have

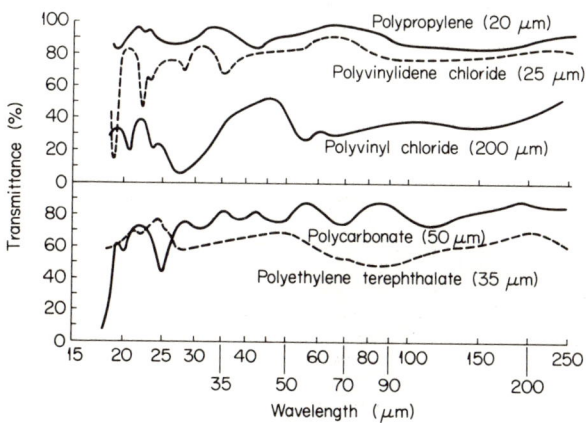

FIG. 3.9. Transmission spectra of polypropylene, polyvinyl chloride, polyvinylidene chloride, polyethylene terephthalate, and polycarbonate (after Mitsuishi *et al.*[13a]).

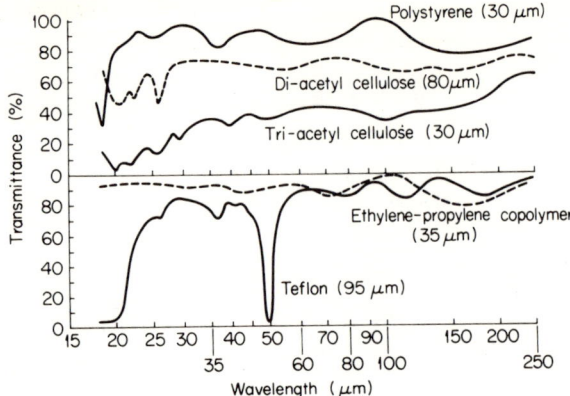

Fig. 3.10. Transmission spectra of polystyrene, di-acetyl cellulose, tri-acetyl cellulose, ethylene-propylene copolymer, and Teflon (after Mitsuishi et al.[13a]).

obtained the improved transmittance shown in Fig. 3.8. Selenium has a weak absorption band below 100 μm which contributes to the poor performance observed at short wavelengths. Di-acetyle cellulose films (refractive index 1.9) cemented on Ge have given improved transmission at wavelengths above 200 μm. These Japanese workers have investigated the effects of polymer films cemented on silicon. A list of polymers and their static dielectric constants are given in Table 3.2, while some transmission spectra of thin films are shown in Figs. 3.9 and 3.10. Figure 3.11

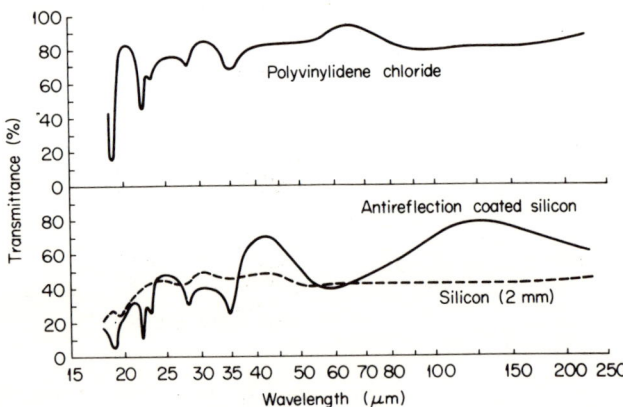

Fig. 3.11. Anti-reflection coating on silicon. Upper half: transmission spectrum of polyvinylidene chloride film. Lower half: dotted curve gives the transmittance of silicon plate 2 mm thick, and the solid line is the transmittance of the plate when coated with polyvinylidene chloride films (after Mitsuishi et al.[13a]).

3.2. WAVES IN MATERIAL MEDIA 99

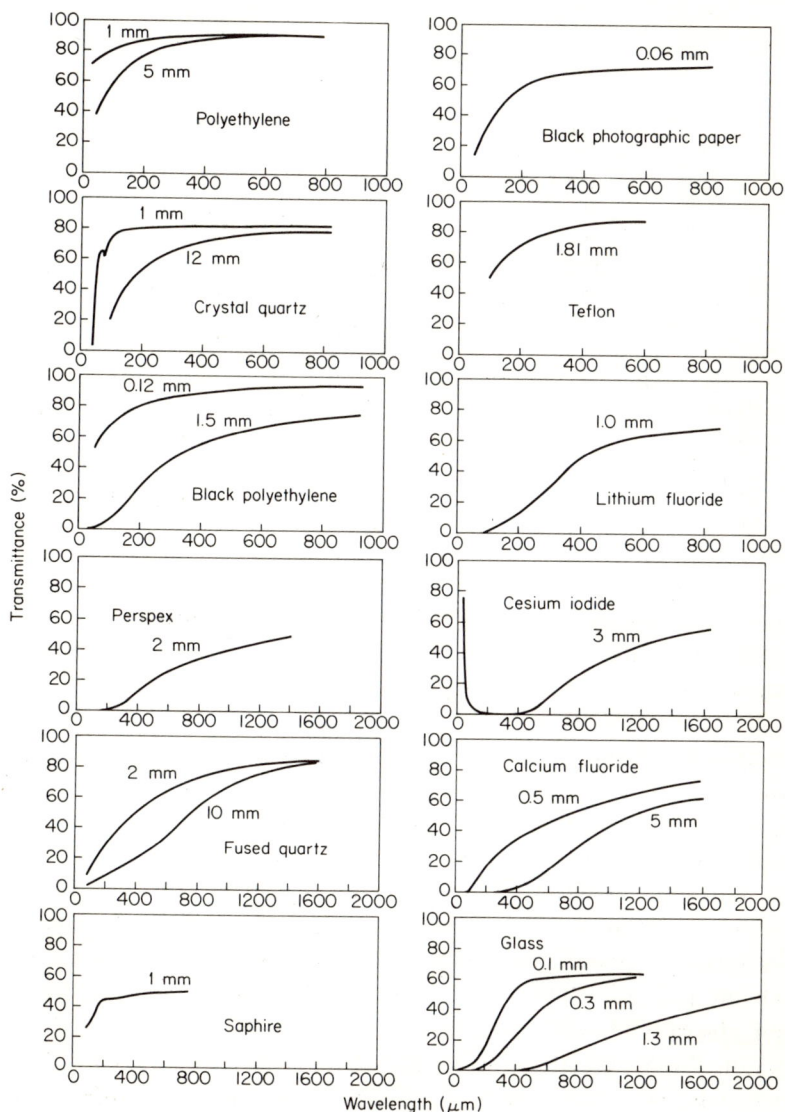

FIG. 3.12. Transmittance of materials in the far-infrared. (After Kimmitt.[19a] Crown copyright. Reproduced by permission of the Controller of HM Stationery Office.)

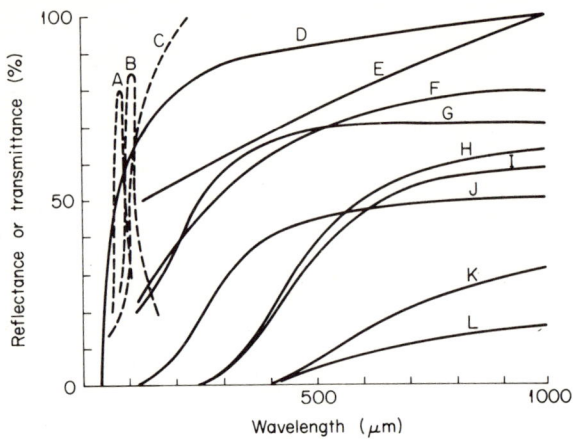

FIG. 3.13. Approximate reflection and transmission curves of filters used in the far-infrared; dashed curve: reflectivity; solid curve: transmittance. A, Pressed plate of KBr; B, pressed plate of RbBr; C, aluminium plate ground with grade 150 carborundum powder; D, 1-mm crystal quartz; E, 0.015-mm mica; F, 1-mm soot in polyethylene; G, 1-mm 5% soot in paraffin; H, 1-mm KBr or 1-mm RbCl; I, 1-mm RbBr; J, 1-mm 15% soot in paraffin; K, 1-mm RbI; L, 1-mm TlCl. (After Bloor et al.[14] Reproduced by permission of the Royal Society.)

shows the transmission spectrum of polyvinylidene chloride film and its effect on the transmittance of a 2-mm slab of silicon. The transmission maxima shown centered at 125, 42, and 20 μm correspond to $p = 0$, 1, and 2, respectively, in Eq. (3.27). The zeroth order peak is about 80% transparent, but the other peaks are somewhat reduced due to absorption in the polymer.

Reasonable transparency for wavelengths greater than 50 μm can be obtained with crystalline quartz and thin sheets ($\sim$1 mm thick) of polymers and other materials. The transmission characteristics of a number of useful far-infrared materials are shown in Figs. 3.12 and 3.13. Because of its relatively good transmittance throughout most of the far-infrared spectrum, polyethylene is extensively used for thin windows. It has three major absorption bands and these are in the infrared at 3.5, 6.8, and 13.6 μm. Black polyethylene (polyethylene loaded with carbon black) is also widely used particularly to prevent visible and near infrared from entering sensitive detectors. Sooted quartz may also be used for this purpose.

[14] D. Bloor, T. J. Dean, G. O. Jones, D. H. Martin, P. A. Mawer and C. H. Perry, *Proc. Roy. Soc. Ser. A* **260**, 510–522 (1961).

In Fig. 3.14 the far-infrared transmittance of black polyethylene manufactured by several companies is shown. The scatter in data from different sources is due to differences in carbon particle size, the concentration of the carbon in the polyethylene, and the form of the carbon.[15]

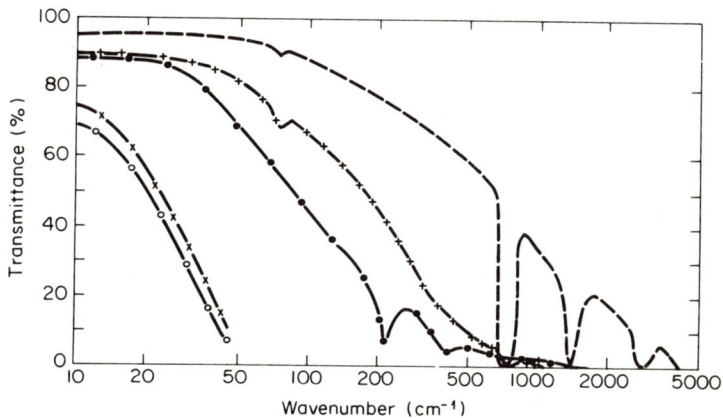

FIG. 3.14. Transmittance of black polyethylene as a function of wavenumber. Dashed curve: U.S. Industrial Chemical Co. (0.150 ± 0.007 mm); +−+−+ Dow Chemical Co. (0.157 ± 0.005 mm); —·—·— Monsanto Chemical Co. (0.165 ± 0.007 mm); -×-×-×- Dow Chemical Co. (2.57 mm); -O-O-O- Dow Chemical Co. (3.40 mm) (after Blea et al.[15]).

A new polyolefine based on a poly 4 methyl pentene-1 has recently become available under the trade name TPX. It is one of the best far-infrared window and lens materials.[16] Its transmission, as shown in Fig. 3.15, is good out to about 400 cm^{-1} with an average absorption coefficient ∼3 neper cm^{-1} (compared with 2 neper cm^{-1} for good samples of polyethylene). Below 50 cm^{-1} the transmission becomes very good indeed. It has a refractive index of only 1.43, so that it reflects relatively little energy. In contrast to polyethylene it is hard and therefore more suited to conditions involving stress; it is more resistant also to deformation by heat. It can be molded under pressure at elevated temperature to form lenses which lend themselves to optical alignment techniques because they are relatively transparent in the visible, and the refractive index is similar in the visible and the far-infrared.

[15] J. M. Bled, W. F. Parks, P. A. R. Ade and R. J. Bell, *J. Opt. Soc. Amer.* **60**, 603–606 (1970).
[16] G. W. Chantry, H. M. Evans, J. W. Fleming and H. A. Gebbie, *Infrared Phys.* **9**, 31–33 (1969).

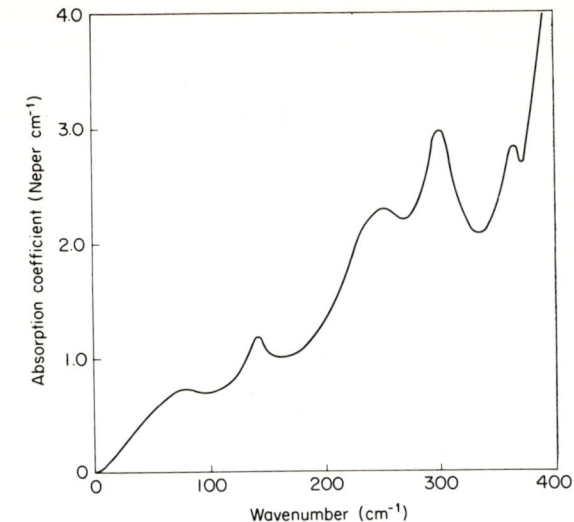

FIG. 3.15. Transmission of TPX as a function of frequency (courtesy of G. W. Chantry).

Effective low (frequency)-pass filters with cuton wavelengths in the range 30–200 μm can be made by the process of dissolving reststrahlen crystals in polyethylene, a process first developed by Yamada et al.[17,18] The filters absorb or scatter in the region of the natural resonances of the bound ion pairs, and they can give a fairly sharp cuton of transmission on the longer wavelength side of the resonance. On the high-frequency side they again transmit so that, in order to get the desired low-pass characteristic, it is necessary to use another filter to eliminate the high frequencies. The loaded polyethylene filters are made in the form of thin sheets by mixing the powders with softened polyethylene stuck to heated rollers rotating at 20 rpm.[17] The thin sheets are finally cut from the rollers. In the Japanese work described in 1962, the sheet thickness was 0.3 mm. It has been found that the sharpness of the cut-on regions increases with thinner sheets and with finer powders. Between 15 μm and the cuton wavelength they are quite opaque, with transmittance generally below 0.1% and never greater than 1%. Short wavelength radiation (below 20 μm) can be eliminated with a black polyethylene filter or with a chopper filter of NaCl or KCl crystals. The chopper ampli-

[17] Y. Yamada, A. Mitsuishi and H. Yoshinaga, J. Opt. Soc. Amer. **52**, 17–19 (1962).
[18] H. Yoshinaga, Jap. J. Appl. Phys. Suppl. 1 **4**, 420–427 (1965).

tude modulates the radiation above 20 μm so that it can be selected from the unmodulated short wavelength radiation.

Transmission characteristics of a number of reststrahlen crystal-polyethylene filters are shown in Fig. 3.16. Powder mixtures were chosen to give cut-on wavelengths in the range 30–200 μm. For wavelengths beyond 200 μm there are no reststrahlen crystals available.

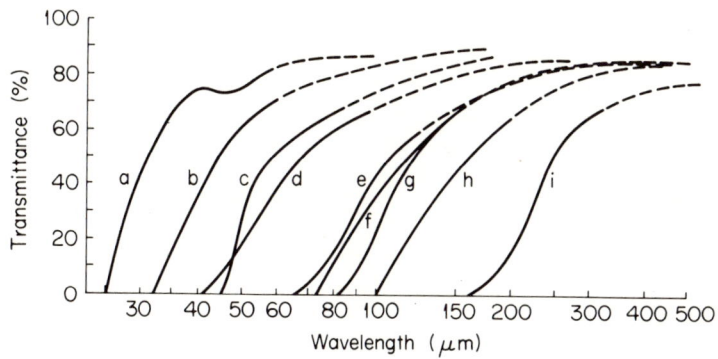

FIG. 3.16. Transmittance of far-infrared filters made from the following reststrahlen crystals dissolved in polyethylene: (a) BeO + ZnO; (b) LiF; (c) SrF_2 + LiF; (d) CaF_2 + LiF; (e) BeO + ZnO + NaF + KCl; (f) LiF + SrF_2 + KCl; (g) KBr + BaF_2 + LiF; (h) $CaCO_3$ + MgO + KCl; (i) TlCl + TlI + NaF (after Yaroslavsky and Stanevich[10]).

Low-pass transmission filters for wavelengths throughout the far-infrared can be fabricated by the polyethylene replica grating technique described by Möller and McKnight.[19,19a] They are made by impressing diffraction grating contours into the surface of polyethylene sheet about 0.2–1 mm in thickness. Radiation with wavelengths long compared to the periodic spacing S of the grating is transmitted through the material with little loss. However, wavelengths of the order of and shorter than S are scattered and experience a sharp decrease in transmission (see Fig. 3.17). Waves polarized parallel and perpendicular to the grooves are scattered a little differently to that replica gratings molded on only one side have a polarizing effect. This can be eliminated by molding both sides of the sheet with orthogonal grooves.

The transmission curve of a double moulded polyethylene filter is shown in Fig. 3.17. Filters of this type are particularly useful for wave-

---

[19] K. D. Möller and R. V. McKnight, *J. Opt. Soc. Amer.* **53**, 760–761 (1963).
[19a] M. F. Kimmitt, *Roy. Radar Establ. Tech. Note* No. 716, December 1965.

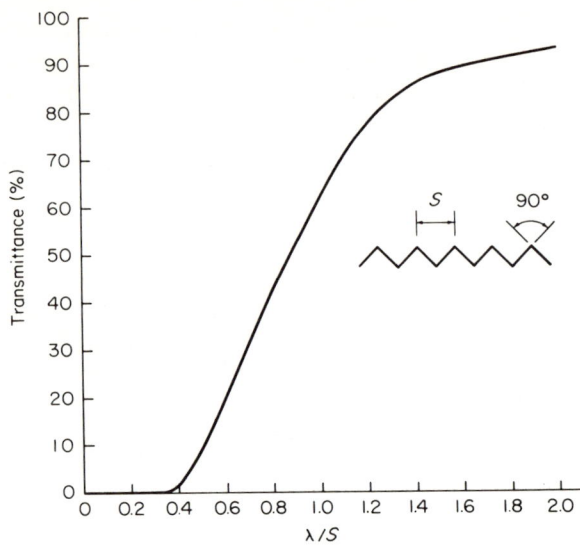

FIG. 3.17. Cutoff of a polyethylene grating transmission filter.[18,19] Inset is the surface profile. λ is the wavelength and S the grating spacing.

lengths longer than 200 μm, that is, beyond the range of reststrahlen filters. Below 100 μm the filtering capabilities decline due to the increased absorption of polyethylene. From 25 up to 100 μm reststrahlen crystal-polyethylene filters are preferable because they provide sharper cuton characteristics.

Mitsuishi *et al.* have made a Brewster angle polarizer for the region 30–200 μm from sheets of polyethylene.[20] The index of refraction of polyethylene is about 1.46 in the whole far-infrared region, so the angle of incidence on the polarizer should be about 55°. With 15 sheets of polyethylene with thicknesses ranging from 20–50 μm, 97% polarization was obtained except in regions where multiple interference effects occurred within the films. The transmission exceeded 75% of the appropriate linearly polarized component of the incident radiation.

Linear polarization can also be achieved with waveguides since, if the dimensions are chosen appropriately, only one plane of linear polarization can propagate. This property follows from the principles discussed in Sections 3.4.1 and 3.4.2. In practice, the waveguide must be very small indeed—of the order of the wavelength—and this makes the techniques

---

[20] A. Mitsuishi, Y. Yamada, S. Fujita and H. Yoshinaga, *J. Opt. Soc. Amer.* **50** 433–436 (1960).

3.2. WAVES IN MATERIAL MEDIA

satisfactory only at longer wavelengths, greater than 700 μm, or so. Losses will occur through resistive dissipation of waveguide wall currents, and care is necessary in preserving the continuity and symmetry in sections of waveguide tapering down to the polarizer section.

Wire grid polarizers are commonly used in far-infrared laboratories at wavelengths greater than about 100 μm. They are linear strips of metal which may be fabricated by vacuum evaporation of gold on polyethylene sheet. We may think of the grid as "shorting out," and hence reflecting, the component of the electric field vector parallel to the direction of the metal strips, while transmitting the component perpendicular to the strips. For wavelengths greater than the grid constant (i.e., the distance between the centers of neighboring strips) by a factor of 3 or more they can transmit with about 98% linear polarization. While the influence of the grid depends primarily on the polarization of the incident wave and the ratio of the grid constant to the wavelength, the strip width and the properties of the metal itself play a role. The various influencing factors are brought out in the Survey by Larsen.[21]

Band-pass interference filters can be made with Fabry–Perot interferometers working in low orders of interference. Such filters have been made by Ulrich et al.[22,23] with reflecting metal meshes of parallel strips and grids made of orthogonal set of strips.

The important characteristics of interference filters are the Q, for this determines the bandwidth $\Delta\lambda$, and the peak transmission τ through the filter. The resolving power (or Q-value) for a given order of interference,

$$l = \phi/2\pi, \qquad (3.29)$$

ϕ being the total phase shift between two successive beams in the multiple beam interferometer, is

$$Q = \lambda/\Delta\lambda = l\mathscr{F}, \qquad (3.30)$$

where λ is the wavelength of the resonance, and $\mathscr{F}$ is the finesse. If the power reflectance $\mathscr{R}$ at each grid is greater than 0.6 the finesse is closely approximated by the expression

$$\mathscr{F} = \pi\mathscr{R}^{1/2}/(1 - \mathscr{R}). \qquad (3.16)$$

---

[21] T. Larsen, *IRE Trans. Microwave Theory Tech.* **10**, 191–200 (1962).

[22] R. Ulrich, K. F. Renk and L. Genzel, *IEEE Trans. Microwave Theory Tech.* **11**, 363–371 (1963).

[23] K. F. Renk and L. Genzel, *Appl. Opt.* **1** 643–648 (1962).

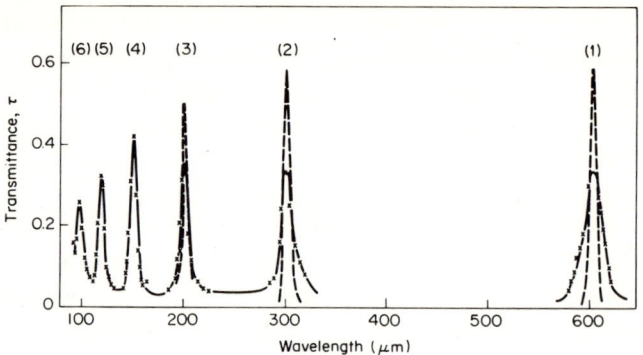

Fig. 3.18. Measured characteristic of an interference filter (grids on crystal quartz plates, each 5 mm thick). −×−×−×− measured; − − − − corrected to zero spectrometer slit-width and parallel beams. The orders of interference are given in parentheses (after Renk and Genzel[23]).

The peak transmission is given by [see Eq. (3.14)]

$$\tau = [1 + A/(1 - \mathscr{R})]^{-2}, \qquad (3.31)$$

where A is the power absorptance. From Eqs. (3.16) and (3.31) we see that high $\mathscr{F}$ and high τ require large reflectivity $\mathscr{R}$ and small absorption.

Measurements of the performance of interference filters consisting of unsupported grids of nickel, and metallic grids on plates of crystal quartz and polyethylene have been made by Renk and Genzel in the region 100–800 μm. They measured coefficients of finesse from 10–60, and peak

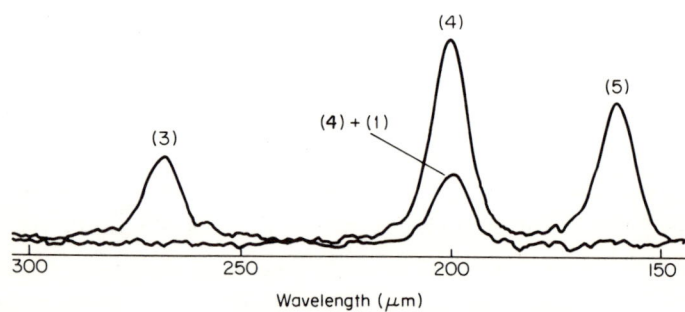

Fig. 3.19. Recording of the transmission lines of a combination of two interference filters. Upper curve: interference filter with the fourth-order peak transmission at 200 μm. Lower curve: interference filter with first-order peak transmission at 200 μm in addition to the fourth-order 200 μm interference filter. The orders of interference are given in parentheses (after Renk and Genzel[23]).

transmission up to 0.9. The best interference filters were made with parallel strips without substrate.

In Fig. 3.18 transmission is plotted as a function of wavelength for grids on a 5-mm thick crystal quartz plate. Figure 3.19 shows the effect of increasing the order of interference by increasing the spacing between the two grid reflectors, and the effect of combining a first- and fourth-order filter. Higher-order interference gives higher Q [see Eq. (3.30)] but, of course, neighboring pass-bands occur for different orders of nearby wavelengths. Two filters working in different orders can be used to select certain coincident resonances and reject others.

3.2.3. Reflection Filters

Reststrahlen crystals[24] can be used in reflection to give selective band-pass filter characteristics. From Eq. (3.27) it can be seen that when the refractive index n and the absorption coefficient K rise to large values in

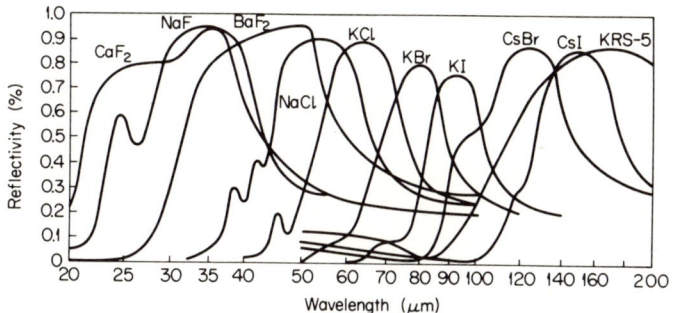

FIG. 3.20. Reflectivity of a number of reststrahlen crystal filters. The measurements were made[25] with angles of incidence of 12° and 15°.

the region of the lattice resonance, the reflectivity will be large. It can be between 80–100%. Away from resonance K falls to very small values but the off-resonance value of n can give 10–30% reflection.

Figure 3.20 shows some collected data for a variety of crystal reflection filters. The off-resonance reflection, particularly for long wavelengths, can be quite high, and it is therefore often necessary to use several re-

---

[24] E. Fermi, "Molecules, Crystals and Quantum Statistics." Benjamin, New York, 1966.

[25] A. Mitsuishi, Y. Yamada and H. Yoshinaga, *J. Opt. Soc. Amer.* **52**, 14–16 (1962).

flections in order to obtain satisfactory rejection. The waves need not be normally incident on the crystal. This is shown by the measurements of Mitsuishi et al.[25] for incident angles of 12° and 15°.

Figure 3.21 shows the observed reflection spectra of CdSe and CdS, together with values of n and K derived from dispersion analyses of the

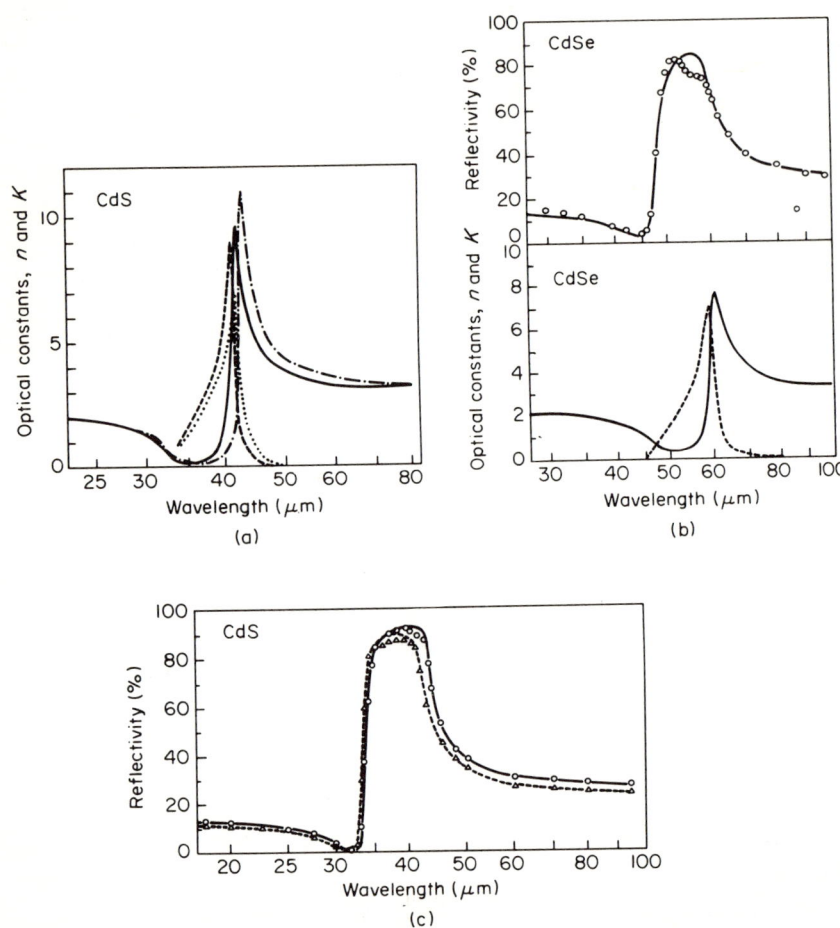

FIG. 3.21. Curves (a) and (b) show the refractive index n and absorption coefficient K near resonance in the reststrahlen crystals CdS and CdSe. [In (a), solid curve is n, dashed curve is K, for $C_\perp$; —·— is n, ··· is K, for $C_\parallel$; in (b) (top), solid curve is calculated values, o is observed values; (bottom) solid curve is n, dashed curve is K; in (c), o is observed values, solid curve is calculated values, for $C_\parallel$; △ is observed values, dashed curve is calculated values, for $C_\perp$.] (b) and (c) were made at room temperature (after Mitsuishi et al.[13a]).

measured data. As Mitsuishi *et al.* point out, the reflection behavior of CdS is associated with two slightly separated resonances—one produced by the radiation component polarized perpendicular to the *C*-axis of the crystal and one produced by the component polarized along the *C*-axis. A further interesting effect is the temperature dependence of the reflectivity and resonance frequency shown for NaCl, KCl, and KBr in Fig. 3.22.

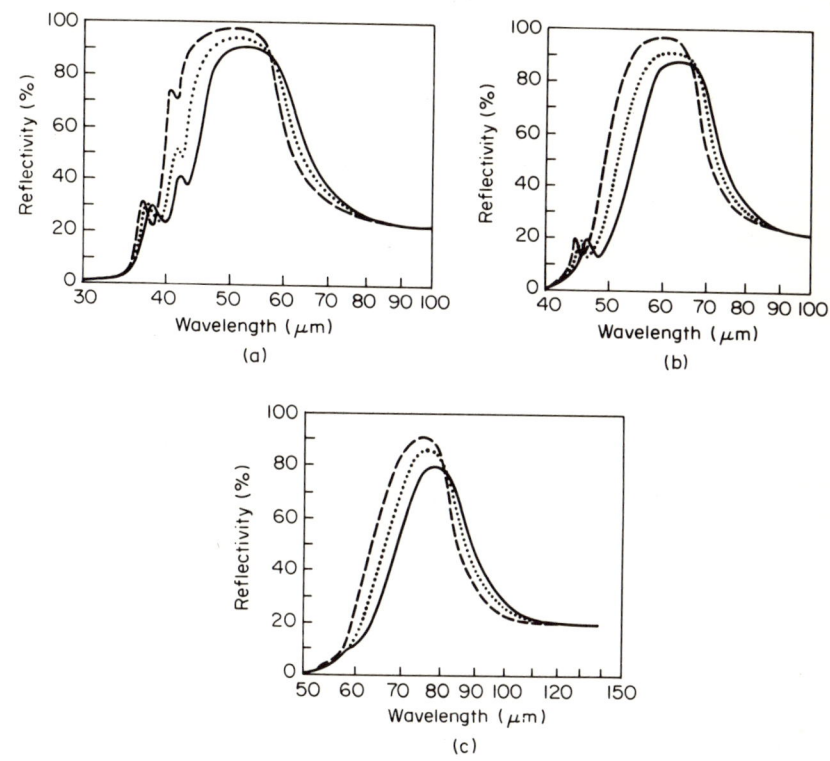

FIG. 3.22. Temperature dependence of reflectivity at the reststrahlen resonance in the following ionic crystals: (a) NaCl; (b) KCl; (c) KBr [--- 100°K, ··· 200°K, ——— 300°K] (after Mitsuishi *et al.*[25]).

The reflectivity increase with a temperature decrease can be interpreted as due to a decrease in the damping constant γ in the Drude–Lorentz dispersion formula.[25]

Grooved metal surfaces can be used as low (frequency)-pass filters. They can take the form of a diffraction grating used in zero order, or a metal mesh or abraded surface. Wavelengths short compared to the depth and width of the surface irregularities are scattered, while wavelengths longer than the irregularities are specularly reflected.[14,26]

3.2.4. Metallic Reflection

Metallic reflection is described reasonably well by classical theory in which the conduction electrons are treated as a "gas" of free electrons capable of oscillating in an applied field with collision-damped motion.[27] The reflectivity is described in terms of a complex refractive index $n - iK$, and can be determined by specializing the Drude–Lorentz dispersion equations. The free electron case is obtained from Eqs. (3.25) and (3.26) by writing $\varkappa_\infty = 1$ and reducing the restoring force to zero, that is, take $\omega_0 = 0$. The damping constant γ is to be interpreted as the collision frequency for scattering, N is the volume density of conduction electrons. Equations (3.25) and (3.26) then become

$$n^2 - K^2 = 1 - \omega_p^2/(\omega^2 + \gamma^2), \qquad (3.32)$$

$$2nK = \omega_p^2 \gamma/\omega(\omega^2 + \gamma^2), \qquad (3.33)$$

where

$$\omega_p^2 = Nq^2/\varepsilon_0 m. \qquad (3.34)$$

The reflectivity is given by Eq. (3.27). It depends on the relative magnitudes of the wave frequency ω, the collision frequency γ, and the plasma frequency ω_p. The plasma frequency can be determined from the electron density. Take the case of copper, which has a density of 8.9 gm cm^{-3} and atomic weight 63.5. The number of atoms in a gram atom is given by Avagadro's number, 6.02×10^{23}, so there are 8.5×10^{22} atoms per cm^3. The copper atom has, in addition to closed shells, only one electron in its unfilled 4S state. If we assume, then, that there should be one electron per atom the electron density will be 8.5×10^{22} cm^{-3}. In the case of silver, which has a density of 10.5 gm cm^{-3} and atomic weight 107 gm, the electron density would be 5.9×10^{22} cm^{-3}.

The electrons in a metal are in the periodic field of the lattice and are not completely free. Only in the case of zero lattice field are Eqs. (3.32) and (3.33) strictly valid. Quantum-mechanical treatment of the problem of electron motion in the periodic field of the crystal shows that the electron acceleration is different from the acceleration of a perfectly free

---

[26] J. U. White, *J. Opt. Soc. Amer.* **37**, 713–717 (1947).
[27] R. K. Wangsness, "Introduction to Theoretical Physics." Wiley, New York, 1963.

electron. The effect can be taken into account, and Eqs. (3.32) and (3.33) used, if N is replaced by an "effective" number of free electrons N_{eff}. This matter is discussed by Mott and Jones.[28] On the basis of optical measurements it has been concluded that copper, silver, and gold have 0.37, 0.89, and 0.73 effective free electrons per atom, respectively. For copper, then,

$$N_{\text{eff}} = 3.1 \times 10^{22} \quad \text{cm}^{-3}.$$

For silver,

$$N_{\text{eff}} = 5.2 \times 10^{22} \quad \text{cm}^{-3}.$$

The plasma frequencies are, accordingly,

for copper,

$$\omega_p = 0.98 \times 10^{16} \quad \text{sec}^{-1},$$

for silver,

$$\omega_p = 1.27 \times 10^{16} \quad \text{sec}^{-1}.$$

It must be pointed out that there is some uncertainty as to the values of N_{eff}. The factors taken here have been derived from optical measurements and may be too low because of the effects at these high frequencies of oxidation. Calculations by Wigner and Seitz suggest that the factor is very close to unity, and this conclusion is consistent with the values of electronic specific heat at low temperatures.[29]

The collision frequency in a metal can be estimated from the Drude expression for the low frequency conductivity,

$$\sigma_0 = N_{\text{eff}} q^2 / m\gamma. \tag{3.35}$$

For copper and silver the conductivities are 5.8×10^7 ohm^{-1} m^{-1} and 6.1×10^7 ohm^{-1} m^{-1}, respectively, and the approximate collision frequencies are for copper,

$$\gamma = 1.5 \times 10^{13} \quad \text{sec}^{-1},$$

and for silver,

$$\gamma = 2.3 \times 10^{13} \quad \text{sec}^{-1}.$$

---

[28] N. F. Mott and H. Jones, "The Theory and Properties of Metals and Alloys." Oxford Univ. Press (Clarendon), London and New York, 1936.

[29] N. F. Mott and H. Jones, "The Theory and Properties of Metals and Alloys," p. 316. Oxford Univ. Press (Clarendon), London and New York, 1936.

When these parameters are substituted in Eqs. (3.32)–(3.34), and (3.27) it is found that n and K are of the same order and the power reflection coefficient is

$$\mathscr{R} \simeq 1 - 2/n. \tag{3.36}$$

This is the Hagen–Rubens relation[30] which applies to very good conductors ($\sigma_0 \gg \varepsilon_0 \omega$). Its applicability is discussed by Mott and Jones.[28] For copper, silver and gold Hagen and Rubens found it a good approximation for wavelengths as short as 12 μm.[30]

Equation (3.36) predicts more than 99% reflection of power by good conductors throughout the far-infrared. Aluminum is also a good reflector. Silver, of course, can be an excellent reflector provided it is free of tarnish.

3.3. Quasioptical Components

Many components useful for the handling of far-infrared radiation have been taken from optics. Mirrors, reflection diffraction gratings, and light pipes get widespread use because the reflectivity of most metals is very high. Lenses, Fresnel zone plates, and prisms are less widely used largely because transmitting materials are not very transparent. With all of these components diffraction is an important consideration, particularly if the dimensions are less than an order of magnitude larger than the wavelength.[31,32] At the aperture of a light pipe with diameter $D = 1$ cm, for example, the angular width of the principal lobe of the diffraction pattern for 500 μm radiation is $\theta \sim 2.44\lambda/D \sim 8°$. Mirrors and diffraction gratings used in monochromators and interferometers (see Sections 3.5 and 3.6) generally have linear dimensions $\sim$20 cm, and for these ray optics is a reasonably good approximation.

At room temperatures, lenses and zone plates can be made of polyethylene or quartz. A polyethylene lens 6 cm thick has $\sim$50% transmission for wavelengths longer than 50 μm.[33] However, if one is prepared to tolerate the inconvenience of maintaining certain crystals at low temperatures much better transmission can be obtained.[34]

---

[30] M. Born and E. Wolf, "Principles of Optics," 3rd ed. Pergamon, Oxford, 1965.
[31] J. Brown, *Advan. Electron. Electron Phys.* **10**, 107–152 (1958).
[32] R. Tremblay and A. Boivin, *Appl. Opt.* **5**, 249–278 (1966).
[33] A. Hadni, "Essentials of Modern Physics Applied to the Study of the Infrared," p. 643. Pergamon, Oxford, 1967.
[34] L. Genzel, *Jap. J. Appl. Phys. Suppl. 1* **4**, 353–356 (1965).

For further information about quasioptical components and systems the reader is referred to Section 3.4.3 where light pipes are considered in conjunction with oversize waveguides, Section 3.5.1 for the diffraction grating, Sections 3.7 and 3.4.4 for the Fabry–Perot interferometer. For a discussion of the system of phase transforming lenses known as the "beam waveguide" reference should be made to the literature.[32,35]

3.4. Waveguides and Quasimicrowave Components

Radiation can be guided by means of hollow metal pipes with highly conducting walls. Within the guide it can be divided in power, stored in resonant cavities, shifted in phase, attenuated, matched, rotated in polarization, and so forth, by a variety of components based on principles of microwave electromagnetic theory.[6,36] The properties of such guiding systems are well understood for centimeter and millimeter waves, where the dimensions are of the order of the wavelength; beyond this range wave handling components have been developed in larger metal pipes where the bounded fields increasingly tend to take on the characteristics of waves in free space and where some of the powerful techniques of microwaves (e.g., stub matching) are compromised.

3.4.1. Standard Waveguides

The cross-sectional dimensions of standard waveguides are close to the free-space wavelength, and so chosen that only one wave of the form $e^{i(\omega t - kz)}$ satisfies Maxwell's equations and the boundary conditions at the excitation angular frequency ω. This wave is called the *principal* or *dominant* mode of propagation. A knowledge of its field distribution and wavelength within the guide enables the experimenter to control the various procedures of wave manipulation mentioned above.

To obtain the important properties of fields in waveguides one must solve Maxwell's equations subject to the boundary conditions that the tangential component of **E**, and the normal component of **H** are zero at the perfectly conducting walls. We take a region of vacuum enclosed by the metal boundaries, thereby eliminating problems of dielectric and atmospheric loss, and consider the conductor geometries usual in microwaves, namely, rectangular and cylindrical. Following the usual proce-

---

[35] G. Goubau and F. Schwering, *IRE Trans. Antennas Propagat.* **9**, 248–256 (1961).
[36] E. L. Ginzton, "Microwave Measurements." McGraw-Hill, New York, 1957.

dure we consider separately *transverse electric*, or TE, modes, for which the longitudinal components of electric field E_z are zero, and *transverse magnetic*, or TM, modes, for which $H_z = 0$. Any arbitrary field distribution excited in a lossless guide may be expressed as the superposition of a number of TE and TM modes with suitable amplitudes and phases.

The analysis proceeds as follows: For fields varying with time as $e^{i\omega t}$, solutions are to be found for the homogeneous wave equation

$$\nabla^2 \begin{bmatrix} \mathbf{E} \\ \mathbf{H} \end{bmatrix} + k_0^2 \begin{bmatrix} \mathbf{E} \\ \mathbf{H} \end{bmatrix} = 0, \qquad (3.37)$$

where $k_0 = \omega/c = 2\pi/\lambda_0$ is the free-space propagation constant. The wave equation in **H** (and similarly in **E**) represents three equations, one in each of the components of field. However, they are not independent because the field components must be related through Maxwell's equations. Thus we can solve the wave equation for the z-component of field and derive the other components via Maxwell's equations.

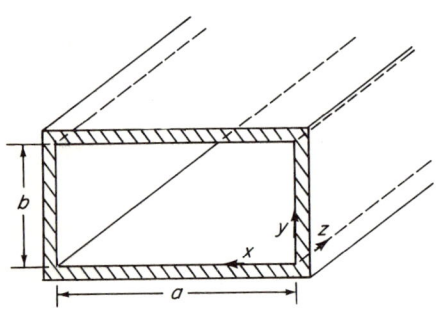

FIG. 3.23. Rectangular waveguide and cartesian coordinate axes.

For TE modes in the rectangular waveguide (Fig. 3.23) the wave equation in H_z is

$$\frac{\partial^2 H_z}{\partial x^2} + \frac{\partial^2 H_z}{\partial y^2} + \frac{\partial^2 H_z}{\partial z^2} + k_0^2 H_z = 0. \qquad (3.38)$$

Following the method of separation of variables, substitute

$$H_z = XYZ,$$

X being a function of x only, Y of y only, and Z of z only. We obtain

$$\frac{1}{X}\frac{\partial^2 X}{\partial x^2} + \frac{1}{Y}\frac{\partial^2 Y}{\partial y^2} + \frac{1}{H_z}\frac{\partial^2 H_z}{\partial z^2} + k_0^2 = 0. \qquad (3.39)$$

Since the last two terms taken together are equal to the negative of the first and second terms, they are independent of z and must be equal to a constant. Equating to the (as yet unspecified) separation constant k_c^2 we have

$$\frac{1}{H_z}\frac{\partial^2 H_z}{\partial z^2} + k_0^2 = k_c^2. \tag{3.40}$$

The solution (including the time dependence) is the traveling wave

$$H_z = H_{z0}\exp[-i(\omega t - k_g z)], \tag{3.41}$$

where

$$k_g^2 = k_0^2 - k_c^2. \tag{3.42}$$

Now from Maxwell's equations the following relations can be derived[37] which enable the determination of the transverse field components associated with H_z:

$$H_x = \frac{\partial}{\partial x}\frac{1}{k_c^2}\frac{\partial H_z}{\partial z}, \tag{3.43}$$

$$H_y = \frac{\partial}{\partial y}\frac{1}{k_c^2}\frac{\partial H_z}{\partial z}, \tag{3.44}$$

and

$$\frac{E_x}{H_y} = -\frac{E_y}{H_x} = \frac{\omega\mu_0}{k_g}. \tag{3.45}$$

Returning now to Eq. (3.39) to impose the field constraints at the walls of the guide shown in the Fig. 3.23, we can equate the first and second terms to separation constants $-k_x^2$ and $-k_y^2$. Thus

$$\frac{1}{X}\frac{\partial^2 X}{\partial x^2} = -k_x^2, \tag{3.46}$$

$$\frac{1}{Y}\frac{\partial^2 Y}{\partial y^2} = -k_y^2, \tag{3.47}$$

and

$$k_c^2 = k_x^2 + k_y^2. \tag{3.48}$$

Solutions of Eqs. (3.46) and (3.47) satisfying the boundary conditions that

[37] R. V. Langmuir, "Electromagnetic Fields and Waves." McGraw-Hill, New York, 1961.

at $x = 0$ and a, $H_x = 0$ [or by Eq. (3.43), $\partial H_z/\partial x = 0$], and at $y = 0$ and b, $H_y = \partial H_z/\partial y = 0$, are cosine functions of $k_x x$ and $k_y y$, and k_x and k_y are given by

$$k_x = \pi l/a, \qquad l = 0, 1, \ldots, \qquad (3.49)$$

$$k_y = \pi m/b, \qquad m = 0, 1, \ldots. \qquad (3.50)$$

Hence the field traveling along the $+z$-axis has the z-component

$$H_z = A \cos(\pi l x/a) \cos(\pi m y/b) \exp[-i(\omega t - k_g z)], \qquad (3.51)$$

and components H_x, H_y, E_x, and E_y are easily found from Eqs. (3.43)–(3.45). A is a constant.

We can interpret the behavior of the TE waves in rectangular waveguide from the foregoing equations. From Eq. (3.42) the significance of the constant k_c can be seen. Wave propagation occurs for k_g real, that is, when $k_0 > k_c$, but forward propagation is terminated and the field declines exponentially when $k_0 < k_c$ and k_g is imaginary. Under these conditions all the wave energy is reflected back along the guide. Writing Eqs. (3.42) in terms of the wavelengths: $k_0 = 2\pi/\lambda_0$, $k_g = 2\pi/\lambda_g$, λ_g being the wavelength in the guide, and $k_c = 2\pi/\lambda_c$, we have

$$1/\lambda_g^2 = (1/\lambda_0^2) - (1/\lambda_c^2). \qquad (3.52)$$

We see that propagation is possible only for $\lambda_0 < \lambda_c$. That is, λ_c is a cut-off wavelength; it is the maximum value permitted for the free-space wavelength if guide propagation is to be possible. In terms of waveguide dimensions we have, from Eqs. (3.49), (3.50), and (3.52),

$$\lambda_c = \frac{2}{[(l/a)^2 + (m/b)^2]^{1/2}}. \qquad (3.53)$$

As the constants l and m take on the allowed integer values, the field patterns are described by an equation such as Eq. (3.51), and Eq. (3.53) gives the cutoff wavelengths for the various TE modes. For rectangular waveguide with $a > b$, the longest wavelength mode is designated the TE_{10} mode, where the subscripts specify $l = 1$, $m = 0$, and indicate one variation of the field across the broad dimension of the guide and constant field across the narrow dimension:

$$TE_{10}: \quad \lambda_c = 2a.$$

The next mode is the TE_{01} with a cutoff wavelength

$$TE_{01}: \quad \lambda_c = 2b.$$

Clearly, from Eq. (3.53), successively higher modes have decreasing cutoff wavelengths.

A procedure similar to that above can be carried out for the TM modes of propagation. The cutoff wavelengths are again given by Eq. (3.53), and it transpires that the lowest wavelength mode is the TM_{11} mode:

$$TM_{11}: \quad \lambda_c = \frac{2}{[(1/a^2) + (1/b^2)]^{1/2}}.$$

It is usual in the design of rectangular waveguide to choose $a = 2b$ so that the cutoff wavelength of the TE_{10} mode is twice that of its nearest neighbor. With this side ratio, the waveguide can be excited over a frequency range of 2 : 1 with the certainty that only the TE_{10} field pattern can propagate. For the appropriate frequency range, the TE_{10} mode is thus a nondegenerate mode; it is the principle mode mentioned earlier. Figure 3.24 illustrates the fields of the TE_{10} mode and the wall currents associated with these fields. A number of the lower modes supported in rectangular waveguide are given in Table 3.3.

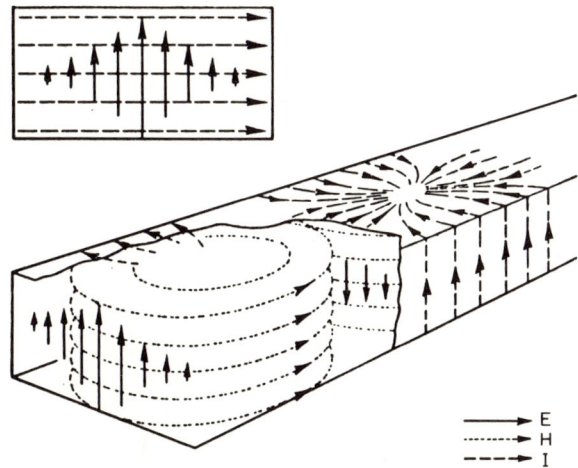

FIG. 3.24. Electric and magnetic field patterns, and wall currents, in the TE_{10} mode in rectangular waveguide (after Mariner[37a]).

[37a] P. F. Mariner, "Introduction to Microwave Practice." Heywood, London, 1961.

TABLE 3.3. Cutoff Wavelengths of the Five Lowest Frequency Modes in Rectangular Waveguide with Sides in the Ratio $a/b = 2$

Mode	Cutoff wavelength
TE_{10}	$2a$
TE_{01}	a
TE_{20}	a
TE_{11}	$0.89a$
TM_{11}	$0.89a$

The variation with frequency of wave attenuation due to finite conductivity for TE and TM modes in rectangular waveguide has the general form of Fig. 3.25. There is a minimum with steeply rising losses near the low frequency cutoff and rather gradual increasing attenuation for frequencies above the minimum.

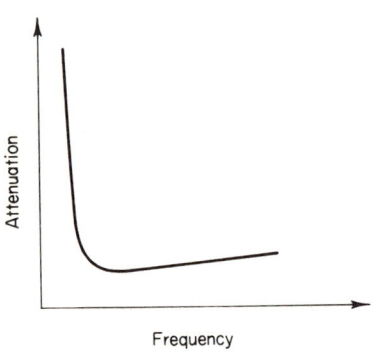

FIG. 3.25. Form of the attenuation versus frequency characteristic of the modes of propagation in rectangular waveguide. The attenuation has a minimum, and beyond this it increases with increasing frequency. Most modes in cylindrical waveguide have similar attenuation–frequency characteristics, the exception being the TE_{0m} modes.

The problem of propagation in circular–cylindrical guide can be analyzed by the same procedure. The modes of propagation can be designated TE_{lm} and TM_{lm}, where now l describes the number of circumferential variations of the field and m the number of radial variations. Some of the lower modes and their cutoff wavelengths are given in Table 3.4.

The principal mode is the TE_{11} mode illustrated in Fig. 3.26. It is closer to its nearest neighbor (the TM_{01} mode) than is the principal mode in rectangular waveguide, and can be excited nondegenerately over a

3.4. WAVEGUIDES AND QUASIMICROWAVE COMPONENTS

TABLE 3.4. Cutoff Wavelength of the Seven Lowest Frequency Modes in Cylindrical Waveguide of Radius a

Mode	Cutoff wavelength
TE_{11}	$3.41a$
TM_{01}	$2.61a$
TE_{21}	$2.06a$
TE_{01}	$1.64a$
TM_{11}	$1.64a$
TE_{31}	$1.49a$
TM_{21}	$1.22a$

1.31:1 frequency range. For most cylindrical waveguide modes, attenuation versus frequency has the same form as the curve given in Fig. 3.25 for rectangular waveguide, and, for many, the losses are comparable in order of magnitude. However, the TE_{0m} modes in cylindrical guide are exceptional in that they have monotonically decreasing attenuation as the frequency is raised. For these particular modes the only magnetic field tangential to the conductor is the longitudinal component H_z, and there are therefore no longitudinal currents. As the frequency increases, H_z and the wall currents decrease indefinitely, and the loss approaches

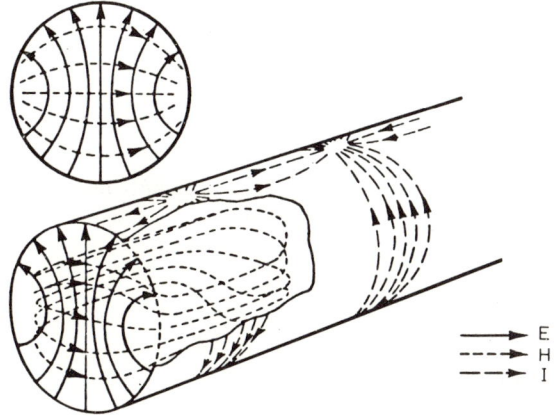

FIG. 3.26. Electric and magnetic field patterns, and wall currents, in the TE_{11} mode in cylindrical waveguide (after Mariner[37a]).

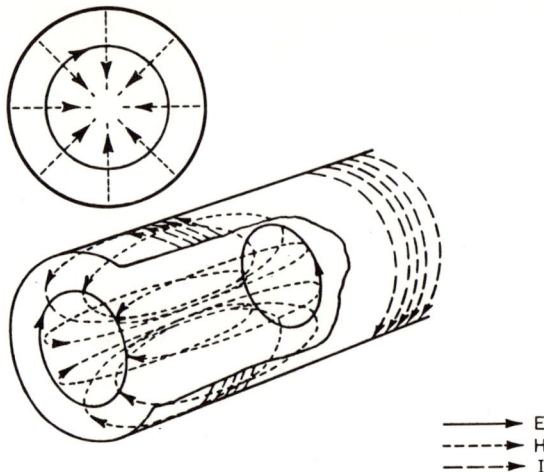

FIG. 3.27. Electric and magnetic field patterns, and wall currents, in the TE_{01} mode in cylindrical waveguide (after Mariner[37a]).

zero.[38] The TE_{01} field pattern in cylindrical waveguide is drawn in Fig. 3.27; the attenuation-frequency characteristic is illustrated in Fig. 3.28.

Microwave guidance techniques can be extended down to about 1 mm, but at this short wavelength the guide dimensions and tolerances become very small while the wall losses are exceedingly large. For example, RG-139 standard rectangular waveguide can propagate waves between 1.36 and 0.92 mm in the TE_{10} mode. Its dimensions are 0.034×0.017 in and, when fabricated from high conductivity silver, has theoretical attenuation

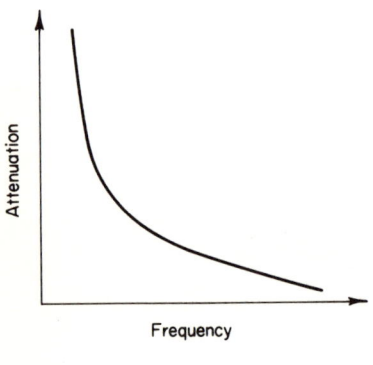

FIG. 3.28. Attenuation versus frequency characteristic of the TE_{0m} modes in cylindrical waveguide.

[38] S. Ramo and J. Whinnery, "Fields and Waves in Modern Radio." Wiley, New York, 1953.

of about 17–11.5 dB m^{-1}. At 1 mm the theoretical attenuation is 14 dB m^{-1}.[39] In practice, surface oxidation, strains, and roughness may well double these loss figures. For this mode, and also for the other modes in rectangular and circular guide, the theoretical attenuation increases as $\omega^{3/2}$ as we scale the waveguide size in proportion to the wavelength in going to higher frequencies.

Cylindrical silver waveguide with a diameter of 0.094 in can operate between 1.61–1.17 mm in the TE_{01} mode with theoretical attenuation of 2.5–1.1 dB m^{-1}. The TE_{01} cylindrical waveguide mode has been investigated with a view to millimeter wavelength communication systems, but it has serious drawbacks. Any asymmetry or bending of the guide produces wall currents and a corresponding increase in losses. The cross section should be of uniform area, precisely circular, and quite straight, otherwise field distortions arise which excite other modes of propagation.

3.4.2. High-Pass Waveguide Filters

Cutoff waveguides are frequently used as high (frequency)-pass filters. They can be fabricated by making a gradual transition from a standard waveguide down to a short section of very small guide and then another gradual transition back to the original size guide.

As an illustration of the use of a cutoff waveguide filter, suppose we wish to separate the higher harmonics of an electron cyclotron maser from the fundamental of wavelength 1000 μm. Let the filter be a section of guide 2 mm long with a cutoff wavelength of, say, 700 μm. For simplicity, we assume the guide to be perfectly conducting, with the understanding that a numerically correct result would require inclusion of wall resistance. This short section will propagate all higher harmonics but not the fundamental. From the results formulated in Section 3.4.2, the propagation constant for the fundamental is imaginary and has magnitude

$$|k_g| = (2\pi/\lambda_0)[(\lambda_0/\lambda_c)^2 - 1]^{1/2} \simeq (2\pi/\lambda_0),$$

and the factor by which the amplitude is reduced is

$$\exp(-2\pi z/\lambda_0) = \exp(-4\pi) \sim 109 \quad dB.$$

---

[39] W. Culshaw, *Advan. Electron. Electron Phys.* **15**, 197–263 (1961).

3.4.3. Oversize Waveguides and Light Pipes

One of the best means of transmitting millimeter and submillimeter waves is in oversize waveguide.[40] This technique is easy to use successfully, and it improves both the attenuation and power handling capabilities. RG-96 guide which is designed for principal mode operation in the range 26.5–40 GHz can carry submillimeter power largely in the TE_{10} mode if the field is launched through a carefully tapered section from smaller waveguide. According to Taub et al.,[41] a linear taper should be at least 50 wavelengths long. A too abrupt taper will cause a large curvature of the wave front and the increased possibility of exciting other modes. A 750-μm wave (frequency 400 GHz) in this guide has a theoretical attenuation of approximately 2.3 dB m^{-1}, which is one order of magnitude lower than in standard guide fabricated from silver. In practice, the losses will be several times larger than the theoretical value but still well below those in small-size guide.

Whereas the fields in standard waveguide have a large longitudinal component, in oversize guide the longitudinal component diminishes and in the limit of infinitely distant walls the fields reduce to the purely transverse waves of free-space propagation. This transition can be seen from the results in Section 3.4.1. From Eq. (3.51) for the TE_{10} mode, we have

$$H_z = A \cos(\pi x/a) \exp[-i(\omega t - k_g z)],$$

and from Eqs. (3.43) and (3.44),

$$H_y = 0,$$
$$H_x = -i(k_g/k_c^2)(A\pi/a) \sin(\pi x/a) \exp[-i(\omega t - k_g z)].$$

Thus the magnitude of the ratio of the maximum values of the longitudinal and transverse fields is

$$(H_z)_{max}/(H_x)_{max} = ak_c^2/\pi k_g. \qquad (3.54)$$

From Eqs. (3.42), (3.49), and (3.54), we can determine the magnitude of the longitudinal field relative to the transverse field. Consider, for example, the RG-96 guide mentioned above. It has dimensions 0.28 × 0.14 in and

---

[40] D. J. Kroon and J. M. Van Nieuwland, in "Spectroscopic Techniques" (D. H. Martin, ed.), Chapter 7. North-Holland Publ., Amsterdam, 1967.

[41] J. J. Taub, H. J. Hindin, O. F. Hinckelmann and M. L. Wright, *IEEE Trans. Microwave Theory Tech.* **11**, 338–345 (1963).

a cutoff wavelength of 1.42 cm. It is readily calculated that the longitudinal magnetic field of a 40 GHz wave in this guide is about 0.6 times as strong as the transverse field. However, for a 400-GHz wave the longitudinal field is only about 5% of the transverse field. For the frequency 400 GHz, RG-96 guide is ten times oversize. Guides more than ten times oversize will, of course, have still weaker longitudinal fields and hence will support essentially transverse electromagnetic waves.

For large rectangular waveguides with conducting walls of pure silver, the theoretical attenuation constant is given by[39]

$$\alpha = 5.60 \times 10^{-9} \nu^{1/2}/b \quad \text{dB m}^{-1},$$

where ν is the frequency in Hz, b is that dimension of the tube parallel to the electric field. Table 3.5 compares the attenuation in oversize and standard waveguide for the same wavelength.

TABLE 3.5. Theoretical Loss in Transmission by the TE_{10} Mode in Oversize and Standard Waveguide Made from Pure Silver[a]

Wavelength (mm)	Large tube		Standard waveguide		
	Dimensions (in)	Attenuation (dB m^{-1})	Number	Dimensions (in)	Attenuation (dB m^{-1})
8.6	6×6	0.69×10^{-2}	RG-96	0.280×0.140	0.50
3.75	2.84×1.34	0.047	RG-99	0.122×0.061	1.75
1.00	0.90×0.40	0.30	RG-139	0.034×0.017	14.0

[a] After Culshaw.[39]

Light pipes are essentially oversize waveguides, but with the difference that no attempt is made to control or limit the modes of propagation. Each mode tends to be largely transverse, and the superposition of the many modes supported gives a nearly plane wave in the pipe. Losses tend to be quite low provided the walls are clean and free from oxide layers and the pipe is evacuated. Figure 3.29 and Table 3.6 shows some measured values of transmittance through a number of pipes.[33,42]

[42] R. C. Ohlmann, P. L. Richards and M. Tinkham, *J. Opt. Soc. Amer.* **48**, 531–533 (1958).

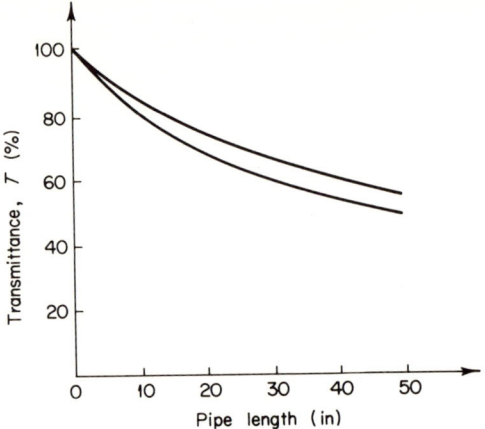

Fig. 3.29. Measured transmittance of 140-μm wavelength radiation through pipes of diameter 0.43 in. The upper curve corresponds to a polished brass pipe, and the lower curve fits measurements in copper, aluminum, and silvered glass (after Ohlmann et al.[42]).

TABLE 3.6. Measured Transmittance, at 100 μm and 500 μm, of Light Pipe 1.1 cm in Diameter, 91 cm Long, with Polished Brass Condenser Cones[a]

	100 μm	500 μm
Degreased hard drawn brass	0.78	0.83
Polished brass	0.79	0.84
Copper-plated brass	0.83	0.87
Stainless steel	0.50	0.63
Two cones, back to back	0.81	0.77

[a] After Richards.[43]

3.4.4. Resonant Cavities and Wave Handling Devices

We have seen the importance of resonant cavities in Chapter 2 in connection with coherent wave generators. They are also important as wavemeters for the measurement of frequency. From the microwave approach we can make enclosed cavities by placing metal plungers in rectangular and cylindrical waveguides, and from optics we have the

[43] P. L. Richards, *J. Opt. Soc. Amer.* **54**, 1474–1484 (1964).

3.4. WAVEGUIDES AND QUASIMICROWAVE COMPONENTS

Fabry–Perot resonator composed of two parallel reflecting plates in free-space.

Just as waveguides are increasingly troubled by wall losses as the frequency is raised, so too are resonant cavities. When the cavity dimensions are of the order of the wavelength, as they must be if we wish to limit the number of supported modes, the Q will be small. In general, the Q is roughly proportional to the ratio of the cavity volume to surface area because stored energy depends upon volume and losses depend upon the area. Thus improvements in Q may be obtained by using large cavities in higher modes, but in this case problems can arise in the identification of a particular mode among the many possible overlapping and degenerate modes. The TE_{01n} modes in circular cylindrical pipe, like their traveling wave constituents the TE_{01} waves in cylindrical waveguide, have exceptionally low loss at very high frequencies. It is possible to locate coupling irises in cylindrical cavities to favor the preferential excitation of this low-loss mode, but because it is identically degenerate with the TM_{11n} mode [see Table 3.4 and Eq. (3.52)] cross-coupling can readily occur, which acts to lower the Q of the TE_{01n} mode.

The Fabry–Perot interferometer has been investigated by Culshaw at wavelengths around 6 mm and has shown high Q capabilities. It is made with silver plates perforated with small holes, and may be regarded as a resonant cavity with multiple reflection of plane waves between the parallel plates. With plates 30 cm square perforated with an array of small holes, power reflectance $\mathscr{R} = 0.999$ has been both calculated and measured and loaded Q values in excess of 50,000 obtained. These are the sharpest Fabry–Perot fringes ever observed—sharper by one order of magnitude than optical fringes. The holes are analogous to the coupling irises of enclosed cavities and thus provide the means of coupling energy into and out of the resonance region between the plates. Loading on the cavity is controlled by the adjustment of hole diameter. Diffraction losses out the sides are small and will become smaller as the wavelength is decreased.

Culshaw has extended his calculations into the submillimeter wavelength region, and has predicted extremely high Q performance in this region. The considerable potential of the Fabry–Perot resonator in the far-infrared is illustrated by the theoretical results given in Table 3.7.

The high-Q Fabry–Perot resonator has been investigated by Valkenburg and Derr[44] from wavelengths of 3–1 mm. Rather than perforated

---

[44] E. P. Valkenburg and V. E. Derr, *Proc. IEEE* **54**, 493–498 (1966).

TABLE 3.7. Computed Unloaded Q Values for Silver Plates Spaced 2.5 cm Apart in a Millimeter Wave Interferometer[a,b]

λ (mm)	Order of interference, l	Power reflectance, $\mathscr{R}$	Cavity, Q
3.125	16	0.99917	60,300
2.0	25	0.99896	75,300
1.0	50	0.99852	106,500
0.5	100	0.99792	150,000
0.1	500	0.99533	333,900

[a] After Culshaw.[39]
[b] The conductivity of silver is taken as $6.139 \times 10^7 \, \Omega^{-1} \, m^{-1}$.

plates, these workers used a semiconfocal mirror system (one plane and one spherical mirror) with coupling into and out of the cavity via open-ended waveguide placed in the plane mirror. With mirror separation of 61 cm, mirror radius of curvature 122 cm, and mirror diameter of 22.86 cm, Q values of 3×10^5 were measured at 1 mm.

Many components based on optical and microwave handling techniques have been developed for short millimeter and submillimeter wavelengths. For a description of direction couplers, phase shifters, attenuators and bends in oversize waveguides,[40,45,46] and for crystal mounts,[47] and mode filters,[46] the reader is referred to the literature. A number of devices based essentially on the concept of plane waves in oversize waveguides are illustrated in Figs. 3.30–3.32. The principles are elementary: In the directional coupler (or, in optical language, the wave amplitude divider) a thin dielectric sheet separates a fraction of the wave by reflection into a side arm. Another principle is the technique of "frustrated total internal reflection" with two prisms closely spaced so that the evanescent wave is coupled into a propagating wave. Variable spacing of the two prisms facilitates variable directional coupling. The four-port junction with a diagonal reflecting film can also be used as a variable attenuator and phase shifter, in the manner illustrated in Fig. 3.31. Wave energy in arm 4 is varied by changing the relative phases of the waves reflected in arms 2 and 3.

[45] J. Bled, A. Bresson, R. Papoular and J. G. Wegrowe, *Onde Elec.* **44**, 26–35 (1964).
[46] H. J. Butterwecke and F. C. de Ronde, *Philips Tech. Rev.* **29**, 86–101 (1968).
[47] D. L. M. Blomfield, *J. Sci. Instrum.* **41**, 517 (1964).

3.4. WAVEGUIDES AND QUASIMICROWAVE COMPONENTS

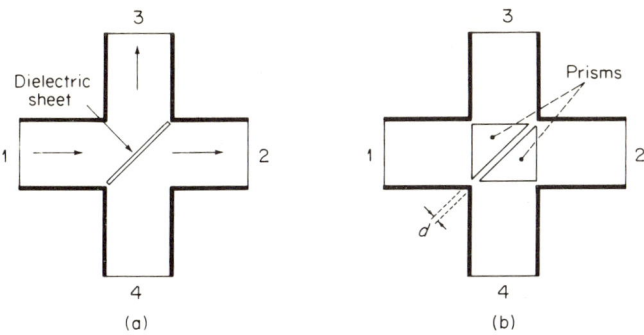

FIG. 3.30. Directional couplers for millimeter and submillimeter waves in oversize (3 cm) waveguide. (a) A fixed coupler in which radiation from arm 1 is split by the dielectric sheet and coupled into arms 2 and 3. (b) A variable coupler in which the power division is controlled by the separation d between the prisms (after Kroon and Van Nieuwland[40]).

Other useful wave handling devices have been based on interactions in "optically" active materials. Circular polarization can be produced in this way, and the plane of polarization of a linearly polarized wave can be rotated by a "circulator." Far-infrared "isolators" have been devised. Isolators are, of course, commonly used at microwave wavelengths where, in the form of magnetized ferrites with nonreciprocal propagation characteristics, they protect or "isolate" the generator from variations in the load.[36]

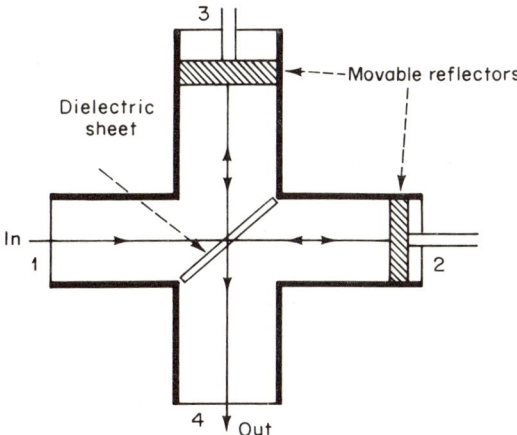

FIG. 3.31. Variable attenuator in oversize (3 cm) waveguide. The output power to arm 4 is determined by the relative phases of waves reflected from the movable plungers (after Kroon and Van Nieuwland[40]).

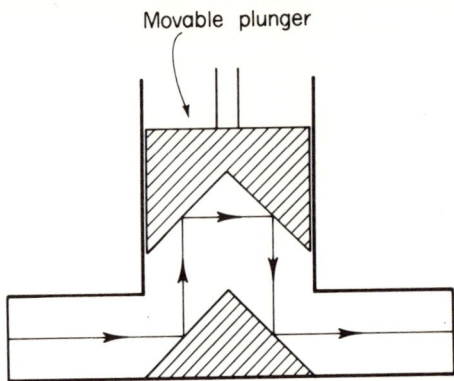

Fig. 3.32. Phase shifter based on the lengthening of the propagation path in oversize (3 cm) waveguide (after Kroon and Van Nieuwland[40]).

In solids where the plasma frequency is well below the lowest frequency for which interband transitions occur, there is a spectral window between these two frequencies for which the solid is nearly transparent (see Section 4.3.2). If a magnetic field is applied so that the cyclotron frequency lies in this transparent band, effects similar to those in ferrites at microwave frequencies occur. The efficiency of wave processing effects in this region is limited by lattice absorption and reflection, especially near reststrahl resonances.

Circular polarizers, isolators, and circulators based on cyclotron resonance in magnetorotary solids have been developed by Richards and Smith[48] for the far-infrared; microwave nonreciprocal devices have been extended to short millimeter wavelengths. Richards and Smith used n-type InSb with a carrier concentration of 5×10^{13} cm^{-3} in the form of a 1 mm-thick slab placed normal to both the direction of propagation and the magnetic field. Conversion to circular polarization can be simply achieved through the absorption of one circularly polarized component (that which rotates in the right-hand sense, in the same sense as the motion of carriers about the field) of a plane polarized or unpolarized incident wave. Results of calculations of the relative absorption of the two counter rotating waves, based on well-known equations[49] describing wave propagation in magnetized InSb, are plotted in Fig. 3.33. The capability of large selective absorption is indicated in the region of the cyclotron frequency ω_c.

[48] P. L. Richards and G. E. Smith, *Rev. Sci. Instrum.* **35**, 1535–1537 (1964).
[49] G. Dresselhaus, A. F. Kip and C. Kittel, *Phys. Rev.* **100**, 618–627 (1955).

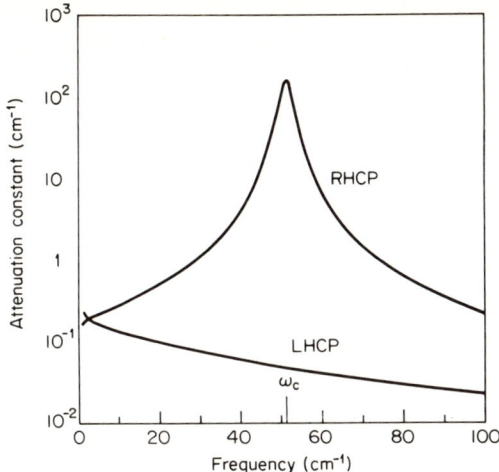

FIG. 3.33. Calculated attenuation constant versus frequency for the right-hand circularly polarized (RHCP) and left hand circularly polarized (LHCP) waves in InSb magnetised with a field of 7.12 kOe in the direction of wave propagation. The relaxation time $\tau = 3 \times 10^{-12}$ sec (after Richards and Smith[48]).

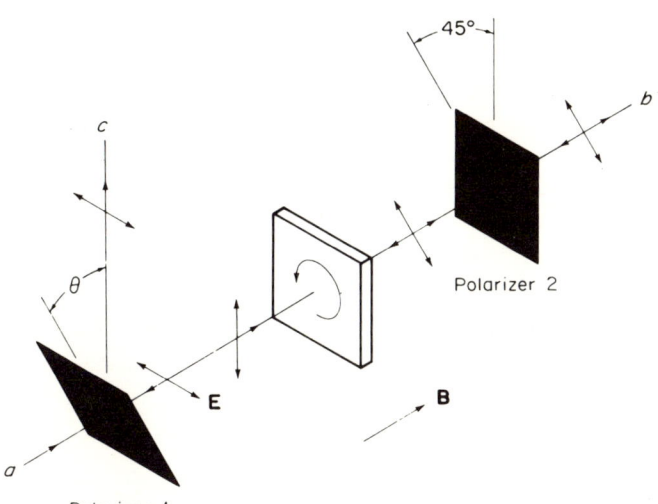

FIG. 3.34. Three-port far-infrared circulator using wire grid polarizers and a magnetorotary medium near its cyclotron resonance frequency. The incident radiation, a, is transmitted, while returning radiation is reflected in the direction c (after Richards and Smith[48]).

Figure 3.34 illustrates the circulator and isolator of Richards and Smith. A grid of parallel wires produces a linearly polarized wave incident on the slab of magnetized InSb. When the frequency of interest lies on the wings of the cyclotron resonance, a Faraday rotation of 45° in the plane of polarization can be obtained with only about 5% carrier absorption. The rotated beam is then passed through a second parallel grid polarizer tilted at 45° with respect to the first polarizer grid. In this form we have a two-port circulator. The system becomes an isolator with no radiation getting back to the input when polarizer grid 1 is tilted through some angle θ. Reflected radiation entering the magnetorotary slab from polarizer grid 2 will be rotated a further 45° and will reach grid 1 with the direction of linear polarization that the grid reflects. This reflected wave, with linear polarization orthogonal to the incident polarization can be absorbed at c.

Although Richards and Smith had used these techniques at low temperatures (4.2°K), InSb polarizers could be used at much higher temperatures, since the carrier concentration remains constant and the mobility increases with temperature up to about 100°K.

3.5. Grating Monochromators

Spectral separation in the far-infrared is commonly achieved with spectrometers based on the diffraction grating as the dispersive element. The layout of a suitable instrument—a Czerny–Turner monochromator—is illustrated in Fig. 3.2. Polychromatic radiation illuminates the entrance slit located at the center of focus of a collimating mirror. Parallel radiation is then directed towards the grating, and the diffracted wave is focused by a second mirror to pass through the exit slit to a detector.

If the angle of incidence on the grating is θ_i, the diffracted waves will reinforce in the direction θ given by

$$S(\sin \theta_i + \sin \theta) = m\lambda. \tag{3.55}$$

This is known as the grating equation; the *grating spacing* S is illustrated in Fig. 3.35. The integer m is the *order* of the interference maxima. The zeroth order ($m = 0$) corresponds to specular reflection; $m = 1, 2, \ldots,$ give the first, second, and higher orders which occur at increasing angles of diffraction θ.

3.5. GRATING MONOCHROMATORS

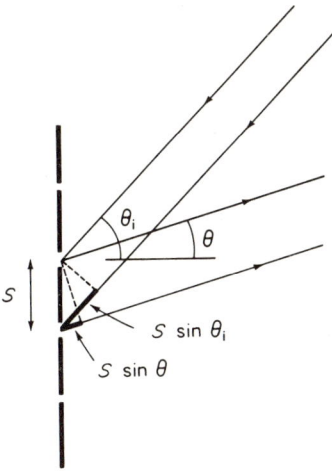

FIG. 3.35. Line illustration of a wave incident on a diffraction grating at an angle θ_i, and the wave diffracted at an angle θ. S is the grating spacing.

3.5.1. Radiation Dispersion and Spectral Resolution

We are interested in the spectral resolution of the grating monochromator and in the energy transmitted through the instrument from the source to the detector. The resolution is determined essentially by the width of the entrance and exit slits. High resolution requires narrow slits and these in turn reduce the energy transmittance. The energy-limited resolution can be calculated as follows: From differentiation of the grating equation, the dispersion is

$$d\theta/d\lambda = m/(S \cos \theta). \tag{3.56}$$

If the maximum spectral spread transmitted by the system is $\Delta\lambda$, this will be dispersed in angle by $(d\theta/d\lambda) \Delta\lambda$, and in space by the width of the exit slit s, if

$$s = f(d\theta/d\lambda) \Delta\lambda, \tag{3.57}$$

where f is the focal length of the mirror. Thus, from Eqs. (3.56) and (3.57), the energy-limited resolution is

$$\Delta\lambda = (s/f)(S \cos \theta)/m. \tag{3.58}$$

S/m can be estimated from the grating equation. For typical values of the angles used in monochromator designs, $\theta_i \sim 40°$, $\theta \sim 20°$,

$$S = \lambda m/(\sin \theta_i + \sin \theta) \gtrsim \lambda m. \tag{3.59}$$

Although the dispersion increases as θ approaches 90°, the maximum diffraction angle is kept below 60° because the intensity falls off as $(1 + \cos\theta)^2$. This is the *obliquity factor* which shows itself in the refined treatment of diffraction due to Kirchhoff.[30] The resolving power of an energy-limited grating monochromator is

$$\lambda/\Delta\lambda = \nu/\Delta\nu \approx f/s. \tag{3.60}$$

Secondary to the limitations of resolving power set by the slit-width is a diffraction limitation set by the width of the principal maxima of the diffraction pattern and determined by the grating itself. To find this diffraction limit consider the wavefronts shown in Fig. 3.36, where normal incidence is taken without loss of generality. A principal maximum in the diffracted wave occurs in a direction θ for which the waves from corresponding elemental strips of successive slits add in phase. That is, the path difference for points separated by the grating spacing S is $m\lambda$. For a

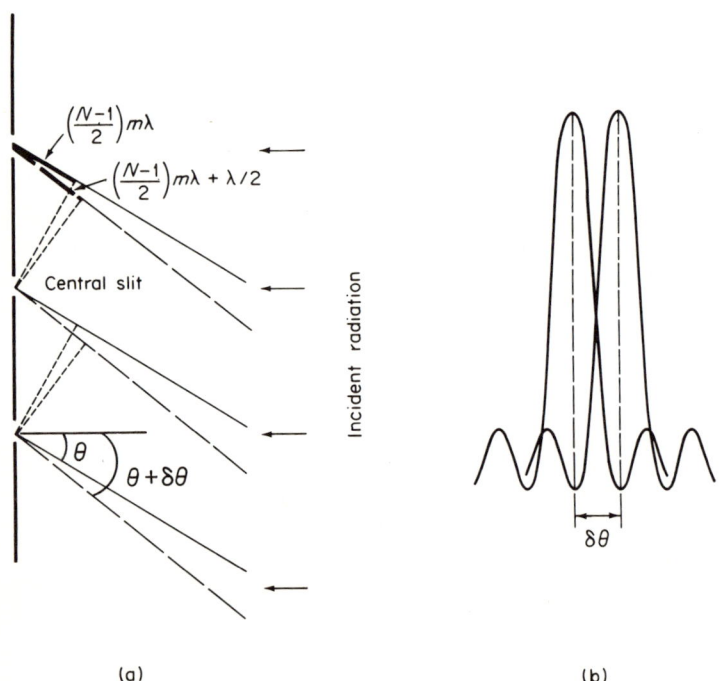

FIG. 3.36. (a) Line illustration of waves diffracted by a grating at the angles θ and $\theta + \delta\theta$, where $\delta\theta$ is the angular half-width of the principal maximum of the diffraction pattern. (b) Illustration of the limit of resolution of two spectral lines, according to the Rayleigh criterion.

3.5. GRATING MONOCHROMATORS

grating of N slits, there will be a path difference $(N-1)m\lambda/2$ between the wave originating at the central and extreme upper slits. At a slightly larger angle of diffraction $\theta + \delta\theta$ the path difference will increase to $(N-1)m\lambda/2 + (\lambda/2)$, and the radiation from the central slit and extreme upper slit will now be opposite in phase and will cancel. Similarly, radiation from corresponding slits taken successively below the central and upper slits will annul, so that in the direction $\theta + \delta\theta$ there will be a minimum of intensity. The cancellation is not quite complete in this direction, for if N is odd, there will remain one slit which is not annulled, and if N is even there will be no slit exactly at the center. From Fig. 3.36 the angular separation between the maximum and minimum of the diffraction pattern is given by

$$(B/2)\,\delta\theta = [(N-1)/2]S\cos\theta\,\delta\theta = \lambda/2,$$

where B is the breadth of the plane wave diffracted in this direction. N is large so that $N-1$ can be replaced by N. Thus the angular half-width of the principal maximum is[†,50]

$$\delta\theta = \lambda/(NS\cos\theta). \tag{3.61}$$

The resolving power is given by the Rayleigh criterion that two spectral lines are resolved when their diffraction patterns are displaced by at least the separation between the principal maximum and the first minimum. This condition is illustrated in Fig. 3.36. The smallest wavelength interval resolved is, by Eqs. (3.56) and (3.61),

$$\Delta\lambda = \delta\theta\,\frac{d\lambda}{d\theta} = \frac{\lambda}{NS\cos\theta}\,\frac{S\cos\theta}{m} = \frac{\lambda}{mN}. \tag{3.62}$$

The resolving power in the diffraction-limited case is

$$R = \lambda/\Delta\lambda = mN. \tag{3.63}$$

Resolution is normally quoted in terms of wave number $\sigma\ (= 1/\lambda)$ rather than wavelength differences. Differentiation of the relation

---

[50] F. A. Jenkins and H. E. White, "Fundamentals of Physical Optics." McGraw-Hill, New York, 1937.

---

† This expression is exact. A more rigorous treatment derives $\delta\theta$ from the well-known diffraction function $\sin^2 N\gamma/\sin^2 \gamma$, γ being half the phase difference for corresponding points in any two adjacent slits.

gives the conversion from $\Delta\lambda$ to $\Delta\sigma$

$$|\Delta\sigma| = \Delta\lambda/\lambda^2, \qquad (3.64)$$

where $|\Delta\sigma|$ is expressed in units of cm^{-1}.

3.5.2. Separation of Overlapping Orders. Echelette Gratings

From the grating equation [Eq. (3.55)] it is apparent that at the angle θ for which the first order maximum occurs at a wavelength λ, maxima also occur in the second, third, and higher orders for the respective wavelengths $\lambda/2$, $\lambda/3$, etc. This is the problem of overlapping orders which must be overcome in the design of a grating spectrometer.

Gratings for far-infrared work are invariably of the *echelette type*.[51] Such gratings were first suggested by Rayleigh and used by Wood[51a] to increase the energy transmittance of the spectrometer. They are shaped in a saw-toothed profile as shown in cross section in Fig. 3.37, and oriented

FIG. 3.37. Cross section of an echelette or blazed diffraction grating.

so that specular reflection from the long faces of the grooves coincides with the direction of a chosen order of the diffracted wave. The grating is said to be "blazed" for the wavelength of that order. The distribution of intensity in the various orders of diffraction is vastly altered from that for a plane grating. Instead of maximum intensity in the zeroth order, most of the power is concentrated in the direction of the blaze (the direction for which specular reflection coincides with the chosen lobe of the diffraction pattern).

[51] J. Strong, "Concepts of Classical Optics." Freeman, San Francisco, 1958.
[51a] R. W. Wood, *Phil. Mag.* **20**, 770 (1910).

3.5. GRATING MONOCHROMATORS

In the direction of the blaze the various higher orders of harmonics of the blazed wavelength still overlap and indeed all are strengthened by the effect of specular reflection. The relative intensities of the overlapping orders depend on the spectral distribution of the radiation source, and, as theoretical studies have shown,[20,52] on just how closely the direction of observation is to the direction of blaze. Since the spectral brightness of the hot body source increases with increasing frequency, the higher orders in the blaze can be very strong, indeed, and difficult to remove with filters. However, calculations and experiments show that the angular widths of the various overlapping lobes decrease with increasing order number. Thus, if the angular setting of the grating is altered so the direction of observation varies from the direction of the blaze, the relative intensities of the various orders change. This is illustrated in Fig. 3.38 for the first, second, and third overlapping orders of a grating blazed at 10°. It is found in practice that considerable rejection of unwanted orders can be achieved when the angle of observation is between 1.25–2 times the blaze, that is, between 12.5–20° for the grating of Fig. 3.38.

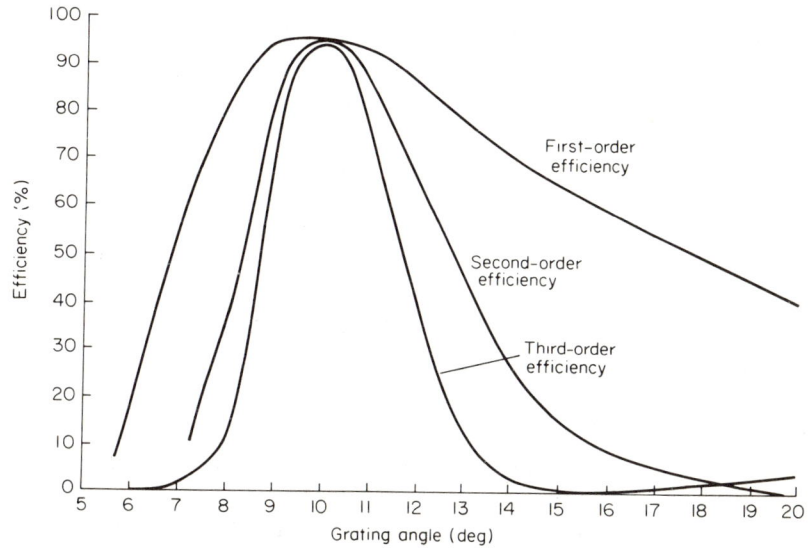

FIG. 3.38. Efficiency of an echelette diffraction grating blazed at 10°, for the first, second, and third orders of interference. (After Kimmitt.[19a] Crown copyright. Reproduced by permission of the Controller of HM Stationery Office.)

[52] G. R. Wilkinson and D. H. Martin, in "Spectroscopic Techniques" (D. H. Martin, ed.), Chapter 3. North-Holland Publ., Amsterdam, 1967.

It is usual to work with low orders ($m = 1$ or 2) because of their relatively large separation from other interfering wavelengths. Orders higher than the fourth are particularly difficult to isolate either with grating techniques or with filters.

3.5.3. Energy Transmittance

A major task in grating spectrometry in the far-infrared is to ensure the delivery of sufficient energy to the detector to give an adequate signal-to-noise ratio. A satisfactory compromise must be obtained between the observing time or rate of scanning of the spectral region of interest and the spectral resolution required.

The energy illuminating the grating is clearly proportional to the power B_ν radiated by the source and the area ls of the entrance slit. While s is chosen on the basis of resolution, the height l of the entrance slit is ultimately limited by the aperture and optics of the detector. It is normally about 1–2 cm. The fraction of the input energy which illuminates the grating is proportional to the angle Ω_0 subtended at the entrance slit by the grating (via the collimating mirror), and, after diffraction, a fraction $s(f\, d\theta/d\lambda)^{-1}$ will pass through the exit slit to the detector. Thus, taking into account the overall reflectivity R_ν of the grating and mirrors, the energy transmittance in a given spectral interval is

$$R_\nu B_\nu \Omega_0 ls^2 (f\, d\theta/d\lambda)^{-1}. \tag{3.65}$$

To make best use of the rather feeble far-infrared emission from the source, the grating should be large to give a large solid angle Ω_0. However, the product of the grating area and the angle subtended at the collimating mirror by the entrance slit should be consistent with the area–solid angle product $A\Omega$ throughout the entire optical system. The mirror should gather a large amount of radiation from the source. The collimating and condensing optics, the aperture of the detector and the solid angle over which it is capable of receiving radiation, should be consistent with the invariance of the étendue along the path of the beam. That is, along a beam where rays are not destroyed (by reflection, attenuation, etc.):

$$A\Omega = \text{invariant}. \tag{3.2}$$

Normally it is the solid angle over which the detector can receive radiation and the size of its entrance aperture which limits the étendue of the system.

3.5.4. Features of Practical Designs

Some of the general features of far-infrared grating monochromators can be seen from considerations of particular instruments. Let us consider the monochromator designed by Kimmitt[19a] and used extensively in England, and one developed by Ginsberg, Richards, and Tinkham[43,53] in the U.S.A. The layouts of these instruments are shown in Figs. 3.39

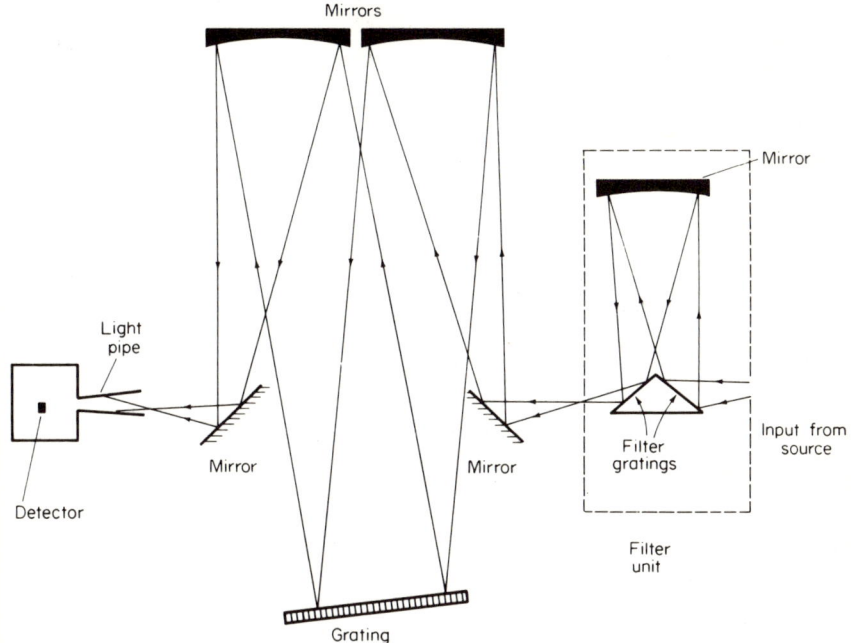

Fig. 3.39. Layout of a diffraction grating spectrometer developed in England. (After Kimmitt.[19a] Crown copyright. Reproduced by permission of the Controller of HM Stationery Office.)

and 3.40. The English design uses two spherical mirrors to collimate and condense the radiation and is of the type first used by Czerny and Turner.[54] It has an F-number of 2.5, and combines good light gathering power with reasonable compactness. The American monochromator has a single spherical mirror to collimate and condense the radiation. An F-number 1.5 mirror is used to minimize the overall length and give good light gathering power.

[53] P. L. Richards and M. Tinkham, *Phys. Rev.* **119**, 575–590 (1960).
[54] M. Czerny and A. F. Turner, *Z. Phys.* **61**, 792–797 (1930).

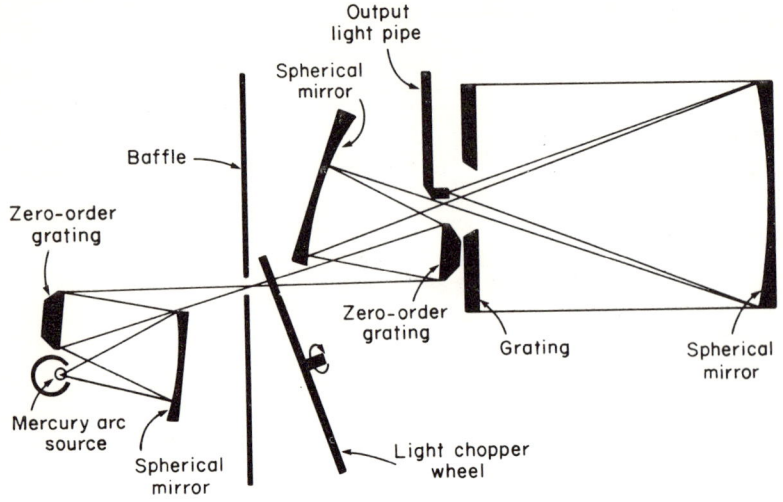

Fig. 3.40. Layout of a diffraction grating spectrometer developed in the U.S.A. (after Richards and Tinkham[53]).

In both instruments the input radiation has its short wavelength components removed by scattering from reflecting grating filters. Alternatively, polyethylene replica gratings may be used in transmission. The functioning of grating filters has been discussed in Sections 3.1.2 and 3.1.3. As shown by Fig. 3.17, an echelette grating is an effective low-pass filter, for it scatters radiation with wavelengths equal to and shorter than the grating spacing. It is used when the shortest wavelengths desired in the monochromator are one hundred or more micrometers, and can be cut with a 45° profile as in Fig. 3.17, and conveniently made about 10 cm square. Still shorter wavelengths can be removed with reststrahlen crystals or polyethylene filters. For the removal of wavelengths shorter than 50-μm crystal quartz or a thin sheet of carbon filled polyethylene is placed in the path of the beam.

After initial filtering the radiation is collimated and directed onto the diffraction grating. In the instrument illustrated in Fig. 3.39 the gratings are 20×25 cm² with a blaze angle of 10°. They are used in first order and, by Eq. (3.59), the grating spacing is approximately three times the wavelength in the blaze. Kimmitt's instrument has worked over the range 2–2000 μm with a variety of prefilters appropriate to the various parts of the spectral range, and with a range of diffraction gratings. Each diffraction grating used above 100 μm or so has an appropriate pair of grating filters with a suitable cutoff wavelength. In one set of investiga-

TABLE 3.8. Diffraction Gratings and Filter Gratings for the Wavelength Range 100–2400 μm with Blaze Angle of 10°[a]

Grating spacing (in)	Wavelength range (μm) (12.8°–20°)	Polyethylene filter grating spacing (in)
0.009	100–150	0.004
0.0135	150–225	0.006
0.018	200–300	0.008
0.027	300–450	0.012
0.036	450–600	0.016
0.054	600–900	0.024
0.072	800–1200	0.032
0.108	1200–1800	0.048
0.144	1600–2400	0.064

[a] After Kimmitt.[19a] Crown copyright. Reproduced by permission of the Controller of HM Stationery Office.

tions, Kimmitt used only three diffraction gratings to work throughout the wavelength range 120–2300 μm. Ruled with 20, 8, and 3 lines per cm they gave wavelengths in the blaze of 175, 440, and 1150 μm, respectively. To cover this range it was necessary to work over the wide angular range 0.7–2.0 times the blaze angle, and to tolerate troubles from overlapping orders, particularly near the blaze. However, with nine gratings, each working from 1.28–2.0 times the blaze angle, adequate rejection was obtained throughout the range 100–2400 μm. The properties of these gratings are given in Table 3.8 together with the spacings of polyethylene replica gratings used as transmission filters in this monochromator.

Kimmitt's instrument employed mirrors 20 cm in diameter with 50-cm focal length. The slits were 2.5 cm high and could open to a width of 2 cm. With these slits and with a mercury arc lamp source, the resolution was better than 0.5 cm^{-1} from 200–10 cm^{-1}. With a water vapor laser source, the slit width could be decreased to give an accuracy of 1 part in 1000 in the measurement of the laser wavelength. To measure wavelength accurately, the procedure is to measure the angle of the grating from the position of the zero-order spectrum carefully, and calculate λ from the grating equation.

In the monochromator of Fig. 3.40 the mirror is 30 cm in diameter with a focal length of 45 cm. The echelette diffraction gratings are 30 cm square with holes approximately 5×10 cm^2 cut in the centers; they are used in first order. For the spectral range 100–2000 μm, the gratings are cut with about 3–20 lines per cm.

The ultimate resolution of a monochromator is determined by the measurement of the signal power reaching the detector in a background of noise. The signal will be limited by the width of the entrance and exit slits and will accordingly be related to the resolution of the instrument. This rms noise amplitude will depend on the time of observation, as $t^{-1/2}$. Thus for a given signal-to-noise ratio there will be interdependent limits to the resolution and the rate at which a spectral range can be scanned. We may use narrow, high-resolution slits and scan the spectrum very slowly, or we may scan rapidly with wider slits and corresponding reduced resolution.

3.6. Multiplex Spectrometry

Consider the output from a Michelson interferometer when radiation containing a range of frequencies is beamed into the input and one mirror is moved with constant velocity. For each of the spectral components of the two beams periodic interference will occur, the period being the time taken for the mirror to move through a distance of a half wavelength. The different spectral intervals will, of course, be modulated with different frequencies. Thus one might pass the output through a system of narrow-band filters and thereby effectively Fourier analyze the spectral composition of the incident radiation. Alternatively, the output for the entire spectral range as a function of the path difference (the "interferogram") may be recorded and decoded by Fourier transform analysis to yield the spectrum. This technique records information about all spectral intervals simultaneously and is appropriately called multiplex spectrometry. It is also frequently called Fourier transform spectrometry, and has been the subject of many excellent reviews.[43,55–57]

In practice, the technique of recording and mathematically analyzing interferograms has generally proven preferable to the technique employing filters.[43] It was used by Michelson[57a,57b] in his determinations of profiles

---

[55] E. V. Lowenstein, *Appl. Opt.* **5**, 845–854 (1966).
[56] J. Connes, *Rev. Opt.* **40**, 45–79, 116–140, 171–190, 231–265 (1961).
[57] J. Strong and G. A. Vanasse, *J. Opt. Soc. Amer.* **49**, 844–850 (1959).
[57a] A. A. Michelson, *Phil. Mag. Ser. 5.* **31**, 256 (1891).
[57b] A. A. Michelson, "Studies in Optics," p. 124. Univ. of Chicago Press, Chicago, 1927.

3.6. MULTIPLEX SPECTROMETRY

of spectral lines, although the full power of the method was restricted at that time by deficiencies in detectors necessary for the recording of interferograms and in computers necessary for performing Fourier transforms.[55] Multiplex spectrometry was used in the far-infrared by Rubens and Wood in 1911[57c] for investigations of the radiation emitted by a Welsbach mantle. In 1948 Jacquinot and Dufour[58] pointed out that multiplex spectrometry enables high resolution without the necessity of a limiting aperture as in dispersion spectrometry. Consequently there can be much greater energy transmittance, and this is most important with weak incoherent sources in the far-infrared.

The full power of multiplex spectrometry was realized by Fellgett in 1951.[59a] He not only calculated numerically the Fourier transform of an interferogram, but he pointed out an important advantage in signal-to-noise ratio obtainable with multiplex spectrometers over monochromators. The "Fellgett advantage" or "multiplex gain" is operative in detector-noise limited spectrometry in the far-infrared. In such systems the signal integrates in direct proportion to t, the time available for the observation, while the noise integrates in proportion to $t^{1/2}$ (see Section 4.7). In a monochromator the spectrum is explored sequentially in (say) m spectral intervals. The average time available for the observation of each is t/m, and the signal-to-noise ratio is thus proportional to $(t/m)^{1/2}$. In the multiplex system where the entire spectrum is observed simultaneously for the same total time t, the signal-to-noise ratio is proportional to $t^{1/2}$, so there is a gain of $m^{1/2}$ over the monochromator.

3.6.1. Fourier Transform Principle

To formulate the principle let us consider the relation of the radiation entering a two-beam interferometer to the output obtained as the path difference is changed. For monochromatic incident radiation of wave number σ, split into two interfering beams each of intensity $S(\sigma)/2$, the intensity of the emerging beam is

$$I(\Delta) = 2S(\sigma) \cos^2 \pi\sigma\Delta = S(\sigma) + S(\sigma) \cos 2\pi\sigma\Delta, \qquad (3.66)$$

where Δ is the path difference between the two beams.

[57c] H. Rubens and R. W. Wood, *Phil. Mag.* **21**, 249 (1911).
[58] P. Jacquinot and C. Dufour, *J. Rech. Cent. Nat. Rech. Sci. Lab. Bellevue (Paris)* **6**, 91 (1948).
[59a] See, for example, P. Fellgett, *J. Phys. (Paris)* **28**, 165 (1967).

For polychromatic radiation with intensity $S(\sigma)\,d\sigma$ in the interval between σ and $\sigma + d\sigma$, the output intensity is

$$I(\Delta) = \int_0^\infty S(\sigma)\,d\sigma + \int_0^\infty S(\sigma) \cos 2\pi\sigma\Delta\,d\sigma. \tag{3.67}$$

The first integral on the right-hand side is half the intensity $I(0)$ for zero path difference between the beams. Thus

$$I(\Delta) - \tfrac{1}{2}I(0) = \int_0^\infty S(\sigma) \cos 2\pi\sigma\Delta\,d\sigma. \tag{3.68}$$

The quantity $I(\Delta) - \tfrac{1}{2}I(0)$ is called the interferogram function; it is the Fourier cosine transform of the spectral distribution of the input radiation $S(\sigma)$. By the Fourier integral theorem for the even function $I(\Delta)$, the spectrum can be obtained from the interferogram by the relation

$$S(\sigma) = 4 \int_0^\infty [I(\Delta) - \tfrac{1}{2}I(0)] \cos 2\pi\sigma\Delta\,d\Delta. \tag{3.69}$$

Equation (3.69) is the basic relation of Fourier transform spectroscopy. As the path difference of a two-beam interferometer is changed progressively the output is registered with a detector. The Fourier cosine transform of the recorded interferogram is calculated to yield the spectrum $S(\sigma)$.

3.6.2. Resolving Power and Apodization

The resolving power in all spectrometers is determined by the maximum path difference between two interfering waves. In a diffraction grating, for example, the separation between the central maximum of the interference pattern and the first minimum is determined by the condition for interference between waves emanating from the central and extreme slits of the grating (see Section 3.5.1). Through the Rayleigh criterion this separation defines the resolving power.

In the case of the two-beam interferometer the movable reflector can be displaced only over a limited range 0 to $\Delta_{\max}$ and this determines the resolving power. Restricting the range on the limits of integration of Eq. (3.69) gives the approximate spectrum

$$S'(\sigma) = 4 \int_0^{\Delta_{\max}} [I(\Delta) - \tfrac{1}{2}I(0)] \cos 2\pi\sigma\Delta\,d\Delta. \tag{3.70}$$

3.6. MULTIPLEX SPECTROMETRY

To find the resolving power, consider monochromatic incident radiation of wavenumber σ and unit intensity, that is, $S(\sigma) = \delta(\sigma - \sigma_0)$, δ being the Dirac delta function. The interferogram function is then $I(\Delta) - \tfrac{1}{2}I(0) = \cos 2\pi\sigma_0\Delta$, and the spectrum determined is

$$S'(\sigma) = 4 \int_0^{\Delta_{\max}} \cos 2\pi\sigma_0\Delta \, \cos 2\pi\sigma\Delta \, d\Delta$$

$$= \left[\frac{\sin 2\pi(\sigma - \sigma_0)\Delta_{\max}}{2\pi(\sigma - \sigma_0)\Delta_{\max}} + \frac{\sin 2\pi(\sigma + \sigma_0)\Delta_{\max}}{2\pi(\sigma + \sigma_0)\Delta_{\max}} \right] 2\Delta_{\max}. \quad (3.71)$$

In practice, the path difference between the two interfering beams is varied over many wavelengths so for all frequencies of interest $\sigma\Delta_{\max} \gg 1$, and the second term on the right-hand side of Eq. (3.71) is negligibly small.

The function

$$\frac{\sin 2\pi(\sigma - \sigma_0)\Delta_{\max}}{2\pi(\sigma - \sigma_0)\Delta_{\max}} \quad (3.72)$$

is the *instrumental* or *scanning function*. It is the function recorded at the output when an infinitely sharp spectral line of wave number σ_0 illuminates the interferometer. It is analogous to the interference pattern that appears in the amplitude distribution for a diffraction grating, and it plays an analogous role in determining the resolution.

As it stands, the instrumental function in the form Eq. (3.72) is not entirely satisfactory from the point of view of resolution, because the secondary maxima are so large that they would tend to hide weak lines in the vicinity of a strong line. It is desirable to reduce them considerably, and this can be done by a process called *apodization*. This process consists of introducing an additional function $A(\Delta)$ so that the truncated integral of Eq. (3.70) takes the form

$$S'(\sigma) = 4 \int_0^{\Delta_{\max}} [I(\Delta) - \tfrac{1}{2}I(0)]A(\Delta) \cos 2\pi\sigma\Delta \, d\Delta, \quad (3.73)$$

where $A(\Delta)$ is the apodization function. For example, it is found that the linear apodization function $A(\Delta) = 1 - (\Delta/\Delta_{\max})$ for $0 < \Delta < \Delta_{\max}$, gives the instrumental function

$$\left[\frac{\sin \pi(\sigma - \sigma_0)\Delta_{\max}}{\pi(\sigma - \sigma_0)\Delta_{\max}} \right]^2. \quad (3.74)$$

This is identical with the intensity diffraction pattern of a grating, and accordingly gives the same resolution as a grating spectrometer.

The Rayleigh criterion that two spectral lines are resolved when their diffraction patterns are displaced by the separation between the principal maximum and the first minimum gives the resolving power for the unapodized case [Eq. (3.72)]:

$$R = \sigma/d\sigma = \sigma \Delta_{max} = \Delta_{max}/\lambda. \tag{3.75}$$

This condition, that two frequencies are resolved if $d\sigma\, \Delta_{max} > 1$, means that the maximum displacement of the movable reflecting element of the interferometer must produce a differential phase shift of 2π or more between the two frequencies.

The instrumental functions for the unapodized case and for a linear apodizing function are seen from Eqs. (3.72) and (3.74) to be different. It can be seen that apodization can decrease the secondary maxima; however, the price paid for this improvement is that the width of the central peak widens, thus degrading the resolving power.

Apodization is applied mathematically to the interferogram during the computation of the truncated integral of Eq. (3.73) rather than by cumbersome experimental intensity modulation techniques. Once an interferogram is recorded, different apodizing functions can be put into the computer program to improve the resolution. For discussions of computational procedures for multiplex spectrometry the reader is referred to papers by Connes[56] and Richards.[59b]

3.6.3. Permissible Solid Angle. Étendue Advantage

In the foregoing considerations we have assumed beams of parallel radiation within the interferometer, ignoring the spread that must occur in practice between the various rays. The rays within the solid angle accepted by the interferometer will be shifted differently in phase as the mirror in the Michelson interferometer is displaced, and the resolution will accordingly be reduced.

To find the connection between the solid angle and the resolution assume the limiting aperture admits radiation within a solid angle Ω. The extreme ray will then be at an angle i to the axial rays where $\Omega \approx \pi \sin^2 i \approx \pi(1 - \cos^2 i)$, and the ratio of the path differences between

---

[59b] P. L. Richards, in "Spectroscopic Techniques" (D. H. Martin, ed.), Chapter 2. North-Holland Publ., Amsterdam, 1967.

these two rays will be $\cos i$. That is, for a path difference Δ_{max} for the axial ray, that for the extremal ray will be

$$\Delta_{max}/\cos i = \Delta_{max}[1 + (\Omega/2\pi)].$$

The difference is thus

$$\delta\Delta_{max} = (\Omega/2\pi) \Delta_{max},$$

and the effect is the same as would be obtained with parallel radiation with frequency spread $\delta\nu$, where

$$\nu/\delta\nu = \Delta_{max}/\delta\Delta_{max} = 2\pi/\Omega.$$

The limit of resolving power R set by the effect of a solid angle Ω is therefore given by

$$R\Omega = 2\pi. \tag{3.76}$$

This expression is due to Jacquinot.[58,60] It gives the ultimate limit of resolution attainable with interferometers with division of amplitude, and these, according to Jacquinot and Dufour,[58] necessarily have circular symmetry. It is applicable to the Michelson instrument for here the beam splitter "divides" the beam amplitude, and the circular symmetry is well known from optical interference fringes. The Jacquinot limitation on solid angle is not at all a serious restriction; with the largest solid angles used in practical instruments the limits of resolving power are several thousand.

For a required resolution, the solid angle of radiation permitted by the interferometer is large compared with that attainable with a monochromator where the angle of beam acceptance is restricted in one dimension by the narrow slit. Jacquinot has shown[1,60] that if the length of the slit subtends an angle β at the condensing mirror, the gain in solid angle is very approximately $2\pi/\beta$ and, if the two instruments have the same area of cross section, the gain in étendue is also $2\pi/\beta$. The signal-to-noise ratio is therefore increased by a factor $(2\pi/\beta)^{1/2}$.

3.6.4. Effect of Solid Angle on the Frequency Determination

For the average of the rays within the solid angle Ω of reception discussed in Section 3.6.3, a path difference $\Delta[1 + (\Omega/4\pi)]$ occurs when the

---

[60] P. Jacquinot, *Rep. Progr. Phys.* **23**, 267–312 (1960).

axial ray is changed in path by Δ. Consequently, if we use the axial path difference in Eq. (3.73) the frequencies in the spectrum will be overestimated by the factor $1 + (\Omega/4\pi)$. In the reduction of data the spectrum must be corrected by this factor.

3.6.5. Interferometers for Multiplex Spectrometry

The Michelson interferometer (Fig. 3.41) is widely used in multiplex spectrometry, but its performance is somewhat affected by the limitations of the beam splitter. Wire meshes have been used but their capability

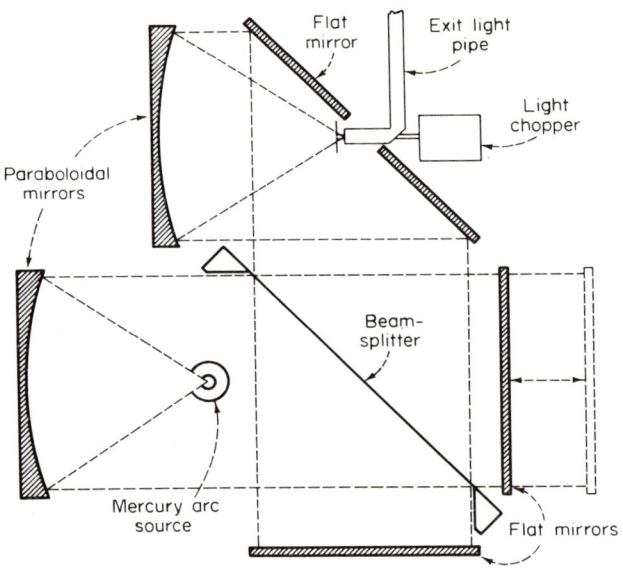

FIG. 3.41. Diagram of a Michelson interferometer for the far-infrared (after Richards[43]).

of dividing the wave amplitude is restricted to wavelengths in the vicinity of the wire spacing. Mylar† polyester sheet beam splitters, introduced by Gebbie, are now widely used. Its reflectivity is rather too low for optimum division and it suffers from interference effects within the film itself.

In traversing a Michelson interferometer each beam is reflected once from the beam splitter and transmitted once through it. The intensity of the resulting output radiation is therefore proportional to RT, where

---

† "Mylar" and "Melinex" are trade names for polyethylene terephthalate.

3.6. MULTIPLEX SPECTROMETRY

R and T are the reflection and transmission coefficients. Neglecting loss within the film, we have $R + T = 1$, and the output intensity varies as $R(1 - R)$. If we equate the derivative of the intensity with respect to Δ to zero, we can find the optimum value of R to give maximum intensity. This occurs for $dR(1 - R)/dR = 0$, or $R = 0.5$, and gives for an ideal beam splitter $RT = 0.25$.

The efficiency of an actual beam splitter, defined as that percentage of 0.25 its RT product is, has been plotted in Fig. 3.42 for a sheet of 0.003-in thick mylar and for a wire mesh grating. As the frequency increases the efficiency oscillates due to interference within the film, and damping

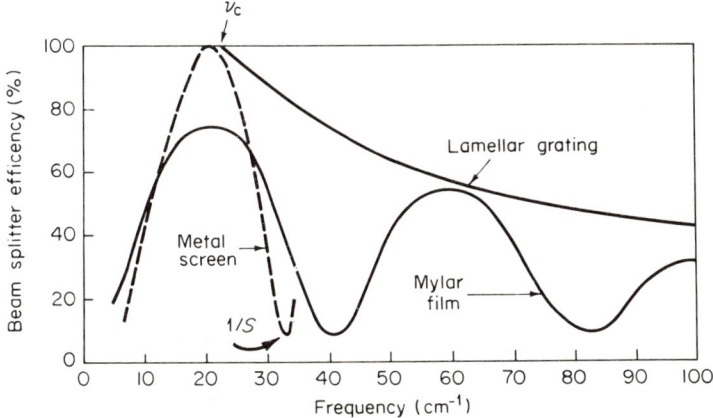

FIG. 3.42. Efficiency of mylar, metal screen, and lamellar grating beam splitters as a function of wavenumber. The mylar is 0.003 in thick and the wire screen has ~250 lines per cm. The frequency at which cancellation begins to occur in the lamellar grating interferometer, and the frequency at which the wavelength equals the grating spacing of the metal screen are marked ν_c and $1/S$, respectively (after Richards[43]).

occurs because of increasing absorption. It is clear that reasonable efficiency is obtainable with a single mylar sheet only over a restricted frequency range. At the frequencies corresponding to the minima of efficiency the signal-to-noise ratio becomes low and some of the Fellgett advantage is lost. The effectiveness of the mylar beam splitter can be improved by baking it at temperatures in the range 130–160°C for a few hours. This results in crystal growth in the mylar and increased flatness. The mylar may also have its reflectivity increased by coating with a thin film of germanium, but if this is done, it is necessary to have a compensator in the interferometer. By using these techniques and choosing mylar sheets with thicknesses appropriate to the various spectral bands, the far-

TABLE 3.9. Film Thickness and Frequency Intervals for Grubb–Parsons Interferometer

Mylar beam-splitter thickness (μm)	Spectral interval (cm^{-1})
3.75	120–660
6.25	80–400
12.5	40–200
25.0	20–100
50.0	10–50

infrared region can be covered with reasonable satisfaction. In an interferometer marketed by Grubb-Parsons in England, five beam-splitters are used. The film thicknesses and corresponding frequency intervals are given in Table 3.9.

At wavelengths longer than about 500 μm the decline in emission from the source means that low efficiency in the beam splitter becomes a serious drawback. Under these conditions the lamellar grating of Strong and Vanasse[61,62] may be preferred. This instrument operates not by amplitude division but by wave-front division. It is illustrated in Fig. 3.43. The grating is illuminated with radiation from a slit source. Partial reflection with a given phase occurs at the upper set of rectangular laminations, while the rest of the incident wave propagates between the upper laminations and is reflected from the lower set. Relative movement of the two interleaving sets of laminations enables the phase variation necessary to give the interferogram. The efficiency of a lamellar grating beam-splitter is compared with that of a mylar sheet and wire mesh in Fig. 3.42. The decline in efficiency of the lamellar grating beyond $v_c = f/sS$, where f is the focal length of the collimating mirror, s is the slit width, and S is the grating spacing, has been discussed by Richards[43] and Vanasse and Strong.[61,62] This experimental comparison was made by Richards[43] and shows a superiority of the lamellar grating for wavelengths longer than about 100 μm. The lamellar grating interferometer does have a reduced étendue relative to the Michelson, but this is largely counteracted by the increased efficiency of the beam splitting. It is also somewhat more complicated mechanically.

[61] J. Strong and G. A. Vanasse, *J. Phys. Radium* **19**, 192–196 (1958).
[62] J. Strong and G. A. Vanasse, *J. Opt. Soc. Amer.* **50**, 113–118 (1960).

3.6. MULTIPLEX SPECTROMETRY

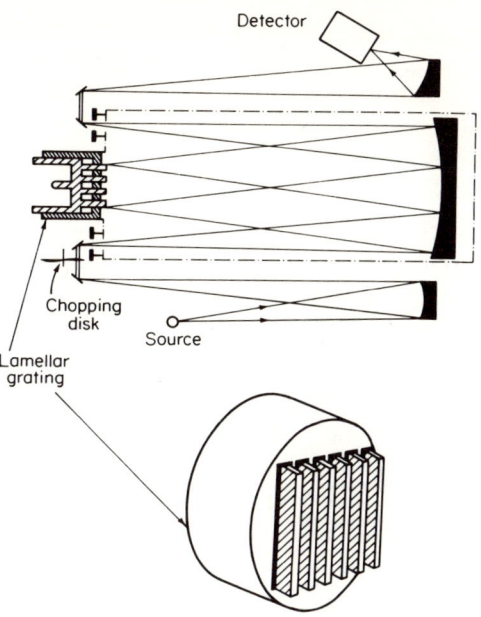

FIG. 3.43. Diagram of the lamellar grating interferometer (after Strong and Vanasse[61]).

3.6.6. Periodic Mode of Operation

We have mentioned the possibility of changing the path length between the two beams linearly with time, thereby "marking" the various spectral intervals with an amplitude modulation proportional to the frequency. With a system of narrow-band low audio frequency filters, a Fourier analysis of the signal would be carried out.

This method has been studied with lamellar grating interferometers with one set of the reflecting laminations moved backwards and forwards with near-constant speed and with the minimum turnaround time. The filter can be a single lockin amplifier, but to gain the multiplex advantage a large number of lockin amplifiers set to the frequencies of the different spectral elements is desirable.

Low resolution periodic interferometers have been built with the reciprocation action given by a cam.[60,63] Major problems in such instruments are the mechanical tasks of moving the mirror with sufficient speed and amplitude, and with abrupt reversal at the extremes of motion, and also the deriving of the necessary range of reference frequencies for the lockin amplifiers from the cam shaft drive.

---

[63] L. Genzel, *J. Mol. Spectros.* **4**, 241–261 (1960).

3.7. Extensions of the Techniques of Microwave Spectroscopy

Techniques of microwave spectroscopy differ markedly from the optical type methods described in Sections 3.5 and 3.6. The differences arise essentially because of the nature of the radiation sources. Microwaves are generated monochromatically, and there is accordingly no need for dispersive or interference techniques. Waveguide methods of propagation and measurement are appropriate, for the losses are low and the mechanical dimensions and tolerances reasonable. In consequence of the high degree of frequency stability attainable, and the existance of high-precision frequency measurement techniques, microwave spectrometers are very high resolution instruments. Resolving powers are typically many orders of magnitude larger than those obtained with grating or multiplex spectrometers.

It is true that in the early days of microwave spectroscopy the methods were largely quasioptical. The experiments of Cleeton and Williams[64] in the early 1930s used a split-anode magnetron, but parabolic mirrors were used for collimation and echelette gratings for wavelength measurements. However, the understanding of microwave electronics derived in the intensive development of the early 1940s changed the situation, and resulted in the type of experimental system illustrated in Fig. 3.44. In essentially this form, microwave spectrometers have been used at centimeter and millimeter wavelengths, and their lower wavelength limit has reached below 500 μm. It is interesting to note[65] that in the same year (1954) Burrus and Gordy extended microwave spectroscopy down to 770 μm,[66] Genzel and Eckhardt[67] working upwards in wavelength from the infrared made measurements to 990 μm. The painstaking development revealed by this overlapping of optical and microwave techniques indicated the considerable progress in experimental techniques from the days in 1923 when Nichols and Tear struggled with the impractical spark gap oscillator (see Chapter 2).

The simple microwave spectrometer of Fig. 3.44 is suitable for gas spectroscopy. It consists of an electron tube oscillator, such as a klystron, electronically tuned in frequency by a saw-tooth voltage on the repeller electrode[36] derived from the X-sweep of the oscilloscope. The signal is passed through a length of waveguide enclosed by windows and filled

---

[64] C. E. Cleeton and N. H. Williams, *Phys. Rev.* **45**, 234–237 (1934).
[65] C. A. Burrus and W. Gordy, *Phys. Rev.* **101**, 599–602 (1956).
[66] C. A. Burrus and W. Gordy, *Phys. Rev.* **93**, 897–898 (1954).
[67] L. Genzel and W. Eckhardt, *Z. Phys.* **139**, 592–598 (1954).

3.7. EXTENSIONS OF MICROWAVE SPECTROSCOPY

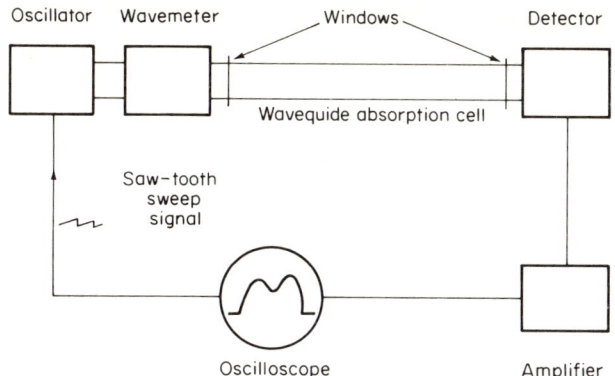

FIG. 3.44. Simplified block diagram of a microwave spectrometer used in gas spectroscopy.

with the gas under study. It is then detected with a video crystal diode whose output current is displayed as the y-deflection of the oscilloscope. A square-law diode gives a power versus frequency display on the oscilloscope screen. As the frequency sweeps through the resonance of the gas the line shape is displayed as an absorption curve on the bell-shaped mode of the klystron output power.

Many variants of the microwave spectrometer have been devised. For very broad spectral lines, as may occur with gases at high pressure, the absorption line may be plotted point by point at a succession of fixed frequencies. With this method the attenuation is measured with and without gas in the cell, the difference giving the gas absorption characteristic. One may also replace the long waveguide cell by a resonant cavity and observe the absorption line either in reflection or transmission. The multiple bouncing of waves within the cavity gives a long effective path for absorption [see Eq. (3.78)]. More sensitive means of detection can also be employed. Heterodyne detection can give much greater sensitivity. Phase-sensitive detection with lockin amplifiers can also be used to increase sensitivity. Here the signal is modulated either with radio frequency modulation superimposed on the slowly varying saw-tooth wave used to sweep the klystron, or the gas absorption line is itself modulated by the Stark or Zeeman effects.[68,69]

---

[68] W. Gordy, W. V. Smith and R. F. Trambarulo, "Microwave Spectroscopy." Wiley, New York, 1953.

[69] C. H. Townes and A. L. Schawlow, "Microwave Spectroscopy." McGraw-Hill, New York, 1955.

One of the most serious problems in microwave spectroscopy is that of distinguishing absorption due to the material under study from unwanted resonances set up within the waveguide system itself. Such standing waves will arise from reflections between slight discontinuities in the waveguide, particularly at the coupling joints, windows, etc. They are often relatively broad (low Q) resonances and can sometimes be distinguished from narrower gas lines on this basis. However, for weak absorption lines in gases, the waveguide cell may be many meters long and the enclosing windows imperfectly matched. In such a case, as Fig. 3.45 illustrates, a small frequency sweep can excite many closely spaced

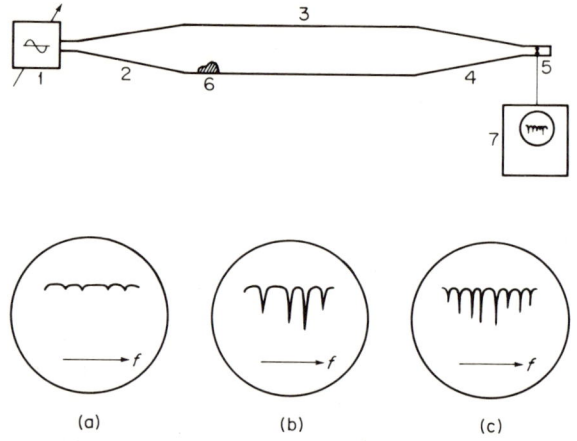

FIG. 3.45. Trapped-mode resonances. Microwave power from a swept-frequency generator (1) with standard-size output is fed through a taper (2) into oversized waveguide (3) and by a second taper (4) into a broadband detector (5) in standard-size waveguide. The tapers are assumed to be free from mode conversion. Mode conversion will take place at any irregularity (6). A higher-order mode will be reflected in the tapers and be trapped, giving resonance losses at certain frequencies, which can be determined from the power-frequency plot on the oscilloscope (7). (a) Low mode conversion at 6; (b) high mode conversion; (c) density of resonances increases if the length of the oversized waveguide (3) or its degree of oversizing is increased (after Butterwecke and de Ronde[46]).

and sharp resonances. In a straightforward microwave spectrometer such as that in Fig. 3.44 this sets a serious limitation and demands maximum care in the elimination of discontinuities within the system. Stark modulation of the molecules or Zeeman modulation of paramagnetic molecules offer effective means of overcoming this limitation, for they produce a modulation that is characteristic of the molecules alone and the output is almost completely insensitive to systematic power variations due to anything but the spectral lines.

Short millimeter and submillimeter wavelength extensions of conventional microwave spectrometry use klystron oscillators with the output multiplied in frequency by crystal harmonic generators. High harmonic output extending to wavelengths below 1000 μm can be directed into an oversize waveguide absorption cell via a section of tapered waveguide. As Fig. 3.46 indicates, the wave, after transmission through the cell, is again passed through a tapered section to a crystal diode detector. Crystal multipliers and detectors for this application are described in Chapters 2 and 4.

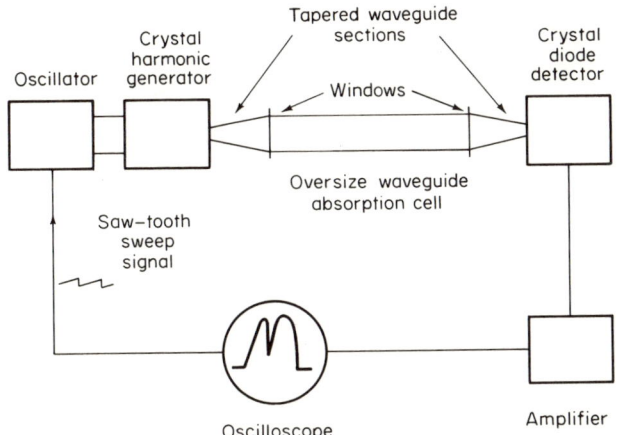

FIG. 3.46. Simplified block diagram of a millimeter and submillimeter wave spectrometer. High harmonics of a microwave oscillator are generated in a crystal diode, and detected by a similar diode.

Harmonic radiation frequencies can be measured with moderate accuracy by the use of cavity wavemeters. At the klystron frequency cavity Qs are sufficient to enable reasonably accurate measurements of the fundamental, and higher frequency cavities can indicate the order of the harmonic. More accurate measurements of frequency are required for the highest resolution; for this, methods based on comparisons with broadcast frequency standards have been developed.[68,70] With well-stabilized klystrons, and accurate frequency measurements, resolution somewhat less than 10^{-5} cm^{-1} can be obtained. This resolution is far better than that obtainable with interferometers and grating spectrometers.

[70] M. Cowan and W. Gordy, *Phys. Rev.* **111**, 209–211 (1958).

Another system used for short millimeter wave spectrometry is illustrated in Fig. 3.47. It was used at Columbia University[71] with wavelengths as short as 1.1 mm. The wave generation process, which we have described in Section 2.2.3, consists of a magnetron oscillator operating into a mismatched waveguide. The "phaser" shown in Fig. 3.47 is a section of tapered waveguide capable of sliding inside adjoining sections of large guide to give the best conditions of mismatch for maximizing the harmonic output. In addition, it is tapered down to dimensions so small that the wavelength of the fundamental is beyond cutoff. Most of the unwanted fundamental power is reflected and passes down the side arm of the T-junction where it is absorbed by a matched load.

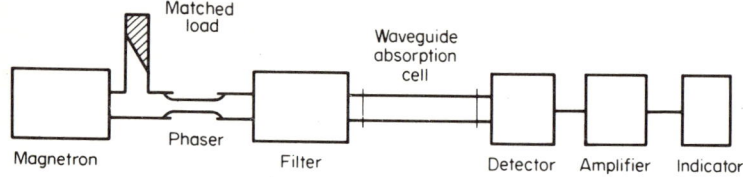

FIG. 3.47. Simplified block diagram of a millimeter wave spectrometer which uses harmonics generated in a magnetron.

Unwanted lower order harmonics must also be filtered out to leave essentially only the desired wavelength reaching the detector. For this separation of harmonics Klein *et al.* have used both tapered cutoff waveguide filters and echelette diffraction gratings. In the former case, experimentation can be carried out using the lowest harmonic transmitted through the filter section, because this wave usually has more power than all other higher harmonics combined.[71] Furthermore, the detector mount can to some extent be tuned to favor the detection of this preferred harmonic. The echelette grating harmonic separator has its collimating and condensing mirrors arranged in the manner discussed in Section 3.5 with waveguide entrance and exit horns placed at the focal points of the mirrors. Problems with overlapping order effects are not serious because of the rapid decrease in power with decreasing wavelength. Pulsed magnetrons have been used with pulse lengths of about one-half microsecond, peak power levels of hundreds of microwatts and a pulse repetition rate of 1000 per second. The magnetrons should be capable of being tuned over a sufficient frequency range to cover the resonance under study. Other-

[71] J. A. Klein, J. H. N. Loubser, A. H. Nethercot and C. H. Townes, *Rev. Sci. Instrum.* **23**, 78–82 (1952).

3.7. EXTENSIONS OF MICROWAVE SPECTROSCOPY

wise a spread of harmonic frequencies can be obtained by using a number of different fixed-frequency magnetrons.

A reflection spectrometer used for studies of cyclotron resonance in semiconductors[72] at 2-mm wavelength is illustrated in Fig. 3.48. Second harmonic of a 4-mm stabilized klystron is directed to an overmoded resonant cavity immersed in a low-temperature cryostat and containing the sample of germanium or silicon. Cavity excitation is in the TM_{016} mode. For a fixed frequency, the magnetic field is swept through the resonance and the absorption is observed in the reflected wave. Carriers are optically excited in the semiconductor by shining light through a quartz light pipe. Chopping of the light beam facilitates phase sensitive detection.

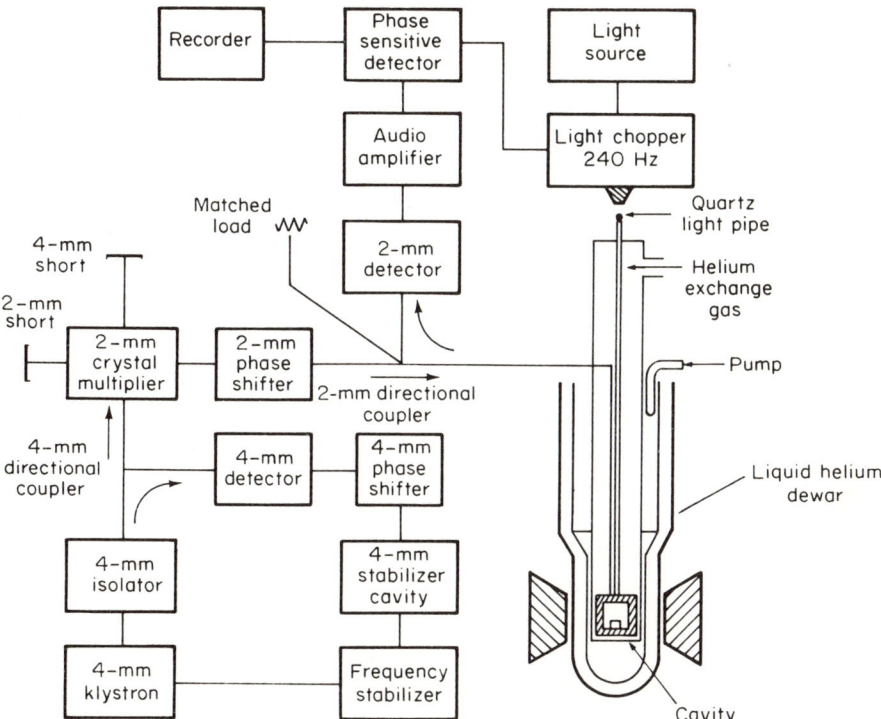

FIG. 3.48. Block diagram of a 2-mm cavity spectrometer used to study cyclotron resonances in semiconductors. Resonances are observed by the cavity reflection technique; the carrier concentration is optically modulated, and the reflected wave is detected by phase-sensitive detection methods (after Stickler et al.[72]).

[72] J. J. Stickler, H. J. Zeiger and G. S. Heller, Phys. Rev. **127**, 1077–1084 (1962).

3.8. The Fabry–Perot Spectrometer

We have seen in Section 3.4.4 that the Fabry–Perot interferometer has been employed as a resonant cavity in the millimeter wavelength region. It is capable of very high Q values and can thus give high resolving power. The resolving power is

$$Q = \lambda/\Delta\lambda, \tag{3.77}$$

where λ is the wavelength of the peak transmission, and $\Delta\lambda$ is the half-width of the resonance.

Within the cavity, energy storage is built up through the multiple passage of radiation between the mirrors. As we have seen in Section 3.1.2, $Q/2\pi$ is the number of periods of oscillation for which a photon is stored, hence the cavity is equivalent to a long absorption cell with "effective pathlength" given by

$$\text{EPL} = Q\lambda/2\pi. \tag{3.78}$$

The desirability of a very high Q resonator is apparent; it implies a greater effective pathlength, and, therefore, greater sensitivity. For example, the Fabry–Perot cavity described in Section 3.4.4 with $Q = 330,000$ at $\lambda = 1$ mm, is equivalent to an absorption cell about 53 m long.

Valkenburg and Derr[44] have measured the absorption of water vapor in the 1–3-mm range with a spectrometer based on this high-Q Fabry–Perot resonator and with a microwave-like measurement system. The signal from a klystron is frequency multiplied in a crossed-guide harmonic generator, and the output is fed into the interferometer via open-ended waveguides in one of the mirrors. The transmitted signal passes out another waveguide to a detector and oscilloscope. The fundamental, and the harmonic output, is frequency modulated when a saw-tooth voltage sweeps the klystron reflector and the oscilloscope displays the Q curve of the resonator. Q values are measured with and without the vapor in the cavity, and from the difference the water vapor absorption is deduced. The sensitivity of this Q reduction technique with video detection is such that gas loss as small as 10^{-7} cm^{-1} can be measured.

4. DETECTION OF FAR-INFRARED RADIATION

Introduction

While it is true that the foremost obstacle to the development of the far-infrared has been the dearth of suitable generators, the limitations of detectors of such radiation have also been a contributive factor. Like the generation problem, the development of detectors has had a long and steady history, and striking recent growth. The bolometer was invented by Langley[1] in 1880, at which time it was the most sensitive detector of infrared radiation. Its most significant rival, the thermocouple, had been in use since 1830. These two detectors were extensively used as the investigations of the spectrum were extended from the 1–2 μm region toward 10s and 100s of microns by the turn of the century. As we will see, the bolometer still continues to be improved and to operate throughout the infrared and far-infrared, while the thermocouple remains prominent among infrared detectors.

Thermometers had been used in 1800 by Sir William Herschel as effective but not very sensitive detectors of infrared. They were displaced for a century by the thermocouple and bolometer, but reemerged in 1947 in the form of a gas-filled bulb called the Golay cell. In this cell, absorbed radiation heats the gas, which expands to deflect a flexible wall of the cell, thereby giving a measure of the absorbed power. This is a slow detector responsive over a considerable spectral band.

This century has seen efforts to extend the rectifier detector of radio waves from millimeter toward submillimeter wavelengths, and the emergence of fast detectors based on the photoelectric effect. Through photoconductivity effects these have been extended upwards in wavelength to cover the far-infrared.

The detectors of present major importance are of four types: point–contact crystal diodes, thermal detectors, photoconductivity detectors, and superconductivity junctions based on the ac Josephson effect. The former are extensions of video and superheterodyne receivers used at

[1] S. P. Langley, *Nature* (*London*) **25**, 14–16 (1881).

microwave frequencies. Thermal detectors are devices in which a temperature rise, following the absorption of radiation, causes some change of physical properties, e.g., a resistance change in a bolometer or the expansion of gas in a Golay cell. In photoconductive detectors, absorbed radiation causes a change in conductivity of a semiconductor, and in the Josephson junction an alternating current through a constriction at the junction of two superconductors results in a dc voltage across the junction.

The essential capabilities of a detector can be specified by four quantities: the *responsivity*, that is, the change in output signal (e.g., voltage) per unit change of input power; the *response time* τ; the *noise equivalent power* (NEP); and the *frequency range* over which it is responsive. The NEP is defined as the quantity of radiation in watts which must fall on the detector to give out a root mean square electrical signal equal to the rms value of the noise. We will sometimes refer to the *minimum detectable signal*, taking this as equal to the NEP, although we must point out that the latter quantity is free from the elements of subjective judgement associated with the former.[2]

Three sources of noise may be distinguished: first, background radiation incident on the detector from thermal emission by objects within the field of view of the detector, from the detector enclosure, etc.; second, noise arising within the detector; and third, noise associated with post-detector amplification.

4.1. Point-Contact Crystal Diodes

Crystal diodes used for millimeter and submillimeter wave detection are usually metal–semiconductor contacts which detect by virtue of their well-known rectification properties. Rectification occurs in a thin barrier-layer region at the interface of the two materials due to the presence of a potential barrier whose magnitude is varied by the incident radiation. We will discuss the behavior in an elementary way for the case of an n-type semiconductor, referring the reader to the literature for other materials and more exhaustive discussions.[3,4]

---

[2] R. C. Jones, *Advan. Electron.* **5**, 2–87 (1953).

[3] C. Kittel, "Introduction to Solid State Physics," Chapter 4. Wiley, New York, 1953.

[4] M. J. O. Strutt, "Semiconductor Devices," Vol. 1. Academic Press, New York, 1966.

4.1. POINT-CONTACT CRYSTAL DIODES

In a metal the conduction electrons are distributed according to Fermi–Dirac statistics. They are located in an energy band, filled from the lowest level up to the *Fermi level*, W_F. The Fermi level is defined as the energy level for which the probability of occupation has the value $\frac{1}{2}$. Its depth below the zero energy level (that is, the energy of an electron free from the metal) is the work function. In a semiconductor, the energy levels are located in bands. As illustrated in Fig. 4.1(a) the conduction band and valence band are separated by an energy gap which is forbidden to electrons in the pure semiconductor. However, impurity atoms are always present, and n-type impurities (atoms with one valence electron more than the atoms of the host semiconductor) can donate electrons to give conduction. For normal, relatively dilute, impurity concentrations, the electron distribution is adequately described by the exponential Maxwell–Boltzmann distribution, with the Fermi level located in the energy gap as indicated by the dotted line in Fig. 4.1(a). When the two materials are brought into contact, equilibrium is established by the principle that the Fermi levels must be equal.[3] The difference between the energy level of an electron outside the crystal and that at the bottom of the conduction band should be approximately preserved at the contact junction.[4] These two conditions can in general be satisfied simultaneously only if the energy bands of the semiconductor are bent in the vicinity of the junction.[4] When the materials are brought into contact, electrons flow from the n-semiconductor to the metal. A negative surface charge density is built up on the metal, and on the n-side there is left a region of net positive charge associated with the ionized impurities. Thus an electrical double layer is formed producing a potential difference which lowers all electron energy levels in the bulk semiconductor to give coincidence of the Fermi levels. This is illustrated in Fig. 4.1(b). The shallow region

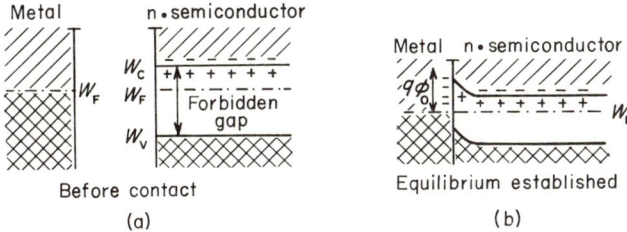

FIG. 4.1. Energy diagram of a rectifying barrier between a metal and an n-type semiconductor. The Fermi levels in the two materials are indicated by W_F; the top of the valence band and the bottom of the conduction band in the semiconductor are indicated by W_V and W_C, respectively. The contact potential difference is ϕ_0.

of the n-material practically stripped of conduction electrons is called the barrier layer or depletion layer. It has low conductivity and behaves like a parallel plate capacitor. At the barrier layer the conduction band bends up to a height $q\phi_0$ above the Fermi level; ϕ_0 is the contact potential difference.

At equilibrium there is a steady current flow I_0 in both directions across the junction from those electrons near the top of the distribution. When an alternating voltage is applied, the process of rectification is accomplished in the following way: The height of the barrier, as viewed from the metal, remains constant at ϕ_0, and the current flow from left to right continues to be I_0. The electron levels in the semiconductor, however, are raised and lowered by the alternating voltage V, so the height of the barrier as seen from the semiconductor alternates. This is illustrated in Fig. 4.2, where two phases of the alternating voltage are

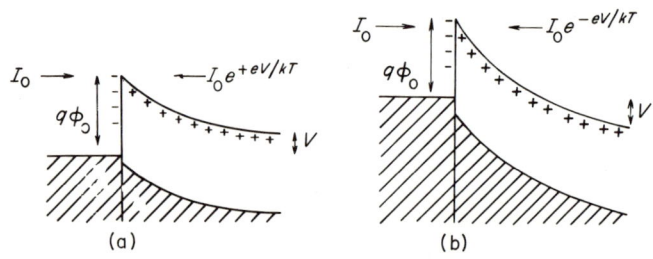

FIG. 4.2. Rectification by a metal–semiconductor barrier. In (a) and (b), two phases of an applied alternating voltage are taken, which respectively raise and lower the height of the potential barrier as seen from the semiconductor, while the contact potential difference ϕ_0 remains constant. The exponential variation of current with voltage is indicated.

taken. Because the electron energy distribution in the semiconductor is exponential above the bottom of the conduction band, the electron current from the semiconductor to the metal varies exponentially with V, as indicated in Fig. 4.2. The net current of electrons which flows across the junction is the difference between the currents from the two sides, namely,

$$I = I_0(e^{qV/kT} - 1). \tag{4.1}$$

The current I_0 arises from the diffusion of electrons in their random equilibrium motion. From the well-known result of kinetic theory,[5] a

[5] E. H. Kennard, "Kinetic Theory of Gases," Chapter 3. McGraw-Hill, New York, 1938.

number $\frac{1}{4}N\bar{v}$, of electrons diffuse into unit area of the barrier per second, where N is the carrier concentration, and $\bar{v}$ their mean speed of random motion. Of these, a fraction $\exp(-q\phi_0/kT)$ negotiate the equilibrium potential barrier. Thus the current I_0 is

$$I_0 = \pi a^2 \tfrac{1}{4} N q \bar{v} \exp(-q\phi_0/kT), \qquad (4.2)$$

where q is the electronic charge, and a is the radius of the area of contact. Thus the current–voltage relation at the rectifying barrier is

$$I = \tfrac{1}{4}\pi a^2 N q \bar{v} \exp(-q\phi_0/kT)[\exp(qV/kT) - 1]. \qquad (4.3)$$

The I–V characteristic is highly nonlinear (see Fig. 4.3) giving substantial conduction in one direction and low conduction in the reverse direction. As the voltage is changed, the charge being pumped into the barrier layer varies, and with it the barrier thickness. The rectifying

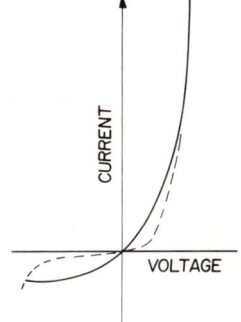

FIG. 4.3. The current through a junction diode as a function of the voltage across it. The full line is the ideal characteristic [Eq. (4.3)]; the dotted line is for a real diode.

barrier can be taken as equivalent to a capacitance which varies with voltage, plus a highly nonlinear parallel resistance R described by Eq. (4.3). An equivalent circuit representation of a diode rectifier is shown in Fig. 4.4.[6] In series with the RC combination, a resistance R_s, the *spreading resistance*, is added to account for ohmic losses in the semiconductor. These losses are proportional to the resistivity of the bulk semiconductor; this dissipation is largely near the junction where the current is concentrated down to the area of contact. The inductance is that of the metal contact-whisker.

[6] W. M. Sharpless, *Bell. Syst. Tech. J.* **35**, 1385–1402 (1956).

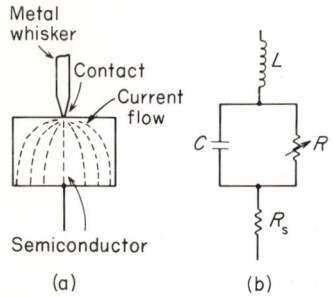

FIG. 4.4. Diagrammatic view of: (a) point-contact rectifier, and (b) its equivalent circuit. The inductance L is that of the contact whisker, C and R are the barrier layer capacitance and resistance, R_s is the spreading resistance of the bulk semiconductor.

The detection properties of point-contact resistive diodes at high frequencies are adversely affected by the inductive reactance, by the bypassing effect of the barrier capacity, and by dissipation in the linear spreading resistance. The design of the diode should aim at achieving a capacitive reactance at the highest frequency approaching or exceeding the barrier resistance, and a spreading resistance much smaller than the barrier resistance. It can be shown that these conditions also serve the interest of a low noise factor.[7] As pointed out by Messenger and McCoy[7] some of the physical parameters must be chosen by compromise between these requirements.

In general, short wavelength diode detectors require small contact areas and high conductivity semiconductors. However, these aims are modified in practice by technological experience of diode fabrication. Consequently, silicon and germanium have been widely investigated, while gallium arsenide has been studied to a lesser extent. Materials with still higher conductivities, such as indium antimonide, are yet to be developed in this application.

When a crystal is used as a video detector its operation will be square-law for sufficiently low power levels. This can be seen from a power series expansion of the exponential function of voltage in Eq. (4.3). The first linear term is, of course, unimportant as far as rectification is concerned, and the higher power terms are negligible for low signal levels P_s. Any characteristic [not necessarily exactly that of Eq. (4.3)] can be expanded by Taylor's theorem about any operating point to show this square-law property. For such a square-law detector, the output current will be

$$I = RP_s, \tag{4.4}$$

where R is the responsivity.

[7] G. C. Messenger and C. T. McCoy, *Proc. IRE* **45**, 1269–1283 (1957).

4.1. POINT-CONTACT CRYSTAL DIODES

For large responsivity at small signal levels the forward knee of the I–V characteristic should occur as near as possible to the origin ($V \simeq 0$), so that the lowest voltages result in a detectable current. The characteristic of p-type silicon particularly has this feature.[8] It is possible to set the operating point near to the knee by biasing a crystal with a dc voltage. In practice, however, the advantages of this are often lost due to the increased "current noise" (see Section 4.5.5) generated by the bias.[8] By so biasing a germanium diode, the VX-3352 developed by the British General Electric Company, Meredith and Warner[9] have detected 2.5×10^{-11} W of 2000-μm radiation with a video radiometer. This figure is for an output bandwidth of 1 Hz, and an input bandwidth to the detector of 10%. Silicon diodes have been used extensively by the spectroscopy group at Duke University[10,8] down to wavelengths of 480 μm.[8] German researchers have reported video detection of incoherent radiation from a heat source down to 100 μm;[11] they record the performance of a germanium diode relative to a bolometer that is shown in Fig. 4.5. There are few figures available on the noise equivalent power of video detectors. At 1000 μm, Meredith and Warner[9] find the NEP is 4×10^{-9} W for an assumed 1-Hz output bandwidth. This appears likely to increase by at least 22 dB per octave as the operating wavelength is reduced.[12]

Point–contact diodes can also be used as mixers in superheterodyne frequency conversion receivers. This is accomplished by supplying local oscillator power, P_L, to mix with the incident signal P_s. The output current of a superheterodyne detector of responsivity R is

$$I = R(2P_s P_L)^{1/2}. \tag{4.5}$$

This result applies provided the local oscillator power is not so high that it takes the operation of the nonlinear element beyond its square-law region.[13] The superheterodyne system can give a current increased by a factor $(2P_L/P_s)^{1/2}$ above that obtained with video detection, and also advantages with respect to noise because of the very narrow bandwidth

---

[8] C. A. Burrus, *Proc. IEEE* **54**, 575–587 (1956); **55**, 1104–1105 (1967).
[9] R. Meredith and F. L. Warner, *IEEE Trans. Microwave Theory Tech.* **11**, 397–411 (1963).
[10] C. A. Burrus and W. Gordy, *Phys. Rev.* **93**, 897–898 (1954).
[11] H. von Happ, W. Eckhardt, L. Genzel, G. Sperling and R. Weber, *Z. Naturforsch. A* **12**, 522–524 (1957).
[12] E. H. Putley, *Proc. IEEE* **51**, 1412–1428 (1963).
[13] H. C. Torrey and C. A. Whitmer, "Crystal Rectifiers" (Radiation Lab. Ser.), Vol. 15, Chapter 1, Sect. 1.3. McGraw-Hill, New York, 1948.

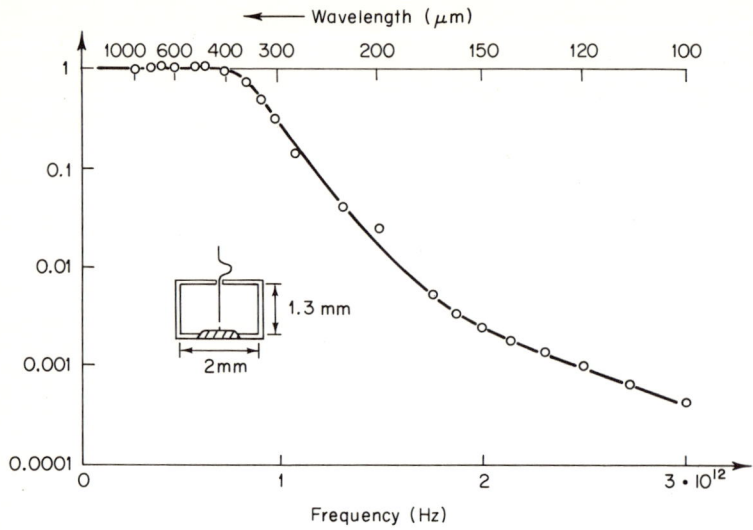

FIG. 4.5. Response of a germanium point–contact diode relative to that of a bolometer, as a function of wavelength (after von Happ et al.[11]).

that can be achieved with these detectors (see Section 4.5). However, the requirement of a high-frequency local oscillator is a serious requirement at far-infrared wavelengths. The rotational–vibrational energy level gas lasers which now exist throughout the far-infrared are not particularly suitable as local oscillators, because they operate with fixed frequencies and therefore do not allow tuning of the receiver.

In the experiments carried out by Meredith and Warner,[9] superheterodyne receivers with Carcinotron local oscillators have operated down to 2000 μm. Their radiometer, with a 60-MHz intermediate frequency bandwidth and 1-Hz output bandwidth, had an NEP of 7×10^{-15} W Hz$^{-1/2}$. The noise figure obtained was 20 dB. These workers have concluded that extensions towards shorter wavelengths would result in sensitivities inferior to that of the Putley photoconductive device discussed in Section 4.3.

An indication of the magnitudes of the equivalent circuit components for millimeter diodes is given by Sharpless[6] and by Meredith and Warner. From work at a wavelength of 5.4 μm with silicon crystals Sharpless has estimated the barrier resistance to be 250 Ω, the spreading resistance 18 Ω, the barrier capacity 5.7×10^{-14} f, and the inductance 3.4×10^{-10} h. Meredith and Warner estimate a barrier resistance of 180 Ω, spreading resistances of 36, 8, and 14 Ω and barrier capacitances of 5.9×10^{-14},

4.3×10^{-14}, and 0.7×10^{-14} f, respectively, for p-type silicon, n-type germanium, and n-type gallium arsenide. We can see from these values of diode capacity and resistance that the time constant of the point–contact diode can be very small, so the diode itself is capable of extremely rapid response. The response time of a receiver is generally set by its bandwidth.

Within the nonlinear diode, harmonics of the local oscillator and signal frequency, and sum and difference frequencies of all the applied signals, will be generated, and these in turn will beat with each other to create still more frequencies, and so on, *ad infinitum*. While in the superheterodyne receiver with fundamental mixing, as discussed above, these higher harmonics can be ignored, it is possible to beat the signal with a higher harmonic of the local oscillator in order to get around the shortage of suitable high-frequency local oscillators. This harmonic mixing technique has higher losses than fundamental mixing but is more sensitive than video detection.[8,9] A radiometer with a 20-MHz input bandwidth developed by Meredith and Warner uses harmonic mixing to give a noise figure of 25 dB and a minimum detectable power of 1.3×10^{-14} W for 1-Hz output bandwidth. Packard[14] has used harmonic mixing down to 500 μm, and its utility to 200 μm has been demonstrated by Strauch *et al.*[15] The mechanisms involved in harmonic mixing have been discussed critically by DeLucia and Gordy.[16]

In this discussion of diode detectors we have spoken only of the point–contact nonlinear resistance junction. This has been a most successful diode detector in the far-infrared. For a discussion of other point–contact junction diodes, and, in particular, nonlinear capacitance "varactor" diodes, the reader is referred to the review written by Burrus.[8]

Payne and Prewer[16a] and Becklake *et al.*[16b] have recently measured and compared the performances of Ge, Si, and GaAs point-contact diodes, and GaAs Schottky-barrier diodes over the wavelength range 4 mm–128 μm. At 337 μm the Ge point-contact performed best with an NEP of about 10^{-8} W Hz$^{-1/2}$ *in video* operation. The Schottky-barrier diodes

---

[14] R. F. Packard, *IEEE Trans. Microwave Theory Tech.*, 13 pp. 211–215 (1965) Symp. Digest.

[15] R. G. Strauch, R. A. Miesch and R. E. Cupp, *IEEE Trans. Microwave Theory Tech.* 13, 873 (1965).

[16] F. De Lucia and W. Gordy, *Proc. IEEE* 54, 1951–1952 (1966).

[16a] C. D. Payne, *Radio Electron. Eng.* 39, 167–171 (1970).

[16b] E. J. Becklake, C. D. Payne and B. E. Prewer, *J. Phys. D Appl. Phys.* 3, 473–481 (1970).

are made with gold or nickel as barrier metals on GaAs using photolithographic techniques. They offer greater stability and a more certain lifetime, and are presently available from E.M.I., England, with NEP of about 10^{-7} W Hz$^{-1/2}$ at 300 μm and a response time of 10^{-8} sec.

4.2. Thermal Detectors

The receiving element in a thermal detector is a piece of radiation-absorbing material thermally coupled to a heat sink at a constant temperature T_0. The temperature response of the element to power absorption will be influenced by its thermal capacity and the thermal conductance between the element and the heat sink. Denoting the thermal capacity and conductance to the sink by C and G, respectively, the energy conservation or heat balance equation for the element is

$$C \, d \, \Delta T/dt = \varepsilon' P - G \, \Delta T, \qquad (4.6)$$

where ΔT is the temperature increase of the element above T_0, ε' the emissivity, and P the incident radiation. Note that the last term in Eq. (4.6) includes conduction, convection, and radiation heat transfer. For example, when radiation is the dominant loss process, it can be described by a term $d(\varepsilon' A \sigma T^4) \, \Delta T/dT$, that is,

$$G = 4\varepsilon' A \sigma T^3, \qquad (4.7)$$

where A is the surface area of the element, and σ is Stefan's constant. When the power flow continues long enough to reach a steady state condition $d \, \Delta T/dt = 0$, the temperature of the element rises to

$$\Delta T = \varepsilon' P/G. \qquad (4.8)$$

If now the power flow is suddenly stopped, the temperature, given by Eq. (4.6) with $P = 0$, at time t later is

$$\Delta T = (\varepsilon' P/G) \, e^{-t/\tau} \qquad (4.9)$$

where

$$\tau = C/G \qquad (4.10)$$

is the thermal time constant of the element.

It is usual to amplitude modulate or "chop" the radiation incident on the thermal detector with some type of mechanical rotating shutter, and to amplify and measure the output at the modulation frequency.[17,18] This modulation of the wanted signal enables it to be discriminated from unchopped radiation; it corresponds in effect to resetting the instrument at every interruption of the signal, thus eliminating drift. Further, it is easier to amplify ac rather than dc signals. If the incident radiation is chopped at an angular frequency ω, we can write $P = P_0 e^{i\omega t}$ in Eq. (4.6). The solution is then

$$\varDelta T = \varepsilon' P_0 e^{i\omega t}/G(1 + i\omega\tau)$$

or

$$|\varDelta T| = \varepsilon' P_0/G(1 + \omega^2\tau^2)^{1/2}. \qquad (4.11)$$

For chopping frequencies fast compared to $1/\tau$, the temperature response falls off as ω^{-1}; for low frequencies such that $\omega\tau \ll 1$, this result reduces to the steady-state response.

From these results it is clear that the requirement for small response time is small thermal capacity and large thermal conductance from the element to the heat sink. A consequence of increased thermal conductance will be reduced responsivity. The requirement of minimum thermal capacity, namely, low density, specific heat, and volume, must be consistent with the need for a surface area and surface emissivity sufficient to enable adequate absorption by the element.

4.2.1. The Golay Cell

The pneumatic thermal detector introduced by Golay[19] in 1947 has, since then, played an important laboratory role as a sensitive broad-band detector. Its receiving element is a radiation absorbing film which heats gas in a small enclosed chamber. A flexible wall (membrane) of the chamber distorts under the pressure of the heated gas, and this gives a measure of absorbed power. High responsivity can be achieved with an optical measurement of wall deflection determined from the deflection of a light beam reflected from the distorted membrane and detected by a photoelectric cell. The response time of the detector is rather long—between 2 and 30 msec. It is convenient, therefore, to chop the incident radiation at about 10 Hz. The NEP is about 10^{-10} W for 1-Hz bandwidth.

---

[17] R. H. Dicke, *Rev. Sci. Instrum.* **17**, 268–275 (1946).
[18] H. M. Randall and F. A. Firestone, *Rev. Sci. Instrum.* **9**, 404–413 (1938).
[19] M. J. E. Golay, *Rev. Sci. Instrum.* **18**, 357–362 (1947).

An outstanding characteristic of the Golay cell is its large frequency range. It can operate from the ultraviolet through the visible and infrared up to microwave wavelengths as long as 7.5 mm. The frequency range is determined by the material of the window used to transmit radiation into the cell; the coverage of a large frequency range requires the use of several different windows. Excluding the absorption effects of windows the response of the cell is largely (but not completely) independent of wavelength.

The Golay cell is a room temperature detector. Because of this, and provided the slow response is acceptable and care is taken not to damage the cell with excess power, it can be a particularly valuable laboratory instrument.

4.2.2. The Bolometer

A radiation bolometer[20] is a slab of electrically conducting material with a high temperature coefficient of resistance through which a steady current is passed. The absorption of radiation causes a change in resistance and thereby a change in voltage which can be observed by a suitable bridge or amplifier. If the resistance is R and the steady current is I, the bolometer will operate at a temperature T given by

$$I^2 R = G(T - T_0). \tag{4.12}$$

The absorption of radiation causes a further temperature rise ΔT and a voltage change

$$\Delta V = IR\alpha\, \Delta T. \tag{4.13}$$

The quantity α is the temperature coefficient of resistance defined by the relation

$$\alpha = (1/R)\, dR/dT,$$

and, by Eq. (4.11),

$$\Delta T = P_0/G(1 + \omega^2\tau^2)^{1/2}, \quad \text{when} \quad \varepsilon' = 1.$$

Thus a high responsivity requires a large resistance and temperature coefficient of resistance. Increased responsivity derived from large current, however, is gained at the cost of increased noise (see Section 4.5.5).

---

[20] R. A. Smith, F. E. Jones and R. P. Chasmar, "The Detection and Measurement of Infra-Red Radiation." Oxford, Univ. Press (Clarendon), London and New York, 1960.

4.2. THERMAL DETECTORS

A suitable circuit arrangement of a bolometer is shown in Fig. 4.6. The load resistance can be another bolometer element, shielded from incident radiation, for this can compensate for ambient temperature changes. If the impedance of the amplifier is large compared to the bolometer resistance R and the load resistance R_L in this circuit, the change in voltage applied to the amplifier when the bolometer resistance changes by ΔR is

$$\Delta V = RV\, \Delta R/(R + R_L)^2. \tag{4.14}$$

The condition for maximum response $\Delta V/\Delta R$ is that the load resistance should equal the bolometer resistance.

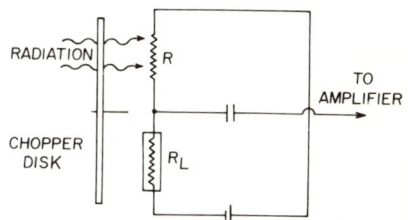

FIG. 4.6. Bolometer circuit: The resistance of the bolometer sensing element is R; the load resistance R_L is another bolometer element shielded from the radiation. The chopper disk modulates the incoming radiation.

In infrared work, vacuum encapsuled elements made from ribbons or thin films of pure metals or semiconductors are used. Blackened platinum has been in use since Langley's first infrared bolometer.[1] This material at room temperature has a value of α equal to $0.003°K^{-1}$. Semiconductor elements can give much higher responsivities because of their larger values of resistivity and α. Bolometers made of "thermistor" materials, that is, sintered mixtures of oxides of nickel, manganese, and cobalt, give both high α values and extremely high resistance values. Infrared thermistor bolometers can achieve response times of the order of 5 msec, responsivities of 100 V W^{-1}, or higher, and NEP values in the region of 10^{-10} W for a 1-Hz bandwidth.

Improved bolometer performance can be obtained by using a sensing element at low temperatures. Not only can noise be reduced in this way, but the decrease of specific heat with cooling can enable improvements in responsivity and/or response time. Boyle and Rodgers[21] in 1959 built a far-infrared bolometer operated at liquid helium temperatures using the unusual resistivity properties of carbon composition resistors. A slab of carbon 0.5 mm thick cut from a commercial resistor has a resistance of about 100 kΩ and the very high temperature coefficient $\alpha = -2$ at

---

[21] W. S. Boyle and K. F. Rodgers, *J. Opt. Soc. Amer.* **49**, 66–69 (1959).

a temperature of 2°K. This element is mounted on a sheet of mylar insulator which in turn is attached to the copper or brass base of the evacuated detector capsule. Such a bolometer is shown in Fig. 4.7. Performance figures quoted are as follows: responsivity 1.4×10^4 V W^{-1}, NEP 6×10^{-12}, and 10-msec time constant. Other designs have given time constants of about 1 msec.[22] An important advantage of carbon is that the resistivity and thickness of this material results in a very large proportion of the incident radiation being absorbed. This is true in the infrared as well as the far-infrared.

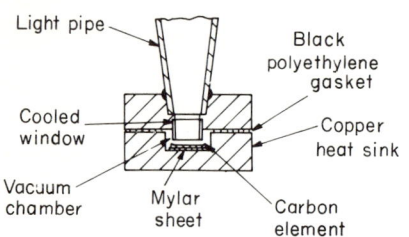

FIG. 4.7. Cross section of a carbon bolometer. The sensing element is thermally coupled to the copper heat sink through a thin sheet of mylar.

Another advance in cooled bolometers has come from the application of the abrupt transition of resistance in superconductors in the vicinity of the critical transition temperature. The transition from finite to zero resistance can occur with a temperature change as small as 0.001°K. Martin and Bloor[23] have used an element of tin 3×2 mm^2 in area and 3 μ thickness, deposited on a mica substrate. This element is maintained at the transition temperature (about 3.7°K) to an accuracy of 10^{-5} °K by control of the vapor pressure in the helium reservoir. The bolometer resistance is between 1 and 10 Ω, and dR/dT lies in the range 100–1000 Ω °K^{-1}. This temperature dependence is several orders of magnitude greater than that of normal metals at room temperature and accounts for the large responsivity. The thermal conductance to the heat sink is via the electrical leads to the sensing film, and these can be designed for suitable response time and responsivity.

Martin and Bloor have developed a detector with approximately constant performance throughout the far-infrared, with a time constant of 1.25 sec and NEP of approximately 10^{-12} W in 1-Hz bandwidth. For a time constant of 10 msec the responsivity is 200 V W^{-1}. It is important to note that the dominant noise contribution here is from the post de-

[22] M. Wang and F. Arams, *Symp. Microwave Theory Tech.*, *Palo Alto, California*, *1966*.
[23] D. H. Martin and D. Bloor, *Cryogenics* **1**, 159–165 (1961).

4.2. THERMAL DETECTORS

tector amplifier, and that, according to Martin and Bloor, an improvement by a factor of 30 would seem possible with a noiseless amplifier.

Relatively fast and sensitive bolometers have been made from germanium. This material has a high temperature coefficient of resistance, low thermal capacity, and high thermal conductivity when it is cooled to low temperatures. Although pure germanium tends to be transparent at long wavelengths, the addition of free carriers tightly coupled to the crystal lattice can produce strong absorption. Doping with 10^{17} cm^{-3} of gallium or indium gives a good absorber. The time constant is controlled by the material and dimensions of the electrical leads which provide thermal coupling to the heat sink, but the very high conductance of the germanium element itself contributes to the attainable fast response. Low[24] has used a gallium doped single crystal of germanium $4 \times 4 \times 0.12$ mm^3, operated at 2°K. His detector has a time constant of 400 μsec, a responsivity of 4.5×10^3 V W^{-1}, and an NEP of 5×10^{-13} W Hz$^{-1/2}$. With a thinner germanium slab, Low has shortened the time constant to 10 μsec. A germanium bolometer has been developed by Richards[25] with a time constant of 100 μsec and an NEP of 3×10^{-12} W Hz$^{-1/2}$. The limit of response time set by thermal factors is perhaps 50–20 μsec.[26,27]

Wang and Arams[22] have developed a useful power measuring room-temperature bolometer responsive from 10 μm–6 mm. They employ a thin film absorber mounted on an alumina substrate in close thermal contact with a flake thermistor bolometer element to measure powers in the range 10^{-3}–10^{-8} W. The responsivity is about 240 V W^{-1}, and the time constant 3 sec. In a more sensitive version, they have reported a NEP of 7×10^{-10} W Hz$^{-1/2}$, response time 6 msec, responsivity 350 V W^{-1}, and spectral range 5 μm–4 mm. The detector may be calibrated for absolute power measurement, and has, in fact, been used to measure the peak and average power output from a HCN laser.[28]

Zwerdling et al.[28a] have developed a pumped liquid helium temperature bolometer for the far-infrared using as an active element a single crystal of p-type compensated germanium containing Ga and Sb impurities. The response time is about 1 msec, and with a chopping frequency in the range 200–500 Hz, the NEP is 2.3×10^{-12} W Hz$^{-1/2}$.

---

[24] F. J. Low, *J. Opt. Soc. Amer.* **51**, 1300–1304 (1961).
[25] P. L. Richards, *J. Opt. Soc. Amer.* **54**, 1474–1484 (1964).
[26] H. E. Stubbs and R. G. Phillips, *Rev. Sci. Instrum.* **31**, 115 (1960).
[27] L. L. Gorelik and E. A. Lobikov, *Sov. Phys. Tech. Phys.* **6**, 90 (1961).
[28] F. Arams, C. Allen, M. Wang, K. Button and L. Rubin, *Proc. IEEE* **55**, 420 (1967).
[28a] S. Zwerdling, R. A. Smith and J. P. Theriault, *Infrared Phys.* **8**, 271–299 (1968).

While bolometers and Golay cells are among the most extensively used detectors of infrared and far-infrared radiation, their most serious disadvantage is their slow response. For detailed measurements of rapidly changing waves, such as those produced by pulsed generators or waves emerging from transient plasmas, they are of little value. For such applications the detectors described in the following sections can be used.

4.2.3. Other Thermal Devices

The thermocouple is extensively used as an infrared detector,[20,29] but little effort has been made to extend it to far-infrared wavelengths. Strong[30] mentions the use of thermocouples between 52 and 152 μm, and Randall and Firestone have reported operation to 100 μm.[18] Descriptions of the design and performance of thermocouple detectors are given by Strong,[30] and by Sutherland and Lee.[31] In the infrared the performance is comparable with that of many bolometers.[20,31] Sutherland and Lee, in their 1947 review of the infrared, quote response times from 140 msec–3 sec and limits of detectability in the neighborhood of 5×10^{-11} W.

Recently thermoelectric effects involving hot carriers in bulk semiconductors[32] have been studied at centimeter and millimeter wavelengths down to 1.5 mm.[33] A dilute concentration of majority carriers is heated nonuniformly by spatially nonuniform absorbed radiation. The effect does not involve lattice heating as in the ordinary thermoelectric effect involved in the thermocouple. The temperature gradient in the carriers gives rise to a thermoelectric voltage proportional to the radiation power.

Pyroelectric detectors have been used in recent years, particularly in applications where room-temperature detection with fast response is called for. Cooper[34] has obtained a time constant of 5 μsec, reducible to about 1 μsec, while in studies of Q-switched CO_2 lasers Shimazu et al.[35] have observed submicrosecond pulses and Kimmitt et al.,[36] have resolved the pulses with a 0.1-μsec response-time detector.

[29] E. H. Putley, *J. Sci. Instrum.* **43**, 857–868 (1966).
[30] J. Strong, "Procedures in Experimental Physics." Prentice-Hall, Englewood Cliffs, New Jersey, 1943.
[31] G. B. B. M. Sutherland and E. Lee, *Rep. Progr. Phys.* **11**, 144–177 (1946–1947).
[32] S. H. Koenig, *J. Phys. Chem. Solids* **8**, 227–234 (1959).
[33] R. I. Harrison and J. Zucker, *Proc. IEEE* **54**, 588–595 (1966).
[34] J. Cooper, *J. Sci. Instrum.* **39**, 467–472 (1962).
[35] M. Shimazu, Y. Suzaki, M. Takatsuji and K. Takami, *Jap. J. Appl. Phys.* **6**, 120 (1967).
[36] M. F. Kimmitt, J. H. Ludlow and E. H. Putley, *Proc. IEEE* **56**, 1250 (1968).

4.2. THERMAL DETECTORS

The pyroelectric effect is exhibited by certain materials that possess a permanent electric moment along a particular axis. Barium titanate, triglycine sulphate, lithium sulphate, and lithium niobate are examples. According to the physical description that we develop in Chapter 7, the polarization is a result of the relative displacement of lattices of positive and negative ions. It is associated with an unstable vibrational mode of the crystal that is temperature dependent. Below a certain temperature (the Curie temperature), which is characteristic of a particular material, the crystal is permanently polarized, but the lattice displacement and hence the electric dipole moment change with temperature. Thus, if the crystal is irradiated with infrared or far-infrared radiation, the temperature and the dipole moment will respond. But the electric dipole moment is generally masked by free charges that accumulate on the surface to neutralize the bound charges of the polarized crystal and, as the dipole moment changes the surface free charge will tend to redistribute. However, the polarization responds quickly to temperature changes, but the free charges cannot redistribute quickly and cannot maintain zero net charge. Thus in response to a temperature change caused by radiation, a voltage appears between the faces of the crystal normal to the axis of polarization.

Very small temperature changes can be measured, for modern electrometers can detect charges of about 5×10^{-16} cb. The temperature coefficient of the dipole moment (measured in cb cm^{-2} °C^{-1}) is 2×10^{-8} for $BaTiO_3$ and triglycine sulphate, 4×10^{-8} for Li_2SO_4, and 4×10^{-9} for $LiNbO_3$, so a change in temperature of less than 10^{-6} °C is measurable.

In the pyroelectric detector developed by Cooper, barium titanate was coated with a gold black layer to absorb radiation. Barium titanate has the merit of being spontaneously polarized at room temperature (Curie temperature = 120°C), and having a linear rate of change of polarization with temperature up to 110°C. In achieving fast response, Cooper compromised sensitivity, which, of course, is not of importance when intense sources are in use, but he has found that a long time-constant detector can be quite sensitive.[37] Ludlow et al.,[38] have reported pyroelectric detection with a response of 2 μsec and NEP of 10^{-8} W Hz$^{-1/2}$. In their experiments with short high-power CO_2 laser pulses, they used a $Li_2SO_4H_2O$ pyroelectric plate 7 mm in diameter and 5 μm thick, while with a similar

---

[37] J. Cooper, *Rev. Sci. Instrum.* **30**, 92 (1962).

[38] J. H. Ludlow, W. H. Mitchell, E. H. Putley and N. Shaw, *J. Sci. Instrum.* **44**, 694–696 (1967).

source the Japanese used triglycine sulphate. In extending pyroelectric detectors to high frequencies, problems arise from mechanical resonances in the ferroelectric. Natural modes of vibration are excited by thermal shock produced by the radiation and give rise to undesirable voltage signals. This, rather than the thermal time constant, was found by Cooper[34] to be the limiting factor.

4.3. Photoconductive Detectors

In recent years, outstanding progress in wave detection has been made with the extension of photoconductive mechanisms from optical and infrared to much longer wavelengths. These developments have resulted in sensitive and fast detectors throughout the far-infrared.

There are many effects suitable for the detection of high-energy photons, but few sensitive to weak photons. Photoemission, for example, is limited to the optical range below 1.2 μm by the work-function of the best materials available; photovoltaic action is also presently limited to a few microns or below.[29] However, there are photoconductive effects in semiconductors which are sensitive to photons from light to the lower energies of microwave quanta. The possible electronic transitions can be divided into three types: Intrinsic photoconductivity, wherein the absorption of a photon can raise an electron across the forbidden gap from the valence to the conduction band of the semiconductor. Conduction by the free electron thus put in the conduction band and the hole in the valence band increases the conductivity and gives a measure of the incident radiation. The extent to which this process can be extended to long wavelengths is determined by the minimum energy gap attainable. This limit is presently in the vicinity of 10 μm. Some results described by Jain[39] and Esaki[40] show energy gaps of 0.01 eV in bismuth–antimony alloys and suggest extensions of intrinsic photoconductivity into the 100 μm region. Small energy intervals do occur, associated with the presence of impurity atoms in semiconductors. Impurity states located in the bandgap can be ionized by photons extending to 140 μm.[29] In germanium, for example, while the bandgap is 0.75 eV, the shallower impurities are only about 0.01 eV below the conduction band.

The two processes, intrinsic and impurity (or extrinsic) photoconductivity, are shown diagrammatically in Fig. 4.8, together with a third

[39] A. L. Jain, *Phys. Rev.* **114**, 1518–1528 (1959).
[40] L. Esaki, *IEEE Spectrum* **3**, 74–86 (1966).

process, free-carrier photoconductivity. This third, and very important, process is the mechanism whereby photoconductive detection can be extended into the microwave range. In this process, photons absorbed by electrons in the conduction band raise the electron energy and thereby the collision frequency of the electron, for this is energy dependent. The consequent conductivity change gives a measure of the absorbed energy.

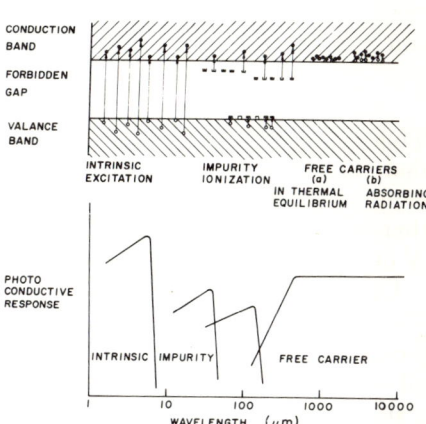

Fig. 4.8. Diagram indicating intrinsic, impurity, and free-carrier photoconductive processes. (After Putley.[29] Reproduced by permission of the Institute of Physics.)

In contrast to intrinsic and impurity transitions with their definite upper wavelength limits, set by the minimum attainable energy gap, free-carrier photoconductivity does not decline as the wavelength increases through the far-infrared. Before we discuss this detector a description of photoconductive detection, particularly impurity photoionization is desirable.

4.3.1. Impurity Photoconductivity

In a semiconductor at temperature $T°K$ with a density N_0 of impurity centers in the forbidden gap there will be a thermal equilibrium distribution of electrons given by the Boltzmann function. With no incident photons, the number of electrons in the conduction band will be

$$N = N_0 e^{-W/2kT}, \qquad (4.15)$$

where W is the depth of the impurity energy level below the conduction band. The material will have a conductivity

$$\sigma = Nq\mu, \qquad (4.16)$$

where q is the electronic charge, and μ is the electron mobility. In this simple treatment, we will not explicitly refer to holes, it being understood that hole conduction is a similar but generally smaller contributing process in n-type semiconductors and a large effect in p-type materials.

When a flux of photons is incident upon the semiconductor with energies $h\nu > W$, more carriers will be raised to the conduction band, and the conductivity will change by an amount

$$\Delta\sigma = q\mu\,\Delta N. \tag{4.17}$$

This change in conductivity can be measured with a circuit similar to that shown in Fig. 4.6 for the bolometer. The resulting voltage change, as indicated by Eq. (4.14) with $R_L = R$, is proportional to $\Delta R/R$ and therefore to $\Delta\sigma/\sigma$. For this to be high, it is necessary to reduce the number N of thermally excited electrons by operating at low temperatures.

Of the incident number of photons only a fraction η will be absorbed. This will be determined by the absorption cross section of the impurity atoms, the density N_0 of impurities, and the thickness of the material, all of which will be factors in the responsivity. For a given flux of photons a steady state will be reached when the number of electrons being raised to the conduction band is equalled by the number recombining with impurities. The lifetime τ of carriers will limit the rate of change of conductivity, thus determining the response time inherent in the detector element. For a flux of Q incident photons per unit area per second, ΔN will be given by the relation

$$d(\Delta N)/dt = \eta Q - \Delta N/\tau. \tag{4.18}$$

The exponential time response $\Delta N \sim e^{-t/\tau}$ is given by this equation when Q is reduced to zero at time $t = 0$. The steady-state response, found by equating $d(\Delta N)/dt$ to zero, is $\Delta N = \eta Q\tau$. Thus the increase in conductivity caused by this photon flux is

$$\Delta\sigma = q\mu\eta Q\tau. \tag{4.19}$$

From this expression, we see the need for materials with high mobilities and adequate photon absorption qualities. It is clear that long carrier lifetimes will give large responsivity.

An insight into the energy levels of impurities in semiconductors can be derived from the application of the quantum theory of the hydrogen atom to impurity atoms. Impurity atoms with a single electron in excess of the electron number in the atoms of the pure semiconductor crystal can be treated in an approximate theory as hydrogenic centers[41] located

---

[41] E. H. Putley, *Phys. Status Solidi* **6**, 571–614 (1964).

4.3. PHOTOCONDUCTIVE DETECTORS

within a uniform dielectric, the permittivity ε being that of the host crystal. An electron deficit can be considered as a hole similarly bound to the impurity atom. The Schrödinger equation for the carrier bound by a Coulomb potential

$$V = -q^2/4\pi\varepsilon r$$

is

$$[(-\hbar^2/2m)\nabla^2 - (q^2/4\pi\varepsilon r)]\psi = W_n\psi, \qquad (4.20)$$

where ψ is the wave function, and W_n the eigenvalue of the energy. The well-known solution is

$$W_n = -m^*q^4/32\pi^2\hbar^2\varepsilon^2 n^2, \qquad n = 1, 2, \ldots. \qquad (4.21)$$

The ionization energy is

$$W_1 = m^*q^4/32\pi^2\hbar^2\varepsilon^2 = (13.6/\varkappa^2)(m^*/m), \qquad (4.22)$$

and the radius of the ground state Bohr orbit is

$$a = 5.29 \times 10^{-9}\varkappa(m/m^*), \qquad (4.23)$$

where m denotes the free electron mass, m^* is the effective mass, and $\varkappa$ is the dielectric constant. The energy given by Eq. (4.22) is in units of electron volts, and the Bohr radius is in centimeters.

This theory, with the effect of the lattice incorporated in the smoothed-out permittivity ε, is a reasonable approximation for orbits large compared to the lattice spacing, provided the spacing between impurities is large compared with the orbital radius, so that we can neglect interactions between impurities. It is applicable to materials like indium antimonide with isotropic effective mass, but modifications are needed for materials such as germanium and silicon with anisotropic effective masses. From the expression for the ionization energy it is clear that shallow impurity levels call for semiconductors with large values of permittivity and small effective masses. However, such loosely bound electrons will describe large orbits so the impurity–impurity interactions will be negligible only for materials with very dilute impurity concentrations. The compromise between low effective mass and purity has restricted the semiconductors thus far used for far-infrared detection. Doped silicon has been used to 10 μm, but the relatively low effective mass and high attainable purity of germanium has enabled this material, with suitable doping, to detect wavelengths approaching 140 μm. It is desirable to reduce the impurities to 10^{15}–10^{13} cm^{-3}, or better. With higher concentrations, the ionization energy decreases, eventually merging with the bands to exhibit metallic behavior at low temperatures. The use of other semiconductor materials with high permittivities and small effective masses is presently limited by the problems of reducing impurities. As we can see from Eq. (4.23), the purity requirements for such materials are more severe. For example, indium antimonide with a dielectric constant comparable with that of germanium but with m^*/m only 0.013, the impurity concentration required to sufficiently reduce interactions between neighboring impurities must be below 10^{13} cm^{-3}. Although indium antimonide is among the purest semiconductors, its impurity concentrations generally exceed 5×10^{13} cm^{-3}.

Germanium detectors have been developed with a number of different types of impurities. With boron, a group III element and therefore different from germanium by a single electron deficit in the outermost shell, the impurity acts as a hydrogenic atom giving shallower energy levels. Boron detectors have operated at wavelengths as long as 140 μm. Non-hydrogenic impurities, with deeper levels, have been introduced to yield a range of germanium detectors covering the spectrum from below 2 μm to about 40 μm.[41] Germanium doped with gold is suitable up to 9 μm; with mercury doping it detects to 15 μm. Germanium–copper and germanium–zinc extend the range to 30 μm and 40 μm, respectively. Beyond this boron, gallium and indium doped germanium are capable of detection up to and beyond 100 μm, and photothermal excitation in epitaxial GaAs has enabled detection to 900 μm.[41a,41b]

We have indicated that the major advantage of photon detectors over thermal detectors is in their fast response and that this is fundamentally determined by lifetimes of carriers in the conduction state. The process of electron capture by impurity centers has been considered by Lax[42] whose calculations give the lifetime and its temperature dependence. In the Lax model of recombination, the electron is first captured into a higher excited state of the impurity atom and cascades down through lower levels to the ground state. The highest state that can capture an electron has an energy of the order of kT below the top of the conduction band. This temperature dependence of the highest state and of the radius of the orbit involved in

---

[41a] G. E. Stillman, C. M. Wolfe and J. O. Dimmock, *Proc. Symp. Submillimeter Waves, Brooklyn, New York, March–April, 1970*, pp. 345–359. Polytechnic Press, Brooklyn, New York, 1971.

[41b] G. E. Stillman, C. M. Wolfe, I. Melngailis, C. D. Parker, P. E. Tannenwald, and J. O. Dimmock, *Appl. Phys. Lett.* **13**, 83–84 (1968).

[42] M. Lax, *Phys. Rev.* **119**, 1502–1523 (1960).

the capture gives a collision cross section which is sensitively temperature dependent. At low temperatures, higher states are involved with wave functions overlapping neighboring impurities to give large and constant cross sections. The lifetime τ is determined not only by the temperature but also by the type and concentration of impurities and can be chosen to have values well below microseconds. It is commonly chosen to give submicrosecond response consistent with reasonable responsivity and noise equivalent power. With some materials the resistance is so high that the RC time constant of the input stage of the postdetector amplifier limits the speed of response.

Fairly typical figures of impurity photoconductive detector performance are response times from 1–0.01 μsec, responsivity 10^3–10^4 V W^{-1}, and NEP $\sim$$10^{-11}$ W Hz$^{-1/2}$.

4.3.2. Free-Carrier Photoconductivity

For the long wavelength region, extending beyond the limits of present impurity photoconductors, a third photoconductive mechanism has been exploited in recent years to yield detectors with submicrosecond time response. This is free-carrier photoconductivity.[41,43–45] It depends on the property of pure, high-mobility semiconductors at low temperatures that the thermal coupling between the conduction electrons and the lattice is very weak, and on the energy dependence of the collision frequency determining the conductivity. The free electrons have extremely small heat capacity, and this contributes to the capability of fast response.

When the electrons absorb photons, electron–electron collisions thermalize the energy; we can thus speak of a free carrier temperature which the "hot electrons" attain above the lattice temperature as a result of photon absorption. The drift mobility of the electrons in a dc field will be determined by the temperature dependence of the collision mechanism limiting the drift, e.g., when ionized impurity scattering is dominant the mobility will vary approximately as the $\frac{3}{2}$ power of the mean carrier energy.[43] The semiconductor conductivity is thus a measure of the energy of absorbed radiation. High purity germanium and indium antimonide have sufficient mobility and weak coupling between the electrons and the lattice when operated at liquid helium temperatures. The device is operated by passing a dc current through the semiconductor element and

---

[43] B. V. Rollin, *Proc. Phys. Soc. London* **77**, Pt. 2, 1102–1103 (1961).
[44] D. W. Goodwin and R. H. Jones, *J. Appl. Phys.* **32**, 2056–2057 (1961).
[45] M. A. C. S. Brown and M. F. Kimmitt, *Brit. Commun. Electron.* **10**, 608–611 (1963).

amplifying the voltage variations. A circuit arrangement similar to that used with the bolometer (Fig. 4.6) can be used.

If the absorption of an amount of power dP by the free carriers changes the mobility by $d\mu$, the relation between incremental changes of voltage, resistance, etc., is

$$dV = I\,dr,$$

or

$$dV/V = dr/r = d\sigma/\sigma = d\mu/\mu,$$

and the responsivity R is

$$R \equiv dV/dP = (V/\mu)\,d\mu/dP, \qquad (4.24)$$

where V is the voltage across the semiconductor element.

For a nondegenerate semiconductor[45a] in the temperature region where momentum scattering is limited by ionized impurities, the mobility μ is related to the temperature by

$$\mu \propto T_{\mathrm{e}}^{3/2}. \qquad (4.25)$$

Thus,

$$R = (3V/2T_{\mathrm{e}})\,dT_{\mathrm{e}}/dP = 3V/2T_{\mathrm{e}}G, \qquad (4.26)$$

where $G = dP/dT_{\mathrm{e}}$ is the coefficient of thermal conductance between the carriers and the lattice. Kinch and Rollins[46,43] have found the responsivity to be in the region of 10^2–10^3 V W^{-1}, and the conductance $G \sim 5 \times 10^{-5}$ joule deg^{-1} sec^{-1}.

For small changes, we can write

$$d\sigma \propto d\mu \propto R\,dP,$$

by (4.24), and this is proportional to E^2, E being the strength of the radiation electric field. The InSb sample can then be described as undergoing a deviation from Ohm's law

$$\sigma = \sigma_0(1 + \beta E^2),$$

where β is a constant, and σ_0 is the conductivity when $E = 0$.

The system of carriers in the semiconductor can be regarded as the sensing element of a "hot electron bolometer." The mean carrier energy

---

[45a] A. S. Grove, "Physics and Technology of Semiconductor Devices." Wiley, New York, 1967.

[46] M. A. Kinch and B. V. Rollin, *Brit. J. Appl. Phys.* **14**, 672–677 (1963).

4.3. PHOTOCONDUCTIVE DETECTORS

W and the temperature T_e will be related by

$$W = CT_e, \qquad (4.27)$$

where C is the electronic heat capacity. The energy of the electron gas will relax to the lattice with a time constant τ,

$$W \sim e^{-t/\tau},$$

and it is this time constant which determines the response time. It is related to the heat capacity and conductance by

$$\tau = C/G. \qquad (4.10)$$

For the indium antimonide sample mentioned above,

$$C \sim 10^{-11} \text{ joule deg}^{-1}.$$

Thus the response time $\tau \sim 10^{-7}$ sec.

The absorption of electromagnetic radiation by the free electrons can be adequately described by the classical Drude–Lorentz theory of conductivity.[47] For the wavelength region of interest, the absorption coefficient is given by

$$\alpha = \frac{2\omega K}{c} = \frac{\sigma}{c\varepsilon_0 \varkappa^{1/2}} = \frac{\sigma_0}{c\varepsilon_0 \varkappa^{1/2}(1 + \omega^2\tau^2)}, \qquad (4.28)$$

when the refractive index $n - iK$ has the condition $n \gg K$. Here $\sigma_0 \, (= nq^2\tau/m^*)$ is the dc conductivity, $\varkappa$ is the dielectric constant (or relative permittivity) of the semiconductor, and τ is the electron scattering time. The term $1/(1 + \omega^2\tau^2)$ in the absorption coefficient gives two spectral regions where the detector has different frequency characteristics. These regions are separated by the critical wavelength $\lambda_c = 2\pi c\tau$, corresponding to $\omega\tau = 1$; for $\omega \gg 1/\tau$, the absorption varies as λ^2. In contrast to intrinsic and impurity photoconductors, there is no long-wavelength threshold; on the contrary, the performance declines with decreasing wavelength. Putley[41] has shown that n-type indium antimonide at 4°K with 5×10^{13} impurities per cubic centimeter and mobility 10 m² V⁻¹ sec⁻¹, has an electron scattering time of 8.5×10^{-13} sec. The critical wavelength λ_c is 1600 μm. At 1000 μm, the absorption coefficient α is

---

[47] R. K. Wangsness, "Introduction to Theoretical Physics—Classical Mechanics and Electrodynamics." Wiley, New York, 1963.

2200 m⁻¹, it falls to 30 m⁻¹ at 100 μm. The thickness t of the semiconductor should be of the order of a millimeter to satisfy the condition $\alpha t \gtrsim 1$ for adequate absorption, at least at longer wavelengths. This device, then, is a broad-band detector, subject to declining performance at short wavelengths. Indium antimonide sensing elements $5 \times 2 \times 2$ mm³, with a resistance of approximately 10 Ω, have been used.

The performance of the detector can be improved by immersing the semiconductor in a magnetic field. This can have two effects. Fields up to 6 or 7 kG can improve the broad-band performance by increasing the resistance of the element. Higher fields tend to diminish the responsivity over most of the far-infrared spectral band by reducing the free-electron absorption, but in the vicinity of the cyclotron resonance frequency the response is enhanced. The mechanism of this latter effect is discussed in Section 4.3.3.

The first effect is associated with the reduction of impurity–impurity interactions by the magnetic field. It can be interpreted in terms of the dilute hydrogenic model discussed in Section 4.3.1. For such a model for n-type indium antimonide, with an impurity concentration of 5×10^{13} cm⁻³, Eq. (4.23) shows that the average separation is only a little over four times the reduced Bohr radius. Sladek[48,49] and others[50] have studied this hydrogenic model in a strong magnetic field and have found that the wave function contracts, particularly in a plane transverse to the field. The effect can be visualized qualitatively in the following way: The magnetic forces are centripetal in a plane perpendicular to the field directions, and they add to the Coulomb forces. The orbital diameter in this plane decreases, and the ground state atom becomes ovoid in shape. As the shape changes the atom shrinks in all dimensions. The shrinkage increases the Coulomb binding force and consequently the ionization energy.

The magnetic field thus causes the impurity energy levels to separate from the conduction band and to descend to an increasing depth (equal to the ionization energy) with increasing field. The change in ionization energy results in a redistribution of the electrons in accordance with Fermi–Dirac distribution statistics. For temperatures such that $kT \ll$ ionization energy, the majority of electrons are bound to impurity atoms. For indium antimonide with 5×10^{13} cm⁻³ impurities in a magnetic field

---

[48] R. J. Sladek, *J. Phys. Chem. Solids* **5**, 157–170 (1958).
[49] R. J. Sladek, *J. Phys. Chem. Solids* **8**, 515–518 (1959).
[50] Y. Yafet, R. W. Keyes and E. N. Adams, *J. Phys. Chem. Solids* **1**, 137–142 (1956).

4.3. PHOTOCONDUCTIVE DETECTORS

of 3–4 kG, temperatures of 4°K or less results in substantial "freeze out" of electrons from the conduction band. At 1.8°K and with 6–7 kG the resistivity increases by about three orders of magnitude, and with it the responsivity increases. Keyes and Sladek[51] have shown that the increase in resistivity is associated not only with a decrease in the number of carriers but also a decrease in their mobility.

Figure 4.9 shows the responsivity measured by Putley[51a] for various values of magnetic field. It can be seen that there is an optimum value of the field in the region of 6–7 kG. This arises because while the decline in free carriers increases the resistivity, it also leads to a reduction in photon absorption. Beyond 7 kG, in the sample used by Putley, this latter effect tends to dominate the responsivity.

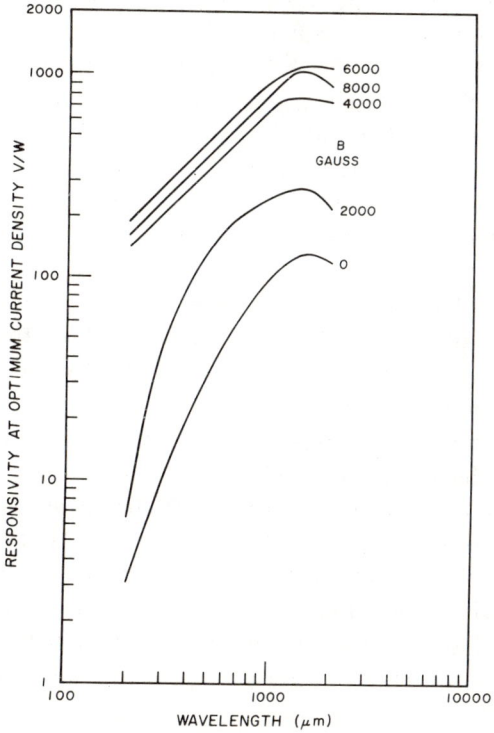

FIG. 4.9. Responsivity of a wide-band indium antimonide photoconductive detector as a function of wavelength and magnetic induction field B. Maximum responsivity is obtained with a field of 6.7 kG (after Putley[51a]).

[51] R. W. Keyes and R. J. Sladek, *J. Phys. Chem. Solids* **1**, 143–145 (1956).
[51a] E. H. Putley, *Appl. Opt.* **4**, 649 (1965).

We have described the operation of the InSb photoconductor as a "video detector," that is, in the square-law regime where the input signal power P_S and output voltage V_S are related as in Eq. (4.24), namely,

$$V_S = RP_S.$$

However, if a local oscillator with power P_L much larger than P_S is applied to the semiconductor we get, as we saw in Section 4.5 [Eq. (4.5)], a larger output,

$$V_S = R(2P_L P_S)^{1/2}.$$

Putley[52] has discussed the improvement attainable with the superheterodyne system at a wavelength of 1000 μm. For a responsivity $R = 1000$ V W^{-1}, he finds that the limit set by equating the signal output voltage to the amplifier noise voltage is $P_S = 1.3 \times 10^{-12}$ W for the video receiver, and, with $P_L = 5 \times 10^{-7}$ W, the superheterodyne limit is $P_S = 1.7 \times 10^{-18}$ W. In experimental comparisons of the performances of a video and a superheterodyne system with both coherent and incoherent radiation, Wort has found the superheterodyne system to be better by about the predicted amount. With a receiver output bandwidth of 1 Hz, Wort has found the limit to be somewhere between 10^{-17} and 10^{-18} W.[53]

Gebbie et al.,[54] have also used the InSb crystal as a mixer. With local oscillator power provided by the 337-μm radiation from a hydrogen cyanide laser, they found a limit of about 7×10^{-13} W. This is well above the theoretical limit.

Arams et al.[54a] have investigated superheterodyne detection at millimeter wavelengths using bulk InSb as the mixing element. At 8 mm they measure a conversion loss of 18 dB, using an IF frequency of 1.5 MHz. This is the upper limit placed on the IF frequency by the relaxation time of the hot electrons to the crystal lattice. In a video system at 150 GHz, they obtained a responsivity as high as 3500 V W^{-1}. Arams et al.,[28] have successfully used the liquid helium cooled InSb detector–mixer to observe the pulse shape of 337-μm emission from an HCN laser.

[52] E. H. Putley, *Proc. IEEE* **54**, 1096–1098 (1966).
[53] D. H. Martin, ed., "Spectroscopic Techniques." North-Holland Publ., Amsterdam, 1967.
[54] H. A. Gebbie, N. W. B. Stone, E. H. Putley and N. Shaw, *Nature* (*London*) **214**, 165–166 (1967).
[54a] F. Arams, C. Allen, B. Peyton and E. Sard, *Proc. IEEE* **54**, 612–619 (1966).

4.3.3. Narrow-Band Photoconductivity in a Strong Magnetic Field

With still higher fields there are interesting developments leading to tunable narrow-band response.[44,45] Over most of the far-infrared frequency range the responsivity of Putley's broad-band detector becomes quite low due to the falloff in free-carrier absorption. This is due to the decline in absorption resulting from carrier freeze-out when the applied magnetic field is large, and also to the ω^{-2} effect shown by Drude theory [equation (4.28) with $\omega\tau \gg 1$]. However, waves with frequencies near the electron cyclotron frequency $\omega_c = qB/m^*$ are strongly absorbed and detected. This phenomena requires a quantum-mechanical explanation. The cyclotron motion of free electrons about the magnetic field can be treated as harmonic oscillators the eigenvalue solutions of which are derived in most elementary textbooks on quantum mechanics.[55] The normally continuous conduction band is split into a series of quantized subbands, or Landau levels, separated by energy intervals $\hbar\omega_c$. The bottom of the lowest Landau level is situated above the normal conduction band minimum by the zero-point energy of the cyclotron motion $\frac{1}{2}\hbar\omega_c$. There is also a similar effect on the electrons bound to impurity atoms. When the magnetic field is so high that the cyclotron orbit is small compared to the Bohr radius, bound electrons can also gyrate about the magnetic field. The cyclotron orbit radius is given by

$$a_c = \left[\frac{2\times \text{Energy}}{m^*}\right]^{1/2}\frac{1}{\omega_c} = \left[\frac{2(n+\frac{1}{2})\hbar}{qB}\right]^{1/2} = 3.63\times 10^{-4}(n+\tfrac{1}{2})^{1/2}B^{-1/2} \tag{4.29}$$

where $n = 0, 1, \ldots$, is the quantum number specifying the eigen energies, B is the field measured in gauss, and the orbit radius is measured in centimeters. The levels of magnetic field required to give impurity cyclotron resonance are apparent from comparison of a_c with the reduced Bohr radius given by Eq. (4.23). These expressions also show that cyclotron resonance in impurity atoms occurs at lower fields (tens of kilogauss) if the effective mass is small. Thus impurity states are also quantized in levels separated by $\hbar\omega_c$; they are (as indicated in Fig. 4.10) located just beneath the Landau levels.

At low temperatures most of the electrons will be in the impurity ground state just below the lowest Landau level—a distance below $\sim 0.1\hbar\omega_c$, equal to the ionization energy. Resonance absorption of a

---
[55] L. Pauling and E. B. Wilson, "Introduction to Quantum Mechanics." McGraw-Hill, New York, 1935.

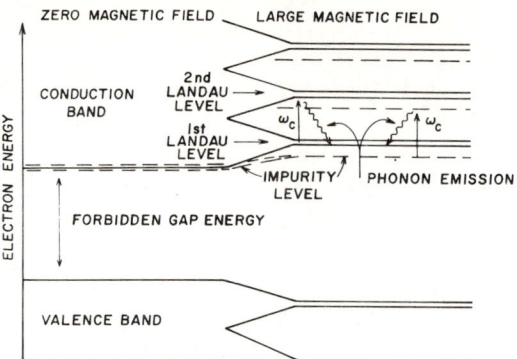

Fig. 4.10. Energy band diagram showing magnetophotoconductive effects in a semiconductor. In a large magnetic field, the normally continuous bands are split into Landau levels separated by $\hbar\omega_c$. Resonance absorption by free and bound electrons is indicated by vertical arrows, and phonon emissions by wavy arrows. (After Brown and Kimmitt.[45] Crown copyright. Reproduced by permission of the Controller of HM Stationery Office.)

photon raises the electron to the first ($n = 1$) impurity Landau level, from which it can proceed to the lowest Landau band by emission of a phonon. The conductivity is thus increased. Furthermore, there is a hot-electron effect with the free electrons. Absorption within the resonance line will raise free electrons to higher Landau energy levels where the scattering time will be longer. This increases the conductivity further. The bandwidth over which there is enhanced absorption is determined by the width of the Landau levels. This is given classically by the Q of the cyclotron resonance. If the average scattering time of an electron is τ, the Q will be $\omega_c\tau/2$. The center of the band tunes linearly with magnetic field. This narrow-band, tunable operation is particularly effective at the higher frequencies of the far-infrared where, as we have seen, free-carrier absorption is weak. At these high cyclotron frequencies there is a sharper Q.

A cross-sectional diagram of the detector developed at the Royal Radar Establishment, England, is shown in Fig. 4.11. A sensing element $5 \times 5 \times 2$ mm³ of indium antimonide is supported in a Dewar of liquid helium, with a niobium–zirconium superconducting solenoid to produce the magnetic field. Radiation is condensed on to the detector by means of a "light" pipe. The high reflectivity of metals in the far-infrared makes tapered hollow metal pipes quite efficient for this purpose. Cooled filters of black polyethylene and crystal quartz remove visible and near-infrared room-temperature radiation which would otherwise energize carriers from the valence to the conduction band to reduce the respon-

4.3. PHOTOCONDUCTIVE DETECTORS

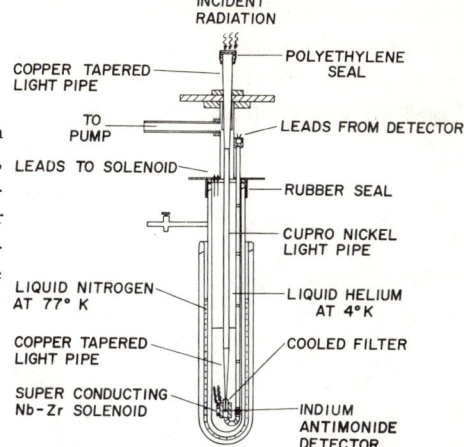

FIG. 4.11. Cross section of an indium antimonide photoconductive detector, complete with a superconducting solenoid and low-temperature Dewar. (After Brown and Kimmitt.[45] Crown copyright. Reproduced by permission of the Controller of HM Stationery Office.)

sivity. The responsivity of the detector when cooled to 4.2°K is shown in Fig. 4.12. With a 15-kG magnetic field, the response is centred on 100 μm with a bandwidth of about 12%. Outside the resonance envelope, the response is orders of magnitude lower. Above 200 μm, the peaked response becomes less well defined as the Landau levels broaden. The resistance of the semiconductor element is about 500 Ω at 5 kG, increasing to 5 kΩ at 25 kG. The response time is less than 1 μsec and the NEP is about 5×10^{-11} W Hz$^{-1/2}$. The dominant noise contribution is not that inherent to the detector; it arises in the postdetector amplifier. If the temperature is reduced (by decreasing the pressure above the liquid helium) the bandwidth is halved, and the responsivity increases four times. At this lower temperature, however, the resistance rises to nearly a megohm with the consequence that detector noise fluctuations now contribute significantly to the total noise. The time constant of this resistance with the capacity of the detector leads and the input stage of the amplifier makes it hard to utilize the inherent speed of the detector.

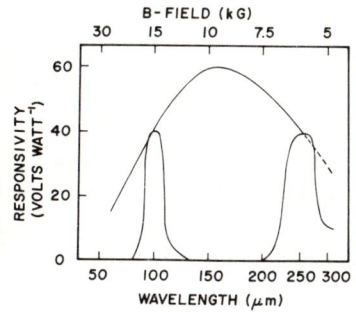

FIG. 4.12. Responsivity of a tunable indium antimonide photoconductive detector as a function of wavelength and magnetic induction field B. The detector, when at 4.2°K, has a bandwidth of about 12% at 100 μm. (After Brown and Kimmitt.[45] Crown copyright. Reproduced by permission of the Controller of HM Stationery Office.)

Brown and Kimmitt[56] have studied the narrow-band photoconductive effect in InSb at 4.2°K with fields up to 75 kG. The photoresponse they observe from 26–1000 μm is shown in Fig. 4.13. At the shorter wavelengths a fine structure is observed which has been attributed to transitions between impurity states bound to conduction band Landau levels.

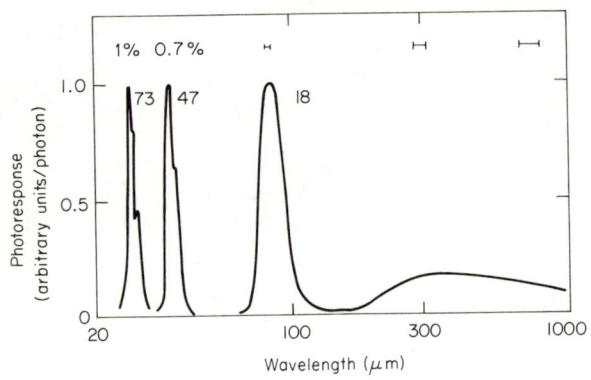

FIG. 4.13. Photoconductive response of InSb as a function of wavelength, for magnetic fields of 18, 47, and 73 kG. The curves have been normalized, and the spectrometer resolution is indicated (after Brown and Kimmitt[56]).

4.4. An Electron Cyclotron Resonance Method of Radiation Detection and Spectral Analysis

The free-carrier photoconductor that we met in Section 4.3.3 is a combined detector–spectrometer below about 200 μm, when a magnetic field is applied to the InSb crystal. The bandwidth of the cyclotron resonance in the solid is set by the relaxation time at about 10^{12} Hz when the temperature is 4.2°K. It is tunable magnetically but, unfortunately, this bandwidth is not sufficiently narrow to make it competitive in spectral resolution with the spectrometers described in Chapter 3. It appears, however, that a tunable spectrometer–detector with a bandwidth narrower than that available with grating and multiplex spectrometers is possible if we use free electrons in cyclotron motion in high magnetic fields.[57] For submillimeter operation, fields greater than 100 kG are required. Such fields are becoming increasingly available and, as we

[56] M. A. C. S. Brown and M. F. Kimmitt, *Infrared Phys.* **5**, 93–97 (1965).
[57] L. C. Robinson, *Infrared Phys.* **10**, 111–123 (1970).

4.4. CYCLOTRON RESONANCE SPECTROMETER–DETECTOR

have seen in Chapter 2, are needed for the electron cyclotron maser and for extensive tuning of Raman spin-flip masers. Thus with coherent generation, and the offer of sensitive detection and high-resolution spectral analysis, high magnetic fields have something quite unique to offer the far-infrared.

In the present section, we discuss the physical principles of a spectrometer–detector; it has not yet been tried experimentally. In essence, the system is an inverse electron cyclotron maser, but nonrelativistic rather than mildly relativistic electrons are used. We choose to develop a classical description of the interaction based on solutions of the equations of motion treated in detail in Chapter 5.

4.4.1. Cyclotron Resonance Energy Exchange

We suppose a system of monoenergetic nonrelativistic free electrons in orbital motion in a homogeneous magnetic induction field **B** interacting with an electric field $E_0 e^{-i(\omega t + \phi)}$ for a short time τ. This field is the circularly polarized component with phase ϕ of a standing-wave electric field in the resonator. Let the **B**-field be in the z-direction and the field E_0 be in the $x - y$ plane. We show in Section 5.1.1 of Chapter 5 that, for an electron with initial kinetic energy $\tfrac{1}{2} m v_0^2$, the kinetic energy after the interaction is

$$\tfrac{1}{2} m v_0^2 - \frac{A m v_0}{\omega_c - \omega} [\cos \phi - \cos\{(\omega_c - \omega)\tau + \phi\}]$$
$$+ \frac{m A^2}{(\omega_c - \omega)^2} [1 - \cos(\omega_c - \omega)\tau], \qquad (4.30)$$

where $A = qE_0/m$, and $\omega_c = qB/m$ is the cyclotron angular frequency.

Consider the second and third terms of expression (4.30). A system of orbiting electrons will have random phases with respect to E_0, so we must average over all ϕ to obtain an expression for the mean energy W_1 absorbed per electron. The result is

$$W_1 = (q^2 E_0^2 \tau^2 / 2m)\left(\frac{\sin^2 \Gamma}{\Gamma^2}\right), \qquad (4.31)$$

where

$$\Gamma = [(\omega_c - \omega)/2]\tau. \qquad (4.32)$$

The quantity $q^2 E_0^2 \tau^2 / 2m$ is the energy absorbed at cyclotron resonance from the radiation field in an interaction time τ, and $\sin^2 \Gamma / \Gamma^2$ is the spectral line shape function, with Γ equal to half the phase shift during the interaction of the particle cyclotron motion with respect to the rotating electric field. It is easily shown that the counterrotating component of the electric field absorbs an amount of energy

$$(q^2 E_0^2 \tau^2 / 2m)(\sin^2 \gamma / \gamma^2),$$

where the phase shift $\gamma = (\omega_c + \omega)\tau/2$ is necessarily so large that negligible transfer results.

While Eq. (4.31) describes the mean energy absorption, the individual electrons are considerably spread in energy. Some are at such phase positions that they acquire energy from the field and spiral in increasing orbits, while others give energy to the field and spiral inwards. We are interested here in the second term of expression (4.30)—the term that averaged to zero in the derivation of Eq. (4.31). Writing W_2 for the energy, additional to that given in Eq. (4.31), of an electron of phase ϕ, it is easily shown that

$$W_2 = qE_0\tau v_0 \sin(\Gamma + \phi)(\sin \Gamma / \Gamma). \tag{4.33}$$

Adding Eqs. (4.31) and (4.33) gives

$$W = W_1 + W_2 = qE_0\tau \frac{\sin \Gamma}{\Gamma} \left[\frac{qE_0\tau}{2m} \frac{\sin \Gamma}{\Gamma} + v_0 \sin(\Gamma + \phi) \right]. \tag{4.34}$$

We note that all values of ϕ from 0–2π are represented homogeneously in the system of electrons and that according to Eq. (4.34), maximum energy is transferred to those electrons for which $\Gamma + \phi = (2n + \frac{1}{2})\pi$, $n = 0, 1, 2, \ldots$. If we suppose that a system of electrons with, say, 20 eV initial energy is brought into resonant interaction for a time $\tau = 10^{-7}$ sec with a field $E_0 = 5$ V m^{-1}, the mean energy W_1 acquired is 0.022 eV, while the maximum energy W reaches 1.522 eV. As we shall argue in the following sections, this relatively large energy increase should be readily measurable, the chosen time of interaction should be attainable, and a field of 5 V m^{-1} can be built up in a resonant cavity excited by low levels of far-infrared radiation.

4.4.2. An Interaction System

We now consider a system capable of bringing a beam of spiraling electrons into interaction with a standing-wave field and measuring, thereafter, the effect of increased particle energy. The system is illustrated

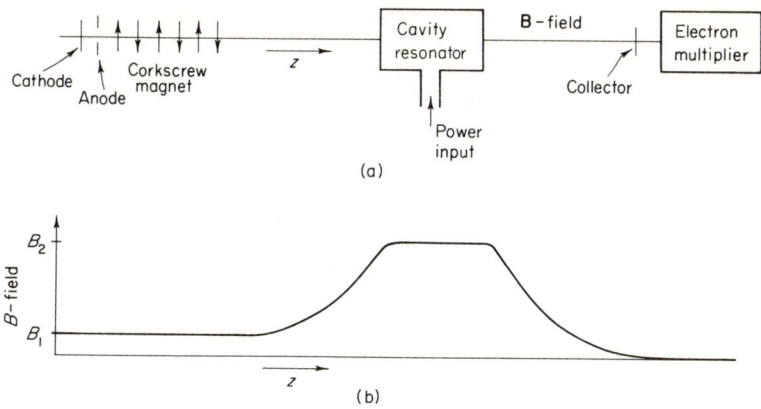

FIG. 4.14. Diagram of a cyclotron resonance experimental system. Electrons from a cathode are wound into helical motion in a helical field magnet and a mirror field. They drift in tight helical motion through a resonant cavity in the uniform field region B_2 shown in (b) and then proceed to the collector. Interaction with radiation in the cavity changes the current reaching the biased collector (after Robinson[57]).

in Fig. 4.14. It is similar to that used by Hirshfield and Wachtel[58] in the electron cyclotron maser. Electrons from an electron gun are fired in the z-direction parallel to a magnetic induction field **B**. In a region where **B** is uniform and of magnitude B_1 they are wound into helical motion by a helical field magnet[59] from which they emerge with a small percentage of their kinetic energy as circular motion in the transverse $x - y$ plane. For motion into the slowly rising **B**-field the magnetic moment of the orbiting particles is an adiabatic invariant—the radius of the motion decreases keeping the flux linkage constant, and the pitch decreases as the electrons acquire energy in their transverse motion at the expense of z-directed motion.[60] If the electrons emerge from the helical field magnet with $p\%$ of their energy in the $x - y$ plane, at a position where the magnetic field has increased to B, the energy in the $x - y$ plane is ($pB/B_1 \%$). Electrons can be reflected in the magnetic mirror when all the energy is translated into the transverse plane, or they may be so controlled as to just reach the field indicated by B_2 in Fig. 4.14 and proceed slowly along the uniform plateau $B = B_2$. It can be shown that a helical field-mirror system such as we have described is capable of velocity selecting a narrow band of electrons from a Maxwellian spread

---

[58] J. L. Hirshfield and J. M. Wachtel, *Phys. Rev. Lett.* **12**, 533–535 (1964).
[59] R. C. Wingerson, *Phys. Rev. Lett.* **6**, 446–448 (1961).
[60] S. Chandrasekar, "Plasma Physics." Univ. of Chicago Press, Chicago, Illinois, 1960.

so that the electrons reaching the plateau region of **B** can be essentially monoenergetic.[61,62]

We suppose B_2 to be in the range 100–200 kG, and we locate a wave-supporting cavity resonant in the wavelength range 1000–500 μm at the **B**-field plateau. The cavity will be a cylindrical structure with its axis along the z-direction and, say, 1 cm long. With no far-infrared radiation fed into the cavity, the electrons will drift unaffected along the **B**-field plateau and unwind as they spiral down the "magnetic hill" towards the collector. With the collector slightly positive with respect to the cathode, electron current will flow in the current meter. However, let us bias the collector slightly negatively with respect to the cathode so that the electrons approach closely to the collector and are then repelled. They return towards the cathode.

Suppose now that a field of strength E_0 is established in the cavity by the inflow of far-infrared radiation. It follows from Eq. (4.34) that approximately half the electrons will acquire additional energy. When this is unwound into z-directed motion towards the collector, some electrons will be energetically capable of overcoming the potential barrier set by the bias voltage. Thus a flow of current will register the presence of radiation in the cavity resonator.

Those electrons which, in accordance with Eq. (4.34), either lose energy to the field or gain insufficient energy to overcome the bias will be reflected back towards the cathode. They will, however, be energetically incapable of reaching the cathode and will, if not removed, be reflected back again in the $+z$-direction to renegotiate the helical field magnet. The helical field magnet is a nonreciprocal device,[59] and it imparts essentially zero transverse energy during the return passage of the electrons. During the third transit (after reflection from the cathode region) it unwinds some electrons and increases the transverse momentum of others thus giving a spread in transverse energies ranging from 0–$4p\%$. Multiple reflections can then follow which give rise to space-charge effects and other complications. Reflected particles must be removed. This can be done with a hollow cylindrical electrode with a small aperture located at the end of the helical magnet nearest the cathode. Electrons from the cathode having nearly zero transverse momentum can pass while the returning electrons with approximately $p\%$ of transverse energy will be intercepted.

[61] L. C. Robinson, *Ann. Phys.* (*New York*) **61**, 265–276 (1970).
[62] L. C. Robinson and G. Szekeres, *Proc. Phys. Soc. London Gen.* **3**, 481–492 (1970).

4.4.3. Spectral Resolution and Response Time

According to Eq. (4.34), the frequency band over which energy is imparted an electron is given by the sinc function (sin Γ/Γ). The response to incoming radiation has a maximum at $\Gamma = 0$ and falls to the first minima at $\Gamma = \pm\pi$ and has successive zeros at $\Gamma = \pm 2\pi, \pm 4\pi$, etc. The frequency resolution, which according to the Rayleigh criterion is the separation of the maximum and first minimum of the sinc function, is given by

$$\Gamma \equiv (\omega_c - \omega)\tau/2 = \pi,$$
$$f_c - f = 1/\tau. \tag{4.35}$$

It is determined by the time of transit of electrons through the resonant cavity. If the cavity length is chosen to be 1 cm and the electrons can be controlled by the helical field-mirror field to have an energy of about 0.07 eV in the z-direction along the **B**-field plateau, $\tau = 10^{-7}$ sec, and the frequency resolution for a perfectly uniform magnetic field will be $\Delta f = 1/\tau = 10$ MHz. It is thus suggested that the cyclotron resonance interaction has the characteristic of a tunable high-resolution spectrometer with resolving power

$$R = f/\Delta f \sim 10^4 - 10^5. \tag{4.36}$$

We note that the sinc function encountered here is that familiar in the theory of spectral resolution of well-known spectrometers, e.g., Fourier transform interferometers and grating spectrometers. Indeed, a physical analogy between the present case and the interference processes operative in classical spectrometry can be seen: In the latter case, the sinc function arises from interference within a superposition of waves after either wavefront or wave-amplitude division—a simple case being the Huygen's construction of phase distributed "wavelets" giving the resultant wave amplitude on a screen in front of a diffraction grating. In the cyclotron resonance spectrometer interference arises from the superposition of elemental vector impulses $qE_0 \, dt$ imparted to the electrons during the succession of subintervals dt of the interaction time τ.[61]

The response time of the detection process will be equal to the transit time τ of the electrons through the cavity. This is considerably longer than the response time $Q/\omega \sim 10^{-11}$ sec of the cavity resonator ($Q < 100$, see p. 159). After the interaction there will be a short delay during which the electrons travel to the collector. For the case taken above the response time is 0.1 μsec. The response time (and, of course, the spectral bandwidth) can be controlled electrically simply by adjusting the current in the helical magnet, and it is clear that very short response times can be realized when the electrons travel quickly through the cavity. However, fast response will be obtained at the expense of responsivity. For example, a reduction in response time from 10^{-7} to 10^{-8} sec will result, according to Eq. (4.34), in a decrease in the peak and mean powers W_2 and W_1 acquired by the electrons by factors of 10 and 100, respectively.

4.4.4. Detection Characteristic

The relationship between the current collected and the radiation field can be found from Eq. (4.34). Consider the resonance condition $\Gamma = 0$, and suppose the bias potential-barrier to be initially greater than $W_1 + W_2$. As the bias voltage is reduced, electrons are first collected when $|qV_{\text{bias}}| = W_1 + W_2$ from those with ϕ in the region of $-\pi/2$ (the negative sign arises from the negative electronic charge). The bias voltage is measured with respect to the cathode. As the electrons are distributed with equal number per unit range of ϕ, the fraction collected will be seen to be

$$(1/\pi) \cos^{-1}[(|qV_{\text{bias}}| - W_1)/W_2].$$

That is, if the total current that can be collected with large positive voltage on the collector is I_0, that in the presence of a negative bias voltage will be I, where

$$I = (I_0/\pi) \cos^{-1}[(|qV_{\text{bias}}| - W_1)/W_2]. \tag{4.37}$$

If $|qV_{\text{bias}}| = 0$,

$$I = I_0[\tfrac{1}{2} - (1/\pi) \sin^{-1}(-W_1/W_2)],$$

and, for the condition $W_1/W_2 \ll 1$, this reduces to

$$\begin{aligned} I &= (I_0/2) + (I_0 W_1/\pi W_2) \\ &= (I_0/2) + (I_0 q E_0 \tau / 2\pi m v_0). \end{aligned} \tag{4.38}$$

On the other hand, if $W_2 \gg |qV_{\text{bias}}| \gg W_1$,

$$I = (I_0/2) - (I_0 |V_{\text{bias}}|/\pi E_0 v_0 \tau). \tag{4.39}$$

4.4. CYCLOTRON RESONANCE SPECTROMETER–DETECTOR

In deriving Eq. (4.38) for the current collected by the biased electrode, we have assumed no spread in electron energies except that imparted by the radiation field. Let us now generalize this by assuming that there is a small spread in energy of the electrons approaching the collector. Suppose that in the absence of radiation the distribution falls off as $\exp[-(v^2 - v_0^2)/2mkT]$ above the energy level $\tfrac{1}{2}mv_0^2$, where T is a "temperature" assigned to the Maxwellian spread, and k is Boltzmann's constant. Writing $(v^2 - v_0^2)/2m = w$, the distribution function giving the fraction of electrons in an energy range dw about w is

$$f(w)\,dw = (1/kT)\,e^{-w/kT}\,dw.$$

Refer now to the distribution as illustrated in Fig. 4.15(a) and consider the electrons within the elemental strip [shaded in Fig. 4.15(a)] to the left of $|qV_{\text{bias}}|$. If n_0 is the total flow of electrons per second in the current I_0, the number in the elemental strip is

$$(n_0/kT)\exp(-|qV_{\text{bias}}|/kT)\,e^{\Delta/kT}\,dw, \qquad (4.40)$$

where

$$\Delta = |qV_{\text{bias}}| - w. \qquad (4.41)$$

Now, when radiation is absorbed there will be a redistribution of the energy of the particles from the strip with some electrons displaced in energy sufficiently to the right to overcome the bias potential. The dashed curves in Fig. 4.15(b) illustrate the redistribution of particles from the strip, W_1 and W_2 being the energies given in Eq. (4.34); the horizontal scale is still energy but now the vertical scale represents ϕ, the phase angle of Eq. (4.34). It is clear that the fraction of electrons collected is $2\theta/2\pi$, and is given by the expression

$$\frac{n_0}{\pi kT}\cos^{-1}\!\left(\frac{|qV_{\text{bias}}| - W_1 - w}{W_2}\right) e^{-w/kT}\,dw. \qquad (4.42)$$

In summing for all elements in the range between $w = 0$ and $w = |qV_{\text{bias}}|$ we note that only electrons for which $\Delta \leq W_1 + W_2$ can reach the collector. Thus the migration of initially low energy electrons to the collector is

$$\frac{n_0}{\pi kT}\int_{w=|qV_{\text{bias}}|-W_1-W_2}^{w=|qV_{\text{bias}}|}\cos^{-1}\!\left(\frac{|qV_{\text{bias}}| - W_1 - w}{W_2}\right) e^{-w/kT}\,dw. \qquad (4.43)$$

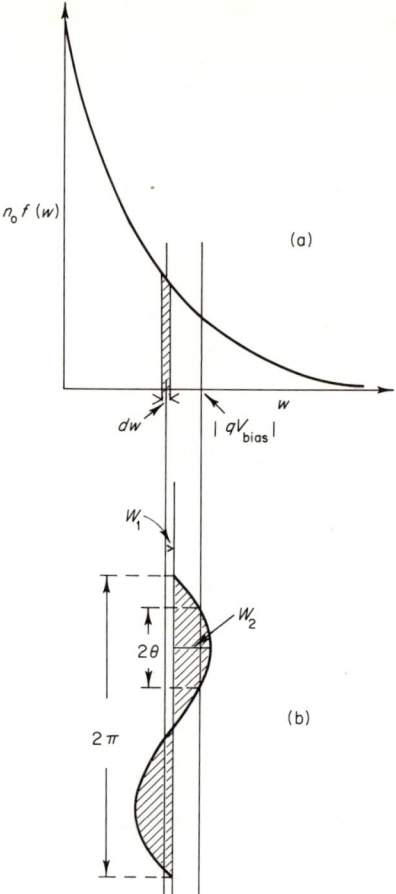

FIG. 4.15. (a) Illustration of an assumed Maxwellian distribution of electrons approaching a biased electrode. In (a) electrons with energy $w > |qV_{\text{bias}}|$ are collected, while those with $w < |qV_{\text{bias}}|$ are repelled from the electrode. In (b) the distribution has been affected in an interaction with a radiation field and the current to the collector is altered. Here the vertical axis represents the phase angle ϕ of an electron with respect to the circularly polarized electric field. The elemental strip dashed in (a) is shifted and spread in energy as illustrated by the dashed region in (b) (after Robinson[57]).

Some electrons initially collected will lose energy to the radiation field and thereafter not be able to overcome the bias potential. We can account for this current decrease by treating an elemental strip to the right of $|qV_{\text{bias}}|$ in the same manner as above. In such a strip, the electron population is

$$(n_0/kT)\exp(-|qV_{\text{bias}}|/kT)\, e^{-\Delta/kT}\, dw,$$

where $\Delta = w - |qV_{\text{bias}}|$. The redistribution due to the radiation field is

$$\frac{1}{\pi}\cos^{-1}\left(\frac{|qV_{\text{bias}}| - W_1 - w}{W_2}\right).$$

In summing over all such strips between $\Delta = |qV_{\text{bias}}|$ and $W_2 - W_1$, we find the loss of electrons per second to be

$$\frac{n_0}{kT}\int_{w=|qV_{\text{bias}}|}^{w=|qV_{\text{bias}}|+W_2-W_1}\cos^{-1}\left(\frac{|qV_{\text{bias}}| - W_1 - w}{W_2}\right)e^{-w/kT}\,dw. \quad (4.44)$$

Subtracting Eq. (4.43) from Eq. (4.44) and multiplying by the electronic charge, we obtain the following expression for the increase in current to the collector when the electrons interact with the radiation:

$$I = \frac{I_0}{kT}\left[\int_{w=|qV_{\text{bias}}|-W_1-W_2}^{w=|qV_{\text{bias}}|}\cos^{-1}\left(\frac{|qV_{\text{bias}}| - W_1 - w}{W_2}\right)e^{-w/kT}\,dw\right.$$
$$\left. - \int_{w=|qV_{\text{bias}}|}^{w=|qV_{\text{bias}}|+W_2-W_1}\cos^{-1}\left(\frac{|qV_{\text{bias}}| - W_1 - w}{W_2}\right)e^{-w/kT}\,dw\right]. \quad (4.45)$$

Integration of Eq. (4.45) is discussed in the literature.[57] In the limit $T \to 0$, the distribution function contracts to a Dirac delta function $\delta(w)$. The second integral on the right-hand side is zero, since the limits of integration exclude $w = 0$. When $W_1 + W_2 < |qV_{\text{bias}}|$, the first integral is also zero, as expected. When $W_1 + W_2 > |qV_{\text{bias}}|$, the range of integration includes zero and Eq. (4.37) is recovered for the current collected.

4.4.5. Space-Charge Effects

In the preceding single-particle analysis, we have ignored space-charge forces on the tacit assumption that the electron beam current can be made small enough to minimize such effects. We have described a method of removing electrons repelled by the collector electrode. Now, it is apparent that the effects of space-charge repulsion will be largest at the high-field plateau where the beam cross section is compressed to a very small size, and where the electron density must increase further in order to satisfy the requirements of current continuity as the forward electron velocity is reduced. Clearly then, space-charge forces can affect the maximum attainable transit time τ; also they will release electrostatic energy to kinetic energy in the $x - y$ plane and adversely affect the energy distribution of electrons approaching the collector. The magnetic field prevents beam expansion but deflects the electrons into azimuthal motion with, of course, the outer electrons acquiring more energy than those near the beam center.

Elementary calculations show that for a narrow beam from the cathode (diameter ~ 0.25 mm) compressed by a magnetic field increase of about 1000 to 1, only a small fraction of a millivolt can be imparted by space-charge forces if I_0 is less than 10^{-12} A.

4.4.6. The Resonator

In considering resonant cavities to build up an intense electric field $\mathbf{E}_0$ from low levels of radiation with wavelengths in, say, the range 500–1000 μm, one might envisage a right-circular cylinder with the axis parallel to the $\mathbf{B}$-field. Higher order modes have been excited in a large cavity of this type in the electron cyclotron maser (ECM), and, because of the close similarity of the present interaction to that in the ECM, it follows from the principle of reciprocity that power fed into the cavity can produce fields suitable for the inverse interaction with power absorption. In work with the ECM carried out by the writer, a copper cavity 2.5 cm in diameter and 16 cm long with partially closed ends exhibited continuous magnetic-field tuning characteristics at wavelengths below 1000 μm, thereby demonstrating the existence of a continuum of overlapping resonant modes with coupling to the electron system.

Let us consider a small fundamental-mode cavity in order to calculate the input power level required to build up the necessary field E_0. We will follow this with considerations of overmoding. To maximize E_0, one naturally thinks of the possibilities of cylinders with walls reentrant along their length, and it transpires that such a geometry has been analyzed theoretically by Hansen and Richtmyer.[63] For bounding surfaces that are sections of confocal ellipsoids and hyperboloids (see Fig. 4.16) curves have been computed for the cavity Q and the shunt impedance R in terms of the dimensionless parameter δ/λ—the ratio of the skin depth to the free-space wavelength—and the cavity dimensions have been calculated for resonance at various frequencies. We choose two geometries for resonance at $\lambda = 1000$ μm, namely $2x_0/a = 3$ and 10 (see Fig. 4.16) and take $\delta/\lambda = 1.41 \times 10^{-3}$ for silver with resistivity 1.58×10^{-6} Ω cm. The results of Hansen and Richtmyer for $2x_0/a = 3$ give $x_0 = 0.19$ and $a = 0.13$ mm approximately, $Q = 96$, and $R = 430$ Ω; for $2x_0/a = 10$, $x_0 = 0.15$ mm, $a = 0.03$ mm, approximately, $Q = 64$, and $R = 141$ Ω.

[63] W. W. Hansen and R. D. Richtmyer, *J. Appl. Phys.* **10**, 189–199 (1939).

4.4. CYCLOTRON RESONANCE SPECTROMETER–DETECTOR

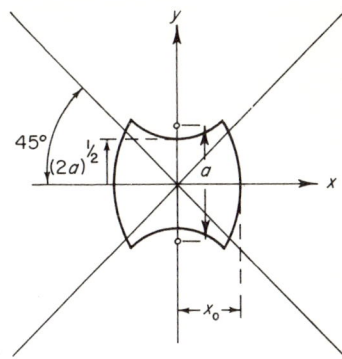

FIG. 4.16. Diagram of a reentrant cylindrical resonant cavity. The surfaces are confocal ellipsoids with equatorial radius x_0, and hyperboloids with interfocal distance a (after Robinson[57]).

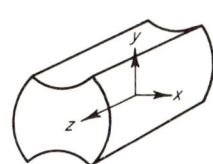

From these parameters, we can determine the integral

$$\int_{(-2a)^{1/2}}^{(+2a)^{1/2}} \mathbf{E}_0 \cdot d\mathbf{y}$$

on the y-coordinate axis implicit in the definition of the shunt impedance for an assumed power P supplied to the cavity. For the first geometry $E_0 \simeq 5.6 \times 10^5 P^{1/2}$ V m^{-1}, and for the second $E_0 \simeq 3 \times 10^5 P^{1/2}$ V m^{-1}. Referring back to Eq. (4.34), and again taking 20 eV electrons and $\tau = 10^{-7}$ sec, we see that a power level of 10^{-10} W raises the energy of the most favored electrons by about 1 eV, while 10^{-16} W can raise the energy by as much as 1 meV. With 500 eV rather than 20 eV electrons, v_0 is increased by a factor of 5 and an input power as low as 10^{-16} W can raise the energy of the most favored electrons by 5 meV.

The question of enlarging the cavity dimensions and operating with higher modes is important because it makes the cavity size more convenient particularly from the point of view of feeding energy into the structure. Furthermore, a spectrum of overlapping modes can broaden the resonator bandwidth, thereby enabling reception at any value of the cyclotron frequency with little or no cavity tuning. We envisage here radially propagating modes as in a Hansen–Richtmyer type resonator but with the equatorial radius x_0 increased to perhaps 1 cm. This would give a sequence of closely-spaced resonances which could be excited by a near-plane wave in an oversize waveguide or light pipe connected through a coupling aperture in the cylindrical wall. The volume of such a cavity increases as x_0^2, the surface area as x_0, and the Q, which varies approximately as the ratio of volume to surface area, increases as x_0. A simple argument based on the constancy of the product of the Poynting vector and area of the wave front for radial wave propagation suggests that, while the number of antinodes of the radial standing-wave pattern increases with x_0, the energy in each antinode is independent of x_0. Thus it follows that the energy stored in an antinode is essentially constant as the cavity is overmoded, and the shunt impedance used in the foregoing estimates applies approximately to overmoded cavities.

In addition to resistive losses in the cavity walls, the electron beam can contribute to the cavity losses.[64] The power dissipation in the beam (P_{beam}) is simply the product of W_1 and the number n_0 of electrons traversing the cavity per second. From Eqs. (4.31) and (4.33), we have, under the resonance condition $\Gamma = 0$, a beam dissipation equal to

$$P_{\text{beam}} = n_0 W_1 = n_0 q E_0 \tau W_2 / 2 m v_0. \qquad (4.46)$$

Beam loading can be made negligible by reducing n_0 to a level such that P_{beam} is much less than the wall losses. This is compatible with the requirements of reducing space-charge forces. In the example taken above where 10^{-10} W is fed into a cavity traversed by 10^7 electrons per second ($I_0 \sim 10^{-12}$ A) P_{beam} is completely negligible.

4.4.7. Further Remarks

The low current levels required for the reduction of space-charge effects lend themselves naturally to an electron multiplier post-detector amplifier. The collector would thus be the first dynode of the multiplier with preceding electrodes biased to select the most energetic electrons and thereafter accelerate them to a suitable energy to give adequate secondary emission. Electron multiplication provides almost noiseless amplification and gives a considerable advantage with respect to other far-infrared detectors presently limited by noise in the postdetector amplifier.

Robinson and Whitbourn have pursued the question of noise in this detector, as well as the description of the interaction in terms of the

---

[64] A. H. W. Beck, "Thermionic Valve." Cambridge Univ. Press, London and New York, 1953.

notions of quantum mechanics.[65] Of the various sources of noise they conclude that the limit of detectivity is set by fluctuations in the flux of photons entering the detector from room-temperature background radiation. As we shall see in Section 4.6, the root-mean-square radiation fluctuation varies as the square root of the input bandwidth. For an input bandwidth of 10 MHz it is found that the minimum detectable signal is close to 10^{-16} W if the detector output bandwidth is 1 Hz. This source of noise varies as $\tau^{-1/2}$, and it is possible that interaction times longer than 10^{-7} sec can be attained and signals much less than 10^{-16} W detected.

Superheterodyne operation of the free-electron cyclotron resonance detector has also been mentioned.[65] If a local oscillator wave is supplied to the resonator, then through the term in E_0^2 in Eq. (4.34) terms in W with sum and difference frequencies occur with amplitudes proportional to the product of the local oscillator and signal amplitudes. This can increase the magnitude of W_1 and its response to the signal, and we can physically interpret this as occurring because the local oscillator increases the velocity and the radius of the electron cyclotron motion so that it becomes more responsive to the input signal. However, the term W_2 in Eq. (4.34) does essentially the same thing through the term v_0, the initial velocity of the electrons. In this sense, then, we can regard the detector as having a superheterodyne principle inbuilt—the "difference" or "intermediate" frequency can be simulated by chopping the radiation, or by chopping the electron beam with an alternating voltage applied to a grid.

4.5. The Josephson Junction

In 1966 a group at Bell Telephone Laboratories reported the operation of point–contact junctions between superconductors as "sensitive, broadband, high-speed detectors of millimeter and submillimeter radiation." This work represented the realization of one of the effects predicted on purely theoretical grounds in 1962 by Josephson.

The Josephson junction is simple to construct and relatively easy to operate. The physics of the effect however, is quite complicated.[66-69]

---

[65] L. C. Robinson and L. B. Whitbourn, *Proc. Phys. Soc. London Gen.* **5**, 263–271 (1972).

[66] R. P. Feynman, R. B. Leighton and M. Sands, "Lectures on Physics," Vol. III, pp. 22. 1–22.18. Addison Wesley, Reading, Massachusetts, 1965.

[67] D. N. Langenberg, D. J. Scalapino and B. N. Taylor, *Sci. Amer.* **214**, 30–39 (1966).

[68] J. M. Blatt, "Theory of Superconductivity." Academic Press, New York, 1964.

[69] See, for example, J. R. Schrieffer, "Theory of Superconductivity." Benjamin, New York, 1964.

The point-contact junction is made by pressing together two pieces of wire with superconductive properties, one with a sharpened end against the other with a flattened end. It is placed in a low temperature cryostat. When electromagnetic radiation falls on the junction alternating currents are induced to flow and a dc voltage occurs across the junction. This voltage is a measure of the incident radiation. Conversely, the application of a dc voltage produces alternating currents, so that the junction generates radiation.

This effect, the ac Josephson effect, is associated with the motion of electrons tunneling through the constrictive barrier of the junction. As we will see, it arises because tunneling pairs of electrons are either transmitted through the junction or reflected according to the relative phases of quantum-mechanical wave functions of the superconducting electrons in the two bulk superconductors. With a dc potential difference between the superconductors, the relative phase increases continuously. As it increases through 0 towards π, then towards 2π, etc., the electron pairs are periodically transmitted and reflected and the tunneling current oscillates. The junction can be a geometrical constriction or we can think of a very thin insulator between the superconductors, as originally envisaged by Josephson.

Let us represent the junction by the diagram of Fig. 4.17, and consider first the electron behavior in the bulk superconductor. Many of the properties of superconductors can be understood in terms of a model consisting of two interpenetrating fluids. One is a "superfluid" composed of

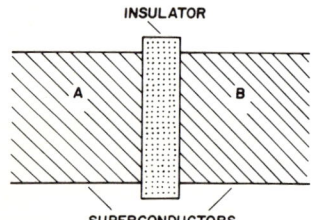

FIG. 4.17. Two superconductors separated by a thin insulator junction.

electrons coupled together in pairs and exhibiting strong pair–pair correlation. The other is a "normal fluid" of weakly interacting single electrons. The two fluids are regarded as two states of the electron system. In the superfluid state the binding between mates of a pair and pair–pair correlations are such as to make this a minimum energy or ground state. The normal fluid is regarded as an excited state separated from the ground state by an energy gap. At zero temperature all electrons are in the super-

4.5. THE JOSEPHSON JUNCTION

fluid state; as the temperature is raised, thermal agitations raise electrons to the excited state until above the critical transition temperature T_c all electrons are in the excited or normal fluid state. The behavior of superconductors is a consequence of the detailed correlations between electrons in the superfluid.

In the theory of metallic conduction[70] electrons are considered to move through the metal lattice, occasionally being scattered by lattice vibrations. The scattering is associated with the absorption or emission of a vibrational quantum (phonon), and might be visualized as a dynamic deformation or polarization of the lattice by the electron, which in turn reacts back on the electron. In the superconducting state, however, the polarization of the lattice can exert an attractive force on another electron. This is a long-range force opposing the Debye-screened Coulomb repulsion. When the attraction exceeds the repulsion (as it can in metals with strong electron-lattice interactions, that is, materials which are normally poor conductors), the two electrons become coupled, the net binding force extending over distances many thousands of times greater than lattice ion separations. It can be shown that a lowest energy state for the coupled pair requires the electrons to have equal and opposite spin and momentum.[68,69] The binding energy of this state is weak as is shown by the ready destruction of superconductivity by the feeble thermal energies kT_c, where the critical temperature is typically only a few degrees above absolute zero.

The wave function describing a well-separated bound pair of electrons is necessarily large in spatial extent. There are usually several electrons for each lattice ion, so it is apparent that the wave function of multitudes of coupled pairs overlap almost completely in space. It was shown by Schafroth[68] and others that all electron pairs can condense into the same state specified by a single pair-wave function. This state is the lowest energy or ground state. Electron pairs are bosons and are not excluded by the Pauli principle from occupancy of a single state. On the contrary, condensation of all electron pairs in the ground state is actually favored in preference to distributions with some pairs in higher states. Thus all the electron pairs in the superfluid or ground state behave as a single matter wave with frequency $v = W/h$, W being the ground state energy and h Planck's constant. The drift velocity of electrons thus locked together may experience local perturbations due to scattering but the cohesion results in an unimpeded supercurrent.

---

[70] F. Seitz, "The Modern Theory of Solids." McGraw-Hill, New York, 1940.

With this picture of the two superconductors shown in Fig. 4.17, each constituted of a single macroscopic matter wave coherent throughout the material, we can consider the tunneling of electron pairs through the thin insulating junction.[71,72] If we connect a battery between the two superconductors we introduce a potential-energy difference $2qV_0$ between electron pairs on either side of the junction. There will be a corresponding frequency difference between the wave functions given by

$$\nu_0 = 2qV_0/h. \qquad (4.47)$$

The waves will thus change in phase relative to each other in a continuous way. If the waves are in phase at time $t = 0$, then the phase difference thereafter will be $\nu_0 t$, that is, the relative phase will change linearly with time. By the process of quantum-mechanical tunneling, electron pairs can penetrate the junction region. If, after tunneling through the junction, the phase of the pair exactly matches that of the wave function it will "fit in" with the matter wave, that is, it will have a large probability of transmission. If, on the other hand, a pair arrives out of phase it will have a large probability of reflection. This, then, is the reason why a dc potential difference across the junction gives rise to an alternating flow of current. The potential difference applied across the junction gives a continuously increasing relative phase and so a periodic occurrence of conditions favorable to electron pair transmission. By this process, a dc voltage gives rise to an alternating supercurrent, and, associated with this, a radiation field. Conversely, a radiation field can induce dc effects.

A simple analytic solution of a two-superconductor system coupled via pair tunneling with a potential difference V applied across the junction, is given by Feynman et al.[66] Wave functions of the form $Ae^{i\theta_A}$ and $Be^{i\theta_B}$ are found for the time-dependent Schrödinger equation. These solutions show that the tunneling current has the form

$$I = I_0 \sin(\theta_B - \theta_A), \qquad (4.48)$$

where I_0 is a constant depending on the degree of coupling between the superconductors. The phase shift is related to voltage by

$$\dot{\theta}_B - \dot{\theta}_A = 2qV/\hbar, \qquad (4.49)$$

[71] B. D. Josephson, *Phys. Lett.* **1**, 251–253 (1962); *Rev. Mod. Phys.* **36**, 216–220 (1964).
[72] B. D. Josephson, *Advan. Phys.* **14**, No. 56, 419–451 (1965).

4.5. THE JOSEPHSON JUNCTION

that is,

$$\theta_B - \theta_A = \delta_0 + (2q/\hbar) \int V \, dt, \qquad (4.50)$$

where the constant of integration δ_0 is the phase difference between the wave functions at time $t = 0$.

Equations (4.48) and (4.50) describe the Josephson effects. For $V = 0$, we have the dc Josephson effect, that is, the existence of a current between $\pm I_0$ (depending on the value of δ_0) for zero voltage across the junction. For constant voltage ($V = V_0$), the phase difference $\theta_B - \theta_A$ increases linearly with time, and the current oscillates sinusoidally with time. This is the ac Josephson effect.

In the radiation detector, a dc voltage V_0 as well as the high-frequency radiation voltage $V_1 \cos(\omega t + \phi)$ are applied to the junction

$$V = V_0 + V_1 \cos(\omega t + \phi).$$

For this voltage, Eq. (4.50) gives

$$\theta_B - \theta_A = \delta_0 + (2qV_0 t/\hbar) + (2qV_1/\hbar\omega) \sin(\omega t + \phi).$$

Substitution into Eq. (4.48) gives

$$I = I_0 \sin[\delta_0 + \omega_0 t + (\omega_0 V_1/\omega V_0) \sin(\omega t + \phi)], \qquad (4.51)$$

where $\omega_0 = 2qV_0/\hbar$.

Now let us simplify by assuming $V_1 \ll V_0$. With this approximation we can expand the sine function to obtain

$$I = I_0[\sin(\delta_0 + \omega_0 t) + (\omega_0 V_1/\omega V_0) \sin(\omega t + \phi) \cos(\delta_0 + \omega_0 t)]. \qquad (4.52)$$

The first term on the right-hand side has no dc component, but the second term has. When the wave frequency ω equals ω_0, the second term gives a dc current proportional to the amplitude of the radiation field. The relation between frequency and junction voltage, Eq. (4.47), gives a tuning sensitivity $2q/h$ or 483.6 MHz μV^{-1}.

More generally, the solution of Eq. (4.51) can be written as a Bessel function expansion:

$$I = I_0 \sum_{n=-\infty}^{\infty} (-1)^n J_n(\omega_0 V_1/\omega V_0) \sin[(\omega_0 - n\omega)t - n\phi + \delta_0]. \qquad (4.53)$$

When $\omega_0 = n\omega$, the current has a dc component given by

$$I_{dc} = I_0(-1)^n J_n(nV_1/V_0) \sin(\delta_0 - n\phi). \qquad (4.54)$$

We see that the dc current may be varied between the limits $\pm I_0 J_n(\omega_0 V_1/\omega V_0)$, and that its sign is determined by the relation of the phase ϕ to the relative pair phase δ_0. Furthermore, as we increase the dc voltage across the junction ω_0 increases and passes through harmonics of ω, giving rise to a succession of steps in the I-V curve. Thus the presence of applied radiation is detected by the step structure that is observed on the I-V curve, and the amplitude of the current in any one of these steps—including that at zero voltage—varies as the radiation power is applied.

We have described only some of the essential features of the Josephson effect, assuming the radiation frequency to be well below the superconducting energy gap (see Chapter 7). There are many interesting properties that we cannot develop here. These include the effect of a magnetic field, the behavior at photon frequencies in the vicinity of the energy gap, the occurrence of maxima of response of some junctions near such frequencies as $\frac{1}{2}$, $\frac{1}{4}$, $\frac{3}{2}$, etc., of the gap frequency while, in others, minima occur in these regions. There are also differences between the behavior of point–contact and thin-film junctions. The reader interested in further details will find the papers of Shapiro,[73] Langenberg et al.,[74] Parker et al.,[75] and Silver[76] of value.

The I-V characteristic observed with a point–contact junction is shown in Figs. 4.18 and 4.19, and in Fig. 4.20 we illustrate the mode of operation used by Grimes et al.,[77] for the detection of broad-band incoherent radiation. The I-V curve of the point–contact shows a zero-voltage current and, at the end of the zero-voltage region, a smooth and continuously increasing current (there are differences between this curve and those for thin-film junctions which Grimes et al.,[77] and Werthamer and Shapiro[78] discuss). They irradiated the junction with modulated radiation from a mercury arc lamp, and with a Fourier transform spec-

---

[73] S. Shapiro, *Symp. Phys. Superconducting Devices, Virginia Univ.*, April 28–29, 1967.
[74] D. N. Langenberg, D. J. Scalapino and B. N. Taylor, *Proc. IRE* **54**, 560–575 (1966).
[75] W. H. Parker, D. N. Langenberg and A. Denenstein, *Phys. Rev.* **177**, 639–664 (1969).
[76] A. H. Silver, *IEEE J. Quantum Electron.* **4**, 738–744 (1968).
[77] C. C. Grimes, P. L. Richards and S. Shapiro, *Phys. Rev. Lett.* **17**, 431–433 (1966); *J. Appl. Phys.* **39**, 3905–3915 (1968).
[78] N. R. Werthamer and S. Shapiro, *Phys. Rev.* **164**, 525 (1967).

4.5. THE JOSEPHSON JUNCTION

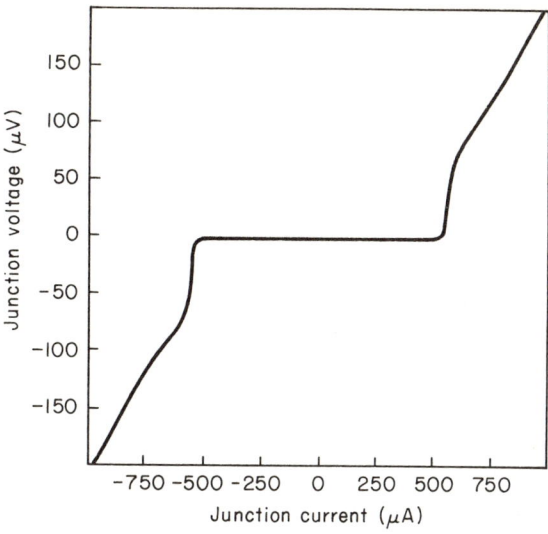

FIG. 4.18. Current–voltage curve for an Nb–Nb point-contact Josephson junction at 4.2°K (after Grimes et al.[77]).

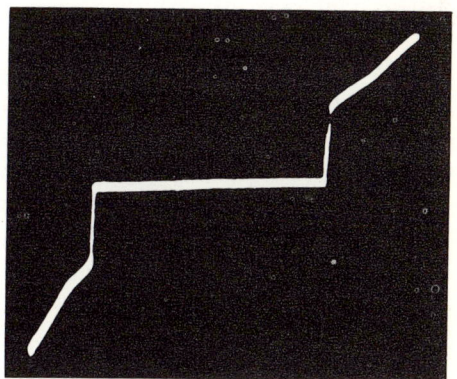

FIG. 4.19. Oscilloscope trace showing the current–voltage curve for an Nb–Nb point-contact Josephson junction at 4.2°K.

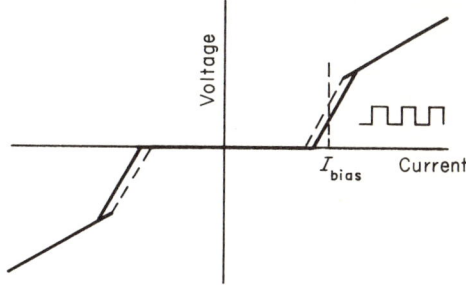

FIG. 4.20. Schematic I–V curve showing how applied radiation changes the zero-voltage current, so that a chopping beam results in a modulated voltage. In the absence of radiation the full curve is observed, but the application of radiation reduces the maximum zero-voltage current and shifts the curve to the dashed position. A constant-current external source sets the operation at a point of high slope.

trometer were able to measure the spectral response. In the spectrometer broad-band radiation of very low power fell on the junction so no current-voltage steps could be expected, but rather the initial decrease of zero-voltage current in the presence of radiation was observed.

The Bell Telephone Laboratory workers used point–contact junctions because their low junction capacitance permits better coupling to radiation at the high frequencies of the far-infrared; they investigated a number of different metals. These they biased with an external source of constant current to a point of high differential resistance. Then, as the modulated radiation caused the I–V curve to shift, the detector output appeared as a similarly modulated voltage, which was rectified and amplified with a lock-in amplifier and plotted on a chart recorder. The spectrometer output interferograms gave the spectral response curves shown in Fig. 4.21.

Monochromatic radiation of sufficient intensity to generate a number of constant-voltage steps in the I–V curve has been used. A record of one such observation is shown in Fig. 4.22, where, with a power level of only 10^{-9} W and a frequency of 150 GHz—higher than half the energy gap, several steps are recorded. Experiments with coherent radiation have been taken above the superconducting energy gap. The 311-μm line of the HCN laser is above the energy gap of niobium, but it has generated steps in the I–V characteristic of a Nb–Nb junction similar to those generated with microwaves. It is clear from the literature that the detector will function at frequencies many times the gap frequency, but, of course, in this regime normal state electrons will be expected to be significant in the tunneling process. It is found that the interval between steps is equal to the expected value for the applied frequency.

Josephson junctions are cryogenic, and are relatively free from noise. Experimental work[77,79] at a number of laboratories has shown the capability of detecting 10^{-14} W or less with a postdetector integration time of 1 sec. Junction capacitance is small, and the response time of the junction is very short indeed. Shapiro[73] has reported the faithful reproduction of pulses with 10 nsec rise times. Richards and Sterling[79a] have used a junction in a cavity and have reported an NEP of about 5×10^{-15} W Hz$^{-1/2}$, while McDonald et al. have extended their frequency response to 78 μm[79b].

[79] Phys. Today **23**, 55–56 (1970).
[79a] P. L. Richards and S. A. Sterling, Appl. Phys. Lett. **14**, 394–396 (1969).
[79b] D. G. McDonald, A. S. Risley, J. D. Cupp and K. M. Evenson, Appl. Phys. Lett. **20**, 296–299 (1972).

4.5. THE JOSEPHSON JUNCTION

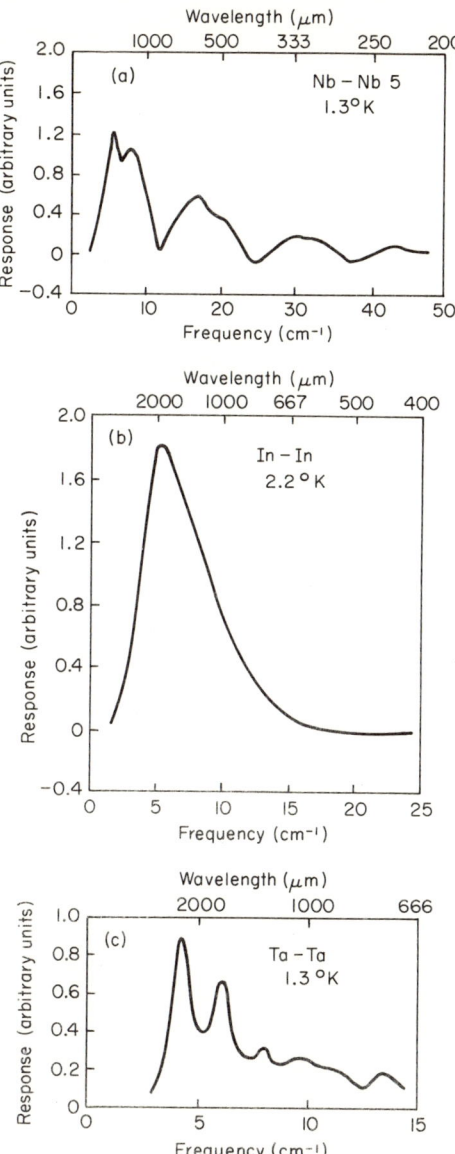

FIG. 4.21. Far-infrared spectral response of (a) Nb–Nb, (b) Ta–Ta, and (c) In–In point-contact Josephson junctions (after Grimes et al.[77]).

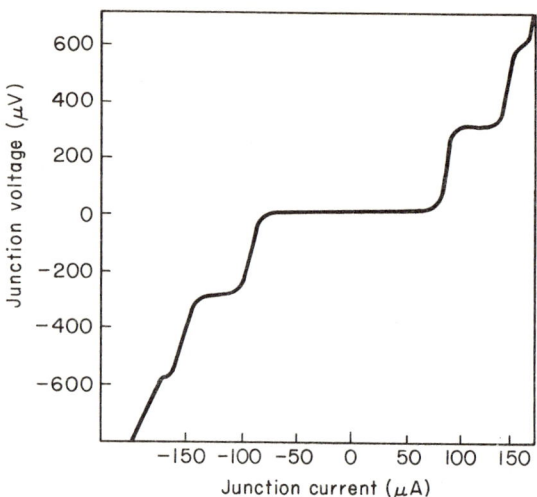

FIG. 4.22. *I–V* curve for an In–In point–contact Josephson junction at 1.3°K in the presence of 10^{-9} W of radiation at 150 GHz (after Grimes *et al.*[77]).

4.6. Limits of Detection Set by Random Fluctuations

Energy and matter are constituted of elementary units (photons, phonons, electrons, molecules, etc.), and there is a natural limit of measurement set by the detection of the statistical effects of these small corpuscular units.[80] Thermal radiation from the general surroundings of the detector element ("background" radiation) and from the detector element itself is composed of photons whose discrete nature will be particularly manifest at low intensities. This radiation arises from the Brownian motion of electrons in material bodies, and it is not surprising that electric field and hence voltage fluctuations will occur across such bodies as a consequence of the random electron movements. Further, thermal lattice vibrations are quantized, so the temperature of a body fluctuates about a mean value; this can lead to changes in resistance and so voltage fluctuations in a detector circuit. The effect of the corpuscular nature of radiation and matter will then be spontaneous fluctuations or noise tending to mask the output signal of a detector caused by a wanted signal. The distinguishability of a signal against this noise can be indicated by the signal-to-noise ratio, this being the ratio of the root mean square amplitude of the signal to the rms noise amplitude.

[80] A. Einstein, *Phys. Z.* **10**, 185–193 (1909).

4.6. LIMITS SET BY RANDOM FLUCTUATIONS

TABLE 4.1. Sources of Noise Significant in Far-Infrared Detection

Noise classification	Kind of noise	Physical mechanism	Detectors concerned
Radiation noise	Fluctuations in background radiation incident on the detector		Affects all detectors
	Fluctuation in radiation emitted by the detector	Bose–Einstein fluctuations of radiation photons	
	Fluctuations in the signal received by the detector[a]		
Detector noise	Johnson noise	Thermal agitation of current carriers	All detectors
	Current noise	Resistance fluctuations, surface contacts, etc.	Bolometers, photoconductive detectors
	Temperature noise	Temperature fluctuations	Thermal detectors
	Generation–recombination noise	Fermi–Dirac fluctuations of current carriers	Photoconductive detectors, semiconductor bolometers
Amplifier noise	Johnson noise	Thermal agitation of current carriers	Affects all detectors
	Flicker noise	Fluctuations of work function	
	Shot noise	Random thermionic emission of electrons	

[a] Significant only at optical and infrared frequencies.[81]

[81] T. P. McLean and E. H. Putley, *Roy. Radar Estab. J.* No. 52, 5–34 (1965).

It is convenient to consider noise in three categories: fluctuations in radiation fields, fluctuations within the detector element itself, and noise originating in the postdetector amplifier. As indicated in Table 4.1 these categories cover many mechanisms of noise generation.

4.6.1. Radiation Noise

The fluctuations in a stream of photon energy are described by Bose–Einstein statistics.[82,83] The distribution of photons as a function of frequency for radiation in thermal equilibrium in an enclosure is derived in books on statistical mechanics.[82] It is found that the average number of photons in the ith mode or quantum state, with energy $h\nu$, is

$$1/(e^{h\nu/kT} - 1).$$

If there are $M\,d\nu$ modes in a narrow frequency interval of width $d\nu$ about ν, the mean volume density of photons $\bar{n}_\nu$ is the sum of the populations of the $M\,d\nu$ states, that is,

$$\bar{n}_\nu = M\,d\nu/(e^{h\nu/kT} - 1). \tag{4.55}$$

By counting the number of normal modes or resonances in which thermal equilibrium radiation is stored in the cavity enclosure[84] one finds

$$M\,d\nu = 8\pi\nu^2\,d\nu/c^3. \tag{4.56}$$

Equation (4.55) is then recognized as Planck's radiation law.

The mean square statistical fluctuation in the number density is[83]

$$\overline{(\Delta n_\nu)^2} = \bar{n}_\nu \{1 + [1/(e^{h\nu/kT} - 1)]\}$$
$$= \bar{n}_\nu e^{h\nu/kT}/(1 - e^{h\nu/kT}). \tag{4.57}$$

The fluctuations in a stream of photons being received by a detector in thermal equilibrium with the radiation can be obtained from the foregoing statistics. The photons in an equilibrium enclosure move isotropically in all directions, so the number moving towards any point within an elemental solid angle $d\Omega = \sin\theta\,d\theta\,d\phi$ is $\bar{n}_\nu\,d\Omega/4\pi$. From a column of length c in the direction θ based obliquely on an area A, there are

---

[82] R. H. Fowler, "Statistical Mechanics." Cambridge Univ. Press, London and New York, 1936.
[83] W. B. Lewis, *Proc. Phys. Soc. London* **59**, Pt. 1, 34–40 (1947).
[84] G. Joos, "Theoretical Physics." Blackie, Glasgow and London, 1949.

4.6. LIMITS SET BY RANDOM FLUCTUATIONS

$(\bar{n}_\nu \, d\Omega/4\pi)cA \cos \theta$ photons which cross A in unit time. The mean square fluctuation in this flux of photons is, from expression (4.57),

$$\frac{2\nu^2 \, d\nu A}{c^2} \frac{e^{h\nu/kT}}{(e^{h\nu/kT} - 1)^2} \cos \theta \, d\Omega. \tag{4.58}$$

If the detector optics restricts the incident photons to within a cone of semiangular aperture α, the total mean square fluctuation is

$$\overline{\Delta J_\nu^2} = \frac{2A}{\lambda^2} \, d\nu \, \frac{e^{h\nu/kT}}{(e^{h\nu/kT} - 1)^2} \int_{\phi=0}^{2\pi} \int_{\theta=0}^{\alpha} \cos \theta \sin \theta \, d\theta \, d\phi$$

$$= \frac{2A}{\lambda^2} \pi \sin^2 \alpha \, \frac{e^{h\nu/kT}}{(e^{h\nu/kT} - 1)^2} \, d\nu. \tag{4.59}$$

Now the effect of the absorption of the individual photons in the stream can be regarded as producing instantaneous delta function electrical output signals. Each has a continuous and uniform Fourier spectrum extending from zero frequency, with a part within the bandwidth of the postdetector amplifier. It can be shown[2] that the noise fluctuations within a bandwidth Δf at the low frequency f of the tuned amplifier is

$$\overline{\Delta J_\nu(f)^2} = \frac{4A}{\lambda^2} \pi (\sin^2 \alpha) \frac{e^{h\nu/kT}}{(e^{h\nu/kT} - 1)^2} \Delta \nu \, \Delta f.$$

The rms fluctuation of this unpolarized radiation is

$$\overline{(\Delta J_\nu(f)^2)}^{1/2} = \left[4\pi(\sin^2 \alpha) \frac{A}{\lambda^2} \frac{e^{h\nu/kT}}{(e^{h\nu/kT} - 1)^2} \Delta \nu \, \Delta f \right]^{1/2}. \tag{4.60}$$

This expression is applicable to detectors which respond to the number of received photons rather than to energy, e.g., impurity photoconductivity. For energy detectors such as bolometers, Golay cells, and free-carrier photoconductors, the corresponding expression for the rms power fluctuation is

$$\overline{(\Delta P_\nu(f)^2)}^{1/2} = h\nu \left[4\pi(\sin^2 \alpha) \frac{A}{\lambda^2} \frac{e^{h\nu/kT}}{(e^{h\nu/kT} - 1)^2} \Delta \nu \, \Delta f \right]^{1/2}. \tag{4.61}$$

Equation (4.61), when integrated over the range of frequencies ν admitted to a detector, and including a multiplying factor equal to the emissivity of the detector surface (which may be frequency dependent), gives the noise power fluctuations due to incident background radiation. The expression is appropriate for detectors sensitive to power in a received beam of radiation. For example, for a thermal detector with a sensing element black and responsive over the entire spectrum, and receptive to incoming radiation over 2π steradians, we have

$$\overline{(\Delta P(f))^2} = \frac{4\pi h^2 A \, \Delta f}{c^2} \int_0^\infty \frac{\nu^4 e^{h\nu/kT}}{(e^{h\nu/kT} - 1)^2} \, d\nu$$
$$= 8k\sigma T^5 A \, \Delta f, \qquad (4.62)$$

where $\sigma = 5.670 \times 10^{-12}$ W cm^{-2} deg^{-4} is Stefan's constant, and $k = 1.38 \times 10^{-23}$ joule deg^{-1} is Boltzman's constant. Equation (4.62) takes account of fluctuations in the incoming radiation only. If the detector element is at the same temperature as the background, there will be similar, but independent, fluctuations due to radiation emission by the element. This will increase the fluctuations to

$$\overline{(\Delta P(f))^2} = 16k\sigma T^5 A \, \Delta f. \qquad (4.63)$$

The rms power fluctuation is

$$[\overline{\Delta P(f)^2}]^{1/2} = 3.56 \times 10^{-17} T^{5/2} (A \, \Delta f)^{1/2}. \qquad (4.64)$$

More generally, if the element is not in thermal equilibrium with the surroundings, the power fluctuations can be written

$$\overline{\Delta P(f)^2} = 8\varepsilon' k\sigma (T_1^5 + T_2^5) A \, \Delta f, \qquad (4.65)$$

where T_1 and T_2 are the temperatures of the element and the surroundings, and we have included the emissivity ε' of the element for generality.

For an ideal[†] power-sensing detector, Eq. (4.61) gives the NEP. Ideal detection can be approached in a detector with sufficiently high responsivity to raise the input signal and radiation noise above other noise sources in the detector and amplifier. For an ideal receiver in thermal equilibrium with the radiation field and responsive over the entire spectrum, we can calculate the minimum detectable power. At a temperature of 300°K, a detector with unit area ($A = 1$ cm^2), unity emissivity and a bandwidth $\Delta f = 1$ Hz, the minimum detectable signal is 5.5×10^{-11} W. In practice, the radiation noise will be somewhat less than this. It will be decreased by cooling the detector and its holder, by limiting the angular aperture of the detector optics, and by reducing the frequency band of the background noise with selective filters. For a detector cooled to

---

[†] An "ideal" receiver is one free from detector and amplifier noise.

4.6. LIMITS SET BY RANDOM FLUCTUATIONS

well below room temperature, with a filter which absorbs all radiation below 100 μm, and with an angular aperture restricted to about 0.2 steradians, the minimum detectable signal will be about 3×10^{-13} W, for a filter with a 10-% bandwidth centered at 500 μm, it is about 2×10^{-14} W. These figures are reasonably consistent with the actual performances of bolometers, Golay cells, and free-carrier photoconductors, in measurements of continuous monochromatic radiation for which the narrow (1 Hz) output bandwidth is suitable. Differences between these figures and actual performances may be due to amplifier noise and Johnson noise in the detector element. These additional sources of noise will be discussed subsequently.

It is of interest to consider at this stage the radiation noise in a superheterodyne indium antimonide detector. In such a detector there will be a relatively broad band of noise incident on the detector. The exact bandwidth might be set by the bandpass of a filter preceeding the detector. However, two bands of radiation noise, each of bandwidth B equal to that of the intermediate frequency amplifier, carry power P_B which will be amplified above noise outside these bands by a factor $(2P_L/P_B)^{1/2}$ in the superheterodyne conversion process. The detector output from this noise will be $R(2P_L P_B)^{1/2}$, where P_L is the local oscillator power and R the detector responsivity. The two bands are centered about the signal frequency ν_S and the image frequency $2\nu_L - \nu_S$, for these mix with the local oscillator signal of frequency ν_L to yield a difference frequency $\nu_L - \nu_S$ equal to the intermediate frequency.

The noise power in this narrow effective bandwidth is found from Eq. (4.61) by taking $\Delta\nu = 2B$. If the radiation is received by an antenna of aperture area $A \gg \lambda$, the free-space wavelength, the solid angle for reception will be defined essentially by the angular spread of the principal diffraction lobe of the aperture. The semiangular aperture set by such considerations is given by $\sin \alpha \sim \lambda/A^{1/2}$, if we neglect a small factor depending on the shape of the aperture.[47] If we assume that only one polarization is received, the radiation noise power in a superheterodyne receiver is

$$P_B = 2h\nu B e^{h\nu/2kT}/(e^{h\nu/kT} - 1). \qquad (4.66)$$

When $h\nu \ll kT$,

$$P_B = 2kTB. \qquad (4.67)$$

This last expression applies to radio frequencies where, if the image band is suppressed, we obtain the familiar expression kTB for the noise power within the bandwidth B of the intermediate frequency amplifier of a perfect receiver. Such a receiver with $B = 1$ Hz operating at room temperature, has an NEP of 4×10^{-21} W Hz$^{-1/2}$. In actual receivers the noise power is larger than that given by Eqs. (4.66) and (4.67) by a factor F, the noise figure of the receiver. This factor takes account of noise arising in the local oscillator, the mixer, and amplifiers. At centimeter wavelengths, low-noise masers can be used to amplify the signal and radiation noise prior to detection to give noise figures close to unity. At short millimeter wavelengths, where low-noise preamplifiers are not yet available, noise figures are somewhat larger. In the superheterodyne receiver designed by Meredith and Warner[9] for operation at 2-mm wavelength, the noise figure is 20 dB. A noise figure comparable with this has been predicted by Putley[85] for a 1-mm indium antimonide detector with a 1-mW local oscillator. In this detector the main sources of fluctuations are due to the postdetector amplifier and Johnson noise in the detector element. These additional sources are discussed subsequently.

The noise in a video radio receiver can be found from Eq. (4.61) by a similar procedure of taking $h\nu \ll kT$ and taking similar account of the aperture diffraction pattern. In this case $\Delta\nu$ is not restricted to $2B$, of course. The result is clearly

$$P_B = kT(2B\, \Delta\nu)^{1/2}. \tag{4.68}$$

Again a factor must be included to take account of amplifier noise.

The limits of detectability for detectors which operate by photoionization is best derived from Eq. (4.60) for the fluctuations in the number of background photons received. The minimum detectable power, or noise equivalent power, for an ideal photoionization detector is set by the criterion that the number of signal quanta n_s must equal the rms fluctuation in the number from the background $\overline{[(\Delta J_\nu(f))^2]}^{1/2}$. For a signal frequency ν_s, this is

$$(\text{NEP})_i = h\nu_s n_s = h\nu_s \overline{[(\Delta J_\nu(f))^2]}^{1/2}. \tag{4.69}$$

The suffix i denotes an ideal detector. Substituting from Eq. (4.60), we obtain

$$(\text{NEP})_i = \frac{2\pi^{1/2}}{ch^{1/2}}\, \nu_s(kT)^{3/2}(A\,\Delta f)^{1/2}\left[\int_{x_0}^{\infty} \frac{x^2 e^x}{(e^x - 1)^2}\, dx\right]^{1/2}, \tag{4.70}$$

[85] E. H. Putley, *Proc. IEEE* **54**, 1096–1098 (1966).

where $x = h\nu/kT$, the integration being taken from the photoionization threshold frequency ν_0, that is, from $x_0 = h\nu_0/kT$. Here the quantum efficiency (the fraction of incident photons that are absorbed by the detector) between ν_0 and infinity is taken as unity, and radiation is incident over a hemisphere. We see immediately from this expression that the noise equivalent power is a minimum when operated near its threshold.

In addition to the noise power given by Eq. (4.70), there is another fundamental source of noise in photoconductors due to fluctuation in the rate of recombination. As discussed in Section 4.6.4, the effect of this process can be accounted for by multiplying Eq. (4.70) by a factor $2^{1/2}$.

4.6.2. Temperature Noise

In this section we present a formulation of noise applicable to thermal detectors in thermodynamic equilibrium with the environment. The formulation is in terms of temperature fluctuations of a detector element coupled by a general conductance G to the surroundings; it is called "temperature noise".[86] These fluctuations are an inherent feature of the statistical nature of thermodynamic phenomena. The element has a statistical interchange of energy with its surroundings, and hence random temperature fluctuations which set a lower limit to the detectable radiation. In Section 4.6.1 we considered the random energy interchange via radiation. The results of this section overlap with those previous results when the conductance G is entirely radiative. However, the results of Section 4.6.1 for this particular case are more general because they are not restricted to equilibrium of the temperature of the element with its environment.

The connection between random power exchange and temperature fluctuations is given by the energy balance Eq. (4.6). If the instantaneous component at frequency f of the power exchange and the corresponding temperature fluctuation are represented by ΔP_f and ΔT_f, respectively, we can see from Eq. (4.11) that

$$\overline{(\Delta T_f)^2} = \overline{\Delta P_f^2}/(G^2 + 4\pi^2 C^2 f^2), \qquad (4.71)$$

where C is the thermal capacity of the element and G is the thermal conductance between the element and its surroundings.

---

[86] J. T. Houghton and S. D. Smith, "Infra-Red Physics." Oxford Univ. Press (Clarendon), London and New York, 1966.

Again, as in Section 4.6.1, we regard the arrival of quanta as producing delta function energy changes at random intervals, and hence giving a flat frequency spectrum. We can thus write

$$\overline{\Delta P_f^2} = K\,\Delta f, \tag{4.72}$$

where K is a constant. The total mean square temperature fluctuation is now given by the integration of Eq. (4.71) over all frequencies. If we treat C and G as remaining constant throughout the spectrum, we obtain

$$\overline{(\Delta T)^2} = \int_0^\infty \frac{K\,df}{G^2 + 4\pi^2 C^2 f^2} = \frac{K}{4CG}. \tag{4.73}$$

With this result established let us turn to other considerations of the temperature fluctuations. We imagine the energy of the entire thermodynamic system (the element and its surroundings) as composed of a large number of energy subintervals. If W_n is the energy in the nth energy interval, Boltzmann statistics gives for the mean energy $\bar{w}$ and the mean square energy of the system

$$\bar{w} = (1/Z) \sum_n w_n \exp(-\beta_n w_n), \tag{4.74}$$

$$\overline{w^2} = (1/Z) \sum_n w_n^2 \exp(-\beta w_n), \tag{4.75}$$

where Z is the partitition function given by

$$Z = \sum_n \exp(-\beta w_n) \tag{4.76}$$

and

$$\beta = 1/kT. \tag{4.77}$$

From Eqs. (4.74) and (4.76), we can write

$$\bar{w} = -(1/Z)\,\partial Z/\partial \beta \tag{4.78}$$

and

$$-\partial \bar{w}/\partial \beta = (1/Z)\,\partial^2 Z/\partial \beta^2 - (1/Z^2)\,(\partial Z/\partial \beta)^2. \tag{4.79}$$

From Eq. (4.75),

$$\overline{w^2} = (1/Z)\,\partial^2 Z/\partial \beta^2. \tag{4.80}$$

Equations (4.78)–(4.80) enable us to obtain the mean square energy fluctuation

$$\overline{(\Delta w)^2} = \overline{w^2} - (\overline{w})^2 = -\partial \overline{w}/\partial \beta = kT^2C, \qquad (4.81)$$

where the thermal capacity of the system is

$$C = \partial \overline{w}/\partial T.$$

The relation between the mean square energy and temperature fluctuations is, of course,

$$\overline{(\Delta w)^2} = C^2 \overline{(\Delta T)^2}.$$

When this is taken in conjunction with Eq. (4.81), we obtain

$$\overline{(\Delta T)^2} = kT^2/C. \qquad (4.82)$$

We now compare the results formulated in Eqs. (4.73), (4,82), and (4.72). The mean square power fluctuation of a thermal detector in equilibrium at temperature T with its surroundings, to which it is coupled by a conductance G, is

$$\overline{\Delta P_f^2} = 4GkT^2 \, \Delta f. \qquad (4.83)$$

This result is appropriate to detectors with sensing elements thermally coupled to the surroundings by any means. As a special case, assume the coupling is entirely through radiation and that the element is black. We saw in Section 4.2 that for this case

$$G = 4\varepsilon' A \sigma T^3$$

where $\varepsilon' = 1$ for a blackbody. Hence

$$\overline{\Delta P_f^2} = 16k\sigma T^5 A \, \Delta f.$$

This is the same as Eq. (4.63) derived in Section 4.6.1.

4.6.3. Johnson Noise

Due to the random motions of electrons within a resistor, fluctuations in voltage arise between its terminals. These fluctuations are called Johnson noise. It is closely related to blackbody radiation, and its spectrum can be determined from considerations of thermal equilibrium between the two.[17] Suppose we have an antenna matched to a coaxial line which in turn is matched to a resistor r. The antenna is located within an enclosure with black walls at temperature T, the same temperature as that of the resistor r at the other end of the line. Radiation from the blackbody is picked up by the antenna and transmitted down the line, where it is absorbed by the resistor. Johnson noise from the resistor passes down the line towards the antenna where it is radiated to the blackbody. If these two powers were unequal, the resistor would either lose or gain energy, resulting in a violation of the second law of thermodynamics. Thus, the available Johnson noise power from a resistor must be equal to the power picked up by an antenna pointed at a blackbody at the same temperature.

On this basis, we can use some of the results of Section 4.6.1 to get a quantitative expression for Johnson noise. From Eq. (4.55), the number of photons per unit volume with frequencies between ν and $\nu + d\nu$ is

$$\bar{n}_\nu = 8\pi\nu^2 \, d\nu / c^3 (e^{h\nu/kT} - 1).$$

If the average absorption cross section of the antenna is σ_{av}, the blackbody power received by the antenna is

$$P = \frac{8\pi\nu^2 \, d\nu \, h\nu\sigma_{\mathrm{av}}}{c^2(e^{h\nu/kT} - 1)}. \tag{4.84}$$

It can be shown[87] that the average absorption cross section of a dipole radiator (or receiver) is

$$\sigma_{\mathrm{av}} = \lambda^2/4\pi. \tag{4.85}$$

Hence,

$$P = \frac{2h\nu \, d\nu}{e^{h\nu/kT} - 1}. \tag{4.86}$$

Now consider the Johnson noise power transferred in the opposite direction. If V_J is the rms noise voltage across the matched terminating resistor r, we know from the power transfer theorem that the power transferred to the antenna resistance (the radiation resistance) is $V_J^2/2r$. Equating this to the received power [Eq. (4.86)], we find the Johnson noise voltage to be

$$V_J = \left(\frac{4rh\nu \, d\nu}{e^{h\nu/kT} - 1} \right)^{1/2}. \tag{4.87}$$

---

[87] S. Ramo and J. R. Whinnery, "Fields and Waves in Modern Radio." Wiley, New York, 1953.

4.6. LIMITS SET BY RANDOM FLUCTUATIONS

For frequencies such that $h\nu \ll kT$,

$$V_J = (4kTr\, d\nu)^{1/2}. \quad (4.88)$$

This derivation reveals the close relationship between Johnson noise and radiation noise, and the difference. The former is based on Eq. (4.55) for the number density of photons in the radiation field, while the latter is based on the fluctuations in this number as given by Eq. (4.57). Thus Johnson noise *is* the power in the radiation field, whereas radiation noise represents the fluctuations in the power in the radiation field.

4.6.4. Recombination Noise

The fluctuations in the rate of arrival of photons described in Section 4.6.1 cause fluctuations in the rate of generation of photoelectrons into the conduction band, and hence in the conductivity. Random ionization effects, however, are not the only cause of conductivity fluctuations; the rate of recombination also fluctuates, and this gives additional noise which is called recombination noise. Recombination noise has been studied by Van Vliet,[88] Petritz[89] and Alkemade.[90] By application of the principle of detailed balancing, it has been found that, for photoconductive detectors, recombination noise is equal to that produced by background radiation. It is thus given by Eq. (4.70). In other words, the combined effect of radiation fluctuations on generation and of recombination noise in a photoconductor is given[29] by Eq. (4.70) multiplied by $2^{1/2}$. This result applies whether the detector element is at the same or at a different temperature from the background.

4.6.5. Current Noise

Johnson noise exists between the terminals of a resistor in thermal equilibrium with its surroundings. When current flows, there is no longer thermal equilibrium, and voltage fluctuations, additional to Johnson noise, appear across the resistor.[20] The origins of this current noise are not completely understood, but they are certainly of a less fundamental nature than Johnson noise in that they are, in principle, avoidable. Nevertheless, in practice the phenomenon is troublesome at low frequencies (particularly below 1 kHz) and in certain materials. In pure metals cur-

---

[88] K. M. Van Vliet, *Proc. IRE* **46**, 1004–1018 (1958).
[89] R. L. Petritz, *Proc. IRE* **47**, 1458–1467 (1959).
[90] C. T. J. Alkemade, *Physica* (*Utrecht*) **24**, 1029–1034 (1959).

rent noise is quite small, but in semiconductors (possibly excepting indium antimonide) and resistors made from compressed powder and deposited films it is most marked.

Empirical studies of current noise show that the mean square voltage fluctuations vary approximately as the inverse first power of the frequency f, and are approximately proportional to the square of the steady current I, that is,

$$\overline{V_i^2} \simeq C(I^\alpha/f^\beta)\, \Delta f, \qquad (4.89)$$

where $\beta \simeq 1$, $\alpha \simeq 2$, and C is a constant. From measurements between 20 and 10 Hz on small carbon resistors carrying about 100 μA, Templeton and MacDonald[91] have found the noise voltage V_i to be given approximately by

$$\overline{V_i^2} = Ar^2(I^2/f)\, \Delta f, \qquad (4.90)$$

where r is the resistance value and A is a constant which lies between 3×10^{-11} and 3×10^{-12}.

Theories of current noise are generally based on fluctuations in resistance at contact surfaces caused by diffusion mechanisms.[92] The surface may be at a contact with the metal electrodes or between microcrystals of the material. The presence of dislocations within single crystals and the state of the surface of a semiconductor are contributing factors.

4.6.6. Shot Noise

Shot noise is a fluctuation arising in electron tubes due to the finite size of the charge on an electron. It was first discussed by Schottky[93] in 1918 in connection with current flow in temperature-limited diodes. When a steady current I is carried by a stream of *free* electrons, a discrete number N of charge carriers pass a given point in t seconds, and these quantities are defined as being related by

$$I = qN/t, \qquad (4.91)$$

where q is the electronic charge. Schottky assumed that under tempera-

---

[91] I. M. Templeton and D. K. MacDonald, *Proc. Phys. Soc. London Sect. B* **66**, Pt. 8, 680–687 (1953).

[92] G. G. Macfarlane, *Proc. Phys. Soc. London* **59**, Pt. 3, 366–374 (1947); *Proc. Phys. Soc. London Sect. B* **63**, Pt. 10, 807–814 (1950).

[93] W. Schottky, *Ann. Phys. (Leipzig)* **57**, 541–567 (1918).

4.6. LIMITS SET BY RANDOM FLUCTUATIONS

ture-limited conditions the electrons are emitted by the cathode in a random way, so that during transit, and when they strike the anode, there is a fluctuation in N given by classical statistics. It is well known that the statistical deviation ΔN of a large random number N from its mean value $\bar{N}$ is given by

$$\overline{(\Delta N)^2} = \bar{N}. \tag{4.92}$$

The mean square fluctuation in the current is therefore

$$\overline{(\Delta I)^2} = \frac{q^2}{t^2}\overline{(\Delta N)^2} = qI/t. \tag{4.93}$$

Fourier analysis of these fluctuations gives the frequency spectrum of shot noise.[2,94] The result is

$$\overline{\Delta I(f)^2} = 2qI\,\Delta f. \tag{4.94}$$

This result is valid for frequencies up to the reciprocal of the transit time of electrons from the cathode to anode. At higher frequencies the noise decreases.[95]

Shot noise has a fundamental and unavoidable origin. It is a consequence of the fluctuations in the discrete (but large) number of charge carriers. It is proportional to the finite unit of charge and would approach zero were the fundamental unit of charge infinitesimally small.

Equation (4.94) applies to diodes operating under conditions of temperature-limited emission. When the cathode emission is space-charge limited, as it usually is in electron tubes, the fluctuations tend to be reduced by space-charge effects. It is found that under these more general conditions[20]

$$\overline{\Delta I(f)^2} = 2\Gamma^2 qI\,\Delta f, \tag{4.95}$$

where Γ is usually between 0.1 and 1. Further, in multielectrode tubes there is partition noise arising from the random effects in the interception of current by the various electrodes.

The noise fluctuations in an electron amplifying tube are commonly all lumped together and described in terms of an equivalent noise resistance which is imagined in series with the grid of the tube and in which is generated only Johnson noise. Thus to shot noise in the tube the

---

[94] J. R. Carson, *Bell Syst. Tech. J.* **10**, 374–381 (1931).
[95] S. S. Solomon, *J. Appl. Phys.* **23**, 109–112 (1952).

Johnson noise in the actual grid circuit resistance (amplified by the gain of the tube) is added. This net output noise is equated to the amplified Johnson noise in the equivalent noise resistance at the operating temperature of the amplifier.

4.6.7. Flicker Noise

In hot cathode electron tubes it is observed that at very low frequencies there is noise above that attributable to shot noise. This is connected with the nature and state of the cathode and is associated with variations in emission from various points on the cathode surface due to the presence of foreign atoms, to fluctuations in the diffusion of ions to the emitting surface, or some such processes.

The variation with current and frequency varies a great deal with the kind of tube, its gain, the cathode type, and whether it operates temperature or space-charge limited. The effect is most pronounced with oxide coated cathodes, and is much less for pure tungsten and thoriated tungsten. Smith et al.[22] have pointed out that, for a wide range of tubes suitable for use in the first stage of a low-frequency amplifier, flicker noise can be represented approximately by an equivalent voltage at the grid. The rms value is about 0.1 μV for a 1-Hz bandwidth at a frequency of 10 Hz and varies approximately as $f^{-1/2}$.

Thus, in very low frequency amplifiers, flicker noise can greatly exceed shot noise and Johnson noise in the grid circuit.

4.6.8. Detectivity

Detector performance is often expressed in terms of *detectivity*, which, in watts^{-1}, is defined as

$$D = 1/\text{NEP}. \tag{4.96}$$

The quantity D, for ideal thermal and photoconductive detectors, varies as $(A \, \Delta f)^{-1/2}$.

Because the areas of detector elements vary somewhat, D is not an entirely satisfactory index for purposes of comparing detectors. An index which reduces the performance to a standard area and bandwidth and called the *specific detectivity* or *detectivity "D star,"* is defined as

$$D^* = D(A \, \Delta f)^{1/2}. \tag{4.97}$$

It is the detectivity of a detector of unit area operating with a postdetector amplifier of unit bandwidth. Unfortunately, the performances of all detectors do not depend on area in the same way, so that the concept of specific detectivity must be used with care.

In Fig. 4.23 the specific detectivities of a number of far-infrared detectors is plotted. Curves for ideal photoconductive detectors are also given.

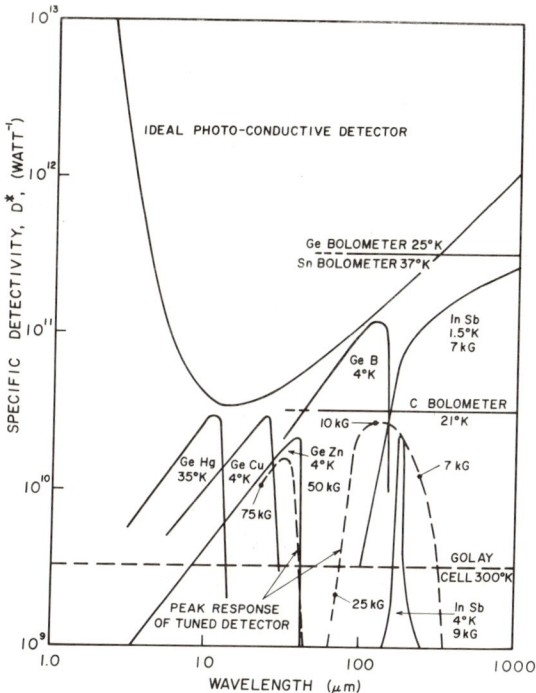

FIG. 4.23. Specific detectivity D^* of a number of detectors in the far-infrared (after Putley[29]).

4.6.9. Postdetector Amplifiers

This section will be restricted to a few general comments about the linking of a detector to the amplifier. For detailed discussion, including amplifier design, the interested reader is referred to Smith *et al.*[20]

Primary requirements of the postdetector circuit are that the amplifier noise should be less than that due to the detector, and that the bandwidth or time constants should be suitable for the particular requirements of response time. As we have seen, the amplifier noise can be represented by an equivalent noise resistance in series with the grid of the first tube. If the detector resistance is larger than this equivalent noise resistance, it can be connected directly to the amplifier input, and detector noise will exceed amplifier noise. Of course, if the resistance is very large (e.g., some impurity photoconductors have resistances ~1 MΩ or more) its product with the input capacity of the amplifier can give a time constant limiting the time response. The detector may have a very low resistance (e.g., in some thermocouples and bolometers, and for the free carried indium antimonide photodetector without a magnetic field, it can be 10 Ω or so, and in some superconducting bolometers it may be less than 1 Ω). In such a case, it is necessary to use an impedance stepup transformer. The turns ratio is chosen to raise detector noise a little above amplifier noise, and, of course, it steps up the detected signal at the same time. Little is gained in using a higher turns ratio; indeed, excessive turns will increase leakage inductance, and this will have an adverse effect on the bandwidth. If the detector is at a low temperature, it may well be that the limit of detectability set by detector noise can only be met if the transformer, too, is cooled to low temperatures.[23,46]

The noise equivalent resistance of an amplifier tube depends very much on the frequency band. For an intermediate frequency amplifier it can be as low as 100 Ω in some of the modern tubes, but at low frequencies $1/f$ noise contributions raise it to much higher values. At frequencies of a few Hertz it can be hundreds of kiloohms.

In slow-response detectors, such as the Golay cell ($\tau \sim$ 2–30 msec) and some bolometers, it is generally preferred to chop (amplitude modulate) the incoming radiation and amplify at the chopping frequency rather than use dc amplifiers. To minimize amplifier noise, the fastest chopping rate should be used, consistent with the response time capabilities of the detector. For a slow thermal detector, the chopping rate may typically be 10 Hz, at which frequency the equivalent noise resistance will approach 10^5 Ω. Photoconductive detectors may be chopped much more rapidly and the amplifier noise reduced accordingly.

4.6.10. Noise Reduction after Rectification

The output from the postdetector amplifier is an alternating signal and must, for convenience of recording, be converted to a dc signal in a rectification stage. The rectification process may be sensitive to the phase of the input signal, or it may be completely independent of phase. Phase-sensitive rectification is frequently used in the far-infrared when the signal is amplitude modulated by a chopping device and fed, along with a reference signal from the chopper, into a *lockin* amplifier. In essence,

4.6. LIMITS SET BY RANDOM FLUCTUATIONS

lockin detection, or *synchronous* detection as it is sometimes called, multiplies the signal plus noise with the reference signal to produce an output dc signal proportional to the input signal and the cosine of the phase difference between the input and reference signals, together with a small amount of noise. Alternatively, the rectification process may be completely independent of the phase of the signal as, for example, with square-law diode rectifiers. The phase-dependent process is frequently called "coherent" detection, and the phase-independent process "incoherent." The relative merits of coherent and incoherent rectifiers have been considered in some detail by Smith,[96] whose results clarify the question of bandwidth and noise reduction after rectification. In this section, we summarize some important features of the interrelationship of observing time, bandwidth, and noise.

The effective bandwidth of the overall receiver can be reduced by averaging or integrating the noise after rectification. A simple integrator might be a condenser which is charged by the signal and noise currents from the rectifier, and which discharges through a suitable resistor and meter. The condenser stores the charge flowing to it with a time constant $\tau = RC$; as the charge is related to the integral of the current flow, the circuit integrates or averages. The time over which the averaging is done —the observing time—is

$$t = 2\tau = 2RC. \tag{4.98}$$

We can also regard the RC integrating circuit as a low-pass filter, passing a narrow band from zero frequency up to a frequency Δf_0, beyond which the capacitive reactance is somewhat less than R. The equivalent bandwidth of the integrator is

$$\Delta f_0 = 1/2t = 1/4RC. \tag{4.99}$$

Thus the circuit increases the signal-to-noise ratio by effectively restricting the bandwidth after rectification. In principle, the noise can be reduced without limit by averaging over an arbitrarily long time or, equivalently, by narrowing the passband to an arbitrarily narrow width.

The effect of postrectifier bandwidth reduction on the recorded signal-to-noise ratio is quite different for coherent and incoherent rectification. Let us denote the signal-to-noise ratio prior to rectification by S_i/N_i and suppose the noise to be rectangular in shape and of width Δf_i determined by input stage bandwidth limitations; after rectification, the signal-

---

[96] R. A. Smith, *Proc. IEE* **98**, Pt. 4, 43–54 (1951).

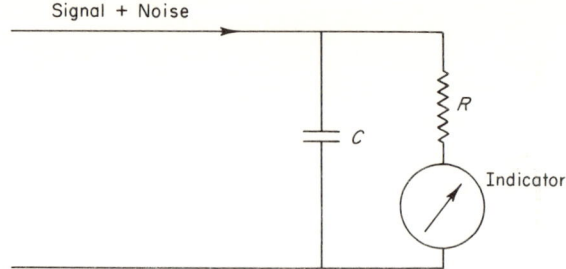

FIG. 4.24. A simple RC integrator circuit which limits the postrectifier bandwidth.

to-noise ratio will be denoted S_0/N_0 and the (output) bandwidth Δf_0. In the case of phase-sensitive rectification, the band of noise about the input signal has the same rectangular shape after rectification as before, but it is shifted to the region of zero frequency, this being the difference between the signal and reference frequencies. The postrectifier filter now selects a fraction $\Delta f_0/\Delta f_i$ of the input noise power with the result that the signal-to-noise ratio at the recorder (the meter in Fig. 4.24) is

$$S_0/N_0 = (S_i/N_i)(\Delta f_i/\Delta f_0)^{1/2}. \tag{4.100}$$

Denoting the rms values of the input signal voltage and noise voltage per unit bandwidth prior to rectification by v and v_n respectively, the input signal-to-noise ratio is

$$(S_i/N_i)^2 = v^2/(v_n{}^2 \, \Delta f_i), \tag{4.101}$$

hence,

$$(S_0/N_0)^2 = v^2/(v_n{}^2 \, \Delta f_0). \tag{4.102}$$

In the case of incoherent rectification, the problem of noise is somewhat more complex. As shown by Smith for a square-law detector, the output noise spectrum is no longer rectangular; it is a function of the input signal and both the pre- and post-rectifier bandwidths. In the limit of small S_i/N_i, Smith obtains the result

$$(S_0/N_0)^2 = v^4/(2v_n{}^4 \, \Delta f_i \, \Delta f_0). \tag{4.103}$$

By comparing Eqs. (4.102) and (4.103), we can see an important advantage of coherent rectification for the detection of small signals. For $S_i/N_i \ll 1$, the output signal-to-noise ratio for coherent detection exceeds that for incoherent detection by the considerable factor

$$(2v_n{}^2 \, \Delta f_i/v^2)^{1/2}.$$

5. CYCLOTRON RESONANCE WITH FREE ELECTRONS AND CARRIERS IN SOLIDS

Introduction

The motion of charged particles in magnetic fields has been studied for some time in a variety of contexts extending from ionospheric physics to elementary particles. However, the possibilities of motions in magnetic fields so high that the cyclotron frequency is in the far-infrared has not long been studied. In 1951 Dorfman[1] in the U.S.S.R. and Dingle[2] in England suggested the possibility of cyclotron resonance in metals. In 1956 Azbel' and Kaner[2a] studied a resonance configuration with the magnetic field parallel to the surface of the metal which proved particularly suited to materials in anomalous skin-effect conditions. It was Shockley[3] who recognized the significance of cyclotron resonance in semiconductors, and from 1953 there followed a decade of intensive work in which the new technique played a considerable role in establishing the dispersion laws for carriers in crystal fields. The relaxation times were short, so that in some materials microwave frequencies were required, while in others far-infrared frequencies were necessary to show the resonances as sharp distinguishable features. The range of applications then covered metals, semimetals, and semiconductors and became intimately connected with plasma effects accompanying dense systems of charges.

Cyclotron resonance of electrons in free-space and in a background of gas atoms and ions has been of interest in recent years in connection with sources of radiation in laboratory and astrophysical plasmas. In these circumstances the cyclotron frequency is normally not as high as the far-infrared, but when the density is high the plasma frequency can raise the region of interesting interactions to this spectral range. For example, some of the waves convenient for the probing of high-density

---

[1] Ya. G. Dorfman, *Dokl. Akad. Nauk SSSR* **81**, 765 (1951).
[2] R. B. Dingle, *Proc. Roy. Soc. Ser. A* **212**, 38–46 (1952).
[2a] M. Ya. Azbel' and E. A. Kaner, *Sov. Phys. JETP* **3**, 772–774 (1956).
[3] W. Shockley, *Phys. Rev.* **90**, 491 (1953).

solid-state (and gaseous) plasmas require a wave frequency ω greater than the plasma frequency ω_p; also in such measurements as the determination of carrier effective mass or electron density by Faraday rotation techniques frequencies not too far removed from ω_p can be necessary.

In the present chapter, we are concerned with the solutions of the classical and quantum-mechanical equations describing the motion of an electron in a homogeneous magnetic field. We introduce the effects of the environment with which the electron interacts through scattering collisions, and we incorporate the effect of the potential of a periodic lattice of neighboring ions by introducing an effective mass of an electron in a solid. The measurement of effective mass by cyclotron resonance methods gives a measure of the curvature of the energy–momentum characteristic, which adds to the understanding of the band structure in semiconductors. Various cyclotron resonance techniques and observations are described. We discuss magnetoplasma effects on wave interactions in semiconductors but we leave until Chapter 6 the derivation of the various plasma formulas. In the final section, we discuss positive and negative cyclotron resonance absorption by weakly relativistic electrons, and in so doing we provide the background description of the electron cyclotron maser discussed in Chapter 2.

5.1. Absorption by Nonrelativistic Free Charges

5.1.1. Classical Motion

Consider the classical motion of a particle of charge q in a uniform magnetic induction field **B** driven by a oscillating electric field $\mathbf{E} = 2\mathbf{E}_0 \cos(\omega t + \phi)$. Suppose the initial kinetic energy of the particle is $\frac{1}{2}m_0 v_0^2$ in the x-y plane normal to the **B**-field. The equations of motion

$$m_0 \ddot{x} = qE_x + qB\dot{y}, \qquad (5.1)$$

$$m_0 \ddot{y} = qE_y - qB\dot{x}, \qquad (5.2)$$

can be solved to give the two cartesian components of velocity. Alternatively, we can calculate the motion produced by the two circularly polarized components of the oscillating electric field. Near cyclotron resonance, the particle interacts strongly with the field component rotating in the same sense and in near synchronism, and only weakly with

5.1. ABSORPTION BY NONRELATIVISTIC FREE CHARGES

the counter rotating component. The equation of motion for the particle and field rotating in the same sense, obtained by multiplying Eq. (5.2) by i and adding Eq. (5.1), is

$$(d\dot{S}/dt) + i\omega_c \dot{S} = Ae^{-i(\omega t + \phi)}, \qquad (5.3)$$

where

$$\dot{S} = \dot{x} + i\dot{y}, \qquad A = qE_0/m_0, \qquad \text{and} \qquad \omega_c = qB/m_0.$$

The particular integral of Eq. (5.3) is

$$-i[A/(\omega_c - \omega)] e^{-i(\omega t + \phi)},$$

and the complementary function is $C \exp(-i\omega_c t)$. The general solution is thus

$$\dot{S} = C \exp(-i\omega_c t) - i \frac{A}{\omega_c - \omega} \exp[-i(\omega t + \phi)].$$

Taking the initial conditions for the interaction to be $\dot{S} = iv_0$, at $t = 0$, the constant C can be determined. The result is

$$\dot{S} = -i \exp(-i\omega_c t) \left[v_0 - \frac{A}{\omega_c - \omega} \{\exp(-i\phi) - \exp(-i\phi) \exp i(\omega_c - \omega)t\} \right] \qquad (5.4)$$

The x and y components of velocity are the real and imaginary parts of Eq. (5.4), and are given by

$$\dot{x} = -v_0 \sin \omega_c t + \frac{A}{\omega_c - \omega} [\sin(\omega_c t + \phi) - \sin(\omega t + \phi)],$$

$$\dot{y} = -v_0 \cos \omega_c t + \frac{A}{\omega_c - \omega} [\cos(\omega_c t + \phi) - \cos(\omega t + \phi)].$$

The kinetic energy is thus

$$W = \frac{m_0}{2}(\dot{x}^2 + \dot{y}^2) = \frac{1}{2} m_0 v_0^2 - \frac{m_0 A v_0}{\omega_c - \omega} \{\cos \phi - \cos[(\omega - \omega_c)t + \phi]\}$$

$$+ \frac{m_0 A^2}{(\omega_c - \omega)^2} [1 - \cos(\omega - \omega_c)t]. \qquad (5.5)$$

The average absorption per particle of a beam of charges initiating their interaction with the field with random phases is found by averaging over ϕ. The result is clearly

$$W = \frac{1}{2} m_0 v_0^2 + \frac{m_0 A^2}{(\omega_c - \omega)^2} [1 - \cos(\omega_c - \omega)t] \qquad (5.6)$$

$$= \frac{1}{2} m_0 v_0^2 + \frac{q^2 E_0^2 t^2}{2 m_0} \left(\frac{\sin^2 \Gamma}{\Gamma^2}\right), \qquad (5.7)$$

where $\Gamma = (\omega_c - \omega)t/2$ is half the phase lag (or lead) of the orbital particle motion with respect to the rotating field during the time t of interaction. The function $\sin^2 \Gamma/\Gamma^2$ may be called the "line shape," "response curve," or "spectral function"; it gives the shape of the absorption line as the wave frequency is tuned through the resonance of the cyclotron oscillator.

Interaction with the counterrotating field $E_0 e^{i(\omega t + \phi)}$ can be analyzed by the same procedure. For this polarization, the energy of a phase-averaged particle is again of the form Eq. (5.6), except that in this case the spectral function is $\sin^2 \gamma/\gamma^2$, where $\gamma = (\omega + \omega_c)t/2$. γ is generally so large that $\sin^2 \gamma/\gamma^2$ is essentially zero, and there is no net energy absorption over the interaction time t.

Equation (5.7) gives the line shape for a system of charges interacting with the field for a definite period of time t. It will, for example, be applicable to a beam of charges drifting in vacuum with uniform velocity along the z-direction in helical motion through a resonator supporting a uniform oscillating field, t being the transit time through the resonator.

However, if the charges undergo collisional scattering with background particles, there will be a spread of interaction times and the line shape will be modified. This will be the case, for example, with electron cyclotron motion in plasmas or solids. In this case, if it is assumed that the probability of a particle suffering a collision between times t and $t + dt$ is dt/τ, and if $P(t)$ denotes the probability of a particle traveling for a time t before a collision,

$$P(t + dt) = P(t)[1 - (dt/\tau)]$$

and hence

$$P(t) \propto e^{-t/\tau}.$$

The average time between collisions is

$$\int_0^\infty t\, e^{-t/\tau}\, dt \Big/ \int_0^\infty e^{-t/\tau}\, dt = \tau. \qquad (5.8)$$

5.1. ABSORPTION BY NONRELATIVISTIC FREE CHARGES

For such collision-damped motion, the line shape can be found from Eq. (5.6), averaged according to the statistical weight $e^{-t/\tau}$. It is

$$\frac{\int_0^\infty \{[1 - \cos(\omega_c - \omega)t]/(\omega_c - \omega)^2\} e^{-t/\tau} \, dt}{\int_0^\infty e^{-t/\tau} \, dt} = \frac{1}{(\omega_c - \omega)^2 + (1/\tau^2)}. \tag{5.9}$$

Thus, for collisional damping, the line shape is Lorentzian.

5.1.2. Quantum-Mechanical Approach

The effect of an electromagnetic field on a charged particle in a homogeneous magnetic field was first discussed from a quantum-mechanical viewpoint by Landau[4] and is treated in a number of texts.[5-7a] The Schrödinger equation (neglecting spin),

$$H\Psi = i\hbar \, \partial\Psi/\partial t, \tag{5.10}$$

is solved with the Hamiltonian operator derived from the classical expression

$$H = (1/2m_0)(\mathbf{P} - q\mathbf{A})^2. \tag{5.11}$$

Here $\mathbf{P}$ is the generalized momentum, related to the ordinary momentum $m_0\mathbf{v}$ by

$$\mathbf{P} = m_0\mathbf{v} + q\mathbf{A}. \tag{5.12}$$

$\mathbf{A}$ is the vector potential of the magnetic field, and, since the field is homogeneous and z-directed, it can be written

$$A_x = -By, \quad A_y = A_z = 0.$$

If we substitute for $\mathbf{A}$, and replace $\mathbf{P}$ by its operator form

$$P_x = (\hbar/i) \, \partial/\partial x,$$

---

[4] L. D. Landau, *Z. Phys.* **64**, 629 (1930).

[5] L. D. Landau and E. M. Lifshitz, Quantum mechanics. "Course in Theoretical Physics," Vol. 3. Pergamon, Oxford, 1958.

[6] R. H. Dicke and J. P. Wittke, "Introduction to Quantum Mechanics." Addison-Wesley, Reading, Massachusetts, 1963.

[7] D. ter Haar, ed., "Selected Problems in Quantum Mechanics." Infosearch, London, 1964.

[7a] S. D. Smith, *in* "Optical Properties of Solids" ed. (S. Nudelman and S. S. Mitra, eds.), Chapter 4. Plenum Press, New York, 1969.

etc., we obtain the form

$$-\frac{\hbar^2}{2m_0}\left[\left(\frac{\partial}{\partial x}+\frac{iqBy}{\hbar}\right)^2+\frac{\partial^2}{\partial y^2}+\frac{\partial^2}{\partial z^2}\right]\psi = W\psi \qquad (5.13)$$

for the time-independent Schrödinger equation. Writing

$$\psi = \exp i(k_x x + k_z z) f(y),$$

the variables may be separated to give

$$\frac{d^2 f(y)}{dy^2}+\frac{2m_0}{\hbar^2}\left[W-\frac{\hbar^2 k_z^2}{2m_0}-\frac{1}{2m_0}(\hbar k_z + qBy)^2\right]f(y) = 0, \qquad (5.14)$$

or

$$\frac{d^2 f(y)}{dy^2}+\frac{2m_0}{\hbar^2}\left[\left(W-\frac{\hbar^2 k_z^2}{2m_0}\right)-\frac{1}{2}m_0\omega_c^2(y-y_0)^2\right]f(y) = 0, \qquad (5.15)$$

where we have defined $y_0 = -\hbar k_x/qB$.

We recognize this equation as being formally identical with that of a harmonic oscillator, oscillating about y_0 with frequency ω_c. Therefore, we know immediately that the quantity $[W - \hbar^2 k_z^2/2m_0]$—which is the energy of cyclotron motion in the transverse plane—has eigenvalues of the form

$$W - (\hbar^2 k_z^2/2m_0) = (n + \tfrac{1}{2})\hbar\omega_c,$$

where $n = 0, 1, 2, \ldots$. If we neglect motion in the z-direction, the kinetic energy in the plane transverse to the homogeneous magnetic field is

$$W = (n + \tfrac{1}{2})\hbar\omega_c. \qquad (5.16)$$

Quantum mechanics thus replaces the continuum of classical energy values by a ladder of discrete levels, the *Landau levels*. The average energy of two neighboring Landau levels, n and $n+1$, is $(n+1)\hbar\omega_c$, and we can equate this to the classical kinetic energy expression $\tfrac{1}{2}m_0 v^2$. The velocity is then

$$v = [2(n+1)\hbar\omega_c/m_0]^{1/2}, \qquad (5.17)$$

and the orbital radius is

$$r = v/\omega_c = [2(n+1)\hbar/m_0\omega_c]^{1/2}. \qquad (5.18)$$

The lowest Landau level, $n = 0$, has zero-point energy $\tfrac{1}{2}\hbar\omega_c$, and the

5.1. ABSORPTION BY NONRELATIVISTIC FREE CHARGES

higher states are separated by the quantum value $\hbar\omega_c$. Drift motion parallel to the magnetic field adds to this an energy continuum.

Absorption and emission of radiation in the vicinity of cyclotron resonance is described in terms of transitions between Landau levels. Transition probabilities are found from quantum theory by regarding an interaction of time duration t between the particle and the oscillating electric field as a "switched-on" perturbation; the results of time-dependent perturbation theory then give the probability for induced transitions in the interaction time t between the nth and $(n+1)$th states[6,8]

$$w_{n,n+1} = w_{n+1,n} = (E_0^2 t^2/\hbar^2)(\mu_{n,n+1})^2 g_\omega(\omega_c), \qquad (5.19)$$

where E_0 is the amplitude of the perturbing electric field. The spectral function is

$$g_\omega(\omega_c) = [(\sin^2 \Gamma/\Gamma^2) + (\sin^2 \gamma/\gamma^2)], \qquad (5.20)$$

with

$$\Gamma = (\omega_c - \omega)t/2, \quad \text{and} \quad \gamma = (\omega_c + \omega)t/2.$$

The first term on the right of Eq. (5.20) is associated with the perturbation by the component of field rotating in the same sense as the charge. The last term is due to the counterrotating component and is negligibly small. The matrix element of the electric dipole moment of the harmonic oscillator for transitions between the nth and $(n+1)$th states has the value[9]

$$\mu_{n,n+1} = q[(n+1)\hbar/2m_0\omega_c]^{1/2}. \qquad (5.21)$$

When the results given in Eqs. (5.19)–(5.21) are substituted into the expression $(w_{n,n+1} - w_{n,n-1})$ for the net probability of absorption by a charge in the nth state, quantitative agreement is obtained with the classical expression for the energy absorbed from the field.

When the orbital motion is damped by collisions, Eq. (5.20) must be averaged in accordance with the statistical weighting factor $e^{-t/\tau}$. As with the classical case, the result is again the Lorentz line shape.

We have thus far shown the origin in quantum mechanics of only the phase average term in the classical energy expression Eq. (5.5). However, as we have seen in Chapter 4 (Section 4.4), the second term on the right

---

[8] R. M. Eisberg, "Fundamentals of Modern Physics." Wiley, New York, 1964.
[9] A. S. Davydov, in "Quantum Mechanics" (D. ter Haar, ed.), pp. 288, 119. Pergamon, Oxford, 1965.

of Eq. (5.5)—which can be readily put in the form $qE_0v_0t \sin(\Gamma + \phi)$ $(\sin \Gamma)/\Gamma$—is of interest. To consider the quantum mechanics of states of motion involving phase ϕ, let us return to Eq. (5.15). The wavefunctions describing the orbiting particle are the well-known Hermite polynomials, and each is, of course, associated with a discrete energy eigenstate specified by the quantum number n. These states may be called "number eigenstates." Phase is unspecified. Now, using number state wavefunctions, we can construct a wave packet from a superposition involving several energy states whose behavior corresponds to the classical motion (see Dicke and Wittke [6, Chapter 8]). As shown by ter Haar,[7] in the plane perpendicular to the magnetic field the wave packet changes its form periodically with a period equal to that of the classical motion of the orbiting particle. If there is momentum in the direction of the magnetic field, the center of the wave packet will move in a helical path precisely as the classical particle.

A useful representation of the motion with phase ϕ with respect to a driving force can be developed by generating a superposition wavefunction from those of many stationary states. The superposition wavefunction is called the coherent state; with this state an operator algebra[10-14] has been developed which enables calculations of transitions between number states beyond the usual range of perturbation theory.[12] To be sure, it is not a state in the sense that it is an eigenstate of H, but rather in the sense that certain operations on the coherent state wavefunction can reproduce the wavefunction multiplied by a (complex) number. The coherent state has the merit that, while it is intrinsically quantum mechanical, it readily leads to the classical limit.

We seek answers to the questions: With what probabilities does a cyclotron harmonic oscillator, initially in the nth energy eigenstate, undergo transitions in the Landau ladder when subjected to the interaction described in Section 5.1.1? How does the quantum description go over to the classical limit? To answer these questions we follow the analysis of Carruthers and Nieto[14] for the coherent state of a forced harmonic oscillator.

[10] E. M. Henley and W. Thirring, "Elementary Quantum Field Theory." McGraw-Hill, New York, 1962.

[11] W. Louisell, "Radiation and Noise in Quantum Electronics." McGraw-Hill, New York, 1964.

[12] R. J. Glauber, *Phys. Rev.* **131**, 2766-2788 (1963).

[13] R. W. Fuller, S. M. Harris and E. L. Slaggie, *Amer. J. Phys.* **31**, 431-439 (1963).

[14] P. Carruthers and M. M. Nieto, *Amer. J. Phys.* **33**, 537-544 (1965).

5.1. ABSORPTION BY NONRELATIVISTIC FREE CHARGES

Coherent states $|\alpha\rangle$ are eigenfunctions of the annihilation operator a in the sense that

$$a\,|\,\alpha\rangle = \alpha\,|\,\alpha\rangle, \tag{5.22}$$

where

$$\alpha = |\,\alpha\,|\,e^{i\phi} \tag{5.23}$$

is a complex number whose argument is the phase angle of the oscillator, and the square of whose modulus gives the mean excitation of the oscillator. The Hamiltonian operator can be written

$$H = \hbar\omega_c(a^+a + \tfrac{1}{2}). \tag{5.24}$$

For a harmonic oscillator in an initial state specified by a (and a^+), and raised by a time-dependent driving force with Fourier transform F to a final state specified by

$$b = a + i\alpha_0, \tag{5.25}$$

where

$$\alpha_0 = F/(2m\hbar\omega_c)^{1/2}, \tag{5.26}$$

Carruthers and Nieto derive an operator S (the S matrix) in terms of which the matrix element $\langle k\,|\,S\,|\,n\rangle$ for transitions from an initial number state n to any final number state k can be found. For the present case of a time-dependent force acting on the harmonic oscillator[14]

$$S = \exp[i(a^+F + aF^*)/(2m\,\hbar\omega_c)^{1/2}]. \tag{5.27}$$

It can be shown[14] that this leads to the transition probability

$$w_{k,n} = |\,S_{kn}\,|^2 = \frac{n!}{k!}\,(\exp -|\,\alpha_0\,|^2)\,|\,\alpha_0\,|^{2(k-n)}[L_k^{k-n}(|\,\alpha_0\,|^2)]^2, \tag{5.28}$$

for $k \geq n$, or by the same formula but with k and n interchanged if $k < n$. $L_k^{k-n}(|\,\alpha_0\,|^2)$ is the associated Laguerre polynomial.

Fuller et al.[13] have summed Eq. (5.28) over upward transitions $k > n$, and downward transitions $k < n$, and have shown that the mean energy transfer to an oscillator is $\hbar\omega_c\,|\,\alpha_0\,|^2$. This is precisely the phase-averaged term in the classical expression Eq. (5.5). Carruthers and Nieto have obtained the same result for the average energy absorbed, but their formulation also yields the phase-dependent term. In the Heisenberg picture of quantum mechanics, the wavefunction remains fixed and the operator representing the observable changes; we can write the energy shift produced by the interaction as

$$W = \langle \Psi_{\text{in}} | H_{\text{out}} - H_{\text{in}} | \Psi_{\text{in}} \rangle. \tag{5.29}$$

Here the subscript "in" refers to the initial state of the oscillator and the subscript "out" refers to the state of the oscillator after the interaction. The Hamiltonian operator of the in state oscillator is

$$H_{\text{in}} = \hbar\omega_c(a^+a + \tfrac{1}{2}), \tag{5.24}$$

and that of the out state is

$$H_{\text{out}} = \hbar\omega_c(b^+b + \tfrac{1}{2}), \tag{5.30}$$

where the annihilation operators are related as in Eq. (5.25). Substitution gives

$$W = \hbar\omega_c[|\alpha_0|^2 + i(\langle \Psi_{\text{in}} | a^+ | \Psi_{\text{in}} \rangle \alpha_0 - \text{complex conjugate})]. \tag{5.31}$$

If Ψ_{in} is a coherent state with phase parameter $\beta = |\beta| e^{i\phi}$, Eq. (5.31) reduces to

$$W = \hbar\omega_c[|\alpha_0|^2 + 2\,\text{Im}(\beta\alpha_0^*)]. \tag{5.32}$$

In going over to the classical cyclotron oscillator, we must have

$$|\beta| = (m\omega_c/2\hbar)^{1/2} r_{c\text{ out}}, \tag{5.33}$$

where $r_{c\text{ out}}$ is the cyclotron orbit radius at the termination of the interaction. For a constant driving field of angular frequency ω and duration t, $F = qE_0 t(\sin\Gamma)/\Gamma$ in Eq. (5.26). When α_0^* and β are substituted from Eqs. (5.26) and (5.33), Eq. (5.32) reduces exactly to the classical expression Eq. (5.5) for the energy imparted by the interaction.

5.2. Cyclotron Resonance and Magnetoplasma Effects in Semiconductors and Metals

5.2.1. Cyclotron Resonance

Experimental observations of cyclotron resonance absorption in the microwave to infrared range provide powerful means for the study of the band structure of solids. Through the measurement of the cyclotron frequency, one has a direct means to determine the mass of the current carriers [from Eq. (5.37)], and from the bandwidth of the resonance the mean collision time can be deduced on the basis of the Lorentz line shape Eq. (5.36). As we shall see, the orbiting particle must be represented by an "effective mass" m^*, which in some materials will be isotropic and in others anisotropic. Cyclotron resonance experiments can also give information regarding the anisotropy.

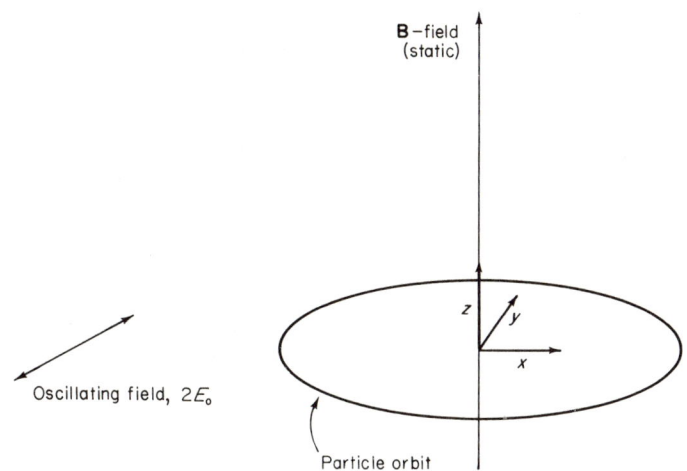

FIG. 5.1. Arrangement of fields in a cyclotron resonance experiment.

For the arrangement of fields shown in Fig. 5.1, and assuming isotropic mass m^* and isotropic relaxation time τ, we can use the conclusions of the preceding section to obtain the cyclotron power absorption. Equations (5.6) and (5.9) combine to give the phase-averaged energy absorption during the time τ between collisions. For n electrons randomly phased in orbital motion with respect to the rotating circularly polarized field, the power absorbed is

$$P = \frac{nW}{\tau} = \frac{nq^2 E_0^2}{2m^*} \frac{\tau}{1 + (\omega_c - \omega)^2 \tau^2}, \qquad (5.34)$$

or

$$\frac{P}{P_0} = \frac{1}{2[1 + (\omega_c - \omega)^2 \tau^2]}. \qquad (5.35)$$

Here $P_0 = \sigma_0 E_0^2$ is the power absorbed at $B = 0$ and $\omega = 0$, $\sigma_0 = nq^2\tau/m^*$ is the dc conductivity, and ω_c is given by Eq. (5.37). Although the result Eq. (5.35) is derived for absorption from a circularly polarized wave

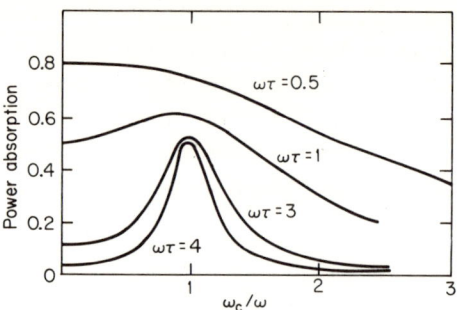

FIG. 5.2. Power absorption in the region of cyclotron resonance, showing the effect of the product $\omega\tau$ of the wave angular frequency ω and the relaxation time τ on the sharpness of the resonance. ω_c is the cyclotron angular frequency (after Lax and Mavroides[17]).

rotating in the same sense as the carriers, it applies equally to a linearly polarized incident wave since (from Section 5.1.1) absorption by the counterrotating component varies as $1/[1 + (\omega_c + \omega)^2\tau^2]$ and is negligibly small. P/P_0 is plotted against ω_c/ω in Fig. 5.2, with $\omega\tau$ as a parameter. At $\omega_c/\omega = 1$, the carriers orbit in synchronism with the field and the resonance absorption is peaked. However, for the resonance to be discernible it is clearly necessary that $\omega\tau \gtrsim 1$. Physically, this means that the carrier must rotate through at least one radian of its orbit between scattering collisions. It is readily seen from Eq. (5.35) that the width of the resonance curve at the points where the power absorption is half the value at resonance is

$$\Delta\omega = 2/\tau. \tag{5.36}$$

In high purity semiconductors at low temperatures, $\tau \sim 10^{-10}$–10^{-11} sec, and for such materials microwaves can show distinct cyclotron resonance absorption. Indeed, millimeter wave spectrometry has been used extensively in the study of germanium, silicon, lead telluride, and the semimetal bismuth. At higher temperatures, however, and in less pure materials, the condition $\omega\tau \gtrsim 1$ means that far-infrared or infrared waves are needed. At room temperature the relaxation times are commonly in the range 10^{-13}–10^{-15} sec.

Now let us consider the origin of the concept of effective mass, which arises because of the interaction of the carriers with the bound atoms of the solid. Electrons in a crystalline solid are not free but are located within the periodic potential of the lattice. The potential has considerable effect on the motion of the electrons, and it is through this effect that

5.2. SEMICONDUCTOR AND METAL MAGNETOPLASMA EFFECTS

information about the form of the potential can be gathered. Such information is central to the problem of band structure determination. For a chosen lattice potential, electron wavefunctions can be calculated from the Schrödinger equation, and it is possible to construct a wave packet representing the motion of the electron. In the equation of motion this representation requires an effective mass m^*, and the expression for the cyclotron frequency becomes

$$\omega_c = qB/m^*. \qquad (5.37)$$

The effective mass is a function of the energy–momentum relation for the carrier electrons and is frequently smaller than the free electron mass. Consider a free electron. In the wave representation of this particle the wave vector $\mathbf{k}$ ($k = 2\pi/\lambda$) has the direction of wave propagation and its magnitude is connected with the momentum $\mathbf{p}$ and kinetic energy W through the de Broglie relation

$$\begin{aligned} \mathbf{p} &= \hbar \mathbf{k}, \\ W &= p^2/2m = \hbar^2 k^2/2m. \end{aligned} \qquad (5.38)$$

Thus for a free electron there is a parabolic W-k relation as shown in Fig. 5.3(a). The second de Broglie postulate relates energy and frequency by $W = \hbar\omega$, and shows that the ω-k dispersion curve for a free electron is a parabola. The velocity v of the free particle must, of course, be the group velocity of the wave packet, and this, by Eq. (5.38), is related to

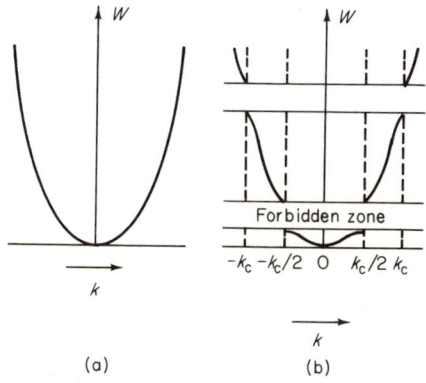

FIG. 5.3. (a) The parabolic energy–momentum relation of a free electron. (b) The energy–momentum relation for an electron in a crystalline lattice. In (b) stop-bands (or forbidden zones) occur when the lattice spacing is equal to an integral number of de Broglie half wavelengths $n\pi/k_c$.

the gradient of the W-k curve,

$$v = (1/\hbar)\, \partial W/\partial k.$$

Furthermore, by differentiation of Eq. (5.38), we have

$$m = \hbar^2/(\partial^2 W/\partial k^2).$$

Next, consider an electron in a solid. For such a particle the dispersion relation will be of the general form characteristic of wave propagation in periodic systems.[15] There will be propagation in certain frequency (or energy) ranges, and these will be separated by forbidden zones that arise from multiple in-phase interference between waves reflected from the sets of lattice planes. When the lattice planes are spaced at multiples of a half de Broglie wavelength, standing-wave resonances build up and propagation is cut off. The resulting W-k relation will be of the form shown in Fig. 5.3(b). Analogously to the case of the free electron, quantum mechanics leads to an expression for the group velocity of the electron and to a definition of effective mass in terms of the curvature of the W-k characteristic:

$$v = (1/\hbar)\, \partial W/\partial k, \tag{5.39}$$

$$m^* = \hbar^2/(\partial^2 W/\partial k^2). \tag{5.40}$$

For the simplest situation, the W-k curve may be given near the minimum of W by

$$W = \hbar^2 k^2/2m^*. \tag{5.41}$$

That is, the surface of constant energy is spherical in k-space. As the momentum value is approached for which constructive interference of Bragg reflections occurs, the W-k curve bends away from the parabola, the slope of the characteristic signifying that the group velocity of the electron has been slowed down by reflections from the lattice planes.

In real crystals a variety of factors arise to complicate the situation. However, while they complicate the necessary analysis, they generally add to the information obtainable from cyclotron resonance experiments. Of particular importance is crystalline anisotropy. In an anisotropic crystal the dispersion relations will, in general, be different for the different directions, and the reciprocal effective mass must be represented

---

[15] L. Brillouin, "Wave Propagation in Periodic Structures." McGraw-Hill, New York, 1946.

5.2. SEMICONDUCTOR AND METAL MAGNETOPLASMA EFFECTS

by a tensor of the form

$$1/m_{ij} = (1/\hbar^2) \, \partial^2 W / \partial k_i \, \partial k_j, \qquad (5.42)$$

where i, j indicate the cartesian coordinate directions. Anisotropy takes a variety of forms that may be characterized in any given material by the symmetry of the constant energy surfaces in k-space. If the surface is spherical, the mass m^* is isotropic, and its sign can be deduced from the sense of rotation of the circularly polarized wave with which it resonates. If the surfaces are ellipsoidal, as they are in the conduction bands of germanium and silicon, resonance occurs for both circular polarizations. Shockley[3,16] has shown that, for the general ellipsoidal surface, the effective mass determining the cyclotron frequency when the static magnetic field makes an angle θ with the longitudinal axis of the energy surface is given by

$$1/m^* = [(\cos^2 \theta / m_t^2) + (\sin^2 \theta / m_t m_l)]^{1/2}, \qquad (5.43)$$

where m_l and m_t are the longitudinal and transverse mass parameters. They are the elements in the simple tensor which, for ellipsoidal energy surfaces, is diagonalized in the principal coordinate system of the ellipsoid. It will be seen from Eqs. (5.43) and (5.37) that ω_c is the same on all parallel sections of the ellipsoid. For a given orientation of the field and crystal, a line of the form predicted by Eq. (5.34) can be observed, but, of course, ω_c will vary with orientation and thereby give information about the axes of the ellipsoid.

In some cases the symmetry of the energy surfaces may be lower than ellipsoidal, and further complications can arise from degeneracy, plasma effects, and geometry to add considerable complexity to the detailed interpretation of band structure. These effects have been discussed in excellent surveys by Lax and Mavroides,[17] Lax,[18] Herman[19] and Fan.[20]

The carriers in semiconductors are distributed in accordance with Fermi–Dirac statistics. The average number of electrons in a state of energy W is

$$2f = \frac{2}{1 + \exp[(W - W_F)/kT]}, \qquad (5.44)$$

---

[16] C. Kittel, "Introduction to Solid State Physics." Wiley, New York, 1956.
[17] B. Lax and J. G. Mavroides, *Solid State Phys.* **11**, 261–388 (1960).
[18] B. Lax, *Rev. Mod. Phys.* **30**, 122–154 (1958).
[19] F. Herman, *Proc. IRE* **43**, 1703–1732 (1955).
[20] H. Y. Fan, *Rep. Progr. Phys.* **19**, 107–157 (1956).

where the factor 2 takes account of the two spin directions allowed by the exclusion principle. The Fermi energy, W_F, is located between the valence and conduction bands. At absolute zero the valence band is completely filled and the conduction band completely empty, and in this situation an external field can produce no current flow because, according to the Pauli exclusion principle, there are no vacant states into which the electrons can jump. Thermal excitation at finite temperatures, or optical, infrared, or microwave irradiation,[18] can excite electrons from the valence band to the minimum (or minima) of the conduction band thereby enabling field interactions. The effective mass of the conduction electrons can then be measured by cyclotron resonance, anisotropy can be studied by varying the direction of the magnetic field, and the results used to interpret the structure of the conduction band minimum (or minima). Furthermore, the valence band is no longer completely filled and cyclotron resonance of the holes from which electrons have been removed can be observed. By this means the structure near the edge of the valence band is explored. In the case of metals and semimetals cyclotron resonance techniques are used to probe the band structure in the vicinity of the Fermi level. Electron resonances in metals have been reviewed by Azbel' and Lifshitz.[21]

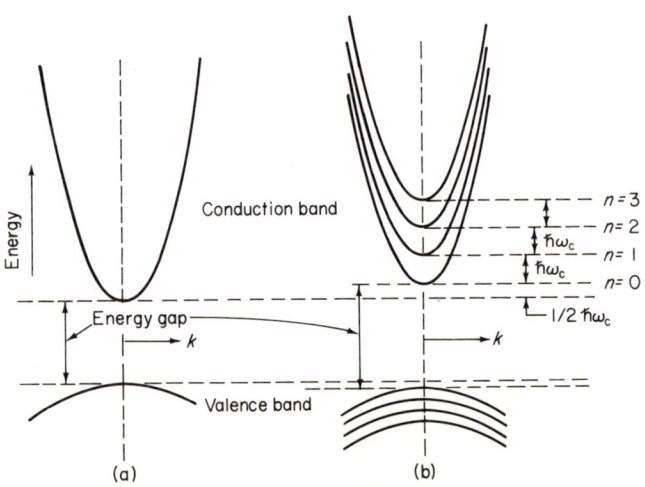

FIG. 5.4. Simple energy bands of a semiconductor (a) in the absence and (b) in the presence of a magnetic field (after Dexter et al.[22]).

[21] M. Ya. Azbel' and E. M. Lifshitz, *Progr. Low Temp. Phys.* **111**, 288–332 (1961).
[22] R. N. Dexter, H. J. Zeiger and B. Lax, *Phys. Rev.* **104**, 637–644 (1956).

5.2. SEMICONDUCTOR AND METAL MAGNETOPLASMA EFFECTS

A useful representation of cyclotron resonance interaction is illustrated in Fig. 5.4 for the case of a simple parabolic energy band. While in the absence of a magnetic field the conduction band and valence band are continua of states, in a magnetic field they split into discrete Landau levels. The minimum of the conduction band and the maximum of the valence band are both shifted by the appropriate zero-point energy, and resonance absorption is associated with interlevel transitions particularly between the states with lower quantum numbers. The interaction is an electric-dipole transition, with transition probability given by Eq. (5.4) with the appropriate Lorentzian line shape function indicated in Eqs. (5.9) and (5.34). This function is, of course, a measure of the width of the Landau levels for finite relaxation times.

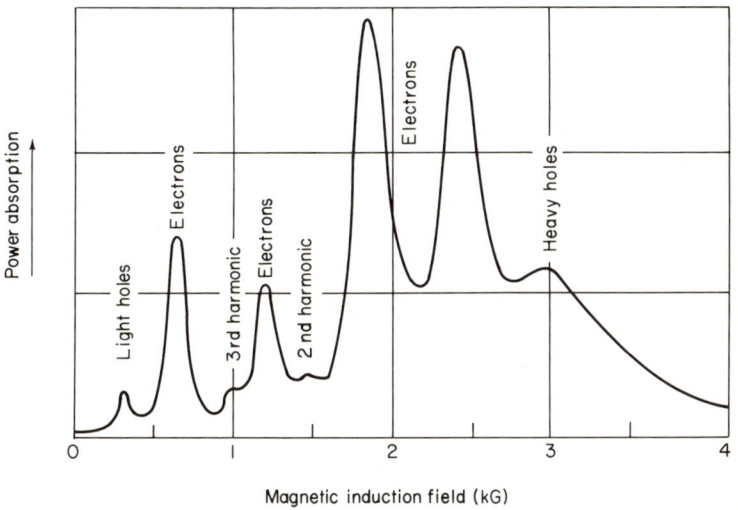

FIG. 5.5. Cyclotron resonance in germanium near 4°K and at 23 GHz. The magnetic field direction (10° out of the (110) plane and 30° from the (100) direction) was chosen to show eight resonances (after Dexter et al.[22]).

Examples of observed spectra for germanium and p-type indium antimonide are given in Figs. 5.5 and 5.6. The electron resonances in germanium arise from a set of ellipsoids oriented along the crystallographic cube diagonals with an effective mass $m_l = 1.64$ m along the principal axis of the ellipsoid and a transverse mass $m_t = 0.082$ m, m being the free electron mass.[17] In addition, the germanium spectrum shows resonances from two kinds of holes and weaker resonances that have been

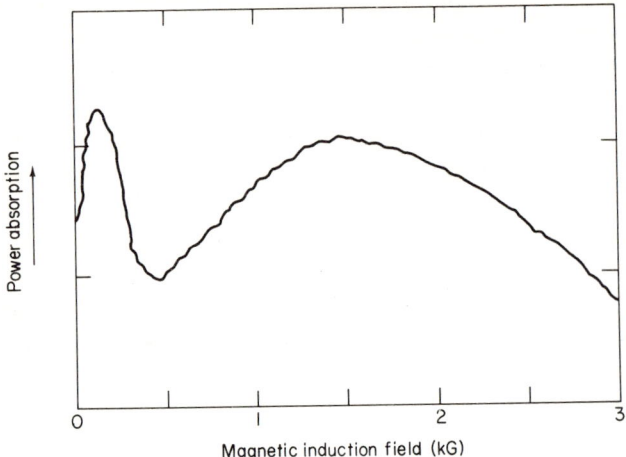

Fig. 5.6. Power absorption (relative scale) versus the static magnetic induction field, in indium antimonide at a frequency of 23,975 MHz (after Dresselhaus et al.[23]).

assigned to harmonic transitions[24] between Landau levels governed by the condition $\varDelta n = \pm 2, \pm 3$, supplementing the usual selection rule $\varDelta n = \pm 1$. From the bandwidth of the sharpest lines in germanium at 4°K Dresselhaus et al.[25] have found approximately isotropic relaxation times for electrons with $\tau \sim 6 \times 10^{-11}$ sec. The sample of p-type indium antimonide had a concentration of less than 5×10^{14} cm^{-3} acceptor impurities and was cooled to 2.2°K. The resonance apparent at low fields showed no anisotropy and was associated with electrons of mass $m^* = 0.013$ m. From the linewidth, the relaxation time is estimated to be $\tau \sim 1 \times 10^{-11}$ sec. In this material broad resonances occur in the higher field regions, but they are not resolved in this record.[17]

A comparison of the germanium and indium antimonide spectra, both recorded at essentially the same frequency, illustrates the importance of high frequencies in the study of the less pure semiconductors. The elements germanium and silicon can be made with a high level of purity and consequently microwave frequencies are high enough to give high resolution spectral data. Microwaves, however, have been less fruitful in the study of alloy semiconductors where impurity scattering is large.

---

[23] G. Dresselhaus, A. F. Kip, C. Kittel and G. Wagoner, *Phys. Rev.* **98**, 556–557 (1955).

[24] B. Lax, *Quantum Electron. Symp.* 1959, pp. 428 (1960).

[25] G. Dresselhaus, A. F. Kip and C. Kittel, *Phys. Rev.* **98**, 368–384 (1955).

5.2. SEMICONDUCTOR AND METAL MAGNETOPLASMA EFFECTS

In such materials it is necessary to use far-infrared, and this means the need for high magnetic fields.

Some far-infrared measurements on indium antimonide are shown in Figs. 5.7–5.9. In Fig. 5.7 the resonance is in n-type indium antimonide at 90 μm and was recorded by Palik et al.[26] for temperatures of 300°K and 80°K. The difference between the two curves emphasizes the change

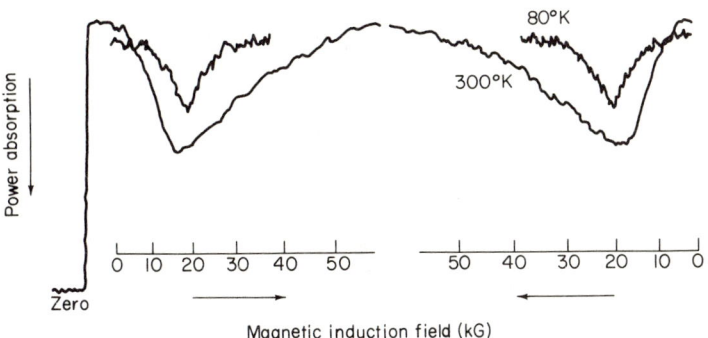

FIG. 5.7. Cyclotron resonance in n-type indium antimonide at 90 μm (after Palik et al.[26]).

in linewidth with temperature. Figures 5.8 and 5.9 show spectra of p-type indium antimonide at room temperature recorded by Burnstein et al.[27] at a wavelength of 41.1 μm. At room temperature the relaxation time is in the region of 5×10^{-13} sec. These records should be compared with that of Fig. 5.6 recorded with a similar material at low temperature. Well-defined effects occur near the cyclotron frequency corresponding to a room-temperature effective mass of 0.015 m. The indium antimonide sample contained the relatively high electron density $n = 2 \times 10^{16}$ cm^{-3}, with a corresponding plasma frequency $\omega_\mathrm{p} = (nq^2/\varkappa_\mathrm{L} m^*)^{1/2} = 1.75 \times 10^{13}$ sec^{-1} or 108 μm wavelength (the lattice dielectric constant $\varkappa_\mathrm{L}$ is 16). At such high concentration levels, plasma effects enter into the interaction and, as we shall see in the next section (see Fig. 5.16), it is desirable that $\omega_\mathrm{c} > \omega_\mathrm{p}$ for the experimental conditions used by Burnstein et al. Burnstein et al. chose a wavelength of 41.1 μm to satisfy this condition and also to avoid the lattice absorption band centered at 54 μm and effective from 42–60 μm.

[26] E. D. Palik, G. S. Picus, S. Teitler and R. F. Wallis, *Phys. Rev.* **122**, 475–481 (1961).

[27] E. Burnstein, G. S. Picus and H. A. Gebbie, *Phys. Rev.* **103**, 825–828 (1956).

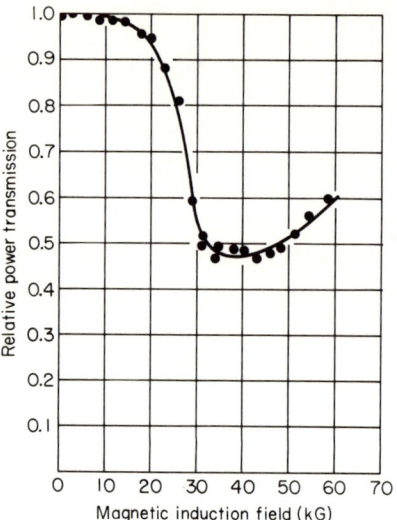

FIG. 5.8. Relative transmission as the magnetic induction field is increased through cyclotron resonance for a slab (0.02 mm thick) of intrinsic indium antimonide at room temperature. The magnetic field is parallel to the surface of the sample and the wavelength is 41.1 μm (after Burnstein et al.[27]).

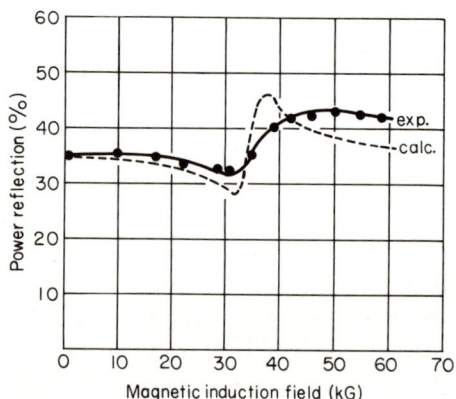

FIG. 5.9. Reflectivity at a wavelength of 41.1 μm from a specimen of intrinsic indium antimonide (0.5 mm thick) at room temperature, with the magnetic field normal to the surface. The reflectivity is that of one surface only, since the sample was strongly absorbing. The calculated curve is based on $m^* = 0.015$ m and $\omega_c \tau = 16$ (after Burnstein et al.[27]).

5.2.2. Magnetoplasma Effects

In metals and semiconductors with high carrier concentrations, a number of problems arise because of the large charge densities; these complicate the interpretation of observations and add requirements additional to $\omega\tau \gg 1$. If the plasma frequency ω_p exceeds ω_c, the effective complex refractive index will have a large imaginary part over a relatively broad range of frequencies adjoining ω_c. The material will be highly reflecting and the cyclotron resonance difficult to distinguish. There are, however, some features of the reflection characteristics for the various possible geometries of interaction and values of ω, ω_c, and ω_p that can give a measure of the plasma parameters. In general, the need is for high magnetic fields and far-infrared radiation as, for example, in the measurements of Burnstein et al.[27]

For the case of metals and semimetals, ω_p is very large, and enormous fields would be required to give the sort of resonance absorption that we have been treating up to this stage. In addition, in metals the interaction is likely to be influenced by the anomalous skin effect, that is, the depth of penetration of the wave into the metal may well be less than the mean free path of the carriers, and less than the radius of the cyclotron orbit. Two geometries are commonly used: in one, the magnetic field is normal to the surface, and in the other, it is in the plane of the surface. With high fields normal to the surface and microwave frequencies of 24 and 72 GHz, Galt[28] and others have been able to determine approximately the cyclotron frequencies in bismuth,[29] graphite,[30] and other semimetals under conditions of normal and anomalous skin effect from broad nonresonant absorption curves. A better arrangement under conditions of highly anomalous skin effect is to apply the magnetic field in the plane of the metal and the **E**-vector of the normally incident wave perpendicular to the magnetic field. This so-called "extraordinary wave" can show sharp cyclotron resonance peaks. Such an interaction was first proposed by Azbel' and Kaner and, as illustrated in Fig. 5.10 for conditions where the skin depth δ is somewhat less than the cyclotron radius, actually makes use of the shallow skin depth. The resonance arises as follows: An electron traveling with the Fermi velocity orbits in a helical path about the magnetic field. Once in each revolution it enters the skin region,

---

[28] J. K. Galt, In "High Magnetic Fields" (H. Kolm, B. Lax, F. Bitter and R. Mills, eds.), Chapter 52. MIT Press, Cambridge, Massachusetts, and Wiley, New York, 1962.
[29] W. S. Boyle, A. D. Brailsford and J. K. Galt, Phys. Rev. **109**, 1396–1398 (1958).
[30] J. K. Galt, W. A. Yager and H. W. Dail, Phys. Rev. **103**, 1586–1587 (1956).

where it interacts with the electric field. Provided successive reentries of the particle into the skin occur for the same phase of the oscillating electric field, resonant absorption of energy takes place. The interaction is somewhat analogous to that in a cyclotron, where particles are accelerated by repetitive impulses in a particular sector of their orbits. For cumulative interactions, i.e., resonance, the frequencies must be related by

$$\omega = n\omega_c, \qquad (5.45)$$

where n is an integer 1, 2, 3, ... Azbel'–Kaner resonance does not give a single peak but rather a succession of peaks as the magnetic field is changed. Constructive phase conditions occur between the oscillating

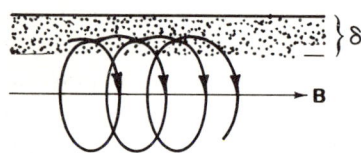

FIG. 5.10. Azbel'–Kaner resonance in a conductor. The particle spirals in orbits with radius larger than the skin depth δ about a magnetic field applied parallel to the surface. The oscillating electric field is normal to the page and loses energy during the succession of in-phase transits of the particle through the skin depth.

electric field and the particles at high magnetic fields when the cyclotron frequency equals the wave frequency, and at successively lower magnetic fields as ω_c passes through subharmonics of ω. Quantitative treatment of this interaction under anomalous skin effect conditions are given in Ziman's book[31] and in the review by Lax and Mavroides.[17]

Azbel'–Kaner cyclotron resonance at centimeter and millimeter wavelengths has been used with marked success in copper,[33] bismuth,[29] zinc and cadmium,[32,32a] and other metals.[17] Figures 5.11 and 5.12 illustrate for the cases of cadmium and copper the type of resonance behavior observed with this configuration of fields. The condition $\omega\tau > 1$ can be satisfied for pure copper at 4°K, for at this temperature τ is between 10^{-9} and 10^{-11} sec. For the copper crystal used in the measurement illustrated in Fig. 5.12, $\omega\tau$ has been estimated to be ~ 50. As the Fermi energy in copper is ~ 7 eV, the Fermi velocity v_F equals 1.6×10^8 cm sec^{-1}

---

[31] J. M. Ziman, "Principles of the Theory of Solids." Cambridge Univ. Press, London and New York, 1964.

[32] J. K. Galt, F. R. Merritt, W. A. Yager and H. W. Dail, *Phys. Rev. Lett.* **2**, 292–294 (1959).

[32a] J. K. Galt, F. R. Merritt, and P. H. Schmidt, *Phys. Rev. Lett.* **6**, 458–460 (1961).

[33] D. N. Langenberg and T. W. Moore, *Phys. Rev. Lett.* **3**, 328–330 (1959).

5.2. SEMICONDUCTOR AND METAL MAGNETOPLASMA EFFECTS

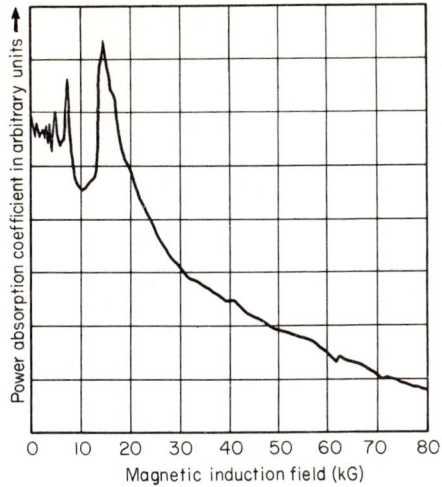

FIG. 5.11. Cyclotron resonance signal of the Azbel'–Kaner type in cadmium at 1.3°K. The wave frequency is 72 GHz and the magnetic field is parallel to the sample surface and directed along a principal axis of the crystal (after Galt et al.[32a]).

(assuming $m^* = m$), so that the mean free path l is between 10^{-3} and 10^{-1} cm. This is considerably larger than the skin depth $\delta \sim 10^{-5}$ cm as given by classical electromagnetic theory, assuming $n \sim 10^{22}$ electrons per cubic centimeter. At a frequency of 23 GHz, the cyclotron orbit radius is $r = v_F/\omega_c \sim 10^{-3}$ cm. Thus the conditions, $\delta \ll r < l$, for the anomalous skin effect prevail. In bismuth at 1.2°K, $n \sim 10^{17}$–10^{18} cm^{-3} and $\tau \sim 10^{-10}$–10^{-11} sec, and in zinc at 1.3°K, $\tau \sim 4 \times 10^{-11}$ sec.

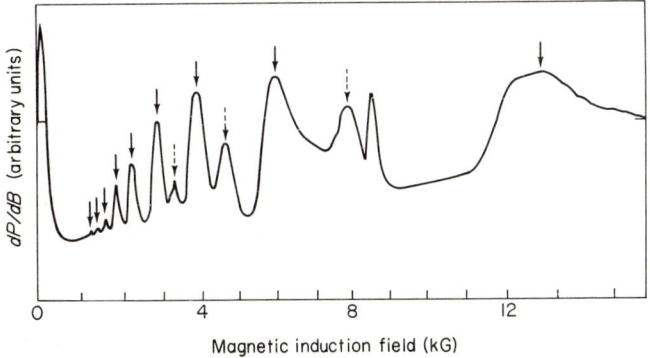

FIG. 5.12. Cyclotron resonance in copper at 24.47 GHz, observed by modulating the magnetic induction field and recording the derivative dP/dB of the absorbed power curve. The magnetic field is in a (110) plane 10° from a (100) axis, and the microwave current is in a (110) plane and 45° from the **B**-field direction, 55° from the (100) axis. The resonances indicated by the solid arrows are associated with carriers of effective mass $m^* = 1.24\,m$, and those indicated by the broken arrows with carrier mass $m^* = 1.29\,m$ (after Langenberg and Moore[33]).

Turning back to the question of magnetoplasma effects, let us assume normal skin-effect conditions and illustrate the main features for the case of normal incidence on the plane boundary of a semi-infinite collisionless plasma. At this point we write down the dispersion relations for the various waves, leaving their justification until Chapter 6 where they occur in a gaseous plasma context. In terms of the effective dielectric constant $\varkappa$, we have

$$\varkappa_{\substack{r\\l}} = \varkappa_L\left[1 - \frac{\omega_p^2}{\omega(\omega \mp \omega_c)}\right], \tag{5.46}$$

$$\varkappa_{ext} = \varkappa_L\left[1 - \frac{(\omega^2 - \omega_p^2)\omega_p^2}{\omega^2(\omega^2 - \omega_c^2 - \omega_p^2)}\right], \tag{5.47}$$

where $\varkappa_L$ is the lattice dielectric constant, and the plasma frequency for charge carriers in a background dielectric is defined as

$$\omega_p^2 = nq^2/\varkappa_L m^*. \tag{5.48}$$

In these equations we denote the right-hand circularly polarized (RHCP) and left-hand circularly polarized (LHCP) waves by the subscripts r and l, respectively, and take the RHCP wave to be that rotating in the same sense as the orbiting electrons. The extraordinary wave propagates transversely across the **B**-field with its **E**-vector perpendicular to **B** and is denoted "ext."

From the standard procedure of imposing continuity of the components of the **E**-vector tangential to the surface of the solid, the amplitude reflection coefficient is found to be

$$R = (1 - \varkappa^{1/2})/(1 + \varkappa^{1/2}). \tag{5.49}$$

On the basis of Eqs. (5.46), (5.47), and (5.49) the reflection behavior can be plotted as a function of ω_c/ω for the RHCP and LHCP waves, and for the extraordinary wave, with ω_p^2/ω^2 as a parameter. In Figs. 5.13–5.15 the variation of the effective dielectric constant with ω_c/ω is shown for each of these three waves. For the RHCP and LHCP waves $\varkappa_{\substack{r\\l}}$ become zero, and propagation is cut off when

$$(\omega)_{\substack{r\\l}} = [(\omega_c^2 + 4\omega_p^2)^{1/2} \pm \omega_c]/2. \tag{5.50}$$

For increasing ω the RHCP wave $\varkappa_r$ remains negative until the singularity $\varkappa_r \to \infty$ is reached at $\omega = \omega_c$. In the intervening frequency range, the square root of $\varkappa_r$ is imaginary, so that $|R| = 1$ and the wave is

5.2. SEMICONDUCTOR AND METAL MAGNETOPLASMA EFFECTS

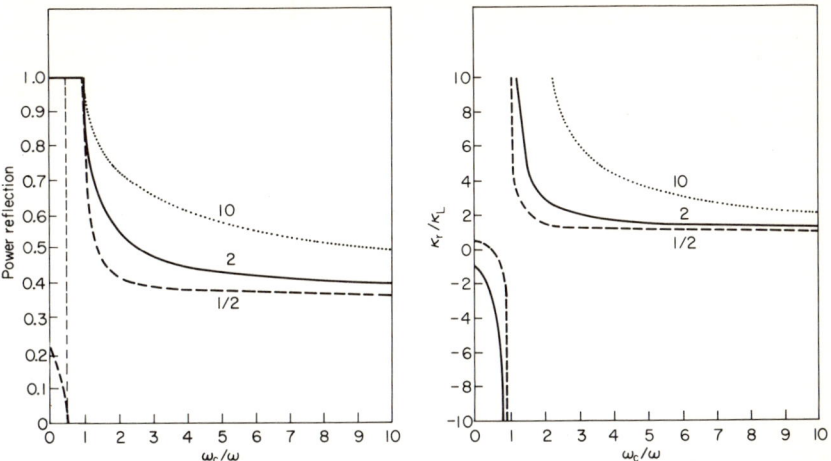

FIG. 5.13. Right panel: ratio $\varkappa_r/\varkappa_L$ of the dielectric constant to the lattice dielectric constant as a function of the ratio of the cyclotron to wave frequency ω_c/ω for a right hand circularly polarized (RHCP) wave in a solid-state plasma. Left panel: power reflection of a RHCP wave traveling parallel to the magnetic field and normally incident on the surface of the solid. The lattice dielectric constant $\varkappa_L = 16$. The dashed curves correspond to $\omega_p^2/\omega^2 = 0.5$, the full curves to $\omega_p^2/\omega^2 = 2$, and the dotted curves to $\omega_p^2/\omega^2 = 10$, where $\omega_p = (nq^2/\varkappa_L m^*)^{1/2}$ is the plasma frequency.

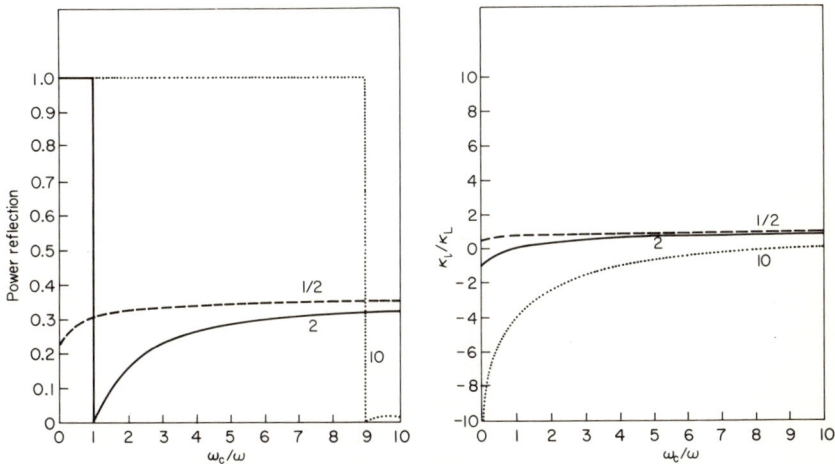

FIG. 5.14. Right panel: ratio $\varkappa_l/\varkappa_L$ of the dielectric constant to the lattice dielectric constant as a function of ω_c/ω for the left hand circularly polarized (LHCP) wave in a solid state plasma. Left panel: power reflection of a LHCP wave traveling parallel to the magnetic field and normally incident on the surface of the solid. The dashed, full and dotted curves correspond to $\omega_p^2/\omega^2 = 0.5, 2, 10$, and $\varkappa_L = 16$.

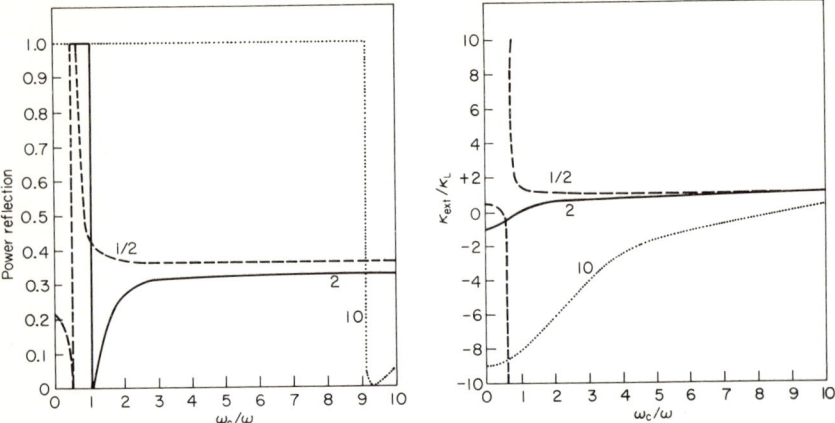

FIG. 5.15. Right panel: ratio $\varkappa_{\text{ext}}/\varkappa_{\text{L}}$ of the dielectric constant to the lattice dielectric constant as a function of ω_c/ω for the extraordinary wave in a solid state plasma. Left panel: power reflection of the extraordinary wave incident normally on the surface of the solid. The dashed, full, and dotted curves correspond to $\omega_p^2/\omega^2 = 0.5, 2, 10$, and $\varkappa_{\text{L}} = 16$.

perfectly reflected. Similarly for the LHCP wave (Fig. 5.14 for ω_c/ω negative) there is a region of ω_c/ω where perfect reflection occurs but, of course, no singularity exists for this wave. The extraordinary wave also has cutoffs at the frequencies given by Eq. (5.50), and a singularity when $\omega_p^2/\omega^2 < 1$ at the upper hybrid frequency:

$$\omega = (\omega_c^2 + \omega_p^2)^{1/2}. \tag{5.51}$$

As the figures indicate, all three wave types show sharp cyclotron resonances only when $\omega_p^2 \ll \omega^2$, that is, in the limit of low charge density. When $\omega_p^2 \gtrsim \omega^2$, the cyclotron resonance frequency may be estimated from such features as the decrease in reflectivity at $|\omega_c/\omega| = 1$, but for large values of ω_p^2/ω^2 this "edge" becomes less clearly defined. In the case of the extraordinary wave, the plasma tends to shield out the cyclotron resonance interaction when $\omega_p^2 \gg \omega^2$. The extraordinary wave is partially transverse and partially longitudinal, and it is through the interaction of the longitudinal component with the carriers that space-charge depolarizing fields are set up.[34] Finite relaxation time has the effect of "rounding off" the sharp edges of the collisionless reflection curves; assuming $\omega \tau = 10$ the reflection coefficient for the RHCP wave is as

[34] G. E. Smith, L. C. Hebel and S. J. Buchsbaum, *Phys. Rev.* **129**, 154–168 (1963).

5.2. SEMICONDUCTOR AND METAL MAGNETOPLASMA EFFECTS

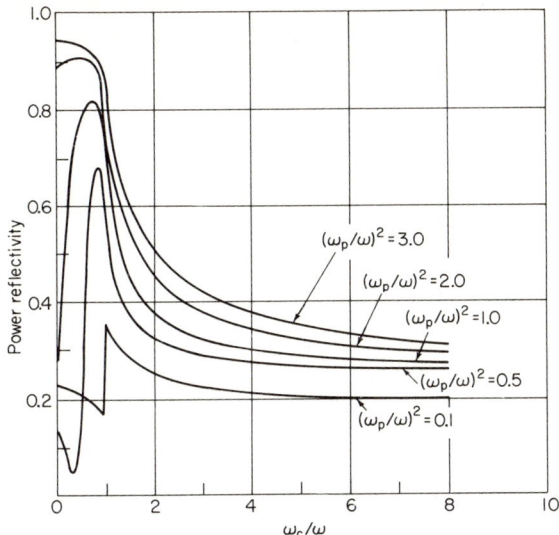

FIG. 5.16. Power reflection from a solid-state magnetoplasma of a right hand circularly polarized wave as a function of ω_c/ω, with ω_p/ω as a parameter. The wave is normally incident on a surface oriented at right angles to the magnetic field. It is assumed that $\varkappa_L = 9$ and $\omega\tau = 10$ (after Zeiger and Hilsenrath[35]).

shown in Fig. 5.16, from which it is clear that a sharp peak near the cyclotron frequency requires $\omega_c > \omega_p$.

Lax and Wright[36] have considered the dispersion relations Eqs. (5.46) and (5.47) in the limit $\omega_c \ll \omega_p$ and have shown that the behavior of the reflection coefficient as a function of ω/ω_p for both the extraordinary wave and unpolarized radiation traveling in the **B**-field direction exhibit two minima in the neighborhood of $\omega/\omega_p \sim 1$ with a separation equal to ω_c. This separation has been used to determine the effective mass. Similarly, the limit $\omega_c > \omega_p$ shows a splitting of the plasma edge by the cyclotron frequency. In both limits infrared or far-infrared waves are needed in order to probe in the general neighborhood of ω_p. In the limit $\omega_c > \omega_p$ very large magnetic fields are called for.

The ordinary wave can also be used in solid-state plasmas to measure the effective mass of carriers provided the carrier density is known. The ordinary wave (symbolised by subscripts "ord" and "//") propagates in

---

[35] H. J. Zeiger and S. Hilsenrath, Lincoln Lab., *Quart. Progr. Rept.*, Group 35, p. 54, February 1957.
[36] B. Lax and G. B. Wright, *Phys. Rev. Lett.* **4**, 16–18 (1960).

an unmagnetized plasma or when a wave propagates across a magnetic field with its **E**-vector parallel to the field. Equation (5.46) gives the dispersion relation if we place $\omega_c = 0$,

$$\varkappa_{\text{ord}} = \varkappa_L[1 - (\omega_p^2/\omega^2)]. \tag{5.52}$$

The effective dielectric constant is equal to unity and the reflection coefficient becomes zero when

$$\omega_p^2/\omega^2 = 1 - (1/\varkappa_L). \tag{5.53}$$

Furthermore, at a frequency only slightly lower, ω_p^2/ω^2 rises above unity, $\varkappa_{\text{ord}}$ is then imaginary and the reflection coefficient becomes unity. This sudden transition from zero to total reflection as the frequency drops below the plasma frequency gives a measure of the ratio of the carrier concentration to the effective mass. Observations of this effect in indium

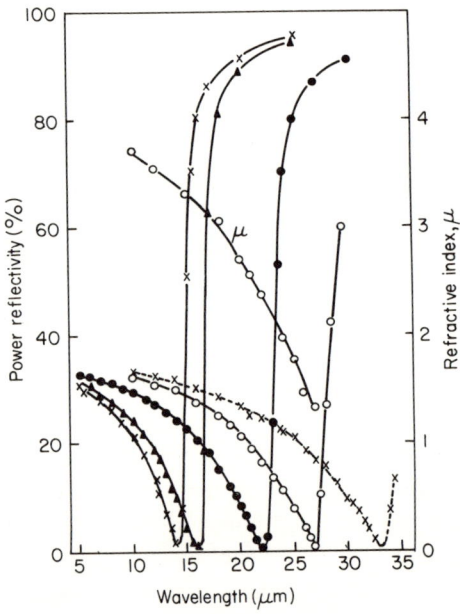

FIG. 5.17. Reflectivity versus wavelength for five n-type indium antimonide samples showing the abrupt transition from zero to nearly perfect reflection as the wave frequency decreases through the plasma frequency. The refractive index curve labeled μ is for the sample with a carrier concentration 6.2×10^{17} cm^{-3}; values of n (cm^{-3}) ×--×, 3.5×10^{17}; ○, 6.2×10^{17}; ●, 1.2×10^{18}; ▲, 2.8×10^{18}; ×——×, 4.0×10^{18} (after Spitzer and Fan[37]).

antimonide by Spitzer and Fan[37] are shown in Fig. 5.17. The dielectric constant $\varkappa_L$ for indium antimonide is 16 and the reflection minimum therefore occurs when $\omega = 1.03\omega_p$. The method is best suited to semiconductors with high carrier densities where the plasma frequency is beyond the range of lattice vibration frequencies.

5.3. Faraday Rotation and Related Magnetooptic Effects in Semiconductors

We have seen in the previous section that the four wave types: the ordinary, the extraordinary, and the two counterrotating circularly polarized waves propagate in a magnetized semiconductor with different phase velocities. Furthermore, losses result in differences in attenuation of the four wave types. Therefore, if a wave incident on a semiconductor plasma has field components that excite two types of wave propagation, phase and attenuation differences occur between them which, in general, produce elliptical polarization in the emerging wave. Such magnetooptic effects when related to the plasma parameters can give the effective mass of carriers and the carrier concentration. Three effects are of particular interest: Faraday rotation of the plane of polarization of a linearly polarized wave propagating along the magnetic field, the magneto-Kerr effect, which is the conversion from linear to elliptical polarization at reflection of a wave incident along the magnetic field, and Voight double refraction —a phase shift between the ordinary and extraordinary waves.

For a transparent material (that is, infinite relaxation time) the angle of Faraday rotation θ of the plane of polarization is determined by the difference in phase change of the RHCP and LHCP components of the linearly polarized wave, namely,

$$\theta = (\omega L/2c)(\mu_l - \mu_r), \tag{5.54}$$

where L is the sample thickness in the direction of propagation. The refractive indices μ_r and μ_l for the RHCP and LHCP waves, respectively, are given by Eq. (5.46):

$$\mu_r = \varkappa_r^{1/2} = \varkappa_L^{1/2}\left[1 - \frac{\omega_p^2}{\omega(\omega - \omega_c)}\right]^{1/2}, \tag{5.55}$$

$$\mu_l = \varkappa_l^{1/2} = \varkappa_L^{1/2}\left[1 - \frac{\omega_p^2}{\omega(\omega + \omega_c)}\right]^{1/2}. \tag{5.56}$$

---

[37] W. G. Spitzer and H. Y. Fan, *Phys. Rev.* **106**, 882–890 (1957).

If $\omega(\omega - \omega_c) \gg \omega_p$, Eqs. (5.54)–(5.56) give

$$\theta = Lnq^3B/2c\varkappa_L^{1/2}m^{*2}(\omega^2 - \omega_c^2). \tag{5.57}$$

Two special cases are of interest:

(i) For $\omega \gg \omega_c$,

$$\theta = Lnq^3B/2c\varkappa_L^{1/2}m^{*2}\omega^2; \tag{5.58}$$

(ii) For $\omega_c \gg \omega$,

$$\theta = -Lnq/2c\varkappa_L^{1/2}B. \tag{5.59}$$

Equations (5.57)–(5.59) thus enable the determination of the effective mass m^* and carrier density n from the measured Faraday rotation. The conditions $\omega(\omega - \omega_c) \gg \omega_p$ and $\omega \gg \omega_c$ call for frequencies in the far-infrared range, while the case $\omega_c \gg \omega$ calls for high magnetic fields.

When losses are included, the dispersion relations have the form (see Chapter 6)

$$\varkappa_{r_1} = \varkappa_L\left[1 - \frac{\omega_p^2}{\omega(\omega \mp \omega_c + j/\tau)}\right]. \tag{5.60}$$

It follows that losses may be neglected when

$$(\omega - \omega_c)\tau \gg 1,$$

a requirement much less stringent as far as the magnetic field is concerned than the condition $(\omega_c\tau > 1)$ demanded for cyclotron resonance.

The Voight effect for an incident wave linearly polarized at an angle of 45° with respect to the magnetic field can also be used to determine effective mass.[38] From Eqs. (5.47) and (5.52) under the conditions $\omega \gg \omega_c$ and ω_p, the phase difference between the components with electric vectors parallel and perpendicular to the magnetic field is given by

$$\delta = (\omega L/2c)(\mu_{\text{ord}} - \mu_{\text{ext}}) \simeq Lnq^4B^2/2c\varkappa_L^{1/2}\omega^3m^{*3}. \tag{5.61}$$

The rotation angle δ of the plane of polarization gives a measure of m^* only if n is known. However, if δ is compared with the Faraday rotation angle θ under the conditions $\omega \gg \omega_c$, the ratio

$$\delta/\theta = \omega_c/\omega \tag{5.62}$$

permits the determination of m^* directly without involving n.

---

[38] E. D. Palik, S. Teitler and R. F. Wallis, *J. Appl. Phys. Suppl.* **32**, 2132–2136 (1961).

In the case of very lossy semiconductors Faraday and Voight effect techniques have limited value because of the heavy attenuation of the waves. Under such conditions the magneto-Kerr effect can give a measure of carrier density and effective mass. The magneto-Kerr effect is a phenomenon, closely related to the Faraday effect, that arises within the skin depth during reflection from a surface oriented at right angles to the magnetic field. A linearly polarized incident wave is converted to an elliptically polarized reflection when the two circularly polarized components are changed in phase by the same process as in the Faraday effect, but they are also differently attenuated to give elliptically polarized reflection. By observing the two orthogonal linearly polarized components of the reflected wave, one can find the plasma parameters.[39] Because the propagation path is into the skin and out again, the net attenuation is not excessive. The technique can be used effectively under conditions where $\omega_p \gg \omega$ and ω_c.

5.4. Positive and Negative Absorption by Weakly Relativistic Electrons

In Section 5.1 we considered the interaction of nonrelativistic charges in a uniform magnetic field with electromagnetic radiation from both classical and quantum-mechanical approaches. Both calculations led to Eq. (5.7) predicting the energy absorbed by a particle in an interaction time t, with a line shape given by the sinc function of half the phase shift of the rotating circularly polarized field with respect to the orbiting particle. Quantum mechanics told us that the spectrum of energy states is not a continuum but rather a ladder of uniformly spaced discrete bands separated by a quantum of the cyclotron frequency radiation $\hbar\omega_c$. For an infinite ladder of uniformly spaced levels, the energy absorption by a large system of cyclotron oscillators is always positive. Classically, positive absorption arose when we averaged over random phases in proceeding from Eq. (5.5) to Eq. (5.7); quantum mechanically, from the expressions (5.19) and (5.21) for the transition probabilities and the matrix element of the dipole moment, the probability for an induced transition from the nth state to the $(n + 1)$th state exceeds the induced transition probability from the nth state to the $(n - 1)$th state, so that for a system of oscillators there is a net positive absorption of energy.

---

[39] L. C. Robinson, *Aust. J. Phys.* **20**, 29–46 (1967).

A question we might ask at this stage is: can we modify the Landau ladder of energy states in such a way as to favor emissive over absorptive transitions thereby raising the possibility of maser action from induced transitions between Landau levels? The answer to this question is yes. In the case of free electrons with relativistic velocities the mass is energy dependent, and this leads to a ladder of states with nonuniform spacings and the possibility of negative absorption. The necessary and sufficient conditions for this mechanism of negative absorption were first considered by Twiss[40] in 1958, and in 1959 Schneider[41] proposed the possibility of a laboratory maser. As we have seen in Chapter 2, the experimental realization of this possibility came in 1964[42-44] and later work with high magnetic fields resulted in cyclotron maser action at far-infrared frequencies. Two other possibilities have been discussed by Lax[45,17] and Wolff,[46] both involving electrons in indium antimonide. In one, the nonparabolicity of the energy–momentum bands is envisaged as a means of attaining unequally spaced Landau levels, and in the other a sudden decrease in relaxation time above a certain level terminates the ladder, thereby effectively eliminating upward transitions by electrons excited (pumped) into the higher levels of the ladder. As these solid-state proposals have not yet been realized in the laboratory, we will confine our attention in this section to the principles of maser action with a system of weakly relativistic free electrons.

Sauter[47] has solved the Dirac equation for a relativistic charged particle in a homogeneous magnetic field and shown that the allowed states of kinetic energy are given by

$$W = m_0 c^2 [1 + 2(n + \tfrac{1}{2})(\hbar \omega_c / m_0 c^2)]^{1/2} - m_0 c^2, \quad n = 0, 1, 2, \ldots, \quad (5.63)$$

where m_0 is the rest mass of the electron, c is the velocity of light in vacuum, and $\omega_c = qB/m_0$ is the cyclotron angular frequency of a nonrelativistic electron. Let us suppose that $n\hbar\omega_c \ll m_0 c^2$, that is, the electron

---

[40] R. Q. Twiss, *Aust. J. Phys.* **11**, 564–579 (1958).

[41] J. Schneider, *Phys. Rev. Lett.* **2**, 504–505 (1959).

[42] J. L. Hirshfield and J. M. Wachtel, *Phys. Rev. Lett.* **12**, 533–536 (1964).

[43] I. B. Bott, *Proc. IEEE* **52**, 330–332 (1964).

[44] J. L. Hirshfield, I. B. Bernstein and J. M. Wachtel, *IEEE J. Quantum Electron.* **1**, 237–245 (1965).

[45] B. Lax, *Proc. Int. Conf. Quantum Electron.*, 2nd, Berkcley, March, *1961* (J. R. Singer, ed.), pp. 465–479. Columbia Univ. Press, New York, 1961.

[46] P. A. Wolff, *Physics* (Long Island City, N.Y.) **1**, 147–158 (1964).

[47] F. Sauter, *Z. Phys.* **69**, 742–764 (1931).

5.4. ELECTRON POSITIVE AND NEGATIVE ABSORPTION

kinetic energy is somewhat less than the rest energy (0.51 MeV) of the electron. An expansion of Eq. (5.63) by the binomial theorem then gives the frequency separation of the nth and $(n-1)$th states,

$$\omega_{n,n-1} \simeq \omega_c[1 - (n\hbar\omega_c/m_0c^2)], \tag{5.64}$$

from which it is apparent that the ladder of Landau states no longer has uniform spacing, but the levels get closer together as the quantum number n increases. If an electron has been accelerated to 5 kV, the ratio of kinetic to rest energy $n\hbar\omega_c/m_0c^2$ is of the order of 10^{-2} and the corresponding levels are 1% closer than the nonrelativistic levels. For far-infrared photon energies, e.g., $\hbar\omega_c = 0.0012$ eV corresponding to a wavelength of 1000 μm, the quantum number n is of the order of 10^6.

Now let us suppose that we inject a beam of monoenergetic electrons into the magnetic field and thereby populate the nth state with N_n oscillators, while leaving all other states empty. The energy absorbed by the system during interaction with an electromagnetic field is

$$W = N_n \hbar[\omega_{n,n+1} w_{n,n+1} - \omega_{n,n-1} w_{n,n-1}], \tag{5.65}$$

where the transition probabilities for absorption and induced emission during the interaction time t ($w_{n,n+1}$ and $w_{n,n-1}$, respectively) are again given by expressions of the form Eq. (5.19) but, of course, with the appropriate frequency separation in the spectral function. Thus

$$w_{n,n+1} = \frac{E_0^2 t^2}{\hbar^2} (\mu_{n,n+1})^2 g_\omega(\omega_{n,n+1}), \tag{5.66}$$

with $\mu_{n,n+1}$ given by Eq. (5.21), and

$$g_\omega(\omega_{n,n+1}) = \frac{\sin^2[(\omega_{n,n+1} - \omega)t/2]}{[(\omega_{n,n+1} - \omega)t/2]^2}. \tag{5.67}$$

Substituting Eqs. (5.66) and (5.67) back into Eq. (5.65), we obtain the energy absorbed from the field in the interaction time t:

$$W = \frac{N_n q^2 E_0^2 t^2}{2m_0} \{n[g_\omega(\omega_{n,n+1}) - g_\omega(\omega_{n,n-1})] + g_\omega(\omega_{n,n+1}) + \text{small term}\}. \tag{5.68}$$

Negative absorption can arise in Eq. (5.68) if $g_\omega(\omega_{n,n-1})$ is sufficiently larger than $g_\omega(\omega_{n,n+1})$. This situation can arise only when $\omega_{n,n+1} \neq \omega_{n,n-1}$, that is, for unequally spaced levels. Conditions under which $g_\omega(\omega_{n,n-1})$

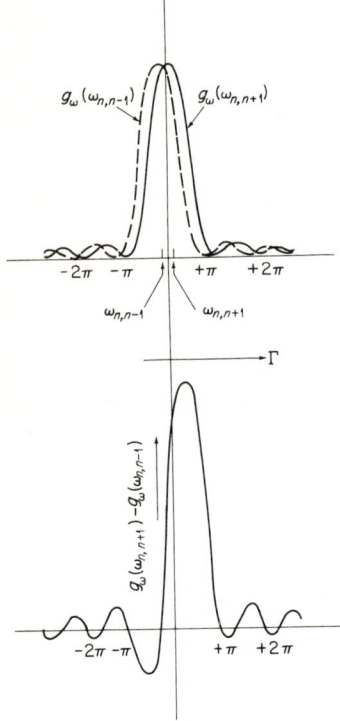

FIG. 5.18. Upper curve: plot of the spectral line shape for cyclotron oscillators with central frequencies $\omega_{n,n-1}$ and $\omega_{n,n+1}$ versus $\Gamma = (\omega_{n,n\pm1} - \omega)t/2$, t being the interaction time and ω the wave angular frequency. The relativistic velocity dependence of mass slightly separates $\omega_{n,n-1}$ and $\omega_{n,n+1}$ and gives regions where $g_\omega(\omega_{n,n-1}) > g_\omega(\omega_{n,n+1})$ and negative absorption arises. Negative absorption regions are shown in the lower curve where the difference between the line shape functions is plotted.

$> g_\omega(\omega_{n,n+1})$ can be seen from a plot of the two sinc functions in Fig. 5.18. Here we plot $(\omega_{n,n+1} - \omega)t/2$ and $(\omega_{n,n-1} - \omega)t/2$ on the horizontal scale taking the centers of the sinc functions $\omega_{n,n-1}$ and $\omega_{n,n+1}$ as separated, in accordance with Eq. (5.64), by

$$\omega_{n,n-1} - \omega_{n,n+1} \simeq \hbar\omega_c^2/m_0 c^2. \tag{5.69}$$

We can regard these curves as giving the behavior, for fixed values of $\omega_{n,n\pm1}$ and t, of $g_\omega(\omega_{n,n+1})$ and $g_\omega(\omega_{n,n-1})$ with ω increasing from right to left. Several regions of ω exist where $g_\omega(\omega_{n,n-1}) > g_\omega(\omega_{n,n+1})$ as can be seen from the lower curves of Fig. 5.18, where the difference between the two sinc functions is plotted. These regions of amplification are slightly displaced from the relativistic cyclotron frequency by an amount determined largely by the interaction time t. As an example, consider a set of oscillators with a cyclotron frequency of 300 GHz interacting with the radiation for a time 10^{-8} sec. We note from Eq. (5.69) that $\omega_{n,n-1} - \omega_{n,n+1} \approx 900$ Hz, and from the plot of Fig. 5.18 that the regions of possible amplification are adjacent to $(\omega_{n,n\pm1} - \omega)t/2 = \pm\pi, \pm2\pi$, etc.

5.4. ELECTRON POSITIVE AND NEGATIVE ABSORPTION

For $t = 10^{-8}$ sec, these occur in bands on either side of the relativistic cyclotron frequency, separated by intervals ~ 100 MHz, with the strongest interaction less than 100 MHz above the relativistic cyclotron frequency.

A more tractable expression for W can be derived from Eq. (5.68) by making the replacement

$$g_\omega(\omega_{n,n+1}) - g_\omega(\omega_{n,n-1}) = (\omega_{n,n-1} - \omega_{n,n+1}) \frac{\partial g_\omega(\omega_{n,n+1})}{\partial \omega}$$

$$= \frac{\hbar \omega_c^2}{m_0 c^2} \frac{\partial g_\omega(\omega_{n,n+1})}{\partial \omega}.$$

The result is

$$W = \frac{N_n q^2 E_0^2 t^2}{2m_0} \left[\frac{\sin^2 \Gamma}{\Gamma^2} + \frac{n\hbar\omega_c^2 t}{2m_0 c^2 \Gamma} \left\{ \frac{2 \sin^2 \Gamma}{\Gamma^2} - \frac{\sin 2\Gamma}{\Gamma} \right\} \right], \quad (5.70)$$

where $\Gamma = (\omega_{n,n+1} - \omega)t/2$. This function is shown in Fig. 5.19 for three values of the factor $n\hbar\omega_c^2 t/m_0 c^2$ which, of course, is a measure of the interaction time and how relativistic the electrons are.

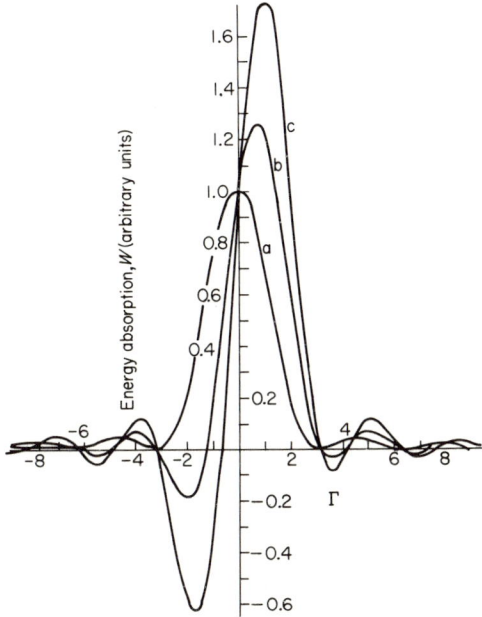

FIG. 5.19. Cyclotron resonance absorption line. W is the energy absorbed in an interaction time t, Γ is half the relative phase shift of the orbital motion with respect to the wave frequency; curve (a) is for nonrelativistic particles and curves (b) and (c) correspond to $n\hbar\omega_c^2 t/m_0 c^2 = 2$ and 4, respectively (after Hsu and Robson[48]).

By a development similar to that of Section 5.1.1., this problem can be treated classically by solving the equations of motion [Eqs. (5.1) and (5.2)], but with the rest mass m_0 replaced by the velocity-dependent relativistic mass m. The resulting nonlinear equation of motion in terms of the position S of a particle rotating in the same sense as a circularly polarized electric field, namely,

$$d\dot{S}/dt + (\dot{S}/m)\, dm/dt + (iqB\dot{S}/m) = qE_0/mv_0, \qquad (5.71)$$

has been solved by Hsu and Robson[48] to yield an expression similar to that given in Eq. (5.70) for the absorbed energy. Again, as in Section 5.1, similar calculations can be carried out for a counterrotating field. Account of this component of a linearly polarized wave simply means an additional term involving $(\omega_{n,n+1} + \omega)t/2$ in the line shape function Eq. (5.67), but its significance in the interaction is small.

---

[48] T. W. Hsu and P. N. Robson, *Electron. Lett.* **1**, 84–85 (1965).

6. WAVE INTERACTIONS IN PLASMAS

Introduction

In this chapter we are concerned with the problem of determining the properties of plasmas from observations of the effects on propagating electromagnetic waves or the emission of electromagnetic radiation.

The dispersion properties of high-frequency waves in plasmas have been studied in ionospheric, plasma, and astrophysical research as a basic investigation of complex wave phenomena and as a means of establishing the dependence of the various wave characteristics on plasma parameters. On the basis of the understanding of wave types thus developed, one can deduce information about the properties and behavior of the plasma itself and relate these to other kinds of measurements.

The plasma may be probed actively with an electromagnetic wave and effects of the plasma on the wave measured, or passive experiments may be performed wherein one observes radiation emission. Developments in the former field have been largely through the use of radio frequencies, microwaves, and optical techniques. However, for many laboratory plasmas the far-infrared wavelength region is about optimum for marked interaction effects but, because of the state of technical development, they are only now coming into extensive use. The experimental methods include interferometry, Faraday rotation, etc., with molecular laser sources, and radiometry with sensitive detectors.

We take as our starting point the equations of electromagnetism and kinetic theory governing wave-particle processes in plasmas. With restrictions on physical processes, and on the orders of magnitude of physical quantities, we derive expressions for those plasma properties which occur in the dispersion equations and establish the dispersion relations themselves. The development is essentially what is known as "temperate" plasma theory, with effects from the more general theory of thermal plasmas stated where necessary. For more intensive analyses of wave-plasma interactions, the reader is referred to the books of Ratcliffe,[1] Allis et al.,[2] Bekefi,[3] and Heald and Wharton.[4]

Radiation from plasmas in the far-infrared is to be expected from transitions between vibrational and rotational energy levels of molecules as well as from the motions of unbound electrons which may or may not be in a state of thermal equilibrium. Little has been done with far-infrared plasma radiation of a molecular origin, but emission from free electrons has been observed. Bremsstrahlung—the incoherent radiation emitted by electrons during encounters with other particles— has been used as a means of measuring radiation temperatures and densities of laboratory plasmas. Cyclotron harmonic radiation in the far-infrared has also been observed. Our discussion of radiation emission in Sections 6.8–6.10 is restricted mainly to an account of the physical processes directly related to the far-infrared laboratory work that has been done. A more complete treatment of the processes of radio and microwave emission from plasmas is given by Bekefi,[3] and Heald and Wharton.[4]

6.1. Plasma Conductivity and Tensor Dielectric Constant

The electromagnetic properties of a plasma are described by the set of Maxwell's equations with the sources of fields, namely the charge and current densities, related to the distribution function which in turn is subject to the kinetic or Boltzmann equation. That is, the distribution function $f_i \equiv f_i(\mathbf{r}, \mathbf{V}, t)$ or number of particles of a given species i per unit volume of velocity space per unit volume of real space, is given in terms of acting forces by Boltzmann's equation

$$\frac{\partial f_i}{\partial t} + \mathbf{V} \cdot \frac{\partial f_i}{\partial \mathbf{r}} + \frac{q_i}{m_i}(\mathbf{E} + \mathbf{V} \times \mathbf{B}) \cdot \frac{\partial f_i}{\partial \mathbf{V}} = \left(\frac{\partial f_i}{\partial t}\right)_{\text{collisions}}, \quad (6.1)$$

where $\mathbf{V}$ is the particle velocity, $\mathbf{B}$ the magnetic induction field, $\mathbf{E}$ the electric field, and $\mathbf{r}$ and t represent position and time.

---

[1] J. A. Ratcliffe, "The Magneto-Ionic Theory and Its Applications to the Ionosphere." Cambridge Univ. Press, London and New York, 1959.

[2] W. P. Allis, S. J. Buchsbaum and A. Bers, "Waves in Anisotropic Plasmas." MIT Press, Cambridge, Massachusetts, 1963.

[3] G. Bekefi, "Radiation Processes in Plasmas." Wiley, New York, 1966.

[4] M. A. Heald and C. B. Wharton, "Plasma Diagnostics with Microwaves." Wiley, New York, 1965.

6.1. PLASMA CONDUCTIVITY AND TENSOR DIELECTRIC CONSTANT

Then in Maxwell's equations

$$\nabla \cdot \mathbf{E} = \varrho/\varepsilon_0, \tag{6.2}$$

$$\nabla \cdot \mathbf{B} = 0, \tag{6.3}$$

$$\nabla \times \mathbf{E} = -\mu_0 \, \partial \mathbf{H}/\partial t, \tag{6.4}$$

$$\nabla \times \mathbf{H} = \mathbf{J} + \varepsilon_0 \, \partial \mathbf{E}/\partial t, \tag{6.5}$$

the source terms are given by

$$\mathbf{J} = \sum_i q_i \int \mathbf{V}_i f_i \, d^3V, \tag{6.6}$$

and

$$\varrho = \sum_i q_i \int f_i \, d^3V, \tag{6.7a}$$

the integrals being carried out over the entire volume of velocity space. In this treatment we will be concerned with only one species, electrons. High-frequency fields interact almost entirely with electrons, the ions being too massive to respond to the field oscillations to any significant extent.

Following Allis *et al.* we now make assumptions to simplify the Boltzmann equation. As the distribution function f is a function of both the thermal velocity v_{th} and the velocity v induced by the oscillating electric field $\mathbf{E}$, the term $\mathbf{E} \cdot \partial f/\partial \mathbf{V}$ is nonlinear in $\mathbf{E}$. It can be linearized by assuming $v_{\text{th}} \gg v$ so that f is essentially independent of $\mathbf{E}$. This assumption also means that the collision frequency for momentum transfer (ν) is temperature dependent but not field dependent. Secondly, we allow only "local" field effects and so avoid the complication of integrating over the spatially varying field $\mathbf{E}$ in the determination of the induced velocity, by assuming that in one period of the oscillating wave the electron travels a distance much shorter than a wavelength. This means that the thermal velocity must be much smaller than the phase velocity v_{ph} of the wave. Our assumptions are thus

$$v \ll v_{\text{th}} \ll v_{\text{ph}}. \tag{6.7b}$$

The inequality $v_{\text{ph}} \gg v_{\text{th}}$ justifies the neglect of alternating magnetic fields thereby completing the linearization of Eq. (6.1). $\mathbf{B}$, then, is the applied static magnetic induction field.

Under these conditions of small amplitude waves in temperate plasma, the Boltzmann equation can be replaced by Newton's equation of motion

$$m \, d\mathbf{v}/dt = q(\mathbf{E} + \mathbf{v} \times \mathbf{B}) - m\mathbf{v}\nu, \tag{6.8}$$

where the three forces on the right-hand side are, respectively, the forces due to the applied electric field, the Lorentz force, and a damping term accounting for the ordered momentum scattered per unit time. In the present model (the "Lorentz model") we treat ν as a constant. Scattering occurs in electron–electron collisions and in electron collisions with neutral atoms and ions, which we assume to be infinitely massive.

If we assume sinusoidal time dependence ($e^{i\omega t}$) of $\mathbf{E}$ and $\mathbf{v}$ and introduce the electron cyclotron frequency vector defined by

$$\boldsymbol{\omega}_c = -(q/m)\mathbf{B}, \tag{6.9}$$

Eq. (6.8) becomes

$$(i\omega + \nu - \boldsymbol{\omega}_c \times)\mathbf{v} = (q/m)\mathbf{E}.$$

Assuming the collision frequency ν to be independent of velocity, this equation can be solved to give

$$\mathbf{v}_\perp = \frac{(i\omega + \nu + \boldsymbol{\omega}_c \times)(q\mathbf{E}_\perp/m)}{(i\omega + \nu)^2 + \omega_c^2}; \quad \mathbf{v}_\parallel = \frac{1}{i\omega + \nu} \frac{q}{m} \mathbf{E}_\parallel, \tag{6.10}$$

where $\mathbf{E}_\parallel$ is the component of the electric field parallel to the $\mathbf{B}$-field which we take to be the z-direction, and the subscript $\perp$ indicates components in the x-y plane.

When multiplied by the charge density nq, Eq. (6.10) gives the nine components of the mobility and hence the nine components of the complex conductivity tensor σ_{ij} relating the current density vector $\mathbf{J}$ to the electric field $\mathbf{E}$ in

$$J_i = \sigma_{ij} E_j, \tag{6.11}$$

where $i, j = 1, 2, 3$, or x, y, z. The result for the conductivity in cartesian coordinates is

$$(\sigma_{ij}) = \frac{nq^2}{m} \begin{bmatrix} \dfrac{i\omega + \nu}{(i\omega + \nu)^2 + \omega_c^2} & \dfrac{-\omega_c}{(i\omega + \nu)^2 + \omega_c^2} & 0 \\ \dfrac{\omega_c}{(i\omega + \nu)^2 + \omega_c^2} & \dfrac{i\omega + \nu}{(i\omega + \nu)^2 + \omega_c^2} & 0 \\ 0 & 0 & \dfrac{1}{i\omega + \nu} \end{bmatrix}. \tag{6.12}$$

6.1. PLASMA CONDUCTIVITY AND TENSOR DIELECTRIC CONSTANT

A more rigorous treatment of conductivity taking account of the velocity dependence of the collision frequency has been given by Margenau[5] and Ginsburg.[6] They show that the Lorentz form of the conductivity can be retained provided that ν is an "effective" collision frequency obtained by averaging over the velocity distribution. The effective collision frequency is given by

$$\nu = \frac{-4\pi}{3} \int_0^\infty \nu(v) \frac{df(v)}{dv} v^3 \, dv,$$

for $\nu \ll \omega$. For the case of a Maxwellian velocity distribution, which is but slightly modified by the imposed high-frequency field,

$$\nu = (8/3\pi^{1/2}) \int_0^\infty \nu(u) u^3 e^{-u^2} \, du, \tag{6.13}$$

where $u = (m/2kT)^{1/2} v$, k being Boltzmann's constant. This matter is discussed by Heald and Wharton.[4]

Spitzer[7] has shown that the collision of electrons with infinitely massive ions leads to an effective collision frequency given by (see Delcroix[8])

$$\nu \approx 2.6 \, nZ \ln \Lambda / T^{3/2}, \tag{6.14}$$

where n is in electrons per cubic centimeter, T is the electron temperature in degrees Kelvin, and Z is the ion charge number. The quantity $\ln \Lambda$ is slowly varying and of the order of 10 for many laboratory plasmas. Values for $\ln \Lambda$, and expressions for $\ln \Lambda$, in terms of T, n, and Z are given by Delcroix.[8]

6.1.1. Conductivity in Rotating Coordinates

Let us transform the physical quantities in Eq. (6.11) to coordinates rotating with the wave frequency ω. Denoting the right-hand sense of rotation by the subscript r and the left-hand sense by l, we have

$$E_r = |E_r| e^{i\omega t} = (1/\sqrt{2})(E_x + iE_y), \tag{6.15}$$

$$E_l = |E_l| e^{-i\omega t} = (1/\sqrt{2})(E_x - iE_y). \tag{6.16}$$

---

[5] H. Margenau, *Phys. Rev.* **109**, 6–9 (1958).
[6] V. L. Ginsburg, *J. Phys. (USSR)* **8**, 253 (1944).
[7] L. Spitzer, "Physics of Fully Ionized Gases." Wiley (Interscience), New York, 1956.
[8] J. L. Delcroix, "Introduction to the Theory of Ionized Gases." Wiley (Interscience), New York, 1960.

In matrix notation, the field and current vectors are transformed by the matrix

$$\mathsf{C} = \frac{1}{\sqrt{2}} \begin{bmatrix} 1 & i \\ 1 & -i \end{bmatrix}.$$

By applying this transformation to Eq. (6.11), the form of the conductivity tensor in the rotating coordinate system can be found. It is a diagonalized matrix whose form can be obtained by considering the 2×2 cartesian matrix formed of the elements σ_{11}, σ_{12}, σ_{21}, and σ_{22} of Eq. (6.12), the third row and column being already diagonalized. Denote the conductivity matrix by $\Lambda = (\delta_{ij}\lambda_k)$, where δ_{ij} is unity if $i = j$ and zero otherwise. We write

$$\begin{bmatrix} J_r \\ J_l \end{bmatrix} = (\delta_{ij}\lambda_k) \begin{bmatrix} E_r \\ E_l \end{bmatrix}.$$

That is,

$$\mathsf{C} \begin{bmatrix} J_x \\ J_y \end{bmatrix} = (\delta_{ij}\lambda_k) \, \mathsf{C} \begin{bmatrix} E_x \\ E_y \end{bmatrix}.$$

Multiplying on the right by C^{-1}, the inverse matrix of C, we obtain

$$\begin{bmatrix} J_x \\ J_y \end{bmatrix} = \mathsf{C}^{-1}(\delta_{ij}\lambda_k) \, \mathsf{C} \begin{bmatrix} E_x \\ E_y \end{bmatrix},$$

from which it follows that

$$\sigma_{ij} = \mathsf{C}^{-1}(\delta_{ij}\lambda_k)\mathsf{C}.$$

Multiplying on the right by C, we obtain

$$\begin{bmatrix} 1 & i \\ 1 & -i \end{bmatrix} \begin{bmatrix} \sigma_{11} & \sigma_{12} \\ -\sigma_{12} & \sigma_{22} \end{bmatrix} = \begin{bmatrix} \lambda_1 & 0 \\ 0 & \lambda_2 \end{bmatrix} \begin{bmatrix} 1 & i \\ 1 & -i \end{bmatrix}.$$

Hence the elements of the diagonalized matrix are

$$\lambda_1 = \sigma_{11} - i\sigma_{12},$$
$$\lambda_2 = \sigma_{11} + i\sigma_{12}.$$

Thus the circularly polarized and longitudinal components of the electric field and current density vectors are related through the diagonalized

6.1. Plasma Conductivity and Tensor Dielectric Constant

tensor conductivity in the following way:

$$\begin{bmatrix} J_r \\ J_l \\ J_z \end{bmatrix} = \frac{nq^2}{m} \begin{bmatrix} \frac{i(\omega + \omega_c) + \nu}{(i\omega + \nu)^2 + \omega_c^2} & 0 & 0 \\ 0 & \frac{i(\omega - \omega_c) + \nu}{(i\omega + \nu)^2 + \omega_c^2} & 0 \\ 0 & 0 & \frac{1}{i\omega + \nu} \end{bmatrix} \begin{bmatrix} E_r \\ E_l \\ E_z \end{bmatrix}. \tag{6.17}$$

It should be noted that our assignments Eqs. (6.9) and (6.15) imply that the right-hand circularly polarized (RHCP) wave has the same sense of rotation as the plasma electrons and the left-hand circularly polarized (LHCP) wave has the opposite sense of rotation. As a consequence the diagonal element of Eq. (6.17) representing the RHCP wave shows the occurrence of a resonance when $\omega = \omega_c$, while the element representing the LHCP wave shows no such resonance.

6.1.2. Effective Dielectric Constant

Instead of explicitly using conduction currents in the formulation of plasma-wave phenomena it is convenient to characterize the plasma by an effective tensor dielectric constant $(\varkappa_{ij})$, which is evolved when the Maxwell Eq. (6.5) is written in the following alternative forms:

$$\begin{aligned} \nabla \times \mathbf{H} &= (\sigma_{ij})\mathbf{E} + \varepsilon_0 \, \partial \mathbf{E}/\partial t \\ &= (\sigma_{ij})\mathbf{E} + \varepsilon_0 i\omega \mathbf{E} \\ &= [(\delta_{ij}) + (\sigma_{ij})/i\omega\varepsilon_0]\varepsilon_0 i\omega \mathbf{E} \\ &= \varepsilon_0(\varkappa_{ij}) \, \partial \mathbf{E}/\partial t, \end{aligned}$$

where ε_0 is the permittivity of free space, and $\varkappa_{ij}$ is called the dielectric constant or relative permittivity.

The effective dielectric constant is thus

$$(\varkappa_{ij}) = (\delta_{ij}) + [(\sigma_{ij})/i\omega\varepsilon_0] = \begin{Bmatrix} \varkappa_\perp & \varkappa_X & 0 \\ -\varkappa_X & \varkappa_\perp & 0 \\ 0 & 0 & \varkappa_\parallel \end{Bmatrix}. \tag{6.18}$$

New symbols for the elements of $(\varkappa_{ij})$ are defined in Eq. (6.18); their values are obtained by substitution from Eq. (6.12).

6.2. Wave Propagation and Dispersion Relations

The wave equation in **E**, obtained by taking the curl of Eq. (6.4) and substituting from Eq. (6.5), is

$$\nabla \times (\nabla \times \mathbf{E}) + \mu_0 \varepsilon_0 (\varkappa_{ij}) \, \partial^2 \mathbf{E}/\partial t^2 = 0.$$

For a plane wave varying as $e^{i(\omega t - \mathbf{k} \cdot \mathbf{r})}$, the wave equation may be written in terms of the propagation constant **k** as

$$\mathbf{k} \times \mathbf{k} \times \mathbf{E} + k_0^2 (\varkappa_{ij}) \mathbf{E} = 0,$$

where k_0 is the propagation constant in free space. In terms of the vector refractive index $\boldsymbol{\mu} = \mathbf{k}/k_0$, we have

$$\boldsymbol{\mu} \times \boldsymbol{\mu} \times \mathbf{E} + (\varkappa_{ij}) \mathbf{E} = 0. \tag{6.19}$$

Written in scalar form, the wave equation in cartesian coordinates becomes the three linear homogeneous equations

$$(\varkappa_\perp - \mu_y^2 - \mu_z^2) E_x + (\varkappa_X + \mu_x \mu_y) E_y + \mu_x \mu_z E_z = 0,$$
$$(-\varkappa_X + \mu_x \mu_y) E_x + (\varkappa_\perp - \mu_x^2 - \mu_z^2) E_y + \mu_y \mu_z E_z = 0,$$
$$\mu_x \mu_z E_x + \mu_y \mu_z E_y + (\varkappa_\parallel - \mu_x^2 - \mu_y^2) E_z = 0,$$

where we have substituted from Eq. (6.18).

It follows from Cramer's rule for the solution of simultaneous linear equations that nontrivial solutions of a homogeneous system can exist only if the coefficient determinant is zero. That is,

$$\begin{vmatrix} \varkappa_\perp - \mu_y^2 - \mu_z^2 & \varkappa_X + \mu_x \mu_y & \mu_x \mu_z \\ -\varkappa_X + \mu_y \mu_x & \varkappa_\perp - \mu_z^2 - \mu_x^2 & \mu_y \mu_z \\ \mu_x \mu_z & \mu_y \mu_z & \varkappa_\parallel - \mu_x^2 - \mu_y^2 \end{vmatrix} = 0.$$

Now specify the direction of the vector refractive index by $\mu_x = \mu \sin \theta$, $\mu_y = 0$, and $\mu_z = \mu \cos \theta$, where θ is the angle between the direction of propagation and the static magnetic field. The determinant now takes the form

$$\begin{vmatrix} \varkappa_\perp - \mu^2 \cos^2 \theta & \varkappa_X & \mu^2 \sin \theta \cos \theta \\ -\varkappa_X & \varkappa_\perp - \mu^2 & 0 \\ \mu^2 \sin \theta \cos \theta & 0 & \varkappa_\parallel - \mu^2 \sin^2 \theta \end{vmatrix} = 0,$$

and reduces to the equation

$$A\mu^4 - C\mu^2 + D = 0, \qquad (6.20)$$

with

$$A = \varkappa_\perp \sin^2\theta + \varkappa_\| \cos^2\theta,$$
$$C = \varkappa_\perp \varkappa_\| + (\varkappa_\perp^2 + \varkappa_\|^2)\sin^2\theta + \varkappa_\perp \varkappa_\| \cos^2\theta,$$
$$D = \varkappa_\| \varkappa_r \varkappa_l,$$
$$\varkappa_r = \varkappa_\perp + i\varkappa_X,$$
$$\varkappa_l = \varkappa_\perp - i\varkappa_X.$$

Here $\varkappa_r$ and $\varkappa_l$ are respectively the effective dielectric constants presented to the right- and left-hand rotating **E**-vectors.

An alternative form of the solution of the wave equation is[9]

$$\tan^2\theta = -\frac{\varkappa_\|(\mu^2 - \varkappa_r)(\mu^2 - \varkappa_l)}{(\mu^2 - \varkappa_\|)(\varkappa_\perp \mu^2 - \varkappa_r \varkappa_l)}. \qquad (6.21)$$

This form gives the (complex) index of refraction as a function of wave frequency and direction of propagation θ. It is easily reduced to the form (6.20). The directions $\theta = 0$ and $\theta = \pi/2$ are simple special cases important from a measurement point of view.

6.3. Propagation across a Magnetic Field

The dispersion relations for wave propagation perpendicular to the applied static magnetic induction field **B** are obtained from Eq. (6.21) by setting $\theta = \pi/2$. Under these conditions, $\tan\theta$ is infinite, implying that the denominator of the right-hand side is zero. There are two possible waves whose refractive indices are

$$\mu_{\text{ord}} = (\varkappa_\|)^{1/2}, \qquad (6.22)$$

and

$$\mu_{\text{ext}} = (\varkappa_r \varkappa_l / \varkappa_\perp)^{1/2}. \qquad (6.23)$$

The subscripts ord and ext designate the well-known "ordinary" and "extraordinary" waves, respectively. They are illustrated in Fig. 6.1.

[9] E. Aström, *Ark. Fys.* **2**, 442 (1950).

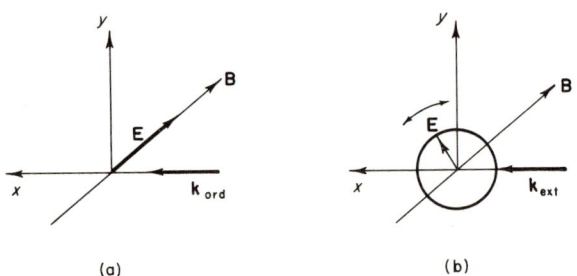

FIG. 6.1. Diagram showing the directions of propagation of the ordinary (a) and extraordinary (b) waves in a magnetized plasma. The E-vector of the extraordinary wave rotates in the x-y plane; it is partly a transverse and partly a longitudinal wave.

The ordinary wave described by the dispersion relation Eq. (6.22) propagates at right angles to **B** with its E-vector parallel to **B**, and its propagation characteristics are the same as those of a wave in an unmagnetized plasma. This wave is commonly used to probe plasmas as, for example, in interferometric measurements of electron density. Its phase velocity and attenuation are independent of the applied magnetic field.

Denote the phase and attenuation constants by γ and α so that

$$k = \gamma - i\alpha \tag{6.24}$$

in the wave propagating as $e^{i(\omega t - \mathbf{k} \cdot \mathbf{r})}$. Now using Eq. (6.22) with the value of $\varkappa_\parallel$ given by Eqs. (6.12) and (6.18), we obtain

$$\mu_{\text{ord}} = \frac{c}{\omega}(\gamma - i\alpha) = \left[1 - \frac{\omega_p^2(\omega + i\nu)}{\omega(\omega^2 + \nu^2)}\right]^{1/2}, \tag{6.25}$$

where $\omega_p^2 = nq^2/\varepsilon_0 m$ is the electron plasma frequency, and where we have substituted $k_0 = \omega/c$, c being the velocity of light in vacuum.

The general properties of the ordinary wave can be seen from Eq. (6.25) taken in the collisionless case $\nu = 0$,

$$\mu_{\text{ord}} = [1 - (\omega_p^2/\omega^2)]^{1/2}, \quad \text{when} \quad \nu = 0. \tag{6.26}$$

The refractive index can be real or pure imaginary according to whether the wave frequency is larger or smaller than the plasma frequency. When $\omega > \omega_p$, the wave propagates in the plasma and its wavelength is

$$\lambda_{\text{ord}} = \frac{\lambda_0}{[1 - (\omega_p^2/\omega^2)]^{1/2}}, \tag{6.27}$$

where $\lambda_0 = 2\pi/k_0$ is the free-space wavelength. Thus the wavelength in the plasma is longer than λ_0 and approaches infinity as ω approaches ω_p. When ω becomes less than ω_p, λ_{ord} is imaginary and wave propagation is "cut off."

The amplitude reflection coefficient at the vacuum–plasma boundary,

$$R = (1 - \mu_{\text{ord}})/(1 + \mu_{\text{ord}}), \tag{6.28}$$

rises to unit magnitude when $\omega \lesssim \omega_p$ and the wave is totally reflected. This phenomena is similar to total internal reflection in optics and waveguide cutoff in microwaves. Within the plasma the wave is "evanescent" with amplitude decaying exponentially with penetration, and with the electric and magnetic field vectors in phase quadrature giving a zero Poynting vector. As we have said, the wave is perfectly reflected. However, although its amplitude remains unchanged for all $\omega_p/\omega \geq 1$, its phase change at reflection increases as ω_p/ω increases beyond unity. These conclusions all follow from Eq. (6.28) with μ_{ord} imaginary.

The collision frequency ν is usually much less than the wave and plasma frequencies when we are working with microwaves or far-infrared radiation. As we see from Eq. (6.25), the inclusion of ν makes μ_{ord} complex. Except when $\omega \lesssim \omega_p$ the imaginary part is much less than the real part which is closely approximated by Eq. (6.26). Collisions have a pronounced effect when $\omega \approx \omega_p$ where they tend to "round off" the step function transition at $\omega = \omega_p$ in the collisionless plasma.

It is usual to speak of a critical cutoff density n_c defined in relation to the wave frequency ω by

$$\omega^2 = n_c q^2/\varepsilon_0 m.$$

In terms of n_c and n, Eq. (6.26) can be written

$$\mu_{\text{ord}} = [1 - (n/n_c)]^{1/2}, \quad \text{when} \quad \nu = 0. \tag{6.29}$$

The properties of the extraordinary wave follow from Eq. (6.23) when $\varkappa_\perp$, $\varkappa_r$, and $\varkappa_l$ are substituted from Eqs. (6.12) and (6.18) using the relations $\varkappa_r = \varkappa_\perp + i\varkappa_X$ and $\varkappa_l = \varkappa_\perp - i\varkappa_X$. The dispersion relation for this wave in the case of no collisions is

$$\mu_{\text{ext}}^2 = \frac{[1 - (\omega_p^2/\omega^2)]^2 - (\omega_c^2/\omega^2)}{[1 - (\omega_p^2/\omega^2)] - (\omega_c^2/\omega^2)}. \tag{6.30}$$

We note first that Eq. (6.30) reduces to Eq. (6.26) when the magnetic field is zero, that is, when $\omega_c = 0$. Propagation of this wave cuts off at the following two conditions:

$$\omega_p^2/\omega^2 = 1 + (\omega_c/\omega),$$
$$\omega_p^2/\omega^2 = 1 - (\omega_c/\omega).$$

If $\omega^2 > \omega_p^2 + \omega_c^2$, the refractive index is negative when

$$[1 - (\omega_p^2/\omega^2)]^2 < \omega_c^2/\omega^2.$$

If $\omega^2 < \omega_p^2 + \omega_c^2$, it is negative when

$$[1 - (\omega_p^2/\omega^2)]^2 > \omega_c^2/\omega^2.$$

It is to be noted that the extraordinary wave can propagate at plasma densities that are above the critical cutoff density n_c for the ordinary wave. If $\omega_c > \omega$ it will propagate at densities more than twice n_c.

There is a resonance when the denominator of Eq. (6.30) becomes zero and μ_{ext} infinite at the frequency $\omega = (\omega_p^2 + \omega_c^2)^{1/2}$. This combination of the plasma and cyclotron frequencies is called the *upper hybrid frequency*.

In Eq. (6.23), the effective refractive index for the extraordinary wave is expressed in terms of $\varkappa_r$, $\varkappa_l$, and $\varkappa_\perp$ and hence it involves "cross product" terms as given in Eqs. (6.12) and (6.18). That is, although it is excited by a plane polarized wave incident on the plasma with its **E**-vector in the y-direction, inside the plasma currents associated with σ_{yy} and σ_{xy} arise. The extraordinary wave is elliptically polarized with the **E**-vector in the plane normal to the static magnetic field. The longitudinal field component will involve space-charge field effects. The electron experiences an ac space-charge force in addition to the magnetic force, and this results in a frequency displacement of the cyclotron resonance from the frequency $\omega = \omega_c$ to the upper hybrid frequency

$$\omega = (\omega_p^2 + \omega_c^2)^{1/2}.$$

6.4. Propagation along a Magnetic Field

When the wave propagates parallel to the **B**-field, $\theta = 0$ and the numerator of the right-hand side of Eq. (6.21) must equal zero. From this condition two dispersion relations result:

$$\mu_r = (\varkappa_r)^{1/2}, \tag{6.31}$$

6.4. PROPAGATION ALONG A MAGNETIC FIELD

and
$$\mu_1 = (\varkappa_1)^{1/2}. \tag{6.32}$$

From our previous discussion it is clear that $\varkappa_r$ and $\varkappa_l$ are the effective dielectric constants corresponding to the first and second diagonal elements of the conductivity tensor in rotating coordinates, Eq. (6.17). They thus represent the dielectric constants "seen" by the right-hand and left-hand circularly polarized waves, respectively. Their polarization properties are illustrated in Fig. 6.2.

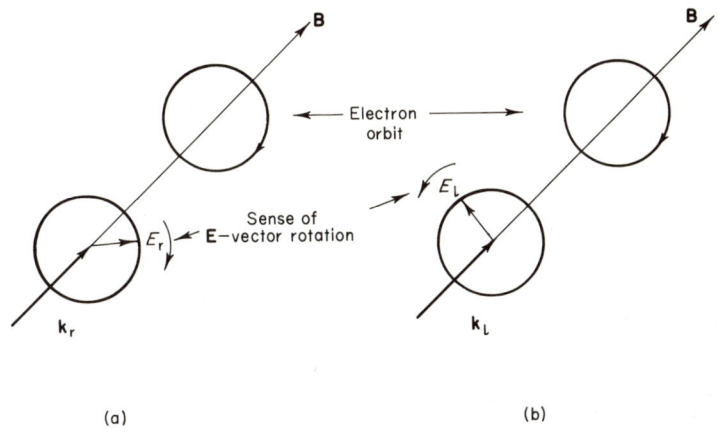

FIG. 6.2. Diagram showing the polarizations of the RHCP and LHCP waves and the direction of propagation parallel to the applied magnetic induction field **B**. The RHCP field rotates in the same sense as the plasma electrons and the LHCP wave rotates in the opposite sense.

For the RHCP wave, the refractive index is

$$\mu_r = \left[1 + \frac{\omega_p^2}{\omega} \left\{\frac{(\omega + \omega_c) - i\nu}{(i\omega + \nu)^2 + \omega_c^2}\right\}\right]^{1/2} \tag{6.33}$$

$$= \left[1 + \frac{\omega_p^2}{\omega(\omega_c - \omega)}\right]^{1/2}, \quad \text{when} \quad \nu = 0. \tag{6.34}$$

For the LHCP wave,

$$\mu_l = \left[1 + \frac{\omega_p^2}{\omega} \left\{\frac{(\omega - \omega_c) - i\nu}{(i\omega + \nu)^2 + \omega_c^2}\right\}\right]^{1/2} \tag{6.35}$$

$$= \left[1 - \frac{\omega_p^2}{\omega(\omega + \omega_c)}\right]^{1/2}, \quad \text{when} \quad \nu = 0. \tag{6.36}$$

We note that the expressions for μ_r and μ_l differ only in the sign of ω_c as we expect, since the electrons orbit in the same sense as the RHCP wave and in the opposite sense to the LHCP wave. The LHCP wave has no resonance, but it is cut off when

$$\omega_p{}^2 \geq \omega(\omega + \omega_c),$$

that is, when

$$\omega \leq [(\omega_c{}^2 + 4\omega_p{}^2)^{1/2} - \omega_c]/2.$$

The RHCP wave has a resonance when $\omega = \omega_c$. Here μ_r approaches infinity, and hence the wavelength in the plasma approaches zero. This is in contrast to the behavior at cutoff, where the wavelength becomes very long. The RHCP wave has a cutoff region in the range of ω given by

$$\omega_c < \omega \leq [(\omega_c{}^2 + 4\omega_p{}^2)^{1/2} + \omega_c]/2.$$

In a low-density plasma ($\omega_p \ll \omega$) the region of cutoff is a narrow band on the high-frequency side of ω_c. When the plasma density is high, as is the case in solids, cutoff occurs over a broad band extending upwards from ω_c.

At frequencies corresponding to the onset of cutoff, perfect reflection occurs abruptly in the collisionless case. When collisions are included, the step-function transition has its sharp edge smoothed out. At frequencies that are cut off the reflection coefficient is now large but not quite unity, and propagation is possible but with very large attenuation.

6.5. Comments on Warm Plasma Effects

In Section 6.1 we have specified the restrictions required for the temperate plasma theory to apply. When thermal velocities are taken into account the equations are modified in form and show additional effects. These effects are discussed by Heald and Wharton,[4] and by Allis et al.[2] A particular modification is that the term $\varkappa_\parallel$ in the numerator of Eq. (6.21) is replaced by a more general term representing a "longitudinal plasma wave" or "space-charge wave" propagating in the $\theta = 0$ direction.

Space-charge waves can be shown in an approximate theory[4] to satisfy the dispersion relation

$$\mu^2 \simeq \frac{c^2}{v_{\text{th}}^2}\left[1 - \frac{\omega_p{}^2}{\omega^2} - \frac{\omega_c{}^2 \sin^2\theta}{\omega^2 - \omega_c{}^2 \cos^2\theta}\right], \quad \frac{v_{\text{th}}}{v_{\text{ph}}} \ll 1, \quad (6.37)$$

6.5. COMMENTS ON WARM PLASMA EFFECTS

for any general direction of propagation θ. The thermal velocity is related to the electron temperature by $v_{th} = (3kT/m)^{1/2}$, and the plasma is assumed collisionless. Here k is Boltzmann's constant. The condition $v_{th}/v_{ph} \ll 1$ requires that the electron velocity distribution function be small at the wave phase velocity.

To satisfy the condition $v_{th}/v_{ph} \ll 1$, it is clear from Eq. (6.37) that space-charge propagation in the $\theta = \pi/2$ direction occurs only at frequencies close to the upper hybrid $\omega = (\omega_p^2 + \omega_c^2)^{1/2}$, and in the $\theta = 0$ direction at the plasma frequency, ω_p.

The properties of these waves in an unmagnetized plasma were studied by Bohm and Gross,[10] and Eq. (6.37) with $\theta = 0$ is called the Bohm–Gross dispersion equation. The wave polarization and the direction of electron motion are longitudinal. The phase velocity is above the thermal velocity and approaches infinity at $\omega = \omega_p$. In the temperate or "cold" plasma approximation there is a singularity at $\omega = \omega_p$ where a longitudinal plasma oscillation with infinite wavelength is excited. This corresponds to the solution $\varkappa_\| = 0$ in Eq. (6.21), a solution which is not a wave and which we have therefore ignored.

In a warm plasma, the refractive indices of the ordinary, extraordinary, RHCP, and LHCP waves are given, to first order, by[4]

$$\mu_{ord}^2 \simeq \left[1 - \frac{\omega_p^2}{\omega^2}\right]\left[1 - \frac{\omega_p^2}{\omega^2}\frac{kT}{mc^2}\right], \tag{6.38}$$

$$\mu_{ext}^2 \simeq \frac{[1 - (\omega_p^2/\omega^2)]^2 - (\omega_c^2/\omega^2)}{[1 - (\omega_p^2/\omega^2)] - (\omega_c^2/\omega^2)}\left[1 - \frac{\omega_p^2}{\omega^2}\frac{kT}{mc^2}\right], \tag{6.39}$$

$$\mu_r^2 \simeq \left[1 + \frac{\omega_p^2}{\omega(\omega_c - \omega)}\right]\left[1 - \frac{\omega_p^2}{\omega^2}\frac{kT}{mc^2}\right], \tag{6.40}$$

$$\mu_l^2 \simeq \left[1 - \frac{\omega_p^2}{\omega(\omega_c + \omega)}\right]\left[1 - \frac{\omega_p^2}{\omega^2}\frac{kT}{mc^2}\right]. \tag{6.41}$$

For normal (nonrelativistic) velocities, then, the temperate plasma dispersion equations are seen to be good approximations under many conditions. Of course, near resonance, where the phase velocity becomes small, we must expect limitations.

---

[10] D. Bohm and E. P. Gross, *Phys. Rev.* **75**, 1851–1863, 1864–1876 (1949).

6.6. Phase Shift and Attenuation Measurements

Measurements of the electron density and collision frequency of a plasma can be made by passing waves whose dispersion equations are known into the body of the plasma, or by observing the reflections of well-understood waves from plasma surfaces. Provided the wave frequency is not too far removed from the plasma frequency one can, on the basis of such dispersion relations as Eqs. (6.25), (6.30), (6.33), and (6.35), deduce information about n and ν. The observations involve phase shift and attenuation measurements or changes in wave polarization as, for example, Faraday rotation.

The phase and attenuation constants of the ordinary wave in a magnetized plasma are particularly simple and independent of the magnetic field, and for this reason the ordinary wave is frequently used for plasma measurements. For this wave, and for waves in unmagnetized plasmas, when $\nu \ll \omega$, Eqs. (6.24) and (6.25) give the phase constant as

$$\gamma = \frac{\omega}{c}\left(1 - \frac{\omega_p^2}{\omega^2}\right)^{1/2} = \frac{2\pi}{\lambda}\left(1 - \frac{n}{n_c}\right)^{1/2}, \qquad (6.42)$$

and the attenuation constant as

$$\alpha = \frac{\omega_p^2 \nu}{2c\omega^2} \frac{1}{[1 - (\omega_p^2/\omega^2)]^{1/2}} = \frac{n\nu}{2n_c c}\left(1 - \frac{n}{n_c}\right)^{-1/2}, \qquad (6.43)$$

where λ is the free-space wavelength.

As defined earlier the critical cutoff density n_c is determined by the wave frequency $f \equiv \omega/2\pi$. When numerical constants for the electron are substituted, the relation is

$$f = 8.98 \times 10^3 n_c^{1/2},$$

where n_c is the number of electrons per cubic centimeter, and f is in Hz. Hence, for an electron density of 10^{15} cm^{-3} only waves with frequencies higher than 285 GHz can penetrate into the plasma. Frequencies lower than this are reflected. Much higher frequencies (e.g., light), on the other hand, make ω_p^2/ω^2 very small and hence reduce the sensitivity of γ and α to the electron density. The frequency for probing a plasma should actually be somewhat above ω_p for the following reason: The measured phase shift ϕ produced by a plasma of thickness L on a propagated wave

6.6. PHASE SHIFT AND ATTENUATION MEASUREMENTS

is expressed in terms of the electron density by

$$\phi = \frac{\omega}{c} \int_0^L \left[1 - \left(1 - \frac{n(x)}{n_c} \right)^{1/2} \right] dx, \tag{6.44}$$

where $n(x)$ is, in general, a function of position along the path of propagation. It is clear that this measured phase change can only be used to give meaningful density information if the dependence on x is known. Frequently, however, the spatial density distribution is not known, and under these circumstances it is necessary to limit the ratio ω_p^2/ω^2 by using a frequency greater than the plasma frequency, so that ϕ is approximately linearly dependent on n. For $\omega_p^2/\omega^2 \lesssim 0.3$, expansion of Eq. (6.44) by the binomial theorem gives

$$\phi \simeq (\pi/\lambda n_c) \int_0^L n(x) \, dx,$$

which yields a meaningful average density along the propagation path

$$\bar{n} = (1/L) \int_0^L n(x) \, dx = (\lambda \phi/\pi L) n_c.$$

We conclude that, in the absence of detailed information of the profile, density measurements in a plasma with $n \sim 10^{15}$ cm^{-3} is best carried out with far-infrared frequencies of the order of 500 GHz.

When the density has been determined from phase measurements, the attenuation can be used to obtain the collision frequency ν through Eq. (6.43). The attenuation along a path length L is $\exp[-\int_0^L \alpha \, dx]$. If the dominant scattering process is electron–ion collisions, one can then obtain estimates of the electron temperature T on the basis of Eq. (6.14). In measuring α, one must, of course, be careful to allow for losses due to reflections at plasma surfaces.

As we have said, most interferometric studies of plasmas have used microwaves and light. With microwave wavelengths as short as two millimeters, plasma densities up to about 10^{14} cm^{-3} can be studied. Light wavelengths are convenient for very high densities and their use to densities as low as, perhaps, 10^{15} cm^{-3} may be possible provided the path length is sufficiently long or fractional fringe shift measurement techniques are developed.

A typical microwave phase-sensitive bridge is shown in Fig. 6.3(a). This is a Mach–Zehnder interferometer with the wave divided into two paths, one through the plasma and one which acts as a phase reference

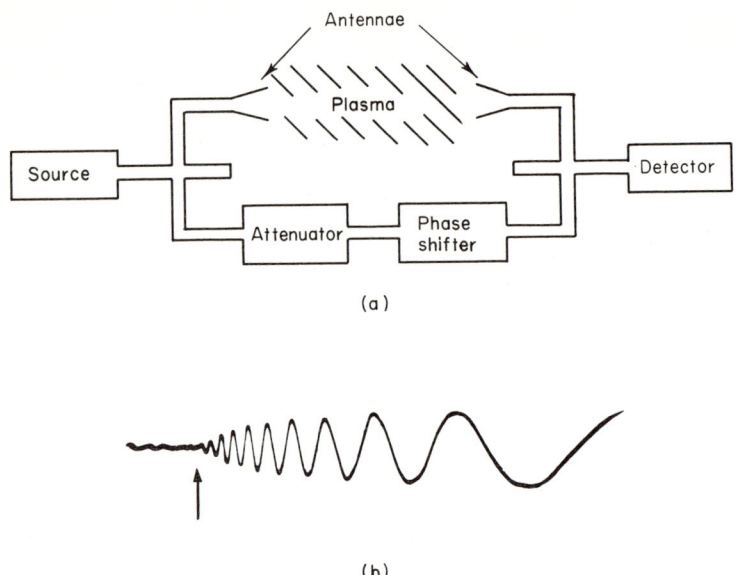

Fig. 6.3. (a) Diagram of a Mach–Zehnder interferometer as used at microwave frequencies for plasma density and collision frequency measurements. (b) Plasma interferogram recorded with a Mach–Zehnder interferometer at 35 GHz. The arrow indicates the time at which the density falls below the critical cutoff density. The period of the interference pattern gives the time-dependent electron density, while the amplitude attenuation is a measure of the collision frequency.

wave. In a transient plasma, for example, as the electron density rises during gas breakdown the mixing of the two waves produces interference beats. When the density rises above the cutoff level n_c, only the reference wave reaches the detector, but, later, when the density has decayed to below n_c, further interference beats occur. A microwave interferogram showing a period of cutoff followed by the onset of propagation is shown in Fig. 6.3(b). The amplitude modulation is due to time-varying attenuation. Bridges of this type are extensively used in plasma laboratories. They are simple and relatively trouble free and with crystal video detectors have very fast (submicrosecond) response times. The reduction of density data from oscilloscope records is quite straightforward. Figure 6.4 shows microwave interferograms produced by an interferometer in which the wave travels twice through the plasma. After the first transit, the wave signal is reradiated from a modulated reflector (a dipole with an alternating current-carrying crystal diode at its center[11]) back through

[11] L. C. Robinson, *Rev. Sci. Instrum.* **35**, 1239–1240 (1964).

6.6. PHASE SHIFT AND ATTENUATION MEASUREMENTS

the plasma. Double passage, of course, doubles the sensitivity to density changes. The interferograms of Fig. 6.4 show periods of cutoff followed by amplitude modulated envelopes produced by interference and attenuation as the density decays. Figure 6.5 illustrates a plasma interferometer based on a helium–neon laser (output lines are produced at 0.6328 μm and 3.39 μm), and an interferogram recorded by Dellis et al.[12] with the

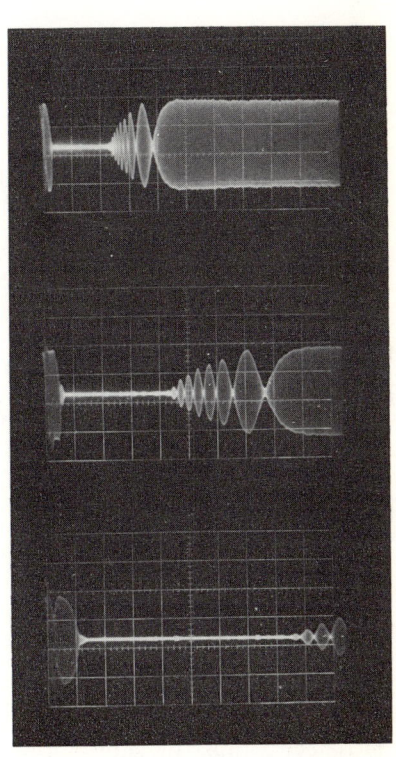

FIG. 6.4. Interference effects recorded with a 35-GHz plasma interferometer in which the wave travels twice through the plasma. After the first transmission it is modulated by a crystal diode carrying a 5-MHz current, and, after the return passage through the plasma it beats with the primary wave. Periods of cutoff are shown, followed by interference envelopes which give a measure of the plasma density and collision frequency. The time scales are 200 μsec cm^{-1} (upper), 100 μsec cm^{-1}, and 50 μsec cm^{-1} (lower).

3.39-μm line. For this line $\omega_p^2/\omega^2 \simeq 10^{-4}$ when $n = 10^{16}$ cm^{-3}. Sensitivity to density changes is enhanced by multiple passage of the light beam through the plasma, and to this end the laser is separated from the plasma by a partially transmitting mirror with a fully reflecting mirror placed at the opposite end of the plasma. Such an arrangement is a coupled cavity system which results in amplitude modulation of the laser output as the changing electron density varies the effective length of the cavity containing the plasma.

[12] A. N. Dellis, W. H. F. Earl, A. Malein and S. Ward, Nature (London) **207**, 56–59 (1965).

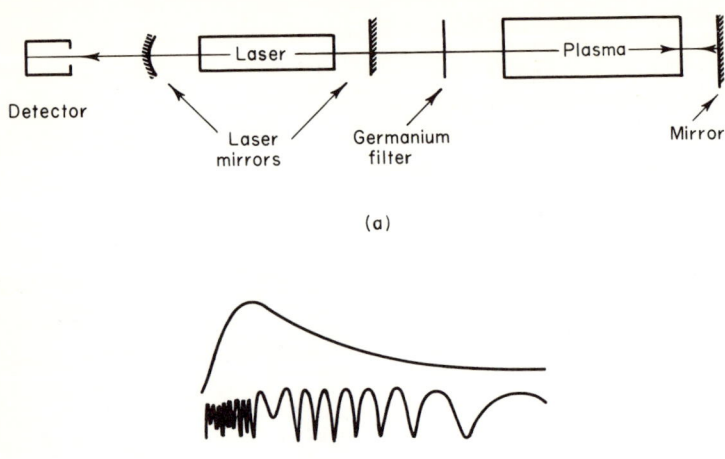

Fig. 6.5. (a) HeNe laser interferometer used in optical and infrared plasma studies. The middle mirror is semitransparent and the two cavities form a coupled system. Density changes in the plasma amplitude modulate the laser output and this modulation gives a measure of the plasma density. The germanium filter separates the 0.6328-μm HeNe line from the 3.39-μm line. (b) Interferogram recorded[12] with the 3.39-μm HeNe line in the interferometer (a). The upper trace shows the gas current pulse. The traces are 1 msec long (after Dellis et al.[12]).

Far-infrared plasma interferometers in present use are largely based on gas laser sources and fast solid-state detectors. Gas lasers are fixed in frequency but this is no disadvantage in studies of transient plasmas where ω_p shifts with respect to ω. The time response requirements of the detector must be shorter than the time taken for the "optical" path through the plasma to change by some small fraction of a wavelength, say, 0.1λ or less depending on the accuracy of fractional fringe counting required. The Putley indium antimonide detector (see Sections 4.3.2 and 4.3.3) when used with an appropriate wide-band postdetector amplifier has a response time of about 0.1 μsec; the gallium-doped germanium detector also has submicrosecond response time when operated in a suitable circuit. These two detectors overlap in the spectral region of 100 μm, the Ge:Ga covering 30–130 μm and the InSb detector extending from below 100 μm up to millimeter and centimeter wavelengths. A far-infrared Mach–Zehnder interferometer with a gas laser source and InSb and Ge:Ga cryogenic detectors is illustrated in Fig. 6.6. The diagram shows a pulsed laser, but it should be noted that some of the emission lines generated by gas lasers can operate continuously (see Chapter 2). Radiation is separated and recombined by Mylar beam splitters (see Section 3.6.5) and a phase shifter (high-density polyethylene sheet) adjusts for a null phase difference between the two waves in the absence of plasma. By this adjustment small phase changes are facilitated.

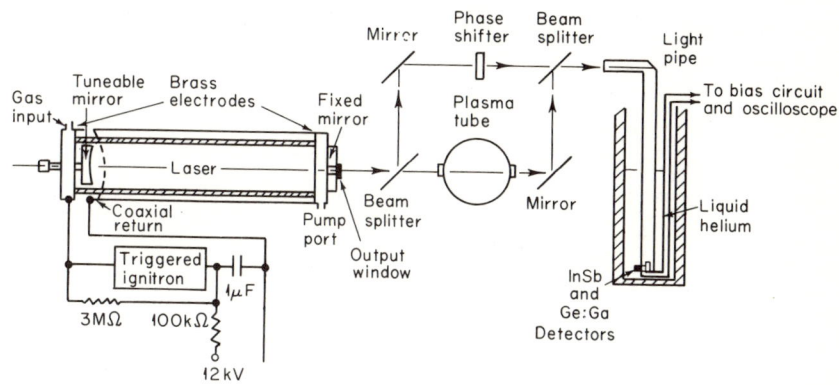

Fig. 6.6. Far-infrared gas laser and Mach–Zehnder plasma interferometer (after Turner and Poehler[13]).

With appropriate gases in the laser, emission lines from the HCN molecule produce a very strong line at 337 μm, a particularly good location in the spectrum for use with many high-density plasmas. Several other lines can be generated with HCN, and of these the 128-μm line has been used. It is generally weaker than the 337-μm line but its higher frequency raises the upper density limit by about one order of magnitude. Shorter wavelength lines can be generated with a water vapor laser discharge, and of these lines those at 119, 79, and 48 μm have been used in plasma experiments. All five lines have been employed by Turner and Poehler[13] in experiments with a transient plasma produced by a "plasma gun." With a small diameter (4 cm), and a density greater than 10^{15} cm^{-3}, the plasma is not suitable for optical or microwave interferometry. Furthermore, the phase shift rate of 2π μsec^{-1} does not come within the response time of most far-infrared detectors. However, with the gas laser lines and InSb and Ge:Ga detectors fast time-resolution measurements proved feasible in the density range 10^{13}–10^{17} cm^{-3}.

[13] R. Turner and T. O. Poehler, *J. Appl. Phys.* **39**, 5726–5731 (1968).

In other experiments with an interferometer similar to that of Fig. 6.5 Chamberlain et al.,[14] have measured phase shifts as small as one tenth of a fringe. By observing phase and attenuation changes of the 337-μm line in a 10-cm thick plasma they have measured densities in the range 7×10^{12}–10^{16} cm^{-3} and collision frequencies between 3×10^{10} and 3×10^{12} sec^{-1}.

Olsen has developed a fractional fringe technique with the 337-μm HCN laser line. He has reported better than 1/100 of a fringe phase shift sensitivity which gives electron density line integrals $\int_0^L n(x)\, dx$ of 3.3×10^{12} cm^{-2} or lower.[15]

As the reader saw in Chapter 2, many gas laser lines are now available throughout the far-infrared to enable plasma experimentalists to probe plasmas over a very large range of densities.

6.7. Changes of Polarization

Faraday rotation, Voight, and magneto-Kerr phenomena in magnetized gaseous plasmas are similar to those discussed in Section 5.3 for semiconductors, and Eqs. (5.54)–(5.62) require only trivial modifications. There are, of course, differences in the information that one seeks. For example, in semiconductors it is the effective mass that is usually to be determined, while in gaseous plasmas the interaction is classical, the mass is known, and it is the parameters n and ν that one determines from magnetoplasma measurements.

The output from a far-infrared laser is polarized with a parallel wire polarizer or with a Brewster angle polarizer made from thin sheets of polyethylene (see Section 3.2.2). After transmission through the plasma, the polarization can be analyzed with a similar parallel wire device or Brewster angle analyzer and the result interpreted in terms of density. Apart from the handling of the polarization, the experimental methods differ little from those of the previous section.

An incident linearly polarized wave, in general, emerges from the plasma with elliptical polarization. In the case of wave propagation parallel to the applied magnetic field, the RHCP and LHCP waves experience different phase shifts and attenuations. If the attenuation is small, the effect of the plasma is to rotate the plane of linear polarization.

[14] J. Chamberlain, H. A. Gebbie, A. George and J. D. E. Beynon, *J. Plasma Phys.* **3**, Pt. 1, 75–79 (1969).
[15] J. N. Olsen, *Rev. Sci. Instrum.* **42**, 104–106 (1971).

If the attenuation is large, one circular polarization can be so attenuated with respect to the other that the emerging polarization is circular. By rotating the polarization analyzer, the two orthogonal linearly polarized components of the wave can be separately recorded and full information about the wave thus built up.

Dellis et al.[12] have measured Faraday rotations in a transient plasma under conditions where Eq. (5.58) applies. The angle of rotation in radians is

$$\theta = 2.63 \times 10^{-25} \, nBL\lambda^2,$$

where n is in electrons per centimeter^3, B is in Gauss, L is in centimeters, and the wavelength λ is measured in micrometers. Dellis et al., used the 27.9-μm line from a water vapor laser. Because of the λ^2 dependence of θ, this line has 68 times the rotation of the 3.39-μm line and 1950 times that of the 0.6328-μm line of a HeNe laser. The input polarizer consisted of 10 sheets of 37-μm thickness and 10 sheets of 52-μm thickness polyethylene inclined at the Brewster angle (55°) to the incident laser beam. For the polarization analyzer, two plates of high resistivity silicon were used, each of thickness 4.2 mm and separated by a thin air gap. After transmission through the plasma, the laser beam was directed to the silicon surface at the Brewster angle for the refractive index of 3.4. One polarization was reflected into a detector, while the orthogonal linearly polarized component was transmitted through the silicon plates to a second detector. Comparison of the signals monitored from the two detectors then gave the Faraday rotation angle.

Over a plasma path of length $L = 200$ cm, and with $B = 6$ kG, Dellis et al. measured rotation angles up to 60°, corresponding to $n = 4.3 \times 10^{15}$ cm^{-3}. The result agreed with interferometry measurements.

6.8. Bremsstrahlung and Blackbody Radiation from Plasmas

It is a well-known result of nonrelativistic classical electrodynamics that an electron undergoing acceleration $\ddot{r}$ in a lossless medium of refractive index μ radiates power[16]

$$P = q^2 \mu \ddot{r}^2 / 6\pi\varepsilon_0 c^3. \tag{6.45}$$

---

[16] W. K. H. Panofsky and M. Phillips, "Classical Electricity and Magnetism." Addison-Wesley, Reading, Massachusetts, 1955.

Acceleration processes of short duration occur when electrons encounter ions and atoms in the plasma, and they produce a spectrum of incoherent emission. The least massive particles, the electrons, experience the largest accelerations, and the effect of collisions is much smaller on the ions and atoms. Thus the source of photon emission is the system of electrons.

If the plasma is "optically thin," so that emitted photons can escape without reabsorption, there is a net output of bremsstrahlung which is the sum of uncorrelated emission events determined by the details of the collisional interaction process. If the plasma has density and dimensions, such that reabsorption of emitted photons is significant, the processes of radiation transport influence the net emission. The emission then has a limit for a plasma many skin depths thick when the radiation and electron system reach thermal equilibrium. We can describe this plasma as a blackbody at the temperature T of the electrons.

To such a system of electron oscillators in equilibrium with the radiation field, we can apply Planck's blackbody radiation law for the energy density of radiation in the frequency interval $d\omega$:

$$\varrho_\omega \, d\omega = \frac{\hbar \omega^3}{\pi^2 c^3} \frac{1}{e^{\hbar \omega / kT} - 1} \, d\omega, \tag{6.46}$$

where $h = 2\pi\hbar = 6.6 \times 10^{-34}$ Joule sec is Planck's constant, T is the electron temperature in °K, and $k = 1.38 \times 10^{-23}$ Joule °K^{-1}. In the far-infrared spectral region, $\hbar\omega \ll kT$, and the Rayleigh–Jeans approximation applies:

$$\varrho_\omega \, d\omega = (\omega^2/\pi^2 c^3) kT \, d\omega. \tag{6.47}$$

For isotropically emitted radiation traveling in space with velocity c, the power density within a solid angle $d\Omega$ is

$$P_\omega \, d\omega = d\Omega (\omega^2/4\pi^3 c^2) kT \, d\omega, \tag{6.48}$$

and the units in the MKS system are W m^{-2}. The *intensity* of the emitted blackbody radiation, defined as the power flow in watts per square meter per steradian per radian frequency interval, and denoted $B(\omega, T)$, is then

$$B(\omega, T) = (\omega^2/4\pi^3 c^2) kT. \tag{6.49}$$

Note that $B(\omega, T)$ comprises two independent polarizations, so that for a receiver sensitive to only one polarized component the power predicted by Eq. (6.49) must be halved.

In the experimental observation of blackbody radiation a light cone or antenna is used to direct the radiation to a detector. The aperture of the antenna can accept radiation only from a solid angle equal approximately to the angular spread of the principal lobe of its diffraction pattern and this, according to diffraction theory, is

$$d\Omega \simeq \lambda^2/A,$$

A being the aperture area. From Eq. (6.48) it follows that the power received with one polarization when the blackbody source subtends a solid angle at the receiver equal to or greater than λ^2/A is

$$P = kT\,df \quad \text{W}. \tag{6.50}$$

Thus a measurement of the power received by a detector of known bandwidth can give the electron temperature of the plasma when its optical thickness is large enough to result in blackbody emission. It is not essential that the entire system of plasma particles be in thermal equilibrium with the radiation. The electrons may be hotter than the ions, or they may be non-Maxwellian, but the system of electrons must be in thermal equilibrium with the radiation they emit. Under these conditions, in the spectral region where the plasma exhibits blackbody characteristics, the emission cannot rise over the blackbody level, and no discrete line in this range can emerge from the plasma.

The conditions under which a plasma may become optically thick can be seen for the case of an unmagnetized plasma from Eq. (6.43). The skin depth or distance in which the plasma attenuates the wave amplitude by the exponential factor e is

$$\delta = 1/\alpha = (2n_c c/n\nu)[1 - (n/n_c)]^{1/2}.$$

When $\omega \gg \omega_p$ only very large bodies of plasma can give sufficient self-absorption of radiation to be opaque. When ω approaches ω_p, δ becomes small and much smaller plasmas can radiate like blackbodies. The further decrease of ω to the near vicinity of ω_p results in increased surface reflection until, as we have seen in Section 6.3, for $\omega < \omega_p$ there is substantial reflection. The plasma has become optically "silver" and, by Kirchhoff's radiation law, it radiates poorly.

Expressions for the radiation from a plasma of moderate optical thickness can be obtained from modifications of Eq. (6.49) which allow for attenuation and surface reflections. For the moment, let us ignore reflections. We require here the equation of radiation transfer[4,17]

$$\mu^2 \frac{d}{dr}\left(\frac{I_\omega}{\mu^2}\right) = j_\omega - 2\alpha I_\omega, \qquad (6.51)$$

and Kirchhoff's radiation law

$$j_\omega/2\alpha = \mu^2 B(\omega, T), \qquad (6.52)$$

where I_ω is the radiation intensity in watts per square meter per steradian per radian bandwidth at the frequency ω; r is distance along the ray path; $2\alpha I_\omega$ is the power absorbed per unit volume along the ray path; and j_ω is the power generated by the medium per unit volume per unit solid angle in the direction **r** in one polarization per unit frequency interval $d\omega$. Equation (6.51) describes the change in intensity along a ray path for all effects except reflections at discontinuities. In a medium with no emission or absorption ($j_\omega = 0$ and $\alpha = 0$) we have $(d/dr)(I_\omega/\mu^2) = 0$. This describes the change of intensity as the solid angle of the flux of rays changes due to refraction in a medium of spatially varying refractive index $\mu = \mu(r)$, and, in particular, it gives $I_\omega/\mu^2 =$ invariant — conservation of total flux. The difference of the terms j_ω and $2\alpha I_\omega$ on the right-hand side of Eq. (6.51) simply gives the net contribution of the medium to the flux of photons.

If we define the *optical depth* as

$$\tau(r_0) = \int_{r_0}^{\text{observer}} 2\alpha \, dr, \qquad (6.53)$$

where r_0 is the position of the source, Eq. (6.51) can then be written:

$$\frac{d}{d\tau}\left(\frac{I_\omega}{\mu^2}\right) = \frac{I_\omega}{\mu^2} - \frac{j_\omega}{2\alpha\mu^2}. \qquad (6.54)$$

The solution

$$(I_\omega/\mu^2)_{\text{obs}} = (I_\omega/\mu^2)_\tau e^{-\tau} + \int_{\text{obs}}^{\tau} (j_\omega/2\alpha\mu^2) e^{-\tau} \, d\tau \qquad (6.55)$$

describes, through the first term on the right, the contribution from the intensity existing at the depth τ, while the last term gives the contribution from emission by the medium between the depth τ and the observer. If we now sum the emission through a plasma of optical depth τ_0, we find

---

[17] G. Bekefi and S. C. Brown, *Amer. J. Phys.* **29**, 404–428 (1961).

for the intensity outside the plasma (where $\mu = 1$)

$$I_\omega = \int_0^{\tau_0} (j_\omega/2\alpha\mu^2)e^{-\tau}\, d\tau. \tag{6.56}$$

Combining Eqs. (6.52) and (6.56) gives for the intensity outside the plasma

$$I_\omega = \int_0^{\tau_0} B(\omega, T)e^{-\tau}\, d\tau. \tag{6.57}$$

When the temperature and power absorption coefficient 2α are constant along the total length L traversed by the rays in the plasma, Eq. (6.57) reduces to

$$I_\omega = B(\omega, T)[1 - e^{-2\alpha L}]. \tag{6.58}$$

When $2\alpha L \gg 1$, $I_\omega = B(\omega, T)$ the blackbody emission, and when $2\alpha L \ll 1$, $I_\omega = 2\alpha L B(\omega, T)$ the emission in the absence of reabsorption.

So far we have ignored reflections at the plasma–vacuum interface. This is justified if the plasma has low density but is nevertheless optically thick by virtue of its physical dimensions, or if the surface density profile is sufficiently gradual to give a matched condition at the interface. Generally, in laboratory plasmas, optical thickness requires ω not far above ω_p, and under these conditions interface reflections can be significant. If the dimensions, density, and attenuation coefficient of the plasma are such that we may ignore multiple bouncing between surfaces, Eq. (6.58) is modified to take the form

$$I_\omega \simeq (1 - \mathscr{R})(1 - e^{-2\alpha L})B(\omega, T), \tag{6.59}$$

where $\mathscr{R}$ is the power reflection coefficient. In the Rayleigh–Jeans limit, we have in place of (6.50)

$$P \simeq (1 - \mathscr{R})(1 - e^{-2\alpha L})kT\, df. \tag{6.60}$$

This result shows the factors involved in the interpretation of observed radiation from a plasma when its emission falls short of that from a blackbody.

In the case of complete transparency, the radiation emission can be calculated from the details of the Coulomb encounters between the electrons and the ions and atoms. Classical calculations of the resulting bremsstrahlung have been based on Fourier analysis of the acceleration described by Eq. (6.45), and the predictions have been compared with those of quantum mechanics. In the spectral region of interest here, there is close accord between the results of the two approaches. In a critical review Oster[18] has summarized the theoretical developments and given expressions for the bremsstrahlung. For electron–ion encounters (which in highly ionized plasmas are more important than electron–atom collisions[17]), and a Maxwellian distribution of electron energies the power emitted from 1 m³ of plasma with one polarization and within a steradian is

$$P_0 = \frac{8}{3} \left(\frac{2\pi}{3}\right)^{1/2} \left(\frac{q^2}{4\pi\varepsilon_0 c}\right)^3 \frac{Z^2}{m^{3/2}k^{1/2}} nn_i T^{-1/2} G \, d\omega, \quad (6.61)$$

where Z is the ionic charge number, q the electronic charge, and n and n_i are the electron and ion concentrations per cubic meter. G is the Gaunt factor, a parameter whose origin lies in the details of the collisional model. For electron temperatures less than 550,000°K, Oster[18] has shown that G has the value

$$G = \frac{\sqrt{3}}{\pi} \ln[(1.12kT/m)^{3/2}(1.12m4\pi\varepsilon_0/Zq^2\omega)], \quad (6.62)$$

provided the inequality

$$10^{-3}T^{-1/2}\omega^{1/3} \ll 1$$

is satisfied. Equation (6.62) applies for $\omega_p^2 \ll \omega^2$. Bekefi and Brown[17] give G for $\omega_p > \omega$.

The power received is obtained by integrating Eq. (6.61) over the volume within the radiation pattern of the receiving aperture. The intensity of the radiation received from a plasma slab as a function of wavelength has the general form of the curves sketched in Fig. 6.7. The two bremsstrahlung curves illustrated have plasma frequencies corresponding to 1000 and 320 μm, and the far-infrared spectrum of blackbody radiation at the same temperature as the source of bremsstrahlung is shown for reference. Bremsstrahlung is below the blackbody level, but approaches blackbody intensity when the plasma becomes optically thick in a narrow region slightly above ω_p. A radiometer measurement of power in the region can be interpreted to give a measure of the electron temperature of the plasma. At $\omega \lesssim \omega_p$, the intensity decreases abruptly as significant reflection occurs at the plasma surface. This feature in the spectrum can be used as a means of estimating the plasma frequency,

[18] L. Oster, *Rev. Mod. Phys.* **33**, 525–543 (1961).

6.8. BREMSSTRAHLUNG AND BLACKBODY RADIATION

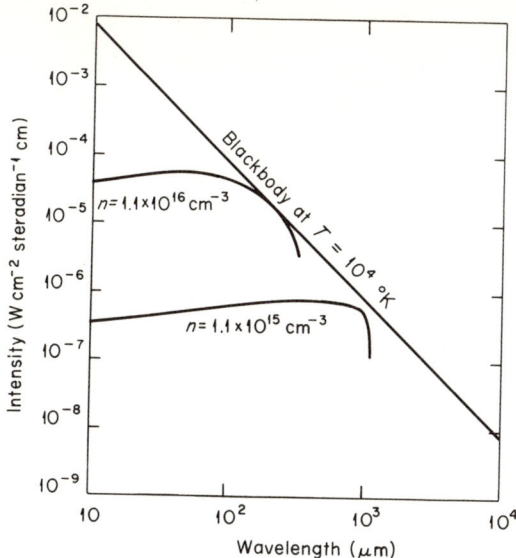

FIG. 6.7. Bremsstrahlung emission intensity from a slab of plasma about 1 cm thick with densities of 1.1×10^{15} and 1.1×10^{16} cm^{-3} and electron temperature 10^4 °K. The corresponding plasma wavelengths are close to 1000 μm and 320 μm. The falloff in intensity near the plasma frequency is due to surface reflection. The blackbody radiation curve at 10^4 °K is also shown (after Kimmitt et al.[19]).

but the shape of the falloff below ω_p is a function of the geometry and density profile of the plasma surface and is not easily interpreted in any detail. The flat region of the emission spectrum is associated with an optically thin plasma and gives the product $nn_i T^{-1/2}$, or $n^2 T^{-1/2}$ for equality of electron and ion densities.

Far-infrared observations of bremsstrahlung spectra have been made on a number of plasmas. We will cite the results of two of these experiments. The first[19] used a plasma compressed to a volume of about 50 cm³ in a cusp configuration of magnetic field, and the second[20] used a plasma in the Zeta machine at Harwell, England.

Figure 6.8 shows far-infrared spectra of the cusp-confined plasma. Measurements were made with a Putley indium antimonide photodetector for two conditions of temperature and density. The lower tem-

---

[19] M. F. Kimmitt, A. C. Prior and V. Roberts, in "Plasma Diagnostic Techniques" (R. H. Huddlestone and S. L. Leonard, eds.), Chapter 9. Academic Press, New York, 1965.
[20] G. N. Harding and V. Roberts, *Nucl. Fusion Suppl.* **3**, 883–887 (1962).

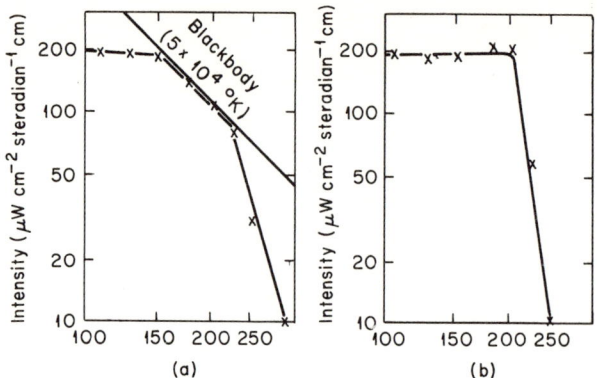

FIG. 6.8. Bremsstrahlung from plasmas produced by shock waves and compressed by a cusp geometry magnetic field. In (a) the electron density and temperature are $n = 1.8 \times 10^{16}$ cm^{-3} and $T = 5 \times 10^4$ °K, while in (b) $n = 2.2 \times 10^{16}$ cm^{-3} and $T = 2 \times 10^5$ °K. At the shorter wavelengths both curves are independent of wavelength as is characteristic of bremsstrahlung from an optically thin plasma. Curve (a) shows a range (150–220 μm) of blackbody behavior when it is optically thick followed by an abrupt decrease as reflections become important. Curve (b) shows a sudden transition from transparency to high reflectivity (after Kimmitt et al.[19]).

perature plasma spectrum shows bremsstrahlung from a plasma that is optically thin up to a wavelength of about 150 μm, and then a wavelength region extending to about 220 μm, where selfabsorption is sufficient to give a close approximation to blackbody emission at 5×10^4 °K. When the temperature is higher, as in Fig. 6.8b, the electron–ion collision frequency is decreased [see Eq. (6.14)] and there is no region where selfabsorption gives emission with blackbody slope. Instead, a sudden transition occurs between a condition where the plasma is optically thin and where it is highly reflecting. This occurs at the plasma frequency and gives a measure of the electron density. Temperatures can then be estimated from the level of bremsstrahlung in the optically thin region.

Zeta is a toroidal shaped plasma which is heated to high temperatures by a pulse of current. The curves of Fig. 6.9 show a flat spectrum corresponding to the wavelength-independent bremsstrahlung equation for a plasma with negligible self-absorption. As the wavelength increases the absorption becomes appreciable and the intensity falls until absorption is so strong that the plasma behaves as a blackbody. The measurements have been analyzed in terms of bremsstrahlung theory. The densities calculated are 1.35×10^{15} and 1.15×10^{15} cm^{-3} for curves (a) and (b), and the corresponding temperatures are 8.5×10^4 and 9.7×10^4 °K.

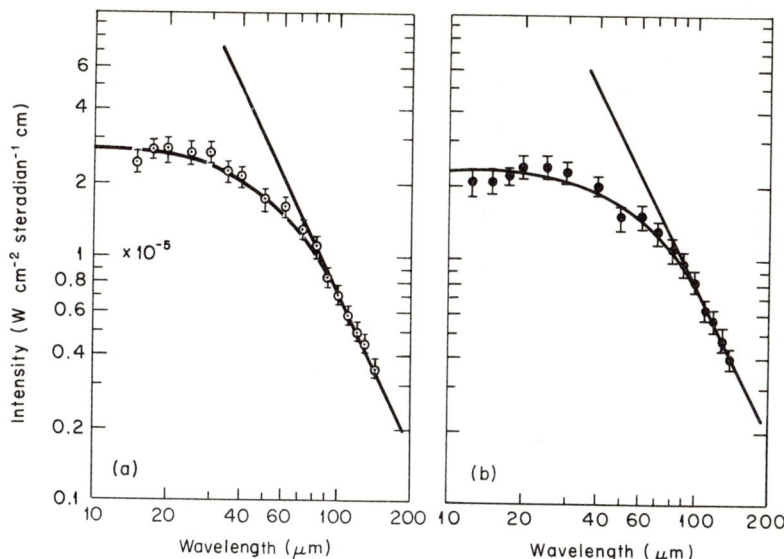

FIG. 6.9. Bremsstrahlung from a pulsed plasma in Zeta. The gas was deuterium and the initial neutral pressure 5 mtorr, and each point represents the average of 3–10 plasma pulses. Analysis of the measurement of (a) gives $n = 1.35 \times 10^{15}$ cm^{-3} and $T = 8.5 \times 10^4$ °K, and for (b) $n = 1.15 \times 10^{15}$ cm^{-3} and $T = 9.7 \times 10^4$ °K (after Harding and Roberts[20]).

6.9. Cyclotron and Synchrotron Radiation

When electrons orbit about magnetic field lines, they are subject to acceleration and therefore emit radiation. If the particle energy is low (less than a few hundred electron volts), the cyclotron emission is concentrated in a single line at the cyclotron frequency ω_c. The radial acceleration of a particle in circular motion is

$$\ddot{r} = (q/m) \, | \, \mathbf{v} \times \mathbf{B} \, | = \omega_c v,$$

and by Eq. (6.45) with $\mu = 1$ the total power radiated is

$$P = q^2 \omega_c^2 v^2 / 6\pi \varepsilon_0 c^3. \tag{6.63}$$

For plasma electrons with a Maxwellian velocity distribution $v = (2kT/m)^{1/2}$. Assuming the plasma depth L and density n to be such as to give negligible selfabsorption, the cyclotron power it radiates in all directions will be

$$P = q^2 \omega_c^2 kTnL/3\pi \varepsilon_0 c^3 m \quad \text{W m}^{-2}. \tag{6.64}$$

We can regard the electron in circular motion as a superposition of two oscillating dipoles, to each of which the well-known classical expression[16] for the radiated power may be applied. The power per unit volume radiated into the solid angle $d\Omega$ from a Maxwellian distribution of electron velocities is then[17]

$$P = (nq^2\omega_c^2/16\pi^2\varepsilon_0 c)(kT/mc^2)(1 + \cos^2\theta)\,d\Omega, \qquad (6.65)$$

where θ is the angle between the direction of observation and the magnetic field.

When the speed of an electron increases towards the velocity of light, the angular spread of the emission decreases and becomes beamed into a forward cone along the direction of motion of the electron. Figure 6.10

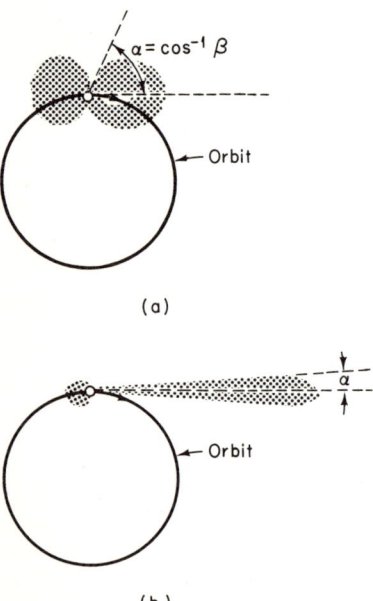

FIG. 6.10. Sketch of the polar diagram of the instantaneous radiation intensity from (a) a nonrelativistic and (b) a highly relativistic electron. In the nonrelativistic case the intensity distribution is that of a dipole. For the relativistic electron the radiation is essentially confined to a narrow cone in the direction of the instantaneous velocity with a semiangle $\alpha = \cos^{-1}\beta$, where β is the ratio of the electron velocity to the velocity of light in vacuum.

illustrates this effect. An observer in the orbital plane receives the energy not in the form of a sinusoidal wave but rather as a sequence of short pulses, one for each period of the electrons orbital motion. The spectrum is the Fourier transform of the received wave-train. As illustrated in Fig. 6.11, it has the form of a series of harmonics of the relativistic cyclotron frequency with an envelope which is the Fourier transform of an individual pulse. For mildly relativistic electrons (kinetic energy $\ll$ rest

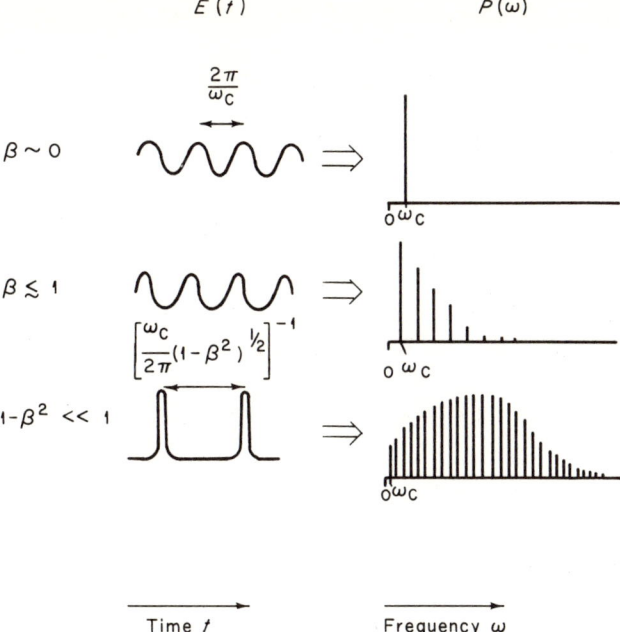

FIG. 6.11. Illustration of cyclotron and synchrotron radiation from an orbiting electron with velocity βc. The left-hand column shows the waveforms of the electric field as seen by a stationary observer in the orbital plane. The emission spectra on the right are the Fourier transforms $P(\omega)$ of $E(t)$. Nonrelativistic motion (β small) gives a line spectrum at ω_c, mildly relativistic motion ($\beta \lesssim 1$) gives a fundamental plus a few harmonics, and highly relativistic motion gives emission rich in harmonics (after Wild et al.[25]).

energy), the energy is concentrated in the first few harmonics and these are sufficiently separated to be resolved experimentally. Highly relativistic energies (total energy ≫ rest energy), however, give a multitude of harmonics which, with line broadening, overlap to form what is essentially a continuous spectrum. In this highly relativistic case the emission is called *synchrotron radiation*.

Calculations of the spectrum of synchrotron radiation have been carried out by Schwinger.[21] The intensity in successive harmonics increases with harmonic number to a maximum and then falls off. The peak of the spectrum occurs at a frequency[17]

$$f_{\max} = 10.7 BW^2. \tag{6.66}$$

[21] J. Schwinger, *Phys. Rev.* **75**, 1912–1925 (1949).

The total intensity radiated by an electron over the entire frequency spectrum is, in watts,

$$P = 6 \times 10^{-22} B^2 W^2, \qquad (6.67)$$

where in both Eqs. (6.66) and (6.67) W is the electron total energy (rest plus kinetic) in MeV, B is in Gauss and the frequency is in MHz. The intensity maximum will occur in the far-infrared (at $\lambda \sim 500\ \mu m$) if $B = 10^4$ Gauss and $W \sim 2.5$ MeV.

The importance of cyclotron harmonic radiation as a loss process in high-temperature plasmas was first pointed out by Trubnikov.[22] For a mildly relativistic electron, Bekefi and others have given the following expression for the power emission in the extraordinary wave (radiation with the polarization of the ordinary wave is negligible) at the frequency ω:

$$P = (q^2 \omega^2 / 8\pi^2 \varepsilon_0 c) \sum_{p=1}^{\infty} A_p\, \delta[p\omega_c (1 - \beta^2)^{1/2} - \omega], \qquad (6.68)$$

where p is the harmonic number. For $p < 3$, they give

$$A_p = [p^{2p}/(2p+1)!]\beta^{2p}. \qquad (6.69)$$

Expressions for A_p when $p \geq 3$ are given by Bekefi.[3] For a Maxwellian distribution of electrons at 50 keV in a plasma where we ignore self-absorption, these results lead to the radiation spectrum of Fig. 6.12.[3,23] The harmonic lines are broadened by the relativistic change of mass, and overlap into a continuum for $p > 4$.

Measurements of the cyclotron harmonic radiation spectrum from a hot plasma have been made by Sesnic et al.[24] They used a plasma confined in a magnetic-mirror configuration and heated by magnetic compression to temperatures of about 80 keV. The cyclotron harmonic radiation measurements were made using a far-infrared monochromator with a 10% bandwidth followed by an indium antimonide photodetector. A magnetic field of 50 kG gave a cyclotron wavelength of 2.14 mm, and the observations scanned the range 4 mm–200 μm. A measured spectrum is shown in Fig. 6.13, together with calculated spectra for temperatures of 75 and 100 keV. The results clearly show the harmonic character of the radiation and are consistent with the theoretical predictions.

---

[22] B. A. Trubnikov, *Sov. Phys. Dokl.* **3**, 136–140 (1958).
[23] J. Hirshfield, D. E. Baldwin and S. C. Brown, *Phys. Fluids* **4**, 198–203 (1961).
[24] S. Sesnic, A. J. Lichtenberg, A. W. Trivelpiece and O. Tuma, *Phys. Fluids* **11**, 2025–2031 (1968).

6.9. CYCLOTRON AND SYNCHROTRON RADIATION

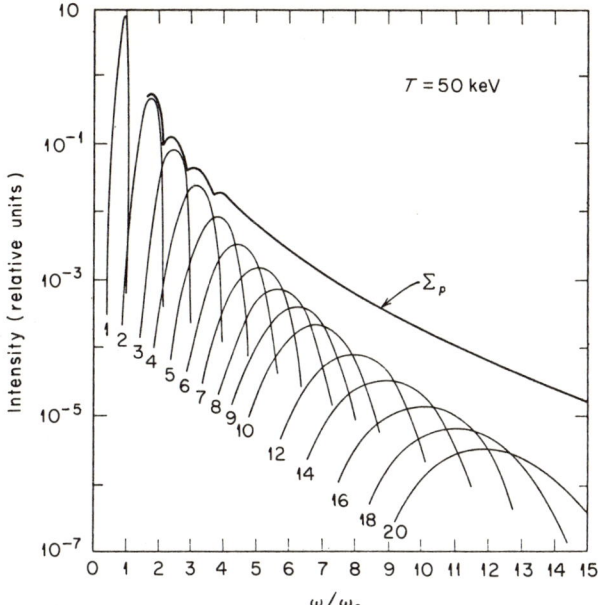

FIG. 6.12. Cyclotron radiation spectrum of the first twenty harmonics from a plasma with a temperature of 50 keV. Self-absorption by the plasma is neglected. The curve Σ_p is the sum of the harmonics (after Bekefi and Brown[17]).

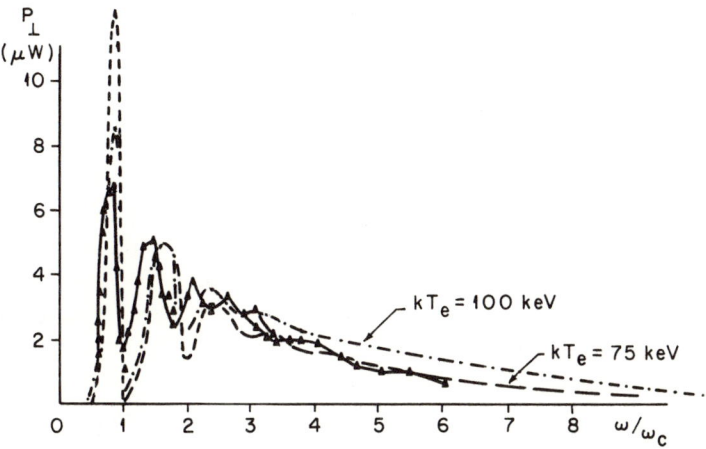

FIG. 6.13. Cyclotron harmonic radiation emitted perpendicular to the magnetic field by a plasma heated to a temperature of about 80 keV. The dashed curves are theoretical and the solid curve experimental (after Sesnic et al.[24]).

6.10. Longitudinal Plasma Waves

A third process by which radiation may be generated in a plasma is via the growth of *longitudinal* or *space-charge* waves, and the subsequent coupling of this wave type to transverse radiation. Radiation emission of this type can occur near the plasma frequency, near cyclotron harmonics, etc., so that for reasonably dense plasmas it is at far-infrared wavelengths. Longitudinal waves have been studied in relatively dilute plasmas at radio and microwave frequencies,[3,25] but, as yet, no far-infrared work appears to have been undertaken.

Longitudinal waves can be generated when electrons stream through the plasma with velocities in excess of the wave phase velocity. For a longitudinal wave, the effective refractive index can exceed unity and the phase velocity in the medium can, therefore, be less than c. It is thus a candidate for wave excitation through the Cerenkov effect of a charged particle passing through the plasma at a speed greater than the phase velocity v_{ph}. Twiss[26] has analyzed the conditions for which electrons acting independently in a stream can stimulate longitudinal growing waves by incoherent Cerenkov emission. For an electron stream with a narrow range of velocities, the condition for wave growth is that the velocity distribution function $f(v)$ of the stream is such that

$$(\partial f(v)/\partial v)_{v=v_{ph}} > 0. \tag{6.70}$$

The coherent growth of longitudinal waves excited by an electron stream has been analyzed by Bohm and Gross.[10] When the electrons travel faster than the phase velocity plasma oscillations are excited, the electron stream becomes bunched in a manner similar to that occurring in traveling wave tubes, and there arises a transfer from the electron kinetic energy to the coherent buildup of wave amplitude. The conditions for wave growth are again given by Eq. (6.70). Energy is concentrated near $\omega \sim \omega_p$, where the dispersion curve is steep. In the presence of a magnetic field, emission can also arise at frequencies near the upper hybrid $(\omega_p^2 + \omega_c^2)^{1/2}$, near harmonics of the cyclotron frequency and other frequencies related to ω_p and ω_c.

Radiation into free space of electromagnetic energy derived from longitudinal plasma waves requires a mechanism of conversion to trans-

---

[25] J. P. Wild, S. F. Smerd and A. A. Weiss, *Annu. Rev. Astron. Astrophys.* **1**, 291–366 (1963).

[26] R. Q. Twiss, *Austr. J. Phys.* **11**, 564–579 (1958).

verse electromagnetic waves. Such coupling is possible in the presence of density or magnetic field gradients under conditions where the two waves have similar electric field directions and velocities, i.e., when their dispersion characteristics happen to be close together. This occurs,[27] for example, in the case of the longitudinal and extraordinary wave in a plasma with a density profile at a position where $\omega = (\omega_p^2 + \omega_c^2)^{1/2}$. Coupling may also occur at sharp plasma boundaries where an effective radiating dipole layer can exist due to the inequality of electron and ion concentrations, or there may be conversion at regions of small-scale fluctuations in plasma density.

[27] H. Dreicer, *Proc. Int. Conf. Ioniz. Phenomena Gases, 6th, Paris, 1963*, (P. Hubert and E. Crémieu-Alcon, eds.), Vol. 3. Achevé D'Imprimer, Paris, 1964.

7. SPECTRA OF GASES, LIQUIDS, AND SOLIDS IN THE FAR-INFRARED

Introduction

In dealing with gases, liquids, and solids we encounter phenomena that are normally separated into three or more specialized compartments of physics. But there are common threads through these topics which tie them loosely together. From a far-infrared spectroscopic viewpoint, one is generally concerned with resonance phenomena, and naturally one frequently encounters, or senses, such things as harmonic oscillator behavior, dispersion theory, and the like. Of course, there are many differences: the particular environment of the oscillating charge, whether it acts as an individual experiencing occasional abrupt encounters, whether it is constrained by coupling to other charges to move in unison with them, etc.; these influences control the precise motion and determine the appropriate representation for a particular material.

In turning to these topics, the order "gases, liquids, solids" has a certain historical basis, but from a far-infrared aspect it is an order of increasing physical complexity. Before we consider rotational molecular spectra in Section 7.1, let us first review some features of molecular spectra and briefly survey some aspects of liquids and solids.

The internal energy of a molecule is made up of the energy of the electrons, the mutual vibrational motion of the nuclei of the constituent atoms, energy of rotational motion, and energy of nuclear orientation. Exact calculations of the quantized energy states of the entire system of interacting particles constituting the molecule are, in general, too difficult to carry out but it is possible to develop an approximation to the real case by treating the various motions separately as noninteracting motions. That is, the wavefunction in the Schrödinger equation,

$$H\psi = W\psi, \qquad (7.1)$$

is taken as the product,

$$\psi = \psi_e \psi_v \psi_r \psi_n, \qquad (7.2)$$

INTRODUCTION

of independent wavefunctions representing electronic, vibrational, rotational, and nuclear spin motions. In this approximation, the Hamiltonian operator H is the sum of four noninteracting terms representing the four motions.

The various reservoirs of energy tend to be well separated for diatomic molecules. However, for polyatomic molecules, some vibrational and rotational frequencies may be comparable, and any accurate calculation of the spectrum must take the mutual interaction into account. Typical energy separations are 0.1–10 eV for electronic states, 0.1–1 eV for vibrational states, less than 0.1 eV for rotational states, and very much smaller still for the energy of nuclear orientation. The corresponding spectra tend to be in the visible to ultraviolet for electronic transitions, infrared for vibrational transitions, at far-infrared and microwave frequencies for rotational motion, and in the radio band for nuclear resonance.

The general polyatomic molecule with N atoms has $3N$ independent modes of motion or degrees of freedom. Of these, three account for translational motion, three are associated with rotation of the molecule (two in the case of a linear molecule), and $3N$-6 with vibrational motion ($3N$-5 in the case of linear molecules). Each mode of vibration and rotation has a set of eigenstates described by wavefunctions and with discrete energy levels. The molecule can be excited in rotation or vibration by electromagnetic radiation through forces exerted by the **E** or **H** field vectors on the electric or magnetic dipole moment to give absorption spectra. Emission spectra can also occur when the molecular dipole undergoes changes which release quanta of energy.

From the theory of dynamics, we know that the rotation of a system of N masses can be calculated in terms of three orthogonal mechanical moments of inertia about axes passing through the center of mass. The molecule can thus be represented by an ellipsoid of inertia centered about the center of mass of the molecule and with three perpendicular principal semiaxes whose lengths equal the inverse square roots of the moments of inertia (I). It is customary to label the principal axes of the ellipsoid A, B, and C in order of decreasing length, so that A is the major axis, B the intermediate axis, and C the minor axis. Since the axial lengths are proportional to $1/\sqrt{I}$, it follows that $I_A < I_B < I_C$. We can classify molecules on the basis of the symmetry of the ellipsoid of inertia as follows:

(i) *Symmetric tops.* These have two moments of inertia equal (i.e., $I_B = I_C$ or $I_A = I_B$). A special case of importance is the linear molecule; for this $I_C = 0$ and $I_A = I_B \neq 0$.

(ii) *Spherical molecules.* In these molecules all three moments of inertia are equal (i.e., $I_A = I_B = I_C$).

(iii) *Asymmetric tops.* These have all three moments of inertia different (i.e., $I_A \neq I_B \neq I_C$).

The existence of a dipole moment is a consequence of asymmetry of the charge distribution within the molecule. It may be an inherent property of the molecule or it may be induced, for example, during collisions and in Raman scattering. Quantum mechanically, the transition rate between energy states is expressed in terms of the matrix element of the electric (or magnetic[†]) moment of the molecule. The x-component of the matrix element for electric dipole transitions between eigenstates k and l is defined as

$$\mu_{x_{kl}} \equiv \langle k | \mu_x | l \rangle = \int \psi_k^* \sum_j q x_j \psi_l \, d\tau. \tag{7.3}$$

The notation $\langle k | \mu_x | l \rangle$ is the Dirac symbol for the matrix element.

The matrix element represents the charge distribution averaged over the molecule with a weighting factor $\psi_k^* \psi_l$, which may be thought of as the probability density of a mixture of the initial and final states involved in the transition. Its existence depends on the symmetry of the wave functions ψ_k and ψ_l. Each component of the electric dipole moment has odd parity, since

$$\sum_j q(-x_j) = -\sum_j q x_j. \tag{7.4}$$

Thus if ψ_k and ψ_l have the same parity (both even or both odd) the integral [Eq. (7.3)] is zero. It will be nonzero only if the initial and final states have different parity. Thus we have a *selection rule* that transitions are allowed between states of different parity. When the wavefunctions for the rotational and vibrational states are examined separately, the selection rules can be stated in terms of changes in the quantum numbers specifying permitted and forbidden transitions between states. When the approximation of separable wavefunctions and energy states is replaced by more exact wavefunctions, it frequently turns out that the matrix elements for forbidden transitions are not exactly zero, and hence, in practice, forbidden transitions occur as very weak lines.

[†] Magnetic dipole interactions tend to be some orders of magnitude weaker than electric dipole interactions (see Section 7.4).

The resulting distribution of spectral lines in the infrared to microwave range are characteristic of the rotational and vibrational origins of the energy, and their study in absorption and emission spectroscopy reveals the internal structure of molecules. In particular, the rotational transition frequencies can give the moments of inertia, the structure, and other properties of the molecules, and the vibrational spectra can give the binding forces, masses of the nuclei, etc.

The response of charges bound in a molecule, or in systems of molecules, will be characterized not only by natural transition frequencies between energy eigenstates but also by line shapes which, in general, reflect the environment of the molecules. An isolated molecule interacting with a radiation field will increase in energy if the wave frequency and the resonance frequency of the molecular oscillator are exactly equal. When the wave frequency and the oscillator frequency differ, the field will excite the motion until such time that a phase difference of half a cycle develops between them and energy begins to be transferred back to the field. Under these conditions the net power transfer to the oscillator is a sinc function of the product of half the difference frequency and the time of interaction, as we have seen in connection with cyclotron resonance absorption. However, the time of interaction is normally set by randomizing processes associated with the interaction of the molecules with the environment. In a gas, molecular collisions determine the mean interaction time and, through the uncertainty principle, the width of the line. The linewidth thus gives a measure of the environment.

When the oscillators become more concentrated, as in a liquid or solid, the line will tend to broaden. In a liquid the motion may be overdamped by viscous restraints and one may, for example, speak of molecular reorientation rather than rotation interrupted by occasional abrupt interactions, and follow the description of dielectric behavior given by Debye.

Usually in a solid the packing of atoms is so close that interatomic or intermolecular forces of attraction come into play and hold atoms in highly ordered crystalline forms against the efforts of thermal motion to disrupt. If the binding forces are ionic, the crystal will have an electric dipole moment and absorb radiation. Coordinated vibrations of atoms linked together in the lattice array are natural to the solid. When optically active they frequently give line-like absorption located in the far-infrared by virtue of the weak restoring forces on displaced charges. Covalently bonded solids are in general transparent, as are ionic crystals outside the frequency regions of optically active natural vibrations. With the introduction of impurity centers, or dislocations, however, the translational symmetry of the crystal is upset and a dipole moment is established by the disturbance of the charge distribution of the perfect crystal. Resonances thus excited can under some circumstances couple to vibrations of the host crystals, thereby giving a tool with which to explore some otherwise inactive modes.

With the highly developed order of the crystalline structure, cooperative ionic movements can give rise not only to stable vibrations but to unstable ones. In ionic crystals, and notably in barium titanate, unstable vibrational motion can result in a slippage in location of atoms of one type relative to atoms of other types. The slippage is not disruptive but is ultimately restrained by nonlinear effects, leaving the crystal in a state of permanent electrical polarization. Unstable modes of this type are part of the phenomenon of ferroelectricity, and the connection with crystal vibration, with extremely high dielectric constant, and with anomalous dispersion in the far-infrared is part of the present chapter.

The dipole moment in certain molecules and crystals is due to non-canceling electron spins, with magnetic resonances and other magnetic properties testifying to the fact. Paramagnetic resonance may be brought into the far-infrared with very large applied magnetic fields, but in ferromagnetic crystals very strong internal magnetic fields may occur naturally. A high degree of order and cooperation produces intricate ferromagnetic and antiferromagnetic resonances which far-infrared waves may probe.

7.1. Rotational Spectra of Molecules

7.1.1. Linear Molecules

To a first approximation, a rotating molecule can be treated as a rigid body with fixed separations between its constituent nuclei. Let us consider linear molecules first. If the molecule is diatomic, its moment of inertia about an axis through the center of mass and perpendicular to the line joining the two nuclei is

$$I = [m_1 m_2/(m_1 + m_2)]r^2, \tag{7.5}$$

where m_1 and m_2 are the masses of the atoms, and r is their separation. If it is polyatomic,

$$I = \tfrac{1}{2} \sum_j \sum_i m_j m_i r_{ij}^2 \Big/ \sum_i m_i. \tag{7.6}$$

In either case it is represented as a simple linear rigid rotator of moment

7.1. ROTATIONAL SPECTRA OF MOLECULES

of inertia I, and its rotational energy eigenstates W and wavefunctions ψ are obtained by solving the Schrödinger equation in spherical polar coordinates θ and ϕ about the center of mass of the molecule

$$\frac{h^2}{8\pi^2 I}\left[\frac{1}{\sin\theta}\frac{\partial}{\partial\theta}\left(\sin\theta\frac{\partial\psi}{\partial\theta}\right) + \frac{1}{\sin^2\theta}\frac{\partial^2\psi}{\partial\phi^2}\right] + W\psi = 0. \quad (7.7)$$

By the method of separation of variables, we may write

$$\psi = \Theta(\theta)\,\Phi(\phi)$$

and separate [Eq. (7.7)] into two equations, one involving the variable θ only, and the other ϕ only. The solutions are

$$\Phi(\phi) = e^{iM\phi}/(2\pi)^{1/2}, \quad (7.8)$$

$$\Theta(\theta) = \left[\frac{(2J+1)(J-|M|)!}{2(J+|M|)!}\right]^{1/2} P_J^{|M|}(\cos\theta), \quad (7.9)$$

where $P_J^{|M|}(\cos\theta)$ are the associated Legendre functions, and J and M are the total angular momentum quantum number and the magnetic quantum number, respectively.

The eigenstates of the energy W are found to be

$$W = (h^2/8\pi^2 I)\,J(J+1), \quad J = 0, 1, 2, \ldots. \quad (7.10)$$

It is usual to write

$$B = h/8\pi^2 I, \quad (7.11)$$

and the energy in the form

$$W/h = BJ(J+1). \quad (7.12)$$

B is called the *rotational constant* and may be expressed in MHz or cm^{-1}. The rigid rotor with fixed moment of inertia, and hence fixed B, is a first approximation to a real linear molecule. A real molecule can vibrate and stretch under centrifugal forces, and the rotational constant will depend on this motion. For a polyatomic linear molecule the value of B in a state of vibrational quantum number v is $B_v = B_e - \sum_i \alpha_i(v_i + \tfrac{1}{2})$, where B_e is the equilibrium value of B and α_i is a small positive constant (of the order of 0.5–1% of B). For a diatomic molecule there is only one mode of vibration (one value of α), and the ground state rotational constant may be written $B_0 = B_e - \alpha/2$.

The total angular momentum is $[J(J+1)]^{1/2}(h/2\pi)$. It is quantized in space and its component along a preferred direction can take the discrete values $Mh/2\pi$, where $M = 0, \pm 1, \pm 2, \ldots$. M can take on $2J + 1$ integral values between $-J$ and $+J$ and is interpreted as meaning that the total angular momentum vector can assume only the discrete spatial orientations given by

$$\cos \theta = \pm M/[J(J+1)]^{1/2}. \tag{7.13}$$

When an electric or magnetic field is present, its direction specifies the preferred direction in space with respect to which the momentum is spatially quantized; the rotational energy levels then have $(2J + 1)$ degeneracy and the rotational energy states split with $(2J + 1)$ multiplicity.

Matrix elements for electric dipole transitions between rotational energy states can be calculated from Eq. (7.3). We have

$$\langle \psi_{JM} | \mu | \psi_{J'M'} \rangle = \int \psi_{JM} \mu \psi_{J'M'} \, d\tau,$$

where

$$\psi_{JM} = \left[\frac{(2J+1)(J-|M|)!}{4\pi(J+|M|)!} \right]^{1/2} P_J^{|M|}(\cos \theta) e^{iM\phi}. \tag{7.14}$$

The first eight normalised wavefunctions are:

$$\psi_{00} = \frac{1}{(4\pi)^{1/2}}, \qquad \psi_{20} = \left(\frac{5}{16\pi}\right)^{1/2} (3\cos^2 \theta - 1),$$

$$\psi_{10} = \left(\frac{3}{4\pi}\right)^{1/2} \cos \theta, \qquad \psi_{21} = \left(\frac{15}{8\pi}\right)^{1/2} \cos \theta \sin \theta \, e^{i\phi},$$

$$\psi_{11} = \left(\frac{3}{8\pi}\right)^{1/2} \sin \theta \, e^{i\phi}, \qquad \psi_{2-1} = \left(\frac{15}{8\pi}\right)^{1/2} \cos \theta \sin \theta \, e^{-i\phi},$$

$$\psi_{1-1} = \left(\frac{3}{8\pi}\right)^{1/2} \sin \theta \, e^{-i\phi}, \qquad \psi_{22} = \left(\frac{15}{32\pi}\right)^{1/2} \sin^2 \theta \, e^{2i\phi}.$$

As we have seen, if a permanent dipole moment exists its parity is odd. Nonzero matrix elements can then occur only if the initial and final states ψ_{JM} and $\psi_{J'M'}$ have opposite parity. The symmetry of the rotational wavefunctions with respect to an inversion operation (see Fig. 7.1) is found by replacing θ by $\pi - \theta$ and ϕ by $\pi + \phi$ in Eq. (7.14). It is clear that the rotational wavefunctions have even parity when $J = 0, 2, 4, \ldots$, and odd parity when $J = 1, 3, 5, \ldots$. We may say that the parity is

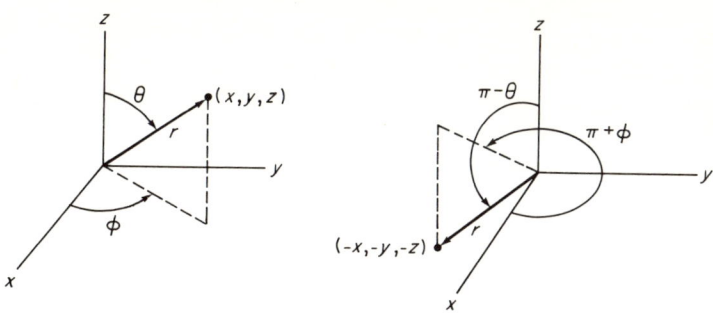

FIG. 7.1. Illustration of parity operation.

determined by $(-1)^J$—it is even if J is even, and odd if J is odd. Thus we have the selection rule for transitions involving permanent dipole moments that *the parity must change*.

On substituting the components of the permanent dipole moment,

$$\mu_x = \mu \sin \theta \cos \phi,$$
$$\mu_y = \mu \sin \theta \sin \phi,$$
$$\mu_z = \mu \cos \theta,$$

into the expression for the matrix elements for electric dipole transitions it is found[1] that the integrals vanish unless

$$\Delta J = J - J' = \pm 1. \tag{7.15}$$

This selection rule governs changes in J for linear molecules. $\Delta J = +1$ indicates absorption, and $\Delta J = -1$ indicates emission of a quantum of rotational energy.

It is clear from the symmetry of homonuclear molecules like N_2 and H_2 that they will have no electric dipole moment, nor will CO_2 for it has the form O—C—O. N_2O, which is linear and arranged as N—N—O, has a permanent dipole moment as, of course, do the unsymmetric linear molecules HCN, HCl, and OCS. They are, therefore, in a position to interact with the electric vector of a radiation field and exhibit pure rotational spectra.

Some linear molecules of particular interest in far-infrared and microwave physics are hydrogen cyanide, hydrogen chloride, and carbonyl

[1] E. Fermi, "Molecules, Crystals and Quantum Statistics." Benjamin, New York, 1966.

sulphide. They are illustrated in Fig. 7.2. HCN has an extremely low moment of inertia for a polyatomic molecule, namely $I = 1.87 \times 10^{-40}$ gm cm², and a dipole moment $\mu = 3$ Debye. Its rotation constant is $B_0 = 44.316$ GHz and, by Eqs. (7.12) and (7.15) or (7.21), the lowest frequency in its pure rotational spectrum is 88.6 GHz, that is, at the millimeter wavelength 3.38 mm. Higher transitions $J1 \to 2$, $J2 \to 3$, $J3 \to 4$, etc., are at wavelengths of 1.69 mm, 1.13 mm, and 845 μm and

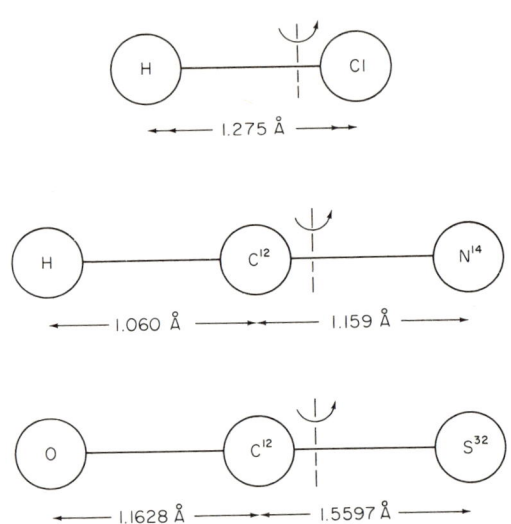

FIG. 7.2. Illustrations of the linear molecules HCl, HCN, and OCS.

extend deeper into the far-infrared with transitions involving large rotational quantum numbers J. The $J0 \to 1$ and the $J1 \to 2$ transitions have not only been observed in absorption but they have been used by Marcuse,[2] and De Lucia and Gordy,[3] to give negative absorption in molecular beam masers. In Marcuse's maser, beams of excited molecules are separated by a field gradient technique and those in the $J = 1$ energy state are fired between the mirrors of a Fabry–Perot resonator. Low power radiation at 88.6 GHz is emitted as the rotationally excited molecules are induced into the $J = 0$ state. Amplification and self-sustained oscillations have been obtained at both 88.6 and 177.2 GHz by De Lucia and Gordy. Copies of oscilloscope tracings of the generated lines are shown in Fig. 7.3. The HCN molecule has also generated laser radiation

[2] D. Marcuse, *J. Appl. Phys.* **32**, 743 (1961); *Proc. IRE* **49**, 1706 (1961).
[3] F. De Lucia and W. Gordy, *Phys. Rev.* **187**, 58–65 (1969).

7.1. ROTATIONAL SPECTRA OF MOLECULES

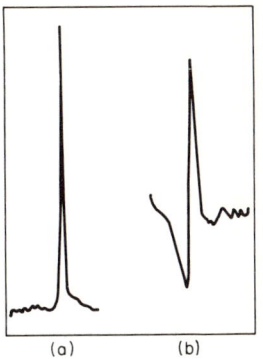

FIG. 7.3. Oscilloscope traces of the output signals from a hydrogen cyanide maser. Signal (a) is the 88.6-GHz $J1 \to 0$ transition, and signal (b) is the 177.2-GHz $J2 \to 1$ transition of HCN (after De Lucia and Gordy[3]).

deeper in the far-infrared (in particular at 337 µm) by transitions between vibrational–rotational states. This topic is discussed in Chapter 2.

The rotational spectrum of the diatomic molecule HCl has historical importance in spectroscopy. It is a molecule with dipole moment $\mu = 1.18$ Debye, moment of inertia $I = 2.7 \times 10^{-40}$ gm cm², and the rotational constant $B_0 = 309.983$ GHz. Pure rotational lines of HCl were measured in absorption by Czerny[4] in 1929, at which time they provided conclusive evidence in favor of the prediction of energy levels proportional to $J(J+1)$ rather than to J^2 as the early form of quantum theory had suggested. Strong[5] later observed HCl emission spectra. Spectral lines measured by Czerny are shown in Fig. 7.4 for transitions down to

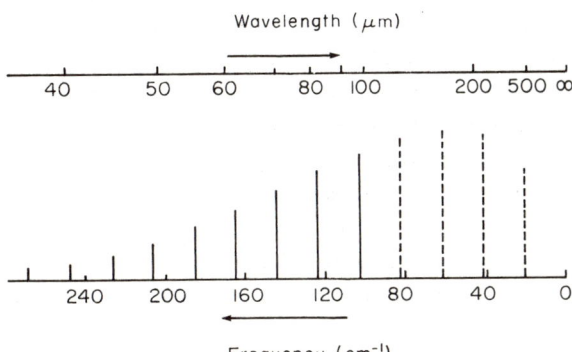

FIG. 7.4. The pure rotation spectrum of HCl. The full lines are the frequencies observed by Czerny[4] (see Sutherland[6]).

[4] M. Czerny, Z. Phys. **53**, 317 (1927).
[5] J. Strong, Phys. Rev. **45**, 877–882 (1934).
[6] G. B. B. M. Sutherland, "Infra-Red and Raman Spectra." Methuen, London, 1935.

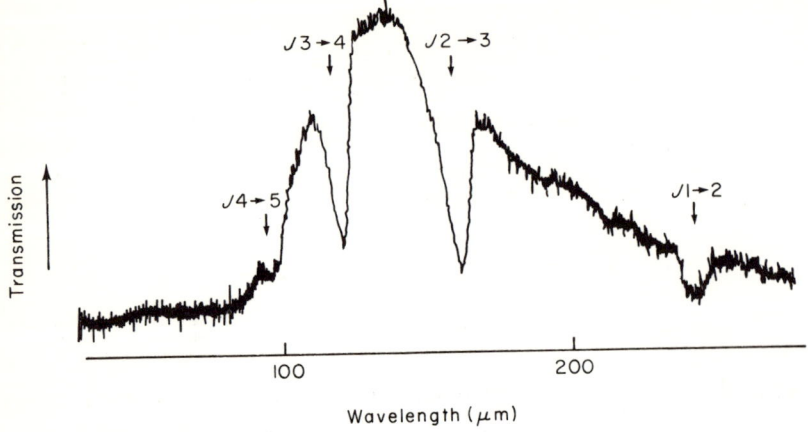

FIG. 7.5. Rotational spectrum of HCl between 100 and 250 μm. (After Bloor et al.[7] Reproduced with permission of the Royal Society.)

$J4 \to 5$. With modern equipment, the longer wavelength transitions have been recorded by Bloor et al.,[7] and Fig. 7.5 shows one of their rotational spectrum recordings.

Like HCl, the small molecules HI and HF also have very large rotational constants. The longest wavelength pure rotational lines for HI[127] and HF[19] occur at about 778.7 and 243.2 μm, respectively. Both have been extensively studied in the far-infrared.[8,9] The two light atoms of HF are separated by less than 1 Å, the moment of inertia is small $(1.32 \times 10^{-40}$ gm cm$^2)$, and B_0 has the value 20.555 cm^{-1}. Through its relatively strong dipole moment (1.91 Debye), it has been successfully observed in absorption to very short wavelengths. Six short wavelength pure rotational transitions of this light molecule between $J = 10 \to 11$ and $J = 15 \to 16$ are shown in Fig. 7.6.[9]

The carbon oxysulphide molecule has been studied extensively at both microwave[10,11] and far-infrared frequencies.[3] Transition frequencies have been measured to seven, and in some cases eight or nine, significant

---

[7] D. Bloor, T. J. Dean, G. O. Jones, D. H. Martin, P. A. Mawer and C. H. Perry, *Proc. Roy. Soc. Ser. A* **260**, 510–522 (1961).

[8] W. Gordy, "Millimeter Waves," Vol. 9, pp. 1–23. Brooklyn Polytech. Press, Brooklyn, New York, 1960.

[9] G. A. Kuipers, D. F. Smith and A. H. Nielsen, *J. Chem. Phys.* **25**, 275 (1956).

[10] C. H. Townes and A. L. Schawlow, "Microwave Spectroscopy." McGraw-Hill, New York, 1955.

[11] D. H. Whiffen, "Spectroscopy." Longmans, Green, New York, 1966.

7.1. ROTATIONAL SPECTRA OF MOLECULES 313

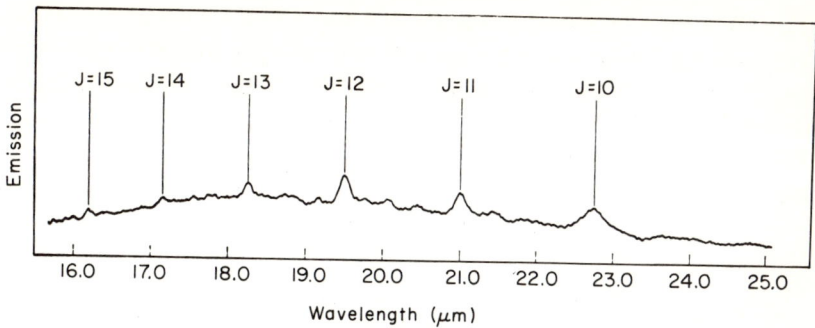

FIG. 7.6. Rotational spectrum of HF in emission (after Kuipers et al.[9]).

figures with the accurate, high resolution instruments of microwave spectroscopy. The isotope $O^{16}C^{12}S^{32}$ has a dipole moment $\mu = 0.71$ Debye, a moment of inertia $I = 137.90 \times 10^{-40}$ gm cm^2, and a rotation constant $B_0 = 6081.480$ MHz. A number of OCS rotational spectral lines observed in microwave and submillimeter wave spectroscopy are given in Table 7.1.

TABLE 7.1. Some Observed $O^{16}C^{12}S^{32}$ Pure Rotational Transitions

Frequency (GHz)	Quantum number assigned
24.325921	$J = 1 \rightarrow 2$
36.48882	$J = 2 \rightarrow 3$
48.65164	$J = 3 \rightarrow 4$
60.81408	$J = 4 \rightarrow 5$
97.30119	$J = 7 \rightarrow 8$
121.62463	$J = 9 \rightarrow 10$
145.94679	$J = 11 \rightarrow 12$
170.26749	$J = 13 \rightarrow 14$
194.58666	$J = 15 \rightarrow 16$
218.90341	$J = 17 \rightarrow 18$
243.21809	$J = 19 \rightarrow 20$
267.52956	$J = 21 \rightarrow 22$
291.83922	$J = 23 \rightarrow 24$
510.4573	$J = 41 \rightarrow 42$
704.437052	$J = 57 \rightarrow 58$
716.546559	$J = 58 \rightarrow 59$
801.259782	$J = 65 \rightarrow 66$
813.553706	$J = 66 \rightarrow 67$

The OCS rotational transitions $J0 \to 1$ and $J1 \to 2$ have been observed in emission by Hill et al., after excitation of the gas by short microwave pulses.[12] Following gas excitation by pulses 10^{-7}–10^{-6} sec long, coherent emission by the "ringing" molecular system occurs, and the frequency of this emission is measured accurately by beating it against a reference signal. Observation of the decay envelope of the radiation has been used to give a measure of the relaxation of the molecular rotation produced by collisions.

Molecular echoes from the OCS transition $J0 \leftarrow 1$ have also been observed to follow irradiation of the gas by two pulses each of length 0.1 μsec separated by an interval of 1 μsec.[12] Echo pulses are emitted at intervals equal to the 1-μsec spacing between the two applied pulses. Molecular echos are a particular case of the echo phenomena now well known[13] for systems of oscillators of several different types: electron spin, electron cyclotron motion, plasma wave oscillations,[13a] etc. Three conditions are sufficient for the production of pulse echoes:

(a) A system of oscillators with a distribution of natural frequencies. Hill et al. used a nonuniform electric field to distribute the frequencies of the OCS molecules by the Stark effect;

(b) Oscillator relaxation time longer than the interval between pulses. The OCS relaxation time exceeds 1 μsec;

(c) Some nonlinear mechanism, as for example, an energy-dependent relaxation time.

With these ingredients the "memory" of the rotating molecules is jolted by the second pulse to recall the effects of its predecessor from phase-mixed storage and regain (at 1-μsec intervals) coherence of phase and the ability to reemit (i.e., echo) the stored energy. Collisional damping reduces the stored rotational energy and hence the magnitudes of the echoes.

Like OCS, the linear molecule N_2O also has its lowest frequency pure rotational transition in the microwave band (at 25,123 MHz). It has a moment of inertia of 66.0×10^{-40} gm cm^2, and a small electric dipole moment of 0.14 Debye. Rotational absorption in N_2O up to $J = 19 \to 20$ has been measured by Bloor et al. whose results are shown in Fig. 7.7.

---

[12] R. M. Hill, D. E. Kaplan, G. F. Herrmann and S. K. Ichiki, Phys. Rev. Lett. **18**, 105–107 (1967).

[13] R. W. Gould, Amer. J. Phys. **37**, 585–597 (1969).

[13a] K. Gentle, Meth. Exp. Phys. **9**, 1–36 (1970).

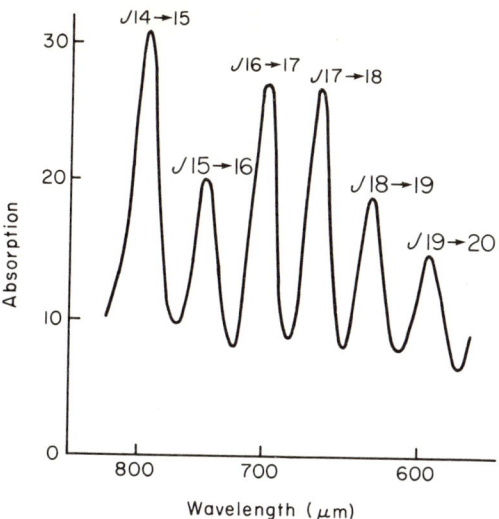

FIG. 7.7. Rotational spectrum of N_2O between 600 and 800 µm. (After Bloor et al.[7] Reproduced with permission of the Royal Society.)

7.1.2. General Symmetric Tops

The linear molecule is a special case of a symmetric top, but the general case is a molecule with two of its principal moments of inertia equal ($I_B = I_C$, say) and different from the moment of inertia I_A. The direction of I_A is called the "figure axis," and this is almost always an axis of symmetry of the molecule.[†] We take this to be the z cartesian direction, and the other two principal axes of inertia to be the x and y directions. If $I_A < I_B$, the moments of inertia are axes of a prolate ellipsoid and the molecule is a prolate symmetric top. If $I_A > I_B$, we have an oblate symmetric top.

The classical expression for the rotational energy of a symmetric top is

$$W = \frac{P_x^2}{2I_x} + \frac{P_y^2}{2I_y} + \frac{P_z^2}{2I_z}, \quad \text{where} \quad I_x = I_y = I_B \quad \text{and} \quad I_z = I_A.$$
(7.16)

The classical motion can be described as a rotation with angular momentum P_z about the symmetry axis and a precession of this axis about

---

[†] It is possible for I_B and I_C to be equal by "accident" in a molecule which is not symmetric in its structure. In such a case, the figure axis is not an axis of symmetry.

the total angular momentum vector **P**. In terms of the cartesian components P_x, P_y, P_z,

$$P^2 = P_x^2 + P_y^2 + P_z^2. \tag{7.17}$$

Quantum mechanically, both P and P_z are quantized, the former with the value $[J(J+1)]^{1/2}(h/2\pi)$ and the latter with the value $Kh/2\pi$, where $K = 0, \pm 1, \ldots, \pm J$. From Eqs. (7.16) and (7.17) it then follows that

$$W = \frac{h^2}{8\pi^2}\left[\frac{J(J+1)}{I_B} + K^2\left\{\frac{1}{I_A} - \frac{1}{I_B}\right\}\right]. \tag{7.18}$$

The rotational energy thus depends on the quantum number J, and also the quantum number K which specifies the component of the total angular momentum $\hbar[J(J+1)]^{1/2}$ along the figure axis of the top. The magnetic quantum number $M = 0, \pm 1, \ldots, \pm J$ occurs in the wavefunctions (as we saw for the special case of the linear molecule) but not in W. The direction of the momentum component $Kh/2\pi$ along the figure axis is specified by the sign of K. The plus and minus signs of K do not affect W, but they give different wavefunctions and so double degeneracy to the energy levels. The level $K = 0$ has $2J + 1$ degeneracy corresponding to the $2J + 1$ values of M, but all levels for which K is not zero have degeneracy $2(2J + 1)$. In terms of the rotational constants $A = h/8\pi^2 I_A$, $B = h/8\pi^2 I_B$, the energy takes the form

$$W/h = BJ(J+1) + (A-B)K^2. \tag{7.19}$$

Spectral lines are given by Eq. (7.19) subject to the appropriate selection rules governing changes of K and J. In a symmetric top, the dipole moment lies along the symmetry axis and there is no component perpendicular to this axis. A radiation field cannot, therefore, exert a torque parallel to the symmetry axis, and such an interaction must leave the dipole moment unchanged. Quantum-mechanical correspondence to this classical argument is simply the requirement that transitions take place only between levels with the same K. It can be shown also that nonzero matrix elements for transitions arise when the total angular momentum quantum number J changes by unity. The selection rules for a symmetric top are then

$$\Delta K = 0, \qquad \Delta J = \pm 1. \tag{7.20}$$

For a symmetric top, linear or otherwise, it follows from Eqs. (7.12), (7.15), (7.19), and (7.20) that the frequency transition $J \leftarrow J - 1$ is given by

$$\nu = B[J(J+1) - J(J-1)]$$
$$= 2BJ. \qquad (7.21)$$

The spectral lines are thus predicted to be evenly spaced with separations $2B$. This is essentially what is observed.

The assumption that a molecule can be treated as a rigid rotor is, of course, not exact and expressions such as Eqs. (7.12), (7.19), etc., must be modified when higher accuracy is required for real (nonrigid) molecules. During rotation centrifugal forces act against the interatomic forces to stretch the dimensions slightly, thereby increasing the moment of inertia. This effect is taken into account by using, in place of Eq. (7.12), for example, the expression

$$W/h = BJ(J+1) - DJ^2(J+1)^2. \qquad (7.22)$$

The constant D is generally very much smaller than B, but for very light, rapidly rotating molecules like hydrogen fluoride this and even higher order corrections may be necessary. Similarly small correction terms must be included in the more general expression Eq. (7.19), and are to be found in the standard texts.[10,14–16]

A symmetric top has a rotational spectrum only if it possesses a permanent dipole moment. By symmetry, if a dipole moment exists it will be along the top axis. When all atoms are in the same plane, there is no dipole moment and no rotational spectrum. However, piramidal symmetric molecules such as NH_3, PH_3, PF_3, and AsF_3 have one atom out of the plane of the three like atoms, and they therefore possess a dipole moment through which rotational motion about an axis perpendicular to the symmetry axis may couple to a radiation field.

The ammonia molecule has played a special role in spectroscopy since Cleeton and Williams[6,17] probed it with 1.25-cm wavelength monochromatic radiation in 1934 in experiments which may be regarded as the beginning of microwave spectroscopy. One of the strongest ammonia

---

[14] W. Gordy, W. V. Smith and R. F. Trambarulo, "Microwave Spectroscopy." Wiley, New York, 1953.

[15] G. W. King, "Spectroscopy and Molecular Structure." Holt, New York, 1964.

[16] G. Hertzberg, "Infrared and Raman Spectra of Polyatomic Molecules." Van Nostrand-Reinhold, Princeton, New Jersey, 1945.

[17] C. E. Cleeton and N. H. Williams, *Phys. Rev.* **45**, 234–237 (1934).

lines was used by Townes in the first maser in the early 1950s,[17a] and, more recently, emission from the "inversion" transition at 1.25 cm has been detected in interstellar NH_3.[17b]

The structure of NH_3 [see Fig. 7.8(a)] has a nitrogen atom at the apex of the shallow pyramid and three hydrogens equidistant from the nitrogen. Its dipole moment is 1.468 Debye, the moments of inertia have the very small values $I_B = 2.82 \times 10^{-40}$ gm cm², and $I_A = 4.44 \times 10^{-40}$ gm cm² giving rotational constants $B_0 = 298.000$ GHz and $A_0 = 189.000$ GHz, and pure rotational spectral lines at submillimeter wavelengths. As indi-

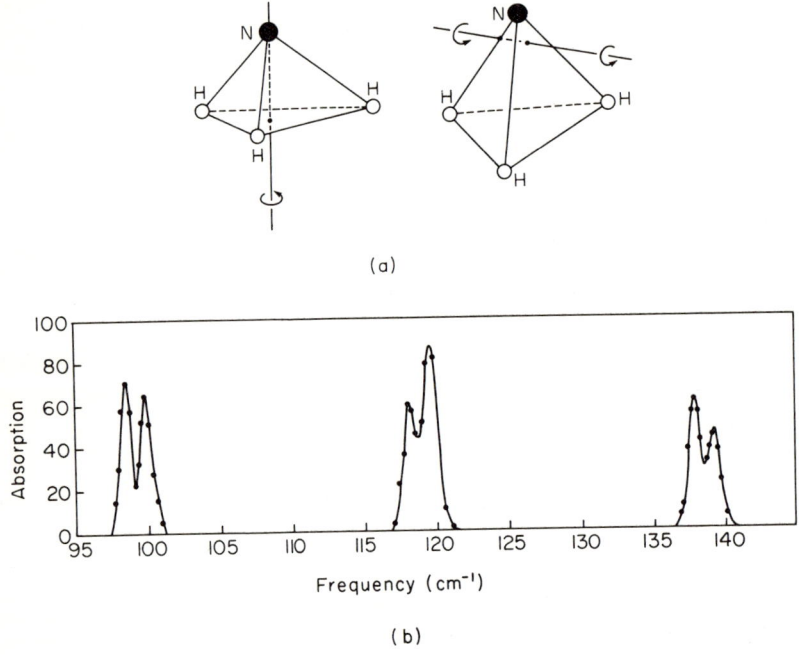

FIG. 7.8. (a) Illustration of the rotation of ammonia (after Von Hippel[17c]). (b) Rotational spectrum of ammonia gas measured with a grating spectrometer (after Wright and Randall[17d]).

[17a] J. P. Gordon, H. J. Zeiger and C. H. Townes, *Phys. Rev.* **95**, 282 (1954).

[17b] A. C. Cheung, D. M. Rank, C. H. Townes, D. D. Thornton and W. J. Welch, *Phys. Rev. Lett.* **21**, 1701 (1968).

[17c] A. Von Hippel, "Dielectric Materials and Applications." Wiley, New York, 1954.

[17d] N. Wright and H. M. Randall, *Phys. Rev.* **44**, 391 (1933).

cated earlier, the subscript zero refers to the ground vibrational state of the molecule. Part of the rotational spectrum of ammonia is shown in Figs. 7.8 and 7.9. The rotational lines of this molecule show splitting into the "inversion doublet." Classically, the nitrogen atom is prevented by a potential barrier from passing through the hydrogenic

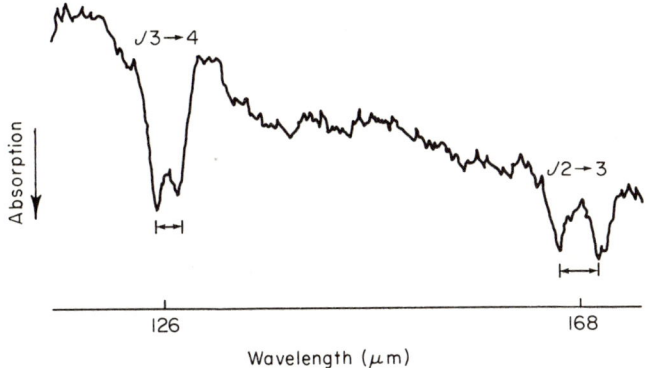

FIG. 7.9. Rotational lines of the ammonia spectrum showing inversion splitting. (After Bloor et al.[7] Reproduced with permission of the Royal Society.)

plane, but quantum mechanically it has a finite probability of inverting the pyramid by tunneling. The inversion frequency with which the nitrogen atom passes from one apex to the other and back again is 23,785.88 MHz, which is a wavelength of about 1.25 cm.

7.1.3. Spherical Tops

The spherical top has all three moments of inertia equal, and its energy levels are found by writing $I_A = I_B = I_C$ in Eq. (7.18). The coefficient of K is zero and the energy,

$$W/h = BJ(J+1), \qquad (7.23)$$

is the same as that for a linear molecule, but there is a difference in degeneracy. The spherical top has an extra $(2J+1)$-fold degeneracy, arising from the $2J+1$ values of K, which has no counterpart in linear molecules.

Because of the high degree of symmetry, spherical tops do not possess a permanent dipole moment and hence they cannot exhibit pure rotational spectra. However, certain of the natural modes of vibration of a spherical molecule can result in a changing dipole moment, and vibrational–rotational spectra can be obtained.[18]

7.1.4. Asymmetric Tops

Unlike the case of symmetric molecules, the problem of developing analytic expressions for the energy levels of molecules with three unequal principal moments of inertia is formidable. Accordingly, general explicit formulas for asymmetric molecules are not available, and it is usual to consider each molecule as a separate problem and to use numerical computation or interpolation methods. Various theoretical approaches are described by Hertzberg,[16] King,[15] and Townes and Schawlow.[10] In the special case of a slightly asymmetric top, where two of the moments of inertia are roughly equal, it is possible to determine the approximate spectrum by considering it as an intermediate case having levels between those of the prolate and oblate symmetric top.[16]

All asymmetric molecules have permanent dipole moments and therefore yield pure rotational spectra. The selection rule is $\Delta J = 0, \pm 1$. Selection of possible transitions is somewhat less restrictive than for linear and symmetric molecules, and the rotational spectral lines are more abundant.

The triatomic molecule H_2O is an asymmetric top that has been investigated in great detail both theoretically and experimentally. It has a dipole moment of 1.94 Debye, and its rotational constants corresponding to the three moments of inertia are $A_0 = 833.2$ GHz, $B_0 = 434.7$ GHz, and $C_0 = 298.5$ GHz. A detailed comparison of observed and theoretically determined spectra has been carried out by Randall et al.[18a] As shown in Fig. 7.10 the agreement is excellent and, as is to be expected for an asymmetric molecule, there are no obvious regularities in the spectrum.

H_2O is a particularly important molecule from the point of view of far-infrared radiation, because it has a multitude of transitions located in this spectral region. As a consequence of this, and of its abundance in the atmosphere, it has a dominant influence on transmission through the atmosphere (see Section 3.2.1) both within and beyond the experimental laboratory. In particular, it is largely responsible for absorption of far-infrared waves emitted by extraterrestrial sources. However, for

---

[18] A. H. Nielsen, *Methods Exp. Phys.* **3**, 57 (1962).

[18a] H. M. Randall, D. M. Dennison, N. G. Ginsburg and L. R. Weber, *Phys. Rev.* **52**, 160 (1937).

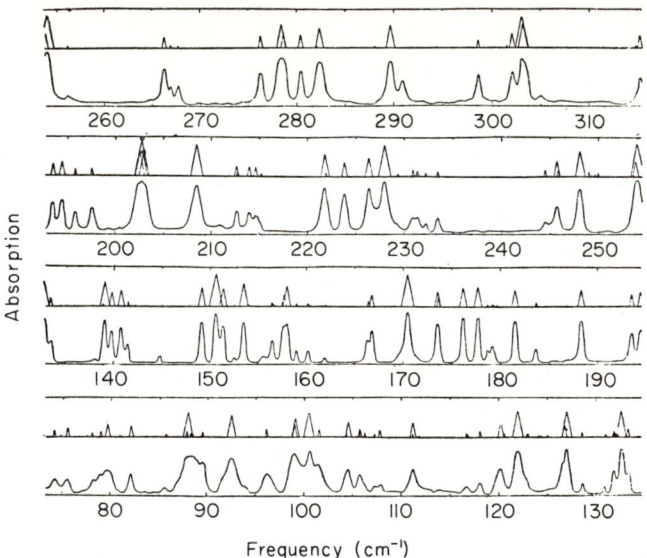

FIG. 7.10. Part of the infrared rotational spectrum of H_2O vapor. The continuous curve represents the observed absorption, and the small triangles above are calculated lines (after Randall et al.[18a]—see Hertzberg, p. 58).

the inconvenience of positive absorption the experimenter has, to some extent, been compensated by the realization of negative absorption by H_2O molecules. This application is discussed in Chapter 2. It has led to the generation of many laser emission lines between 17 and 220 μm and has thereby contributed significantly to a region of the spectrum that has been particularly deficient in sources of coherent radiation.

7.2. Vibration–Rotation Spectra

7.2.1. Vibrational Motion

A molecule containing N atoms can exhibit $3N-6$ modes of vibration if it is nonlinear, and $3N-5$ if it is linear. Each of these vibrations is controlled by an electrostatic force, described by a potential function V, which tends to restore it to an equilibrium position and, in so doing, causes it to oscillate about this position.

Natural vibrational frequencies are determined by the binding forces and atomic masses, and can be calculated by the method of generalized or *normal coordinates* in classical mechanics. In terms of a set of normal coordinates q_i, the quadratic potential approximation is

$$V = \tfrac{1}{2} \sum_i f_i q_i^2, \qquad (7.24)$$

where the f_i are (restoring) force constants, and no cross-product terms $q_i q_j$ are involved. In this formalism the square root of the mass of the nucleus $m_i^{1/2}$ is absorbed in the generalized coordinate q_i, so that the kinetic energy is written

$$T = \tfrac{1}{2} \sum_i \dot{q}_i^2. \qquad (7.25)$$

Using the coordinates q_i, the classical motion can be found from Lagrange's equation in the form

$$\frac{d}{dt}\left(\frac{\partial T}{\partial \dot{q}_i}\right) + \frac{\partial V}{\partial q_i} = 0. \qquad (7.26)$$

Introduction of T and V from Eqs. (7.24) and (7.25) gives simple harmonic equations of motion:

$$\ddot{q}_i + f_i q_i = 0, \qquad (7.27)$$

the solutions of which are

$$q_i = Q_i \sin(f_i^{1/2} t + \delta_i), \qquad (7.28)$$

where $i = 1, 2, \ldots, 3N-6$ (or $3N-5$). Thus each coordinate q_i undergoes harmonic oscillations with amplitude Q_i and phase δ_i. The angular frequencies of the *normal vibrations* are

$$\omega_i = f_i^{1/2}. \qquad (7.29)$$

It is to be noted that the process of absorbing mass into the coordinate q_i means that f_i in Eq. (7.24) is the force per unit mass. ω_i is thus the square root of the force constant divided by the mass.

We have written in Eq. (7.25) an expression for the potential function free from cross-product terms in normal coordinates, which can always be done in the mechanics of normal vibrations. Linear transformations making this possible involve $3N-6$ (or $3N-5$ for the case of linear molecules) simultaneous linear homogeneous equations involving f_i that can be satisfied only if f_i is given by the roots of a secular equation obtained from the determinant of the coefficients of the simultaneous equations. Each of the $3N-6$ (or $3N-5$) roots of the secular equation yields a solution

7.2. VIBRATION–ROTATION SPECTRA

of the classical equations of motion and, through Eq. (7.29), a frequency of a normal vibration. Any arbitrary vibration of the atoms of a molecule may be described as a sum of normal modes. The procedure is relatively simple for the case of a linear triatomic molecule; the interested reader will find a particularly clear normal coordinate analysis of HCN in Dixon's book.[19]

Transformations to normal coordinates that eliminate cross-product terms from the potential energy also enable the quantum-mechanical treatment of molecular vibrations.[20] In normal coordinates the Schrödinger equation takes the form

$$\sum_i \frac{\partial^2 \psi}{\partial q_i^2} + \frac{8\pi^2}{h^2}\left(W - \frac{1}{2}\sum_i f_i q_i^2\right)\psi = 0, \quad (7.30)$$

and this, by writing

$$\psi = \psi_1(q_1)\psi_2(q_2)\cdots\psi_i(q_i), \quad (7.31)$$

separates into the one-dimensional equations

$$\frac{d^2\psi_i}{dq_i^2} + \frac{8\pi^2}{h^2}(W_i - \tfrac{1}{2}f_i q_i^2)\psi_i = 0. \quad (7.32)$$

Each equation is identical with that of a one-dimensional harmonic oscillator, and the eigenvalues for each normal vibration are

$$W_i = (v_i + \tfrac{1}{2})\hbar\omega_i, \quad (7.33)$$

where $v_i = 0, 1, 2, \ldots$.

The wavefunctions of the harmonic oscillator are the well-known Hermite polynomials. When these are used to calculate the matrix elements, one obtains the selection rule for vibrational transitions:

$$\Delta v = \pm 1. \quad (7.34)$$

A diatomic molecule has only one normal vibration, and this can give a vibrational spectrum if there is a dipole moment. Triatomic molecules have four normal modes of vibration. The simplest is the linear triatomic molecule in the symmetric form Y—X—Y of the XY_2 molecule, of which CO_2 is an example. Of its four vibrational modes, illustrated in Fig. 7.11,

---

[19] R. N. Dixon, "Spectroscopy and Structure." Methuen, London, 1965.
[20] L. Pauling and E. B. Wilson, "Introduction to Quantum Mechanics." McGraw-Hill, New York, 1935.

two are degenerate. The ν_1 vibration is a completely symmetric bond-stretching mode, for which the dipole moment remains zero to render it inactive in the infrared. The ν_2 vibration is a bond-bending mode which changes the electric moment perpendicular to the axis and gives rise to an angular momentum of vibration designated by a quantum number l. It is degenerate because this motion is isotropic in two planes. Because the forces opposing bending are small, ν_2 tends to be a low-frequency mode. The mode ν_3 is an unsymmetrical vibrational change in the length of a bond that is infrared active. Bond forces tend to be strong, and ν_3 is consequently a high-frequency mode of vibration.

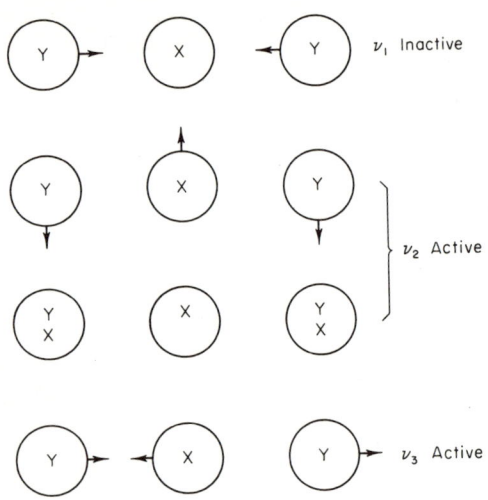

FIG. 7.11. Normal modes of vibrations of the linear triatomic molecule XY_2. The ν_2-mode is doubly degenerate.

Linear molecules of the form XYZ, of which HCN and OCS are examples, vibrate in the normal modes illustrated in Fig. 7.12, which, of course, are the same modes as Fig. 7.11 except that the arrow lengths indicate the magnitudes of the relative motions. The bending mode ν_2 is again low frequency and twofold degenerate, while of the two bond-stretching modes that with the outer atoms moving in opposite directions has the lower frequency ν_1. All three modes ν_1, ν_2, and ν_3 are active to interactions with radiation.

The three vibrational modes of a nonlinear molecule XYZ, such as H_2O, are illustrated in Fig. 7.13. This molecule has a permanent dipole moment and is infrared active in all normal modes.

7.2. VIBRATION–ROTATION SPECTRA

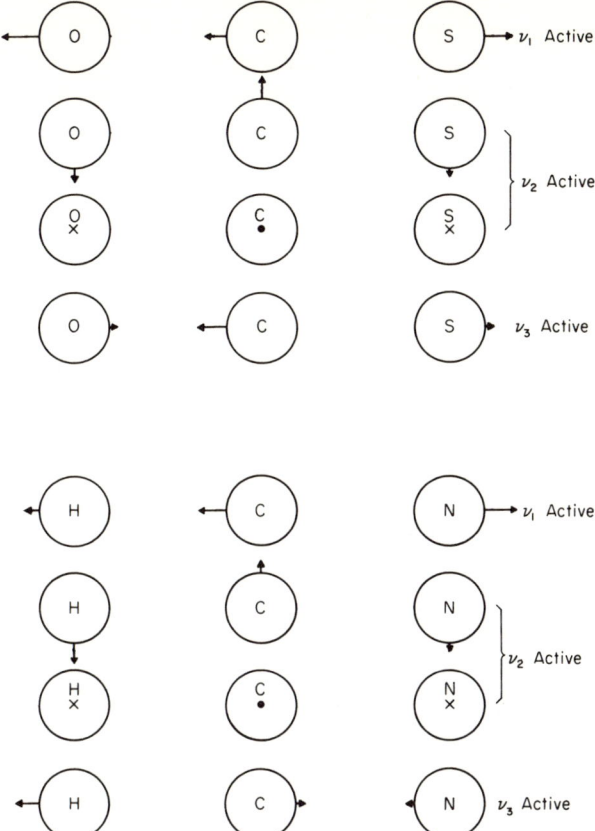

FIG. 7.12. Normal modes of vibration of the linear molecules OCS and HCN. The directions and lengths of the arrows indicate the relative motions of the three nuclei. The ν_2-mode is doubly degenerate.

7.2.2. *l*-Type Doubling and Coriolis Coupling

The degenerate levels of the mode ν_2 of a linear polyatomic molecule can be split by a process called *l*-type doubling which occurs when the molecule has rotational motion. Consider a bond-bending vibration in the y-z plane. Clearly, this will change the moments of inertia about both the x and y axes, but it will change them differently and thereby separate the two otherwise coincident vibrational–rotational states. Linearity of the molecule is perturbed in this mode of vibration, and an additional angular momentum can arise about the z-axis with the quantized values $l\hbar$, where l is a positive integer.

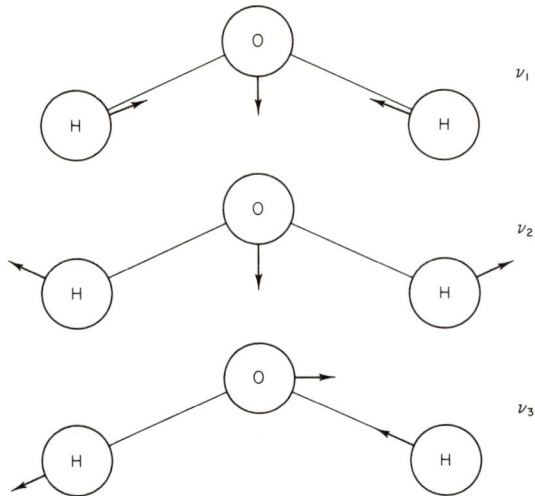

FIG. 7.13. Normal modes of vibration of the nonlinear molecule H$_2$O. All modes are infrared active.

Also, in the presence of both vibrational and rotational motion Coriolis forces $2\mathbf{v} \times \boldsymbol{\omega}$ act on the various atoms in such a way as to couple the ν_2 mode to the ν_3 mode. Coriolis forces act at right angles to the angular velocity. Thus, they arise under the conditions illustrated in Fig. 7.14,

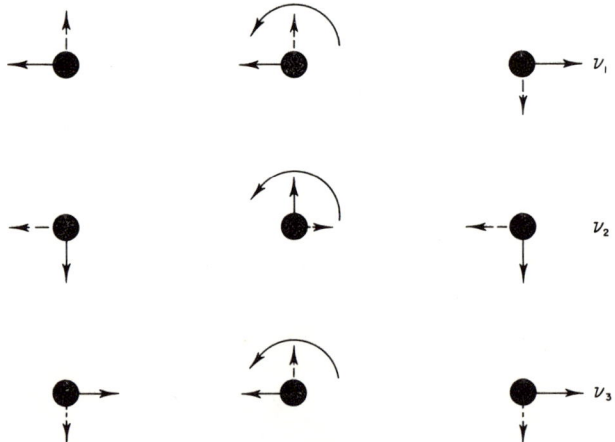

FIG. 7.14. Diagram of the Coriolis forces acting on a triatomic linear molecule. The straight full arrows indicate the three normal modes of vibration ν_1, ν_2, and ν_3, while the curved arrows indicate molecular rotation. The dashed arrows are then the Coriolis forces $2\mathbf{v} \times \boldsymbol{\omega}$.

where the total angular momentum vector and the vibration velocity vectors are at right angles; when these vectors are parallel, the Coriolis interaction force is zero. The effectiveness of the ν_1 mode in exciting the ν_3 mode, and vice versa, will be small except when the two vibration frequencies are close together.

7.2.3. Vibrational–Rotational Bands

In the approximation of noninteracting vibrations and rotations, the total energy of vibrational–rotational motion is the sum of the harmonic oscillator terms W_i in Eq. (7.33) and W in Eq. (7.10) or (7.19), according to whether we are dealing with linear (or spherical) or symmetric molecules. For a single mode of vibration of a linear molecule (other than diatomic) in rotation, the energy is

$$W = (v + \tfrac{1}{2})h\nu + hB_v J(J+1), \tag{7.35}$$

where B_v is the rotational constant of the molecule in the vibrational state v.

For a linear molecule the selection rules for vibration–rotation transitions involving bond-stretching modes are

$$\Delta v = \pm 1, \quad \Delta J = \pm 1, \tag{7.36}$$

and for bond-bending modes, they are

$$\Delta v = \pm 1, \quad \Delta J = 0, \pm 1. \tag{7.37}$$

If rotation were absent, the vibrational spectrum would consist of evenly, and relatively widely, spaced lines. With rotation present a structure of closer, evenly spaced, rotational lines appear in a band about each vibrational frequency. For bond-stretching vibrations the spectrum is called a *parallel* band and the selection rule [Eq. (7.36)] shows that it consists of two branches, a so-called R branch corresponding to $\Delta J = +1$, and a P branch corresponding to $\Delta J = -1$. Bending modes have an additional branch corresponding to $\Delta J = 0$. It is called the Q branch and is at the vibration frequency located midway between the R and P branches (see Fig. 7.15). The spectra associated with bending vibrations are called *perpendicular* bands.

Vibrational lines of a linear triatomic molecule are described by the notation (v_1, v_2, v_3), where the v's are quantum numbers specifying the

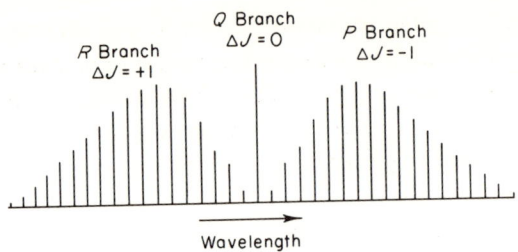

FIG. 7.15. Vibration–rotation band showing the P, Q, and R branches. The lengths of theoretical lines represent the line intensity.

state of excitation in the modes ν_1, ν_2, and ν_3, respectively. To include l-type doubling ν_2 is given the superscript $|l|$, and to specify the rotational state the value of the quantum number J is written in front of the parenthesis. For example, a linear triatomic molecule in the $J = 10$ rotational state of a molecule in the vibrational ground state of the ν_3 mode, and the first excited states of the ν_1 mode and the ν_2 mode with $l = 1$ is denoted $J = 10(11^10)$. Laser action at 337 μm in the HCN laser arises from stimulated transitions between the $J = 10(11^10)$ and $J = 9(04^00)$ vibrational–rotational levels.

Symmetric tops have vibrational–rotational transitions in accordance with the selection rules

$$\Delta J = 0, \pm 1, \quad \text{with } \Delta K = 0, \tag{7.38}$$

$$\Delta J = 0, \pm 1, \quad \text{with } \Delta K = \pm 1. \tag{7.39}$$

The bands corresponding to Eq. (7.38)—the parallel bands of the symmetric top—have P, Q, and R branches and exact resemblance to the perpendicular band of the linear molecule. Perpendicular bands for the symmetric top are complicated somewhat by changes in K. The overall band spectrum is a superposition of P, Q, and R branches associated with each change in K.[6]

7.3. Molecules with Electronic Angular Momentum

In the preceeding discussion we have considered spectroscopic transitions arising entirely from motion of molecular nuclei and have ignored the electrons. In disregarding the possibility of contributions from electronic states, it is implied that the molecule is in the ground state of

7.3. MOLECULES WITH ELECTRONIC ANGULAR MOMENTUM

electronic orbital and spin angular momentum, a state which we will see later is designated by the symbol Σ. This assumption is justified for the great majority of molecules at normal temperatures. However, a small minority of molecules, including O_2, OH, NO, and NO_2, are not in the ground state because their lowest energy electron configurations leave a net electronic angular momentum, or because they are excited by unusual conditions, e.g., high temperatures, to excited states. For example, a molecule may have an odd number of electrons, so that the spins cannot cancel in pairs and there will be a resultant spin. Such a molecule will have a permanent magnetic dipole moment of one Bohr magneton (9.22×10^{-21} erg G^{-1}); it will be paramagnetic and will couple to a radiation field. The homonuclear diatomic molecule O_2 has no electric dipole moment (by symmetry), but it has a net electron spin of 1; it is paramagnetic and is spectrally active at millimeter wavelengths through coupling to the magnetic dipole moment.

By a vector model similar to that used to describe the Stark effect in atomic spectroscopy, spin and angular momentum vectors can be combined with the rotational angular momentum in accordance with certain schemes to give classification models of molecular states. Let us consider the case of a diatomic molecule. Here there is a strong axial field in which a particle moves, and in which, according to classical mechanics, the angular momentum about the symmetry axis remains constant in time. In quantum theory this finds its expression in the fact that the component of angular momentum parallel to the axis of symmetry has discrete values that are integral multiples of $\hbar$, denoted $\Lambda\hbar$. In contrast to the case of an atom, where the central symmetry of the field of force results in the magnitude of the angular momentum remaining constant in the course of time, in the case of an axial field the total angular momentum is not at all constant. In a molecule, then, the magnitude of the electronic orbital angular momentum vector **L** loses any quantized significance, but not so its component Λ along the symmetry axis.

Neglecting spin, let us consider the Schrödinger equation in cylindrical coordinates z, ϱ, and ϕ with potential function $V(z, \varrho)$ symmetrical about the axis $\varrho = 0$ of the molecule. We can separate the equation

$$-\frac{h^2}{8\pi^2 m}\left(\frac{\partial^2 \psi}{\partial z^2} + \frac{\partial^2 \psi}{\partial \varrho^2} + \frac{1}{\varrho}\frac{\partial \psi}{\partial \varrho} + \frac{1}{\varrho^2}\frac{\partial^2 \psi}{\partial \phi^2}\right) + V(z, \varrho)\psi = W\psi \quad (7.40)$$

by writing

$$\psi = \chi(z, \varrho)\Phi(\phi). \quad (7.41)$$

This gives the following equations in χ and ϕ:

$$\frac{d^2\Phi}{d\phi^2} + \Lambda^2\Phi = 0, \qquad (7.42)$$

and

$$-\frac{h^2}{8\pi^2 m}\left(\frac{\partial^2\chi}{\partial z^2} + \frac{\partial^2\chi}{\partial \varrho^2} + \frac{1}{\varrho}\frac{\partial\chi}{\partial\varrho} - \frac{\Lambda^2}{\varrho^2}\chi\right) + V\chi = W\chi, \qquad (7.43)$$

where Λ is the separation constant.

The solutions of Eq. (7.42), namely,

$$e^{+i\Lambda\phi} \quad \text{and} \quad e^{-i\Lambda\phi}, \qquad (7.44)$$

satisfy the physical requirement of single-valuedness only if Λ is real and can take on integer values. That is,

$$\psi = \chi e^{i\Lambda\phi}, \qquad (7.45)$$

where $\Lambda = 0, \pm 1, \pm 2, \ldots$. We see immediately that Λ measures the orbital angular momentum of the electron about the z-axis. We see also from Eqs. (7.43) and (7.44) that for a single eigenvalue of the energy there are two wavefunctions corresponding to the two signs in Eq. (7.44), and these signify the two directions of rotation of electrons around the molecular axis. That is, the states have double degeneracy except for the case $\Lambda = 0$.

If the electronic orbital angular momentum along the axis of a linear molecule is $\Lambda = 0, \pm 1, \pm 2, \ldots$, the molecule is said to be in the $\Sigma, \Pi, \Delta, \Phi, \ldots$ states, respectively. This notation parallels that in atomic vector models where the quantized values $L = 0, \pm 1$, etc., of the total angular momentum of the atom are called S, P, D, etc., states.

As is the case with atoms, strong coupling exists between the spins of the electrons in a molecule and they align themselves parallel and antiparallel to each other to give a total spin **S**. Since an individual spin is $\frac{1}{2}$, the total spin angular momentum **S** (in units of $\hbar$) is either an integer or a half integer, depending on whether the total number of electrons is even or odd. Spins of $0, \frac{1}{2}, 1$, etc., are called singlet, doublet, triplet, etc., states, and this is indicated by a superscript of value $2S + 1$ preceding the Greek letter. The component of spin angular momentum along the

7.3. MOLECULES WITH ELECTRONIC ANGULAR MOMENTUM

interatomic axis is denoted Σ.† It has integer values

$$\Sigma = 0, \pm 1, \pm 2, \ldots, \tag{7.46}$$

and, in the vector model, it is regarded as arising from the various spatial orientations of **S** with respect to Λ.

The components Σ and Λ can be added algebraically to give a total angular momentum with respect to the molecular axis whose magnitude is denoted by Ω (see Fig. 7.16). For any given value of Λ ($\neq 0$) there are $2S + 1$ integer values of Ω between $\Lambda + S$ and $\Lambda - S$. The value of Ω is indicated as a subscript on the right side of the Greek symbol. Thus, as examples, a molecule with $S = \frac{1}{2}$ and $\Lambda = 1$ gives states described by the doublet terms $^2\Pi_{1/2}$ and $^2\Pi_{3/2}$, and a molecule with $S = 1$, $\Lambda = 2$ splits up into triplet terms $^3\Delta_3$, $^3\Delta_2$, $^3\Delta_1$.

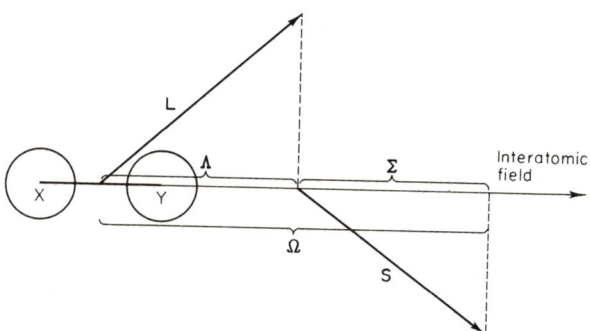

FIG. 7.16. The orientation of the resultant spin and orbit vectors with respect to the axial field in a diatomic molecule XY.

Up to the present we have not considered the rotational angular momentum of the molecule in conjunction with electronic orbital and spin momentum. When this is done several vector coupling schemes may occur which give good descriptions of many linear molecules. We consider two of these, the two extreme cases known as Hund's cases (a) and (b). In case (a) the electronic spin vector couples strongly to the axis of the molecule to give a total angular momentum of magnitude

---

† In molecular spectral notation the letter Σ is used with two different meanings which must not be confused. It denotes both the axial component of the spin and the orbital angular momentum state with $\Lambda = 0$. A double use, similar to this, is made of the letter S in atomic spectral notation.

$\Omega = |\Lambda + \Sigma|$ along the molecular axis, just as we have discussed above. In case (b) this coupling is weak.

That strong coupling is favored by heavy molecules can be easily seen. From the point of view of an electron in orbital motion about the internuclear axis, the nuclei, each with charge Ze, move around it in orbital motion to constitute a current proportional to Z. The consequent magnetic field acts on the spin dipole giving an interaction energy proportional to Z. Furthermore, the spin magnetic moment precesses with a Larmor frequency about the magnetic field also proportional to Z. Proportionality to Z is retained by the interaction energy and precession frequency in a Lorentz transformation to the normal frame of reference.[21] Thus, in the case of a heavy molecule one expects coupling, and rapid precession of S about the axis of the molecule.

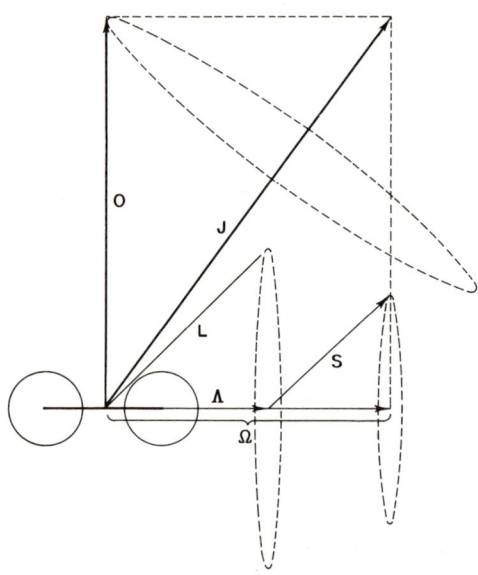

FIG. 7.17. Vector model representing a Hund's case (a) coupling condition.

The situation in Hund's case (a) is illustrated in Fig. 7.17. The rotational angular momentum **O** combines vectorially with the axial angular momentum of magnitude Ω to give the total angular momentum **J**.

Hund's case (b) is illustrated in Fig. 7.18. Here the coupling of S to the molecular axis is negligible, as we may expect for lighter molecules.

[21] R. M. Eisberg, "Fundamentals of Modern Physics." Wiley, New York, 1961.

7.3. MOLECULES WITH ELECTRONIC ANGULAR MOMENTUM

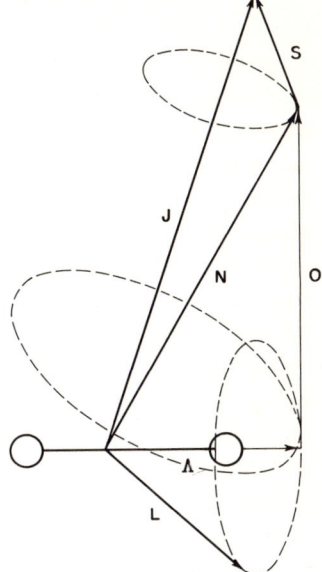

FIG. 7.18. Vector model representing a Hund's case (b) coupling condition.

Vectors Λ and $\mathbf{O}$ combine to give $\mathbf{N} = \Lambda + \mathbf{O}$. The spin S and the angular momentum $\mathbf{N}$ are then oriented with respect to each other, combining to give the total angular momentum $\mathbf{J}$ of the entire molecule.

It can be shown that[10,15,22] for a Hund's case (a) molecule rotational energy states in the millimeter and submillimeter wavelength region can be approximated by the simple form

$$W/h = B_v J(J+1). \qquad (7.47)$$

The rotational energy states of a case (b) molecule is approximately[10,23]

$$W/h = B_v N(N+1), \qquad (7.48)$$

if a small term due to interaction of the electron spin with the magnetic field produced by molecular rotation is neglected. The neglect of spin means, in this case, $J = N$. In both of the above expressions a small centrifugal stretching term may be included.[2,10]

The double degeneracy noted earlier for $\Lambda \neq 0$ in the nonrotating linear molecule is split when the molecule rotates. The splitting is called

[22] J. J. Gallagher, F. D. Bedard and C. M. Johnson, *Phys. Rev.* **93**, 729–733 (1954).
[23] J. S. McKnight and W. Gordy, *Phys. Rev. Lett.* **21**, 1787 (1968).

Λ-type doubling, and its origin may be seen by considering the ϕ-dependent parts, $e^{i\Lambda\phi}$ and $e^{-i\Lambda\phi}$, of the degenerate wave functions. Molecular rotation provides a perturbation that mixes these electron wavefunctions to produce states represented by $e^{i\Lambda\phi} + e^{-i\Lambda\phi}$ and $e^{i\Lambda\phi} - e^{-i\Lambda\phi}$. The two states of a molecule with a Π electron have probability density distributions proportional to $\cos^2 \phi$ and $\sin^2 \phi$, so that for rotation about the z-axis ($\phi = 0$) the electronic moments of inertia are different. The two states, when split in energy, give the Λ-doublet.

Each of the diatomic molecules NO, OH, CH, and SiH has an odd number of electrons, and therefore has electron spin angular momentum $\frac{1}{2}\hbar$ and orbital angular momentum $\hbar$. In the chemically stable nitric oxide molecule the spin of the Π electron is strongly coupled to the molecular axis and it is well described as a Hund case (a) molecule. The two states correspond to $\Omega = \frac{1}{2}$ and $\frac{3}{2}$ and are represented by the notation $^2\Pi_{1/2}$ and $^2\Pi_{3/2}$. In the $^2\Pi_{1/2}$ state the magnetic moments due to the spin and orbital motion cancel, while in the state $^2\Pi_{3/2}$ they add to give a resulting magnetic moment of 2 Bohr magnetons. However, when the molecule rotates, the states mix and the ground state $^2\Pi_{1/2}$ acquires some magnetic moment. The molecule also has a small permanent electric dipole moment of 0.158 Debye through which rotational transitions have been observed. The first rotational transition $J\frac{1}{2} \to \frac{3}{2}$ (in the ground vibrational state $v = 0$) of the $^2\Pi_{1/2}$ state of $N^{14}O^{16}$ occurs at a wavelength of 1.99 mm.[22] The $J\frac{3}{2} \to \frac{5}{2}$ rotational transition of $N^{14}O^{16}$ in the excited state $^2\Pi_{3/2}$ has been observed by Favero et al.,[24] at 1.17 mm, and these workers have found the rotational constant $B_0 = 50848.42$ MHz. Both lines show Λ-doubling with radio frequency separations and hyperfine structure due to nuclear magnetic coupling. A chart recording of the 1.17-mm transition is shown in Fig. 7.19.

Particular interest at radio, microwave, and far-infrared frequencies is associated with the hydrides because their Λ-doublet transitions fall in these bands. Although the free radical OH is chemically very active it can be produced in laboratory ionized gases with electrical discharges, flames, etc., and it is present with relatively high abundance in interstellar space. It is a lightweight molecule with weak coupling between the spin and the molecular axis. At large rotational energies it is very approximately a Hund case (b) molecule, while at small values of J it is intermediate between cases (a) and (b).[10] For no rotation of the nuclei the $O^{16}H^1$ molecule in the state $^2\Pi_{3/2}$ has frequencies associated with Λ-

---

[24] P. G. Favero, A. M. Mirri and W. Gordy, *Phys. Rev.* **114**, 1534–1537 (1959).

7.3. MOLECULES WITH ELECTRONIC ANGULAR MOMENTUM

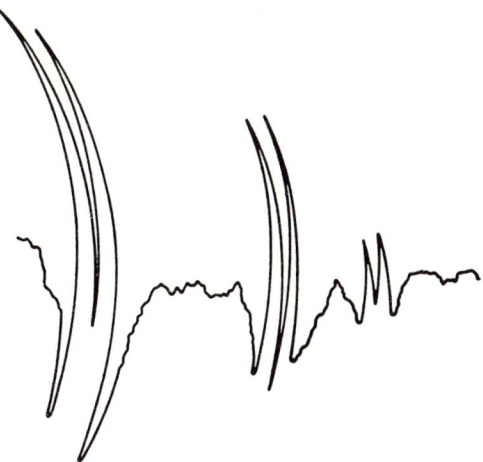

FIG. 7.19. Recording at a wavelength of 1.17 mm of the $J\frac{3}{2} \to \frac{5}{2}$ rotational transition of $N^{14}O$ in the upper $^2\Pi_{3/2}$ electronic state. The gross splitting is due to N^{14} nuclear magnetic coupling, and the small doublet splitting is due to Λ-doubling (after Gordy[8]).

doubling at 1612, 1665, 1667, and 1720 MHz, with four lines occurring instead of two because of hyperfine splitting. Observations of these lines within the galaxy have, within recent years, given considerable impetus to the newly developing field of molecular astronomy.

Like many of the diatomic hydrides, OH has a large rotational constant, $B_0 = 555{,}040$ MHz, and rotational frequencies $2JB_0$ well into the far-infrared. Hydroxyl Λ-doubling transitions have been studied over a large range of rotation energies and, for case (b) coupling, the splitting between the doublet states has been found to be given by

$$W/h = q_\Lambda N(N+1), \qquad (7.49)$$

where q_Λ is very approximately 1 GHz.[10] Expressions for W applicable to case (a) are given by Townes and Schawlow.[10] A comparison between theoretical and experimental determinations of the Λ-doublet separation shows, in Fig. 7.20, an approximately parabolic dependence on N, when N is not too small. For a large range of rotational and hyperfine states in OH produced and raised above the ground state in an electrical discharge through water vapor, Dousmanis et al.[25] have made accurate measurements of Λ-doublet separations. Their observations have been made in the range 7.7–37 GHz.

[25] G. C. Dousmanis, T. M. Sanders and C. H. Townes, *Phys. Rev.* **100**, 1735–1754 (1955).

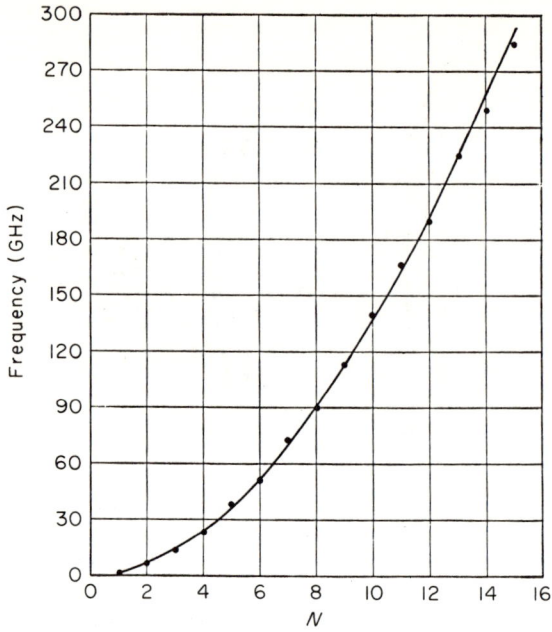

FIG. 7.20. The Λ-type doubling frequency as a function of the quantum number N. The curve is theoretical and the points experimental. (After Townes and Schawlow[10]. Copyright 1955, McGraw-Hill. Used with permission of McGraw-Hill Book Company.)

Oxygen is a molecule of particular interest because it is paramagnetic, and because its abundance in the earth's atmosphere, and possible presence in comets and the atmospheres of other planets, extends its scientific significance beyond the laboratory.[26] It has an even number of electrons, but the chemical bonds are unusual and the electron spins are not all paired. The spin angular momentum is $S = 1$, and the electron orbital angular momentum is zero, so that it is a good Hund's case (b). The ground state of O_2 is accordingly denoted $^3\Sigma$, and its rotational levels are split into triplets by the interaction of the spins of the unpaired electrons and the end-over-end molecular rotation. As $\Lambda = 0$, the rotational quantum number is N and the triplet splitting corresponds to $J = N - 1$, $J = N$, and $J = N + 1$. For the case $J = N$, the rotational energy, if we neglect centrifugal distortion terms, has the simple form[22]

$$W/h = B_v N(N + 1). \tag{7.50}$$

---

[26] A. E. Salomonovic, Highlights of Astronomy, *IAU 13th General Assembly, Prague, 21-31 Aug. 1967*, pp. 148–163. Reidel, Dordrecht, Holland, 1968.

7.3. MOLECULES WITH ELECTRONIC ANGULAR MOMENTUM

TABLE 7.2. Some Millimeter and Submillimeter Molecular Lines

Molecule (electronic state)[a]	Transition	Frequency (GHz)	Comments	Reference
$N^{14}O^{16}(^2\Pi_{1/2})$	$J = \frac{1}{2} \to \frac{3}{2}$	150.372	Associated hfs lines	22
$N^{14}O^{16}(^2\Pi_{3/2})$	$J = \frac{3}{2} \to \frac{5}{2}$	257.867	Associated hfs lines	24
$O^{16}H(^2\Pi_{3/2})$	Λ-doublet for $N = 4, J = 4\frac{1}{2}$; $N = 5, J = 5\frac{1}{2}$	23.827 36.994	Associated hfs lines	27
$O^{16}H^+(^3\Sigma)$	$N = 1, J = 0 \to 1$	133.000		28
$O_2^{16}(^3\Sigma)$	$N = 3, J = 2 \to 3$	62.487	Many nearby lines for other rotational levels	29
	$N = 1, J = 0 \to 1$	118.750		28
$N^{14}H(^3\Sigma)$	$N = 1, J = 0 \to 1$	26.000		28
$CaH(^2\Sigma)$	$J = \frac{1}{2} \to \frac{3}{2}$	254.080		28
$C^{12}S^{32}$	$J = 0 \to 1$	48.991		28
$N_2^{14}O^{16}$	$J = 0 \to 1$ $J = 11 \to 12$	25.123 301.442		28 8
$N^{14}D_3$	$J = 0 \to 1$	309.909		8
$N^{14}H_3$	Inversion, $J = 1, K = 1$	23.694		28
DCl^{35}	$J = 0 \to 1$	323.282	Associated hfs lines	8
$C^{12}O^{16}$	$J = 0 \to 1$ $J = 1 \to 2$ $J = 2 \to 3$ $J = 6 \to 7$	115.270 230.536 345.795 806.651719		30 30 8 31
$CO^+(^2\Sigma)$	$J = \frac{1}{2} \to \frac{3}{2}$	117.980		28
HI^{127}	$J = 0 \to 1$	385.000	Associated hfs lines	8
Li^6D	$J = 0 \to 1 \ (v = 0)$ $J = 0 \to 1 \ (v = 1)$	260.306 254.596		32 32
Li^6F^{19}	$J = 5 \to 6 \ (v = 0)$	538.073		32a
Li^6Cl^{35}	$J = 9 \to 10 \ (v = 0)$	478.893		32a
Li^7Cl^{35}	$J = 11 \to 12 \ (v = 0)$	504.765		32a
$HC^{12}N^{14}$	$J = 0 \to 1$ $J = 1 \to 2$	88.63 177.26		3 3

TABLE 7.2. *(continued)*

Molecule (electronic state)[a]	Transition	Frequency (GHz)	Comments	Reference
$O^{16}C^{12}S^{32}$	$J = 41 \to 42$	510.457		33
	$J = 66 \to 67$	813.354		31
$O^{16}C^{12}Se^{80}$	$J = 38 \to 39$	313.217		8
$Cl^{35}C^{12}N^{14}$	$J = 25 \to 26$	310.366		8
$Br^{81}C^{12}N^{12}$	$J = 55 \to 56$	458.226		8
DF	$J = 0 \to 1$	651.099		31
T Cl^{35}	$J = 0 \to 1$	220.130	Associated hfs lines	34
T Br^{79}	$J = 0 \to 1$	172.366	Associated hfs lines	34
PH_3	$J = 0 \to 1$	266.944	Associated hfs lines	35
DH_3	$J = 0 \to 1$	138.937	Associated hfs lines	35
$Sb^{121}H_3$	$J = 0 \to 1$	176.143	Associated hfs lines	36
$As^{75}H_3$	$J = 0 \to 1$	224.967	Associated hfs lines	36a

[a] If not the $^1\Sigma$ state.

Additional terms occur in the energy expression for the other two states of the triplet. They are given by Townes and Schawlow.[10] For $O^{16}O^{16}$, B_0 is 43.1 GHz and, as the rotational transitions obey the selection rule[10] $\Delta N = \pm 2$, they occur at very short wavelengths indeed. McKnight and Gordy[23] have measured the rotational transition $N = 1 \to 3$, $J = 2 \to 2$,

[27] T. M. Sanders, A. L. Schawlow, G. C. Dousmanis and C. H. Townes, *Phys. Rev.* **89**, 1158 (1953).

[28] C. H. Townes, *Int. Astron. Union Symp., 4th, Manchester, August 1955*. Cambridge Univ. Press, London and New York, 1957.

[29] M. Mizushima and R. M. Hill, *Phys. Rev.* **93**, 745–748 (1954).

[30] F. D. Bedard, J. J. Gallagher and C. M. Johnson, *Phys. Rev.* **92**, 440 (1953).

[31] P. Helminger, F. C. De Lucia and W. Gordy, *Phys. Rev. Lett.* **25**, 1397–1399 (1970).

[32] E. F. Pearson and W. Gordy, *Phys. Rev.* **177**, 59–61 (1969).

[32a] E. F. Pearson and W. Gordy, *Phys. Rev.* **177**, 52–58 (1969).

[33] M. Cowan and W. Gordy, *Phys. Rev. Lett.* **104**, 551–552 (1956).

[34] C. A. Burrus, W. Gordy, B. Benjamin and R. Livingston, *Phys. Rev.* **97**, 1661–1664 (1955).

[35] C. A. Burrus, A. Jache and W. Gordy, *Phys. Rev.* **95**, 706–708 (1954).

[36] A. W. Jache, G. S. Blevins and W. Gordy, *Phys. Rev.* **97**, 680–683 (1955).

[36a] G. S. Blevins, A. W. Jache and W. Gordy. *Phys. Rev.* **97**, 684–686 (1955).

of oxygen at the frequency 424.76 GHz, and the transitions $N = 1 \to 3$, $J = 1 \to 2$, and $N = 1 \to 3$, $J = 2 \to 3$, at 368.5 and 487.25 MHz, respectively.

Table 7.2 lists a number of molecular lines in the millimeter and submillimeter spectrum.

7.4. Intensities and Shapes of Spectral Lines

7.4.1. Perturbation Calculation of Transition Probability

Consider a quantum system subject to a potential that is the sum of a time-independent term $V(x)$ and a small time-dependent term $v(x, t)$. For $V(x)$ alone, the wavefunctions satisfying Schrödinger's equation are

$$\Psi_n(x, t) = \psi_n(x) \exp(-iW_n t/\hbar). \tag{7.51}$$

When the system is perturbed, we assume that the Schrödinger equation for the potential,

$$V'(x, t) = V(x) + v(x, t), \tag{7.52}$$

can be written as the expansion in the unperturbed wavefunctions

$$\Psi'(x, t) = \sum_n a_n(t) \Psi_n(x, t), \tag{7.53}$$

where the coefficients $a_n(t)$ are functions of time. Substituting Eq. (7.53) into the Schrödinger equation

$$-\frac{\hbar^2}{2m} \frac{\partial^2}{\partial x^2} \Psi' + V'\Psi' - i\hbar \frac{\partial \Psi'}{\partial t} = 0,$$

gives

$$\sum_n a_n \left[-\frac{\hbar^2}{2m} \frac{\partial^2}{\partial x^2} \Psi_n + V\Psi_n - i\hbar \frac{\partial}{\partial t} \Psi_n \right] + \sum_n a_n v \Psi_n - i\hbar \sum_n \frac{da_n}{dt} \Psi_n = 0.$$

The term in brackets is the Schrödinger equation in the unperturbed potential V, and it is zero. Thus we have

$$\sum_n a_n v \Psi_n - i\hbar \sum_n \frac{da_n}{dt} \Psi_n = 0.$$

Multiplication of this equation by the complex conjugate of some particular unperturbed wave function,

$$\Psi_k = \psi_k \exp(-iW_k t/\hbar),$$

and integration over x, gives

$$\sum_n a_n \exp\frac{-i(W_n - W_k)t}{\hbar} \int_{-\infty}^{\infty} \psi_k^* v \psi_n \, dx$$

$$= i\hbar \sum_n \frac{da_n}{dt} \exp\frac{-i(W_n - W_k)t}{\hbar} \int_{-\infty}^{\infty} \psi_k^* \psi_n \, dx. \quad (7.54)$$

As the ψ_n are orthogonal and normalized, this equation gives

$$da_k/dt = (-i/\hbar) \sum_n a_n \exp[-i(W_n - W_k)t/\hbar] \int_{-\infty}^{\infty} \psi_k^* v \psi_n \, dx. \quad (7.55)$$

We are interested in harmonic perturbations. When a field $E_x = E_0 \cos \omega t$ interacts with the electric dipole moment $\sum_j q x_j$, perturbing the system between time $t = 0$ and time t, we can integrate Eq. (7.55) to obtain the coefficient a_k:

$$a_k = \frac{-i}{\hbar} \frac{E_0}{2} \int_0^t \sum_n a_n(t) \exp\frac{-i(W_n - W_k)t}{\hbar} (e^{i\omega t} + e^{-i\omega t}) \, dt$$

$$\times \int_{-\infty}^{\infty} \psi_k^* \sum_j q x_j \psi_n \, dx$$

$$= \frac{-iE_0}{2\hbar} \int_{-\infty}^{\infty} \psi_k^* \sum_j q x_j \psi_n \, dx \int_0^t \sum_n a_n(t) \left[\exp i\left\{\frac{\omega - (W_n - W_k)}{\hbar}\right\}t \right.$$

$$\left. + \exp -i\left\{\frac{\omega + (W_n - W_k)}{\hbar}\right\}t \right] dt. \quad (7.56)$$

We suppose that at time $t = 0$ the system is in the state $n = l$. That is,

$$a_n(0) = 0, \quad \text{for} \quad n \neq l,$$
$$a_n(0) = 1, \quad \text{for} \quad n = l.$$

Furthermore, we require that the magnitude of the perturbing force, and the time for which it acts, be so small that the probability of transition from the state l is very small. We can then set

$$a_n(t) \ll 1, \quad \text{for} \quad n \neq l,$$
$$a_n(t) \simeq 1, \quad \text{for} \quad n = l,$$

over the interval 0 to t in the integral Eq. (7.56). This gives

$$a_k = \frac{-i}{2\hbar} E_0 \mu_{xkl} \int_0^t [\exp i(\omega - \omega_{kl})t + \exp -i(\omega + \omega_{kl})t] \, dt, \quad (7.57)$$

7.4. INTENSITIES AND SHAPES OF SPECTRAL LINES

where
$$\omega_{kl} = (W_l - W_k)/\hbar. \tag{7.58}$$

The transition probability to the state k is then

$$a_k a_k^* = \frac{E_0^2}{\hbar^2} |\mu_{x_{kl}}|^2 \left[\frac{\sin^2(\omega - \omega_{kl})t/2}{(\omega - \omega_{kl})^2} + \frac{\sin^2(\omega + \omega_{kl})t/2}{(\omega + \omega_{kl})^2} \right], \tag{7.59}$$

where the matrix element $\mu_{x_{kl}}$ is defined, as in Eq. (7.3), by

$$\mu_{x_{kl}} = \int_{-\infty}^{\infty} \psi_k^* \sum_j q x_j \psi_l \, dx.$$

Equation (7.59) gives significant values of $a_k a_k^*$ through the first term on the right-hand side for $\omega \approx \omega_{kl}$. If we integrate over the width of the line we can express $a_k a_k^*$ in terms of the Einstein coefficient of absorption. Using the standard integral

$$\int_{-\infty}^{\infty} [(\sin^2 x)/x^2] \, dx = \pi,$$

we obtain

$$\begin{aligned} a_k a_k^*/t &= (E_0^2/4\hbar^2) |\mu_{x_{kl}}|^2 \\ &= (\varrho/2\varepsilon_0\hbar^2) |\mu_{x_{kl}}|^2, \end{aligned} \tag{7.60}$$

where $\varrho = \tfrac{1}{2}\varepsilon_0 E_0^2$ is the energy density of the plane polarized field.

The Einstein coefficient of absorption B is defined by

$$a_k a_k^*/t = B\varrho. \tag{7.61}$$

Thus by comparison of Eqs. (7.60) and (7.61), we have

$$B = (1/2\varepsilon_0\hbar^2) |\mu_{x_{kl}}|^2. \tag{7.62}$$

If the radiation is isotropic, we can represent it as composed of three plane polarized components each of amplitude E_0. The Einstein coefficient for isotropic radiation is then

$$B = (1/6\varepsilon_0\hbar^2) |\mu_{kl}|^2, \tag{7.63}$$

where

$$|\mu_{kl}|^2 = |\mu_{x_{kl}}|^2 + |\mu_{y_{kl}}|^2 + |\mu_{z_{kl}}|^2. \tag{7.64}$$

Let us assume that the oscillators have a mean lifetime τ and are distributed as $e^{-t/\tau}$. We average Eq. (7.59) over time. The transition probability averaged over an assemblage of molecules is then

$$a_k a_k^* = \frac{E_0^2}{\hbar^2} |\mu_{x_{kl}}|^2$$

$$\times \left[\frac{\int_0^\infty \frac{\sin^2(\omega-\omega_{kl})t/2}{(\omega-\omega_{kl})^2} e^{-t/\tau} dt + \int_0^\infty \frac{\sin^2(\omega+\omega_{kl})t/2}{(\omega+\omega_{kl})^2} e^{-t/\tau} dt}{\int_0^\infty e^{-t/\tau} dt} \right]$$

$$= \frac{E_0^2}{2\hbar^2} |\mu_{x_{kl}}|^2 \left[\frac{1}{(\omega-\omega_{kl})^2 + (1/\tau^2)} + \frac{1}{(\omega+\omega_{kl})^2 + (1/\tau^2)} \right].$$

(7.65)

For $\omega_{kl} \gg 1/\tau$, the second term on the right is negligibly small, and the transition probability per unit time is

$$\frac{a_k a_k^*}{\tau} = \frac{E_0^2}{2\hbar^2} |\mu_{x_{kl}}|^2 \frac{1/\tau}{(\omega-\omega_{kl})^2 + (1/\tau^2)}.$$

(7.66)

7.4.2. Absorption Coefficient and Linewidth

We have considered a plane polarized wave propagating with velocity c, so the power flux per unit area is $P = \frac{1}{2}\varepsilon_0 E_0^2 c$. We now suppose that of the total number of molecules per unit volume N, a fraction f populate the lower of the two states under consideration, and the number in the higher state is less than this by the Boltzmann factor. The population difference is

$$\Delta N = Nf [1 - \exp(-\hbar\omega_{kl}/kT)] \simeq Nf \hbar\omega_{kl}/kT.$$

The power dP absorbed in a distance dx is then

$$dP = - \left(\frac{Nf\hbar\omega_{kl}}{kT} \right) dx \frac{a_k a_k^*}{\tau} \hbar\omega_{kl}.$$

The absorption coefficient α is defined as $-(1/P)(dP/dx)$ and is given by

$$\alpha = \frac{2\pi^2 Nf \nu_{kl}^2 |\mu_{x_{kl}}|^2}{\varepsilon_0 ckT} \times \frac{1}{\pi} \left\{ \frac{\Delta\nu}{(\nu - \nu_{kl})^2 + (\Delta\nu)^2} \right\},$$

(7.67)

7.4. INTENSITIES AND SHAPES OF SPECTRAL LINES

where $2\pi\nu_{kl} = \omega_{kl}$, and $\Delta\nu = 1/2\pi\tau$. The factor

$$\frac{1}{\pi}\left\{\frac{\Delta\nu}{(\nu - \nu_{kl})^2 + (\Delta\nu)^2}\right\}$$

is the Lorentz line shape function. Its halfwidth is $\Delta\nu$.

Equation (7.67) is a general expression for molecular absorption of plane polarized radiation and, provided the gas pressure is not too high, it gives a reasonable description of the intensity of transitions in the millimeter and submillimeter regions. The expression may be applied to isotropic radiation if a factor of 3 is included in the denominator and the matrix element Eq. (7.64) is used. The total number N of molecules per cubic meter is, of course, related to pressure and temperature. If measurements of pressure are made in torrs and denoted P_{torr}, and temperatures are measured in degrees Kelvin, N is given by

$$N = 9.68 \times 10^{24} \frac{P_{\text{torr}}}{T}. \tag{7.68}$$

The halfwidth of the Lorentz line may be determined by collisions and/or by the Doppler effect. For particular substances and conditions $|\mu_{x_{kl}}|$, the line shape function $S(\nu, \nu_{kl})$, and the population fraction f vary in form to give line strengths and widths characteristic of the molecule and of the conditions of pressure and temperature of the interacting assemblage of molecules.

The factor $|\mu_{x_{kl}}|$ may represent either an electric or a magnetic moment. In the former case, a typical value of the matrix element will be 1 Debye (10^{-18} esu) or 10^{-18} emu, while in the latter case it may be 1 Bohr magneton (0.922×10^{-20} emu). Thus, as the absorption per unit length varies as the square of the matrix element, magnetic dipole interactions tend to be weaker by about a factor of 10^4.

The factors f and $S(\nu, \nu_{kl})$, respectively involving the degeneracy of states and the details of the collisional model, call for a generalization of the expression for α when degeneracy is included explicitly, and the Lorentz line shape function is replaced by a modified form derived by Van Vleck and Weisskopf.[37] The latter can be derived on the assumption that collisions do not completely randomize the orientations of the molecular oscillators but rather they distribute them in accordance with a condition of thermodynamic equilibrium between the oscillators and the

---

[37] J. H. Van Vleck and V. F. Weisskopf, *Rev. Mod. Phys.* **17**, 227 (1945).

existing field immediately after each collision. Van Vleck and Weisskopf obtain the line shape function

$$S(\nu, \nu_{kl}) = \frac{\nu^2}{\pi \nu_{kl}^2} \left[\frac{\Delta \nu}{(\nu - \nu_{kl})^2 + (\Delta \nu)^2} + \frac{\Delta \nu}{(\nu + \nu_{kl})^2 + (\Delta \nu)^2} \right], \quad (7.69)$$

which differs from our previous function [including both terms given by Eq. (7.65)] by a multiplicative factor ν^2/ν_{kl}^2. It differs further from the Lorentz function by the inclusion of the term in $\nu + \nu_{kl}$ [neglected in the foregoing derivation after Eq. (7.65)] which becomes important at higher pressures where $\Delta \nu$ may approach ν_{kl}.

Denoting the degeneracy of the kth state by g_k, the number of molecules N_k with energy W_k will be proportional to g_k times the Boltzmann distribution function. That is,

$$N_k \propto g_k \exp(-W_k/kT). \quad (7.70)$$

The sum over all levels is the total number of molecules N:

$$N = \sum_k N_k \propto \sum_k g_k \exp(-W_k/kT). \quad (7.71)$$

This summation function is the partition function, usually denoted by the symbol Z. Accordingly, the fraction of molecules in the kth state is

$$\frac{g_k \exp(-W_k/kT)}{\sum_k g_k \exp(-W_k/kT)}. \quad (7.72)$$

For isotropic radiation, we can, therefore, write the absorption per unit length in the form

$$\alpha = \frac{2\pi^2 N \nu_{kl}^2 |\mu_{x_{kl}}|^2}{3\varepsilon_0 ckT} \left[\frac{g_k \exp(-W_k/kT)}{\sum_k g_k \exp(-W_k/kT)} \right] S(\nu, \nu_{kl}),$$

where $S(\nu, \nu_{kl})$ is given by Eq. (7.69). This expression gives a reasonably good description of molecular absorption over a wide range of gas pressures. Its limitations are discussed by Townes and Schawlow.[10] For $\nu_{kl} = 0$, it reduces to the Debye formula for the absorption per unit length of a nonresonant medium, namely,

$$\alpha = \frac{\omega N \mu^2}{3\varepsilon_0 ckT} \frac{\omega \tau}{1 + \omega^2 \tau^2}, \quad (7.73)$$

where μ is the dipole moment of a molecule.

7.4. INTENSITIES AND SHAPES OF SPECTRAL LINES

Let us return to Eq. (7.67) which, as we have said, gives a satisfactory description of absorption provided $\Delta\nu$ is not comparable with ν_{kl}. Peak absorption occurs when $\nu = \nu_{kl}$ and is

$$\alpha = 2\pi N f \nu^2 \,|\, \mu_{x_{kl}} \,|^2 / \varepsilon_0 c k T \,\Delta\nu. \tag{7.74}$$

If the linewidth is determined primarily by collisions between molecules, $\Delta\nu$ is proportional to the pressure and is of the order of 10^7 sec^{-1} at a pressure of 1 torr. By Eq. (7.68), the total density of molecules at a pressure of 1 torr is $N = 3.2 \times 10^{22}$ m^{-3}, and so the replacement $\Delta\nu = 3 \times 10^{-16} N$ can be made in Eq. (7.74). As the ratio of gas pressure to temperature drops, as for example, in the dilute gases of interstellar space, or in dense but hot gas discharges, Doppler effects increasingly contribute to line broadening. If the molecular speed is v the observed transition frequency will be $\nu = \nu_{kl}(1 \pm v/c)$, according to whether the particle is traveling towards or away from the observer. A Maxwellian distribution of thermal velocities will give a spread of frequencies, and hence a linewidth given by

$$\Delta\nu/\nu_{kl} = 3.58 \times 10^{-7} (T/M)^{1/2}, \tag{7.75}$$

where M is the molecular weight. For room-temperature gases Doppler linewidths may be 100 kHz or so. In the far-infrared cyanide laser the HCN molecules are above room temperature by some thousands of degrees and the Doppler contribution to the linewidth is about 5 MHz. However, collisions broaden the line also, by about 2 MHz. Interstellar molecular gases may have velocities of, say, 10^4 msec^{-1} which are more or less random and give a Doppler linewidth of about 3×10^{-5} times the molecular line frequency.

7.4.3. Thermal Population Distribution

For vibrational–rotational spectra the factor f in Eq. (7.67) can be expressed as a product, $f = f_v f_r$, of the fraction f_v of molecules in the vibrational state and the fraction f_r of these that are in the lower of the two rotational states involved in the transition. For linear molecules, we have

$$f_r = \frac{(2J+1)\exp[-J(J+1)hB/kT]}{\sum_J (2J+1)\exp[-J(J+1)hB/kT]}, \tag{7.76}$$

and

$$f_v = \frac{g_v \exp(-W_v/kT)}{Z_v}, \tag{7.77}$$

where $W_v = (v + \tfrac{1}{2})h\nu$, and Z_v is the vibrational partition function. Vibrational frequencies occur, in general, above 1000 cm^{-1} where, for $T = 300°$K, $(v + \tfrac{1}{2})h\nu/kT$ is somewhat greater than unity. Thus the vibrational ground state $(v = 0)$ tends to be well populated, and all excited vibrational states relatively unpopulated. This is not generally the case for rotational states, for here the energy increment between levels is smaller. For example, for CO the rotational constant B is 2 cm^{-1} and $hB/kT = 10^{-2}$ at room temperature, while for HCN at 300°K, $hB/kT = 7 \times 10^{-3}$. When the levels are fairly close and the temperature not too low, the rotational partition function can be replaced by the integral approximation

$$\int_0^\infty \exp[-(hB/kT)y]\, dy = kT/hB,$$

where we have let $y = J(J+1)$. We then have

$$f_r = [(2J + 1)hB/kT] \exp[-J(J+1)hB/kT]. \quad (7.78)$$

The fraction of molecules in the various rotational levels is the product of a linearly increasing function and an exponentially decreasing function of J, and this leads to the type of intensity distribution for vibrational–rotational spectra represented by the heights of the lines in Fig. 7.15. Differentiation shows the location of the state of maximum population to be

$$J_{\max} = (kT/2hB)^{1/2} - \tfrac{1}{2}. \quad (7.79)$$

For the HCN molecule at room temperature, the rotational state $J = 8$ has maximum population. The thermal population distribution for this linear molecule is shown in Fig. 7.21.

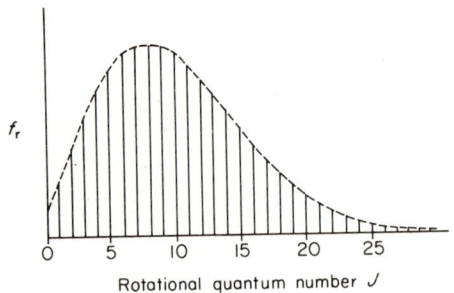

FIG. 7.21. Thermal population of rotational states for the molecule HCN at 300°K.

7.5. Dispersion in Liquids and Solids

As the separations between molecules decrease, as for example when the pressure of a vapor is increased to form condensed phases of matter, the effective time interval between molecular collisions decreases and the resonances broaden in frequency. For liquids and solids, the concept of abrupt collisions used in gas kinetic theory loses its clarity and one tends to speak of a *relaxation time* τ which influences the shape of the absorption versus frequency characteristic according to the relative magnitudes of ω and $1/\tau$. As we have seen previously, resonances are well defined when $\omega\tau \gg 1$ and ill defined when $\omega\tau < 1$. This we can see, in general, from the uncertainty relation

$$\Delta W \, \Delta t \geq \hbar.$$

Writing $\Delta t = \tau$, the time for which a state exists, it is clear that the energy of the state is closely specified if τ is large. For τ small, the uncertainty ΔW broadens and the resonance spreads towards a continuum. For a given relaxation time, such lower frequency resonances as those associated with molecular rotation wash out most readily as their Qs decrease but, of course, this does not mean that the rotational absorption disappears.

In the case of liquids, broad absorption bands arise from damping of the motion of the molecules, and line shapes tend to have the Debye form $\omega\tau/(1 + \omega^2\tau^2)$ given by Eq. (7.73). Molecules in the liquid will be polarized when the E-vector of the radiation distorts the distribution of electronic charge with respect to the nuclei, and when the atomic nuclei are slightly displaced with respect to one another in the normal modes of vibration of the molecule. The sum of the *electronic polarization* (P_E), and the *ionic polarization* (P_I) is called the *distortional polarization*. In the case of polar liquids the molecular orientation in the radiation field increases as the oscillating E-vector increases and relaxes towards the thermodynamic equilibrium distribution when the E-vector decreases towards zero. This is *orientational polarization* and is denoted P_O. Being associated with rotational motion it tends to fall off with increasing frequency in the microwave region when $\omega \approx 1/\tau$. On the other hand, P_I is associated with infrared vibrational modes with $\omega \gg 1/\tau$, which leads to anomalous dispersion characteristics in the frequency region 10^{13}–10^{14} Hz, and a falloff when the nuclear motion diminishes in response to very high frequency waves; the electronic polarization persists into the

TABLE 7.3. Some Typical Relaxation Times[a]

Substance	τ (sec)	Temperature (°C)
Water	9.6×10^{-12}	20
Water	3.2×10^{-12}	75
Nitrobenzene ($C_6H_5NO_2$)	5×10^{-11}	20
Ethyl alcohol (C_2H_5OH)	1.4×10^{-10}	20
Ice	4×10^{-6}	-5
Ice	2×10^{-3}	-45

[a] After Whiffen.[11]

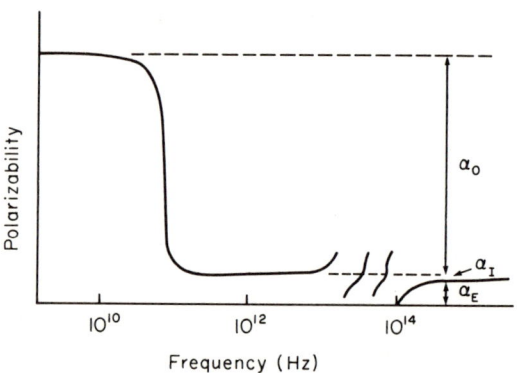

FIG. 7.22. Schematic diagram illustrating the polarizability of matter at a function of frequency. α_O, α_I, and α_E are the orientational, ionic, and electronic polarizabilities, respectively.

visible. Some values of τ are given in Table 7.3, and the frequency dependence of polarization is illustrated in Fig. 7.22 for a material with $\tau \sim 10^{-11}$ sec. In the far-infrared the polarization is $P_E + P_I$.

7.5.1. Debye Theory of Orientational Polarizability

Below about 10^{13} Hz, where P_E and P_I are relatively insensitive to frequency, the behavior of the complex dielectric constant can be described by equations due to Debye. When an electric field E_{eff} acts on an individual molecular dipole, it acquires energy

$$W = -\boldsymbol{\mu} \cdot \mathbf{E}_{\text{eff}} = -\mu E_{\text{eff}} \cos \theta. \tag{7.80}$$

7.5. DISPERSION IN LIQUIDS AND SOLIDS

In thermal equilibrium the distribution of energy and, by Eq. (7.80), the angular orientation will be given by Boltzmann statistics. If the total number of molecules per unit volume is N, the number per unit solid angle about the direction θ will be

$$n(\theta) = \frac{N}{4\pi} \exp \frac{\mu E_{\text{eff}} \cos \theta}{kT} \qquad (7.81)$$

or

$$n(\theta) \simeq \frac{N}{4\pi} \left(1 + \frac{\mu E_{\text{eff}} \cos \theta}{kT}\right), \qquad (7.82)$$

for normal fields and temperatures.

To find the orientational polarization per unit volume, P_0, one simply integrates $n(\theta)\mu \cos \theta$ over the solid angle $2\pi \sin \theta \, d\theta$, with $n(\theta)$ given by Eq. (7.82). That is

$$P_0 = 2\pi \int_0^\pi n(\theta)\mu \cos \theta \sin \theta \, d\theta$$

$$= -\frac{N}{2} \int_0^\pi \left(1 + \frac{\mu E_{\text{eff}} \cos \theta}{kT}\right) \mu \cos \theta \, d(\cos \theta)$$

$$= N\mu^2 E_{\text{eff}}/3kT \qquad (7.83)$$

$$= N\alpha_0 E_{\text{eff}}, \qquad (7.84)$$

where

$$\alpha_0 = \mu^2/3kT \qquad (7.85)$$

is the polarizability per molecule, and $N\alpha_0$ the polarizability per unit volume due to molecular orientation.

In dense materials, the effective field E_{eff} acting on an individual dipole is not the applied field E, but rather it is the resultant of E and the polarization field of atoms in the near neighborhood. It is called the Lorentz local field and, in the case of spherical symmetry, has the well-known form

$$E_{\text{eff}} = E + (P_0/3\varepsilon_0). \qquad (7.86)$$

As the displacement vector is given by

$$\mathbf{D} = \varepsilon_0 \mathbf{E} + \mathbf{P}_0, \qquad (7.87)$$

the field acting on the dipole can be written in terms of macroscopic

parameters as
$$E_{\text{eff}} = (E/3)(\varkappa + 2). \tag{7.88}$$

Using Eqs. (7.87) and (7.88) in Eq. (7.84), we obtain the Clausius–Mossotti relation
$$(\varkappa_s - 1)/(\varkappa_s + 2) = N\alpha_0/3\varepsilon_0. \tag{7.89}$$

The subscript s is now included to remind us that we are dealing with static polarization; furthermore, the reader is reminded of the neglect of electron and atom polarization.

In the classical Debye model of condensed matter the rotation of the polar molecule is dominated by viscous damping. If, for example, the polarizing field is suddenly removed, the polarization will decay as $e^{-t/\tau}$, which is the form we expect for any over-damped oscillator, e.g., an RC circuit. When such a heavily damped oscillator is driven by an oscillating field, $Ee^{-i\omega t}$, we know the frequency response is given by $1/(1 + i\omega\tau)$. Thus one expects the orientational polarizability to be of the form

$$\alpha_0 = \frac{\mu^2}{3kT} \frac{1}{1 + i\omega\tau}.$$

If we include now the electronic and ionic polarizabilities, α_E and α_I, we may plausibly generalize Eq. (7.89) by writing

$$\frac{\varkappa - 1}{\varkappa + 2} = \frac{N}{3\varepsilon_0} \left(\alpha_E + \alpha_I + \frac{\mu^2}{3kT} \frac{1}{1 + i\omega\tau} \right). \tag{7.90}$$

The static value $\varkappa_s$, and the high-frequency value $\varkappa_{IR}$, of the complex dielectric constant are given by

$$\frac{\varkappa_s - 1}{\varkappa_s + 2} = \frac{N}{3\varepsilon_0} \left(\alpha_E + \alpha_I + \frac{\mu^2}{3kT} \right), \tag{7.91}$$

and
$$(\varkappa_{IR} - 1)/(\varkappa_{IR} + 2) = (N/3\varepsilon_0)(\alpha_E + \alpha_I). \tag{7.92}$$

The high-frequency value $\varkappa_{IR}$ is clearly the value at the far-infrared frequencies on the plateau of Fig. 7.22 at about 10^{12} Hz (33 cm^{-1}), below the region of anomalous dispersion associated with ionic vibrational modes. It is often denoted $\varkappa_\infty$ rather than $\varkappa_{IR}$, but we shall be using the symbol $\varkappa_\infty$ later to denote the plateau in the optical region above the ionic resonances in Fig. 7.22, where only electronic polarizability is significant.

By introducing the expressions on the right of Eqs. (7.91) and (7.92) into Eq. (7.90), we obtain the complex dielectric constant with real and imaginary parts[37a]

$$\varkappa' \equiv \mathrm{Re}(\varkappa) = \varkappa_{\mathrm{IR}} + (\varkappa_{\mathrm{S}} - \varkappa_{\mathrm{IR}}) \frac{1}{1 + \omega^2 \tau_e^2}, \qquad (7.93)$$

$$\varkappa'' \equiv \mathrm{Im}(\varkappa) = (\varkappa_{\mathrm{S}} - \varkappa_{\mathrm{IR}}) \frac{\omega \tau_e}{1 + \omega^2 \tau_e^2}, \qquad (7.94)$$

where

$$\tau_e = \tau(\varkappa_{\mathrm{S}} + 2)/(\varkappa_{\mathrm{IR}} + 2). \qquad (7.95)$$

We note that the frequency dependence in Eq. (7.94) is of the same form as in Eq. (7.73), with a modification to the relaxation time brought about by the polarization field which becomes important in dense media. This comparison shows that the frequency response Eq. (7.94) is a reduction of the Van Vleck-Weisskopf line shape[10] in the limit of short relaxation times where the uncertainty of frequency clouds the meaning of a resonance frequency.

7.5.2. Classical Theory of Electronic and Ionic Dispersion

General expressions describing the dielectric properties of material media are given by the classical dispersion theory of Lorentz, Drude, and others, developed in most electromagnetic theory texts.[37b] The equation of motion of a single one-dimensional oscillator is

$$m\ddot{x} + m\gamma\dot{x} + m\omega_0^2 x = qEe^{i\omega t},$$

where $m\omega_0^2$ is the binding force, γ is the damping constant (inverse of the relaxation time), and E is the amplitude of the electric vector driving the bound particle of charge q, mass m. The solution gives the coordinate displacement x and, hence, the dipole moment

$$p = qx = \frac{(q^2/m)E}{\omega_0^2 - \omega^2 + i\gamma\omega}.$$

If there are Nf_j oscillators per unit volume characterized by the constants

---

[37a] A. Von Hippel, "Dielectric Materials and Applications." Wiley, New York, 1964.

[37b] M. Born and E. Wolf, "Principles of Optics." Pergamon, London, 1964.

ω_j and γ_j, the dipole moment per unit volume is

$$P = \sum_j \frac{(Nf_j q^2/m)E}{\omega_j^2 - \omega^2 + i\gamma\omega}. \tag{7.96}$$

Equations (7.96) and (7.87) thus give for the dielectric constant of dilute media

$$\varkappa = 1 + \frac{P}{\varepsilon_0 E} = 1 + \sum_j \frac{Nf_j q^2/\varepsilon_0 m}{\omega_j^2 - \omega^2 + i\gamma\omega}. \tag{7.97}$$

If we suppose the system includes Nf_1 particles with $\omega_1 = 0$ as, for example, free electrons in a metal or heavily damped bound charges in a dielectric, we may write

$$\varkappa = 1 - \frac{Nf_1 q^2/\varepsilon_0 m}{\omega^2 - i\gamma\omega} + \sum_{j \neq 1} \frac{Nf_j q^2/\varepsilon_0 m}{\omega_j^2 - \omega^2 + i\gamma\omega}. \tag{7.98}$$

In the case of a metal, the second term on the right describing free electrons may be written in terms of the conductivity at zero frequency, namely,

$$\sigma_0 = \frac{Nf_1 q^2}{m\gamma}.$$

For dense materials one must replace the applied field **E** in Eq. (7.96) by the effective field given by Eq. (7.86). Assuming one natural frequency of oscillation ω_0, and an oscillator density $N_0 = Nf_0$, one obtains

$$\varkappa - 1 = \frac{N_0 q^2/\varepsilon_0 m}{(\omega_0^2 - \omega^2 + i\gamma\omega)\{1 - (1/3\varepsilon_0)[(N_0 q^2/m)/(\omega_0^2 - \omega^2 + i\gamma\omega)]\}}. \tag{7.99}$$

We introduce a new resonant frequency defined by

$$\omega_0'^2 = \omega_0^2 - (N_0 q^2/3\varepsilon_0 m); \tag{7.100}$$

this enables us to write $\varkappa$ in the form

$$\varkappa = (n - 1K)^2 = 1 + \frac{N_0 q^2/\varepsilon_0 m}{\omega_0'^2 - \omega^2 + i\gamma\omega}. \tag{7.101}$$

Here $n - iK$ is the complex refractive index. Thus the dispersion characteristic of a dense medium can be written in the same form as that in a dilute medium (e.g., a gas), except that we must take the resonant frequency as shifted from the natural frequency of the isolated oscillator towards longer wavelengths.

7.5. DISPERSION IN LIQUIDS AND SOLIDS

The alternative way of using Eq. (7.96) for a dense medium is to replace E by $E_{\text{eff}} = E(\varkappa + 2)/3$, as given by Eq. (7.88), and P by $(\varkappa - 1)\varepsilon_0 E$, which is another form of Eq. (7.87). This gives

$$\frac{\varkappa - 1}{\varkappa + 2} = \frac{(n - iK)^2 - 1}{(n - iK)^2 + 2} = \frac{N}{3\varepsilon_0} \sum_j \frac{f_j q^2/m}{\omega_j^2 - \omega^2 + i\gamma\omega}. \quad (7.102)$$

What we have now is an equation that predicts anomalous dispersion at each oscillator frequency ω_j for the quantity $(\varkappa - 1)/(\varkappa + 2)$ rather than for $\varkappa$. If we assume a (ionic) resonance at ω_1 in the infrared, and another (electronic) resonance well above this in the ultraviolet, then from Eq. (7.102) we can derive the expression

$$\varkappa = \varkappa_\infty + (\varkappa_S - \varkappa_\infty) \frac{\omega_1^2}{\omega_1^2 - \omega^2 + i\gamma\omega}, \quad (7.103)$$

where $\varkappa_\infty$ is the dielectric constant at frequencies above the ionic resonance but below electronic resonances, that is, in the visible spectrum.

In the case of ionic semiconductors, free carriers as well as lattice ions contribute to the effective dielectric properties; from Eqs. (7.98) and (7.103) the dielectric constant is clearly

$$\varkappa = \varkappa_\infty + (\varkappa_S - \varkappa_\infty) \frac{\omega_1^2}{\omega_1^2 - \omega^2 + i\gamma\omega} - \frac{\omega_p^2}{\omega(\omega - i\gamma_{\text{fc}})}.$$

Here γ_{fc} is the reciprocal of the relaxation time for the free carriers, and $\omega_p^2 = nq^2/\varepsilon_0 m^*$ is the plasma frequency of a concentration of n free carriers per unit volume, each of charge q and effective mass m^*.

The equations we have derived provide a satisfactory basis for the interpretation of the main features of spectra of molecules, electrons both free and bound in solids, and ions in liquids and solids. As molecules get closer together in going from a gaseous to a liquid state, coupling between the neighboring oscillators tends to increase. In some substances the coupling may still be weak even in the liquid and solid states, with the result that essentially the same high-frequency resonance can be observed in all three phases. This has been shown by the observations of Ewing for vibrational modes of CO, and by Brecher et al. for cyclopropane[38,39] (see Fig. 7.23). In general, however, one expects the mole-

---

[38] C. Brecher, E. Krikorian, J. Blanc and R. S. Halford, *J. Chem. Phys.* **35**, 1097 (1961).

[39] S. S. Mitra, in "Optical Properties of Solids" (S. Nudelman and S. S. Mitra, eds.). Plenum, New York, 1969.

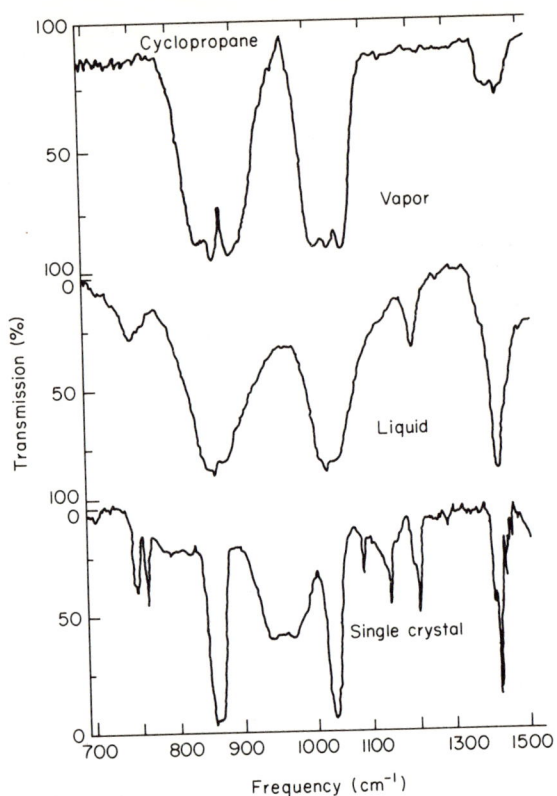

FIG. 7.23. Infrared spectra of cyclopropane vapor (at 25°C), liquid (at −125°C), and crystal (at −190°C), recorded for equivalent absorption paths (after Brecher et al.[38,39]).

cules of a liquid to couple with increased oscillator packing and the spectrum of closely spaced resonances to broaden into bands. We see this in Figs. 7.24 and 7.25.[40] Water shows continuous absorption throughout the far-infrared, increasing as the wavelength decreases from 700 to 50 μm, while the other liquids show resonances at relatively short far-infrared wavelengths, but above 150 μm there is broadening into essentially continuous absorption. Behavior like that of Fig. 7.25 has been found by Chanal et al. for many different liquids.

Of particular importance in the far-infrared are the spectra of crystalline solids, for these have been of long-time interest, especially in the

[40] D. Chanal, E. Decamps, A. Hadni and H. Wendling, J. Phys. (Paris) 28, 165–170 (1967).

7.5. DISPERSION IN LIQUIDS AND SOLIDS

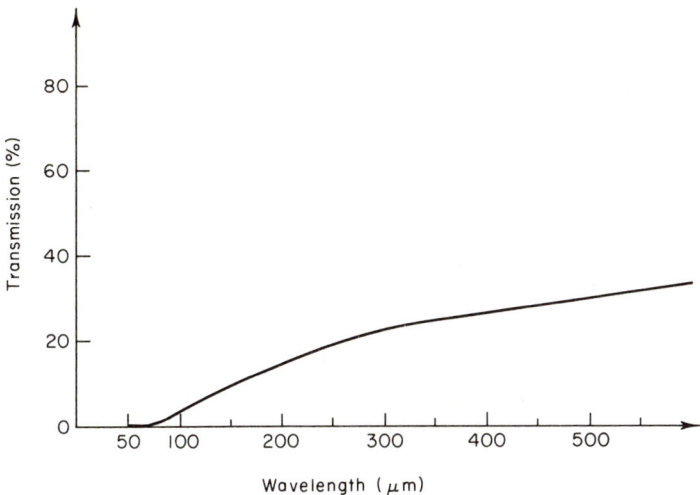

FIG. 7.24. Percentage transmission through a 80-μm thickness of water (after Chanal et al.[40]).

case of alkali hylides such as NaCl, KCl, LiF, and KBr. These ionic crystals couple particularly favorably to a radiation field in the dipole-like mode of vibration where the lattice of negative ions moves in unison with respect to the lattice of positive ions. The strong interaction associated with this particular motion can be interpreted, in its general features, in terms of a dielectric constant, $\varkappa = \varkappa' - i\varkappa''$, with real and imaginary parts given by Eq. (7.103):

$$\varkappa' = n^2 - K^2 = \varkappa_\infty + \frac{(\varkappa_\mathrm{S} - \varkappa_\infty)\omega_1{}^2(\omega_1{}^2 - \omega^2)}{(\omega_1{}^2 - \omega^2)^2 + \gamma^2\omega^2}, \qquad (7.104)$$

and

$$\varkappa'' = 2nK = \frac{(\varkappa_\mathrm{S} - \varkappa_\infty)\omega_1{}^2\omega\gamma}{(\omega_1{}^2 - \omega^2)^2 + \gamma^2\omega^2}, \qquad (7.105)$$

where ω_1 is the lattice vibrational resonance frequency of the mode. For a detailed discussion of this dispersion characteristic, its limitations, and related expressions due to Stigetti and others, the interested reader is referred to the discussions of Mitra,[39] Hadni,[41] and Born and Huang.[42]

---

[41] A. Hadni, "Essentials of Modern Physics Applied to the Study of The Infrared." Pergamon, Oxford, 1967.

[42] M. Born and K. Huang, "Dynamical Theory of Crystal Lattices." Oxford Univ. Press, London and New York, 1954.

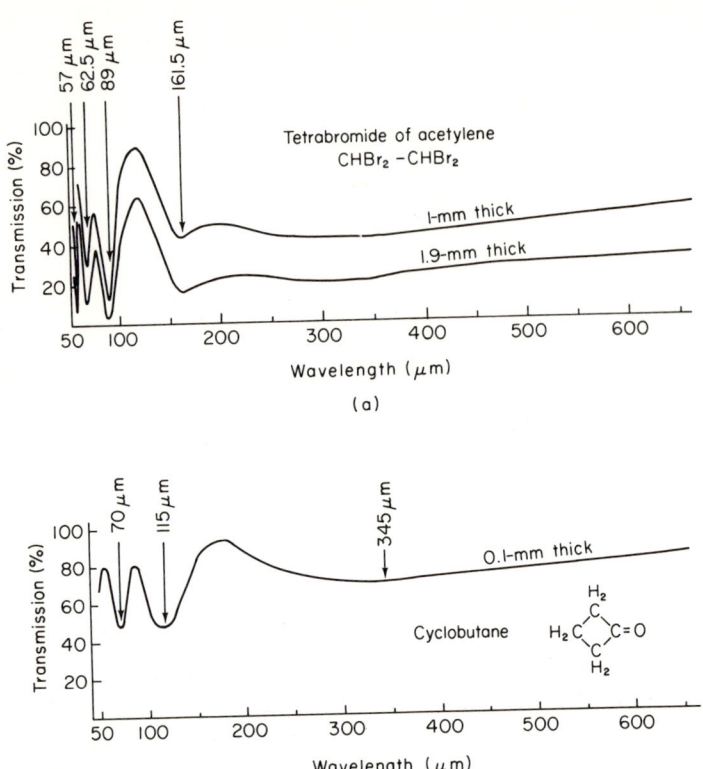

FIG. 7.25. Percentage transmission of far-infrared radiation through tetrabromide of acetylene and cyclobutane. The thicknesses are indicated on the curves (after Chanal et al.[40]).

As we see in the following section, and in Chapter 3, where we discuss crystal transmission and reflection from the point of utility for far-infrared instrumentation, strong lattice vibration modes tend to lie in the range 20–200 μm, where they provide useful filtering capabilities. In addition to the materials discussed in Chapter 3, the results of lattice vibration studies of AgCl, AgBr, AgI, ZnSe, and ZnTe are given by Hadni et al.,[43] and the extensive study by Mitra[39] reviews the state of understanding of lattice resonances in InSb, GaAs, AlSb, InAs, InP, and many others including mixed crystals such as $GaAs_xSb_{1-x}$.

[43] A. Hadni, P. Henry, J.-P. Lambert, G. Morlot, P. Strimer and D. Chanal, *J. Phys. (Paris)* **28**, Suppl. to No. 2, 118–128 (1967).

7.5.3. The Kramers–Kronig Relation

In dispersion theory we treat a dielectric as a system of harmonic oscillators driven by an electromagnetic wave propagating in the medium. The amplitude of an oscillator builds up to a steady-state value set by the condition that the net power absorbed (the difference between the absorbed incident power and the reradiated power) equals the losses. Collisional relaxation influences not only the dissipation but also the phase velocity, for this is determined by the superposition of many reradiated waves, having regard to their amplitudes and the phase shifts incurred between absorption and reemission. As we see from the equations of Section 7.5.2, not only is γ common to the real and imaginary parts of $\varkappa$ but the same set of parameters [ω_j, γ, Nf_j, and ω in Eq. (7.97), or $\varkappa_S$, $\varkappa_\infty$, ω_1, γ, and ω in Eq. (7.103)] determine both. It is not surprising, then, that the real and imaginary parts of the dielectric constant are related rather than independent quantities. Physically speaking they represent two aspects of the same phenomenon, and quantitatively, the relation is so close that a knowledge of the frequency dependence of the real part can be used to yield the imaginary part, and vice versa.

As a simple example consider Eq. (7.97), and suppose there is only one type of oscillator involved with frequency ω_1, damping constant γ, and density $N_1 = Nf_1$. The real and imaginary parts of $\varkappa$ can be written $\varkappa' = fn(N_1, \gamma, \omega_1, \omega)$ and $\varkappa'' = Fn(N_1, \gamma, \omega_1, \omega)$. If the curve $\varkappa'$ versus ω is known, three points on it give three equations and from these N_1, γ, and ω_1 can be determined. Thence we can plot the curve $\varkappa''$ versus ω, given that we accept the functional dependence of the Lorentz theory. This is true quite generally, independently of the structure of any dispersion theory. Mathematically expressed, the calculation of an imaginary part from a given real part, and vice versa, is prescribed by the Hilbert transforms,[44] but to the physicist the connection is known as the Kramers–Kronig relation. For a detailed discussion of this subject the reader is referred to Morse and Feshbach,[44] Cardona,[45] Brown,[46] and Fröhlich.[47] We confine ourselves here to a statement of the relationship.

The amplitude and phase of the complex (amplitude) reflection coef-

---

[44] P. M. Morse and H. Feshbach, *Methods Theor. Phys.* Pt. 1, 370–372 (1953).
[45] M. Cardona, in "Optical Properties of Solids" (S. Nudelman and S. S. Mitra, eds.). Plenum, New York, 1969.
[46] F. C. Brown, "The Physics of Solids." Benjamin, New York, 1967.
[47] H. Fröhlich, "Theory of Dielectrics." Oxford Univ. Press (Clarendon), London and New York, 1958.

ficient, $|R|\,e^{i\theta}$ are related by[47a]

$$\theta(\omega) = \frac{2\omega}{\pi} \int_0^\infty \frac{\ln|R(\omega_1)|}{\omega^2 - \omega_1^2}\, d\omega_1. \tag{7.106}$$

Therefore, from a measurement over the frequency spectrum of the amplitude of the reflected wave, the value of $\theta(\omega)$ can be obtained. One can then find the values of the optical constants n and K from the equation

$$|R|\,e^{i\theta} = (n - 1 - iK)/(n + 1 - iK),$$

and the real and imaginary parts of the dielectric constant from

$$\varkappa' = n^2 - K^2, \quad \text{and} \quad \varkappa'' = 2nK.$$

7.6. Crystal Lattice Vibrations in Solids

7.6.1. Modes of Vibration. Reststrahlen Spectrum

In crystalline solids the atoms are bound in periodic arrays which propagate waves with constraints and behavior imposed by the dynamics of lattice motion allowed by the equations of mechanics. Although ionic crystals are of particular interest in the far-infrared, their behavior in this context is only a particular case of the general phenomena of wave propagation in periodic structures,[48] e.g., filter networks, slow-wave structures in accelerators and wave generators (see Chapter 2), electrons in semiconductors, etc. We recall from Chapter 5 that an electron in a solid when visualized as a de Broglie wave shows a periodic $\omega - k$ dispersion curve describing bands of propagation, and frequency stop-bands where waves cannot propagate, corresponding to wavelengths in the solid in the vicinity of integral multiples of twice the lattice spacing. Precisely the same behavior occurs with acoustic and electromagnetic waves where, between regions of propagation, reflections from the successive planes of atoms add in phase to give constructive interference of reflected waves at the expense of forward wave propagation.

Let us consider as a simplified model of a crystal lattice a linear chain of atoms of two different kinds separated by a massless spring with force

---

[47a] T. S. Robinson, *Proc. Phys. Soc. (London)* **B65**, 910 (1952).

[48] L. Brillouin, "Wave Propagation in Periodic Structures." McGraw-Hill, New York, 1946.

7.6. CRYSTAL LATTICE VIBRATIONS IN SOLIDS

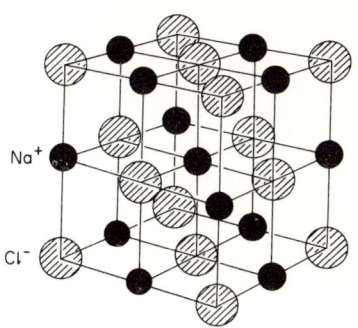

FIG. 7.26. Model of the face-centered cubic crystal structure of sodium chloride. Each Na⁺ ion carries a single positive charge, and each Cl⁻ ion carries a single negative charge.

constant f. With atoms of mass M located at odd lattice points and with nearest neighbors of mass m at the even points we have an approximate representation which shows the essential dispersive characteristics of ionic crystals. For example, in the NaCl crystal shown in Fig. 7.26, M may be the mass of a Cl⁺ ion, m the mass of an Na⁻ ion, and f the force constant determined largely by the Coulomb force between the oppositely charged nearest neighbor ions. Of course, the one-dimensional model represents only waves which propagate in a symmetry direction, such as [111] in NaCl, for then a plane contains only a single kind of ion. In considering the dynamics of the diatomic chain of Fig. 7.27, let the

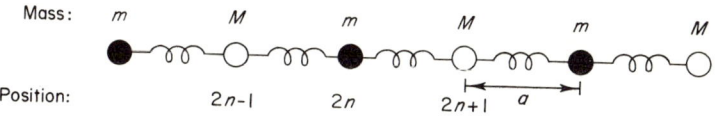

FIG. 7.27. Chain of coupled particles with masses m and M, which is used as a one-dimensional model of a diatomic crystal lattice.

longitudinal displacement from the equilibrium position $2n$ of the atom of mass m be x_{2n}, and that of its neighbor at position $2n+1$ be x_{2n+1}. Newton's second law for the neighboring atoms gives

$$m\ddot{x}_{2n} = f(x_{2n+1} + x_{2n-1} - 2x_{2n}),$$
$$M\ddot{x}_{2n+1} = f(x_{2n+2} + x_{2n} - 2x_{2n+1}).$$

We take trial solutions of the form representing propagating waves

$$x_{2n} = A\, e^{i(\omega t + 2nka)},$$
$$x_{2n+1} = B\, e^{i[\omega t + (2n+1)ka]}, \tag{7.107}$$

and substitute to obtain the simultaneous linear equations in A and B:

$$-\omega^2 mA = fB(e^{ika} + e^{-ika}) - 2fA, \qquad (7.108)$$
$$-\omega^2 MB = fA(e^{ika} + e^{-ika}) - 2fB.$$

Using Cramer's rule we equate the determinant of the coefficients of the unknowns A and B to zero to obtain the nontrivial solutions

$$\omega^2 = f\left(\frac{1}{m} + \frac{1}{M}\right) \pm f\left[\left(\frac{1}{m} + \frac{1}{M}\right)^2 - \frac{4\sin^2 ka}{Mm}\right]^{1/2}. \qquad (7.109)$$

For $ka \ll 1$, the roots of Eq. (7.109) are

$$\omega_1 = \left[2f\left(\frac{1}{m} + \frac{1}{M}\right)\right]^{1/2}, \qquad (7.110)$$

and

$$\omega_2 = ka[2f/(m+M)]^{1/2}, \qquad (7.111)$$

while for $ka = \pi/2$, they are

$$\omega_1 = (2f/m)^{1/2} \qquad (7.112)$$

and

$$\omega_2 = (2f/M)^{1/2}. \qquad (7.113)$$

From Eqs. (7.110)–(7.113) we see there are two separate branches to the $\omega - k$ dispersion characteristic; an upper frequency branch with its highest frequency

$$\left[2f\left(\frac{1}{m} + \frac{1}{M}\right)\right]^{1/2}$$

at $k = 0$, and lowest frequency $(2f/m)^{1/2}$ at $k = \pi/2a$, and a lower branch with $\omega_2 = 0$ at $k = 0$, and its highest frequency $(2f/M)^{1/2}$ at $k = \pi/2a$. The propagation constant k may be positive or negative (corresponding to opposite directions of propagation) and the dispersion symmetrical about $k = 0$; furthermore, from Eq. (7.109), it is clear that the $\omega - k$ dispersion curve over the interval of k-space between $\pm \pi 2a$ is repeated in the zones $\pi/2a$ to $3\pi/2a$, $-\pi/2a$ to $-3\pi/2a$, $3\pi/2a$ to $5\pi/2a$, and so on. The range of k-space $-\pi/2a < k < \pi/2a$ of the one-dimensional lattice is called the *first Brillouin zone*; it follows from the periodicity of the $\omega - k$ curves that waves with propagation constants k beyond this range can be handled by transforming to the first Brillouin zone by an appropriate shift of an integral number of intervals π/a along the k-axis.

7.6. CRYSTAL LATTICE VIBRATIONS IN SOLIDS

According to Eq. (7.107), the ratio of the displacements of the atoms at even and odd lattice positions is

$$x_{2n+1}/x_{2n} = (B/A)\, e^{ika},$$

and, from Eq. (7.108) the ratio of the amplitudes is

$$A/B = (2f - \omega^2 M)/(2f \cos ka). \tag{7.114}$$

Thus, when $ka \ll 1$, we have for the lower branch

$$x_{2n+1}/x_{2n} = e^{ika}. \tag{7.115}$$

This result, which applies to $k \approx 0$, shows in-phase longitudinal periodic motion of neighboring atoms as occurs with sound vibrations, and we accordingly call it the *longitudinal acoustic* (or LA) *branch*. For the upper branch

$$x_{2n+1}/x_{2n} = -(m/M)\, e^{ika}, \tag{7.116}$$

in the limit of small k, and we have what is called the *longitudinal optical* (LO) *branch* with out-of-phase oscillations and hence a dipole moment to which a radiation field can couple.

The dispersion characteristic of a diatomic chain showing the two branches separated by a forbidden frequency (or energy) band is illustrated in Fig. 7.28. Within a branch there is a finite frequency passband but the corresponding range of wavelengths extend from zero to infinity. The binding force affects the curves in that it determines the vertical scale and hence the widths of the frequency regions in which wave propagation is forbidden and allowed.

The one-dimensional representation of a crystal lattice with two kinds of atoms can be extended by considering motion of the particles transverse to the chain in a manner similar to the treatment of longitudinal vi-

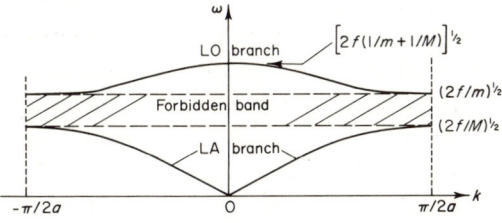

FIG. 7.28. Dispersion diagram of a one-dimensional diatomic chain with masses m and M ($m < M$).

brations. Such an analysis reveals *transverse optical* (TO) and *transverse acoustic* (TA) modes with dispersion characteristics similar to those for longitudinal modes. In both the TO mode and the TA mode there are two orthogonal directions of vibration which give twofold degeneracy in the case of isotropic crystals, but two distinct TO and two distinct TA curves for anisotropic crystals. The situation is illustrated in Fig. 7.29 with the TO mode frequencies located below the LO modes due to the

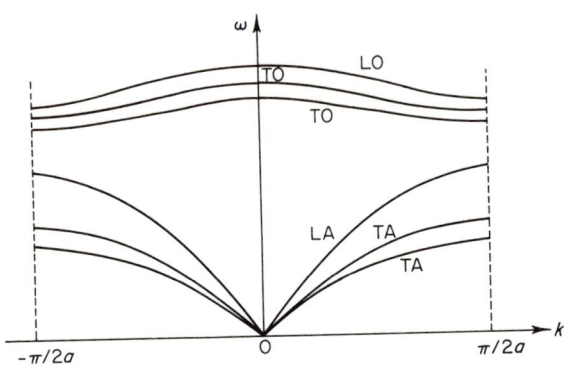

FIG. 7.29. Schematic diagram of the $\omega - k$ dispersion characteristic for a crystal with two atoms per unit cell.

restoring forces being weaker for transverse displacements. Further generalization can be achieved by extensions to two and three dimensions, but these do not change the essential qualitative picture given by the one-dimensional model. A three-dimensional lattice composed of N unit cells each with n atoms has $3Nn$ normal modes of vibration. These $3Nn$ modes are distributed in the dispersion diagram in $3n$ branches, $3n-3$ being optical branches and 3 acoustic.

A crystal is optically active to far-infrared radiation in the region of the top of a TO branch, in that here it represents a dipole moment that can be driven by a transverse radiation field; in addition the phase velocity ω/k of the wave in the region of very long wavelengths $k \approx 0$ will be fast enough to match that of the propagating electromagnetic wave in the crystal, thereby giving large energy exchange. The shape of the spectrum will be described by Eq. (7.103) with the resonant frequency ω_1 given by Eq. (7.110). We can see from Eq. (7.103), or Eqs. (7.104) and (7.105), that the lattice interaction with the electromagnetic field extends upwards from the top (ω_1) of the TO branch of the dispersion curves. Take, for simplicity, the lossless case $\gamma = 0$. Equation (7.104) shows that when ω

7.6. CRYSTAL LATTICE VIBRATIONS IN SOLIDS

reaches the dispersion frequency ω_1, $\varkappa$ and n become infinite and, since the reflection coefficient for normally incident radiation is

$$R = (\sqrt{\varkappa} - 1)/(\sqrt{\varkappa} + 1) = (n-1)/(n+1), \qquad (7.117)$$

the crystal becomes perfectly reflecting. For ω infinitesimally higher than ω_1, $\varkappa$ becomes negative, n pure imaginary, and wave propagation is cut off (just as we saw occur under certain circumstances in gaseous and semiconductor plasmas in Chapters 5 and 6), and $|R| = 1$. Perfect reflection persists with further increase in frequency until

$$\varkappa = 0 = \varkappa_\infty + (\varkappa_S - \varkappa_\infty)\omega_1^2/(\omega_1^2 - \omega^2),$$

that is, up to the frequency $\omega = \omega_L$ given by

$$\omega_L = (\varkappa_S/\varkappa_\infty)^{1/2}\omega_1 = (\varepsilon_S/\varepsilon_\infty)^{1/2}\omega_1. \qquad (7.118)$$

It has been shown by Lyddane *et al.*[49] that the dynamic equations of lattice motion reveal the frequency ω_L to be that of the longitudinal optical mode in the limit of long wavelengths; Eq. (7.118) is accordingly called the *Lyddane–Sachs–Teller* relation, or simply the LST relation. It is discussed in Section 7.7.

We see, therefore, that Lorentz dispersion theory when combined with the dynamics of lattice vibrations predicts cutoff of propagation and reflection over the frequency range extending between the TO and the LO branches at $k = 0$. When damping is included, we get the same type of modification to cutoff that we saw in Chapters 5 and 6, namely, a rounding off of the lossless reflection curves near the frequencies ω_1 and ω_L, and substantial rather than complete reflection between these limits (see Fig. 7.30). This band of high reflection has been commonly called the *reststrahlen spectrum*, since the early days when Rubens and others observed its properties. In addition to the properties delineated for reststrahlen bands, weaker features due to anharmonic effects may arise to alter the pattern in some cases. Anharmonic effects frequently show up as small absorption peaks at the side of the main resonance; they are a consequence of coupling between otherwise independent vibrational modes that occurs when nonlinear terms in the ionic displacement are added to the linear (or Hooke law) restoring forces in the equation of motion.

---

[49] R. H. Lyddane, R. G. Sachs and E. Teller, *Phys. Rev.* **59**, 673 (1941).

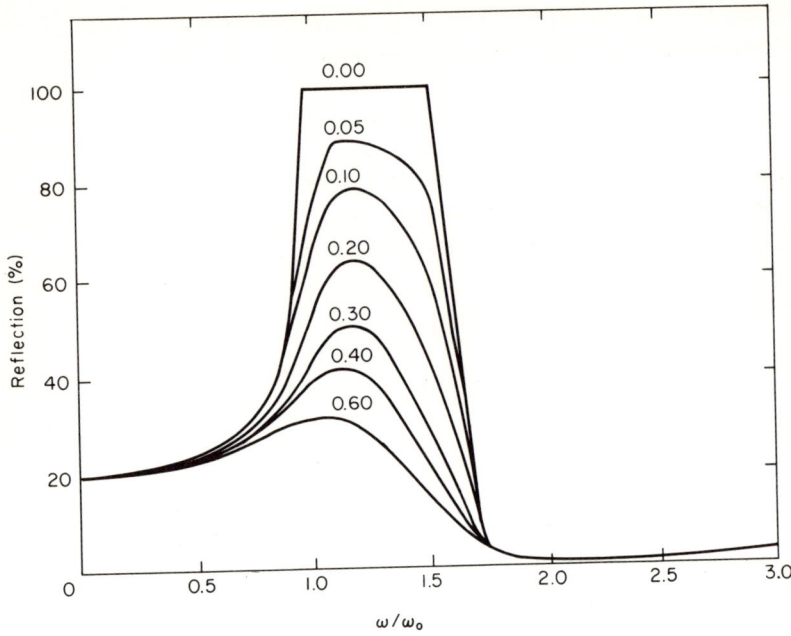

FIG. 7.30. Reflection spectra of a damped oscillator in the vicinity of ω_0 the natural oscillation frequency. The value of γ/ω is marked on each curve, and $\gamma/\omega_0 = 0.00$ is the lossless curve (after Mitra[39]).

In Table 7.4 we record some typical reststrahlen frequencies in the far-infrared, for group III–V and group II–VI compounds, as well as crystals of the NaCl and CsCl type. Data on these and other materials are given by Hadni,[41] Mitra[39] and Brown.[46]

Figures 7.31 and 7.32 show plots of the real and imaginary parts of the complex refractive indices of a ZnSe mixture (commercially known as irtran IV) and AgI in regions of anomalous dispersion in the far-infrared. These measurements were made by Hadni et al.,[43] and based on the use of the Kramers–Kronig relation.

Homopolar crystals such as diamond, silicon, and germanium can also be excited in vibrational modes[50] but rather weakly, and generally at infrared frequencies in the vicinity of 1000 cm^{-1}. When in a pure state, these crystals are relatively transparent in the far-infrared, where they can be used as transmission windows (see Chapter 3). The diamond

[50] J. T. Houghton and S. D. Smith, "Infra-Red Physics." Oxford Univ. Press (Clarendon), London and New York, 1966.

TABLE 7.4. Far-Infrared Reststrahlen Properties of Some Crystalline Solids[a,b]

Material	$\omega_1/2\pi$ (cm^{-1})	$\omega_L/2\pi$ (cm^{-1})	$\varkappa_S$	$\varkappa_\infty$
LiF	306	659	8.81	1.9
LiCl	191	398	12.0	2.7
LiBr	159	325	13.2	3.2
NaBr	134	209	6.4	2.6
NaI	117	176	6.6	2.91
KF	190	326	5.5	1.5
RbF	156	286	6.5	1.9
RbCl	116	173	4.9	2.2
RbBr	88	127	4.9	2.3
RbI	75	103	5.5	2.6
CsCl	99	165	7.2	2.6
TlCl	63	158	31.9	5.1
TlBr	43	101	29.8	5.4
AgCl	106	196	12.3	4.0
AgBr	79	138	13.1	4.6
MgO	401	718	9.64	3.01
NiO	401	580	11.75	5.7
CoO	349	546	13.0	5.3
MnO	262	552	22.5	4.95
InSb	185	197	17.88	15.68
InAs	219	243	15.15	12.25
InP	304	345	12.61	9.61
GaSb	231	240	15.69	14.44
GaAs	269	292	12.90	10.90
GaP	367	403	10.18	8.46
AlSb	319	340	11.21	9.88
AlN	667	916	9.14	4.84
BN	1056	1304	7.1	4.5
ZnS	274	350	8.3	5.0
ZnSe	207	253	9.20	6.10
ZnTe	179	206	10.38	7.8
CdTe	141	168	10.20	7.13

[a] After Mitra.[39]
[b] ω_1 in the angular frequency of the TO mode and ω_L that of the LO mode.

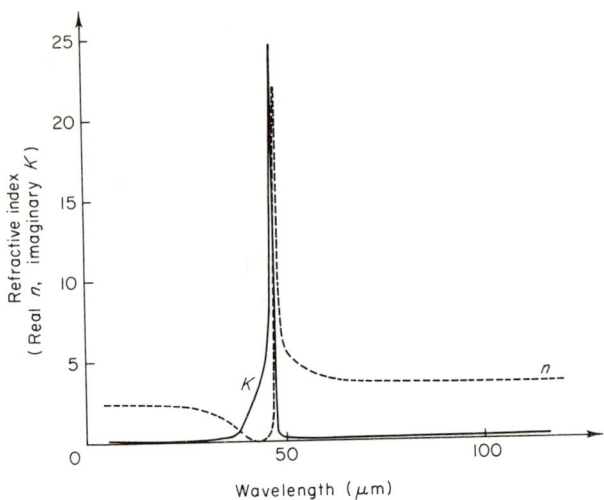

Fig. 7.31. The optical constants, n and K, of ZnSe (commercial name irtran IV) at 90°K, as a function of wavelength (after Hadni et al.[43]).

structure of these crystals can be visualized as two interpenetrating face-centered cubic lattices displaced relative to one another by one quarter of the main body diagonal. There are two atoms of the same kind per unit cell, which means that the crystal possesses no electric dipole moment when in an unperturbed state. However, short wavelength vibra-

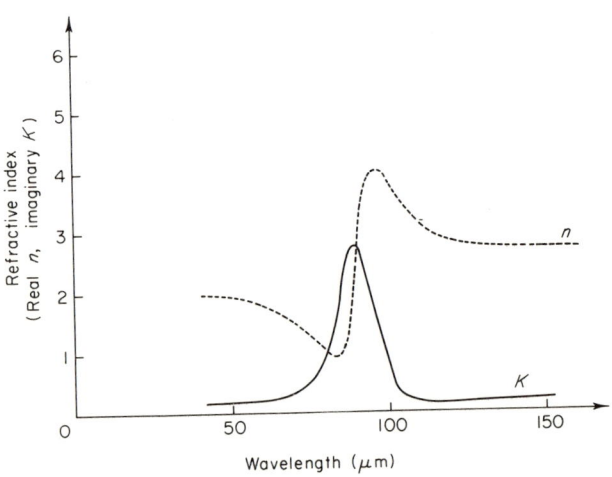

Fig. 7.32. The optical constants, n and K, of AgI at 90°K, as a function of wavelength (after Hadni et al.[43]).

tions can distort the charge distribution by producing unequal displacements in the two lattices and can thereby yield a dipole moment through which infrared radiation may drive the crystal in vibration.

7.6.2. Quantization of Lattice Vibrations: Phonons

By applying boundary conditions to the linear chain of N atoms discussed in Section 7.6.1 (e.g., by "clamping" the first and last atoms), it is easily seen that the number of independent standing vibrational modes on each branch is N, and that each has a particular wavelength. Within the Brillouin zone, then, k is not continuous but rather has N allowed values uniformly spaced with separation $2\pi/Na$, and correspondingly ω has N discrete values. Quantum mechanically, the total of N lattice vibrations in the line of N atoms is formally equivalent to N independent harmonic oscillators, one oscillator for each mode, and we are able accordingly to describe the states of elastic vibration in terms of familiar quantized formalism. Without developing a formal proof, we point out that one can describe the classical motion in terms of normal coordinates (p_i, q_i), and therefore write the Hamiltonian of the one-dimensional crystal as the sum of squared terms representing the kinetic and potential energies and free from cross-product terms. The case of a monatomic chain, namely that of section 7.6.1 with M replaced by m, is particularly simple.[51] With the usual operator replacement $(\hbar/i)(\partial/\partial q_i)$ of the momentum, one obtains the Schrödinger equation for each mode in the form

$$-(\hbar^2/2m)\,\partial^2\psi_i/\partial q_i{}^2 + \tfrac{1}{2}m\omega_i q_i \psi_i = W_i \psi_i. \qquad (7.119)$$

This linear harmonic oscillator equation, with energy eigenstates

$$W_i = (n + \tfrac{1}{2})\hbar\omega_i, \qquad n = 0, 1, 2, \ldots, \qquad (7.120)$$

shows the quantized nature of each of the N independent modes of elastic lattice motion. The energy levels are separated by quanta of vibrational energy, $\hbar\omega_i$, which are called *phonons*.

Transition matrix elements, wavefunctions, creation and annihilation operations, selection rules, etc., may all be taken over from our understanding of the quantum mechanics of the harmonic oscillator. For example, as has been known since the time Einstein applied Planck's quantum concept to the mechanical energy of solids, the specific heat,

[51] A. Yariv, "Quantum Electronics." Wiley, New York, 1967.

density of states, etc., can be calculated analogously to the blackbody radiation problem.

In the language of quantum mechanics, then, the reststrahlen spectra of ionic crystals predicted by the linear theory are $\Delta n = \pm 1$, or one phonon transitions. When anharmonic terms in ionic displacement are introduced, multiphonon transitions are allowed, and one will observe the weak sidepeaks on the reststrahlen bands noted in the previous section. In homopolar crystals two or more phonon processes can act to induce an electric moment of second or higher order in the ionic displacement, so that a photon can excite the crystal oscillators directly between quantum levels separated by $\Delta n = 2$ or more.[48]

7.6.3. Impurity-Induced Lattice Absorption

The introduction of impurity ions, or the occurrence of a lattice defect, can modify the phonon spectrum of the lattice crystal and provide a dipole moment for the excitation of vibrational modes that are infrared inactive in the pure crystal. If, for example, we suppose that in NaCl a Cl^- ion is substituted by a lighter H^- ion,[†] physical intuition leads us to expect a high frequency mode of oscillation of the H^- ion within the cage of Na^+ ions, above the normal lattice vibration spectrum. As a substitute ion will not, in general, contribute precisely the same charge distribution that existed in that vicinity in the pure crystal, it will exhibit an electric dipole moment. Associated with this difference in charge distribution we expect a deformation of the lattice in the region local to the impurity, and a response to the oscillatory motion of lattice ions located within some short range of penetration into the evanescent or "cutoff" crystal. Such an oscillation is called a *local mode*.[49,50] If the impurity ion is heavy and/or weakly bound, its frequency of oscillation may be placed within the passband spectrum of the lattice, in which case its influence will be to excite plane waves propagating through the crystal. One then speaks of a *band mode*. A kind of resonant behavior caused by impurity defects can be observed at far-infrared frequencies lower in the phonon spectrum. These so called *resonant* modes, or *resonant band modes*, are predominantly found in the acoustic branches, and are associated with heavy impurities and weak binding forces. For an impurity that is not too heavy and is bound in a one-dimensional lattice with a force constant f_1 much smaller than that in the host crystal, the resonant mode frequency ω_{RM} can be

---

[†] Substitutional hydride ion impurities are called U centers.

shown to be approximately $(2f_I/m_I)^{1/2}$. This is the natural frequency of vibration of the impurity in a stationary host lattice.

The problem of local modes can be put on a formal basis by representing the impurity substitute ion as a mass m_I replacing either the lighter mass m or the heavier mass M in a simple linear diatomic chain with nearest neighbor force constant. The simpler case of a replacement ion with $m_I < m$ on a monatomic chain is treated by Kittel.[52] The essential conclusions of an analysis of the perturbed diatomic chain are illustrated by the band diagram of Fig. 7.33. In the case of an impurity m_I replacing

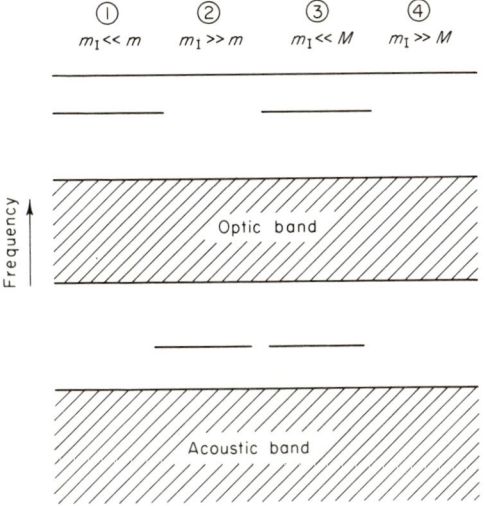

FIG. 7.33. Energy diagram showing positions of local modes due to ion replacement in a diatomic linear chain. In columns (1) and (2) one of the lighter molecules (m) in the chain is replaced, and in (3) and (4) a heavier molecule (M) is replaced (after Sievers[53a]).

one of the lighter ions m of the crystal, there are two solutions. For $m_I \ll m$ a single mode exists above the optical branch, while for $m_I \gg m$, there is a single mode in the band gap between the acoustic and optical phonon branches. Falling as they do in frequency regions where wave propagation is forbidden in the host crystal, they are local modes. In the second case, m_I replaces M; two local modes arise when $m_I \ll M$, one in the forbidden gap and one above the optical branch. Local modes such as these, well separated from the band frequencies of the crystal,

---

[52] C. Kittel, "Introduction to Solid State Physics," 4th ed. Wiley, New York, 1971.

cause sharp absorption at far-infrared frequencies, similar in shape to the resonance line of an isolated molecule. Provided the impurity concentration is not so high as to give impurity–impurity interactions and, of course, provided they are optically active, the absorption will be proportional to the concentration. For $m_I \gg M$, no local modes exist. Heavy impurities may vibrate in the acoustic or optic bands, and "drive" the host crystal vibrations.[53a–c]

TABLE 7.5. Some Impurity-Induced Local Mode Absorption Lines in Alkali Halide Crystals[a]

Crystal: impurity combination	Local mode frequency (cm^{-1})	Temperature (°K)	Linewidth (cm^{-1})
LiF : H$^-$	1015	295	19
	1027	20	4
NaF : H$^-$	846.7	295	12
	859.5	20	0.6
NaF : D$^-$	607.5	295	10
	615.0	20	1.3
NaCl : H$^-$	565	90	5.6
NaCl : D$^-$	408	90	4.0
NaBr : H$^-$	498	90	17
NaBr : D$^-$	361	90	10
KCl : H$^-$	496.5	300	26.2
	502	90	2.3
KCl : D$^-$	357.5	300	30.8
	360	90	2.3
KBr : H$^-$	446	90	6.0
KI : H$^-$	446	90	6.0
RbCl : H$^-$	476	90	4.8
RbCl : D$^-$	339	90	3
RbBr : H$^-$	425	90	8
CsBr : H$^-$	363	80	10

[a] After Genzel.[54]

[53a] A. J. Sievers, NATO Advanced Study of Elementary Excitations and Their Interactions, Cortina d'Ampezzo, Italy, July 11–23, 1966.

[53b] A. J. Sievers, in "Elementary Excitations in Solids." Plenum Press, New York, 1969.

[53c] A. J. Sievers, in "Far-Infrared Spectroscopy" (K. D. Möller and W. G. Rothschild, eds.), Appendix I. Wiley, New York, 1971.

[54] L. Genzel, in "Optical Properties of Solids" (S. Nudelman and S. S. Mitra, eds.). Plenum, New York, 1969.

7.6. CRYSTAL LATTICE VIBRATIONS IN SOLIDS

TABLE 7.6. Some Impurity-Induced Resonant Mode Frequencies in Alkali Halide Crystals[a]

Crystal: impurity combination	Resonant mode frequency (cm^{-1})	Linewidth (cm^{-1})
NaCl : Li$^+$	44.0	0.7
NaCl : F$^-$	59.0	—
NaCl : Cu$^+$	23.7	0.6
NaCl : Ag$^+$	52.5	10.5
NaBr : Ag$^+$	48.0	—
NaI : Ag$^+$	36.7	—
KCl : Li$^+$	42	—
KCl : Ag$^+$	38.8	5.5
KBr : Li^{6+}	17.9	1
KBr : Li^{7+}	16.3	0.8
KBr : Ag$^+$	33.5	4.5
KI : Ag$^+$	17.4	0.9
RbCl : Ag$^+$	21.4	—
	26.4	—
	36.1	—

[a] After Sievers.[53a–c]

It is apparent that defects which destroy the translational symmetry of the host crystal offer a means of probing the vibrational spectrum with far-infrared radiation. In the case of perfect ionic crystals, only the resonance associated with long wavelength vibrations at $k = 0$ on the TO branch can be excited. Beyond this, in the region of the phonon band where absorption by the crystal is normally low, the introduction of impurities of appropriate types can enable activation of, and absorption by, a number of one phonon transitions normally inaccessible in the band spectrum. In the case of homopolar crystals, such as germanium and silicon, the pure lattice is essentially transparent throughout the far-infrared, and the additional absorption due to the introduction of impurities can be detected and interpreted to give information about the phonon spectrum.

Far-infrared impurity-induced absorption in crystals has been an active field of research for the past ten years, and during this period the alkali halides have been particularly studied. Summarizing some of the observational data, Tables 7.5 and 7.6 list resonant frequencies of some local modes and resonant modes produced by various impurities. Figure 7.34 shows line shapes and their temperature dependence for Ag in KI.

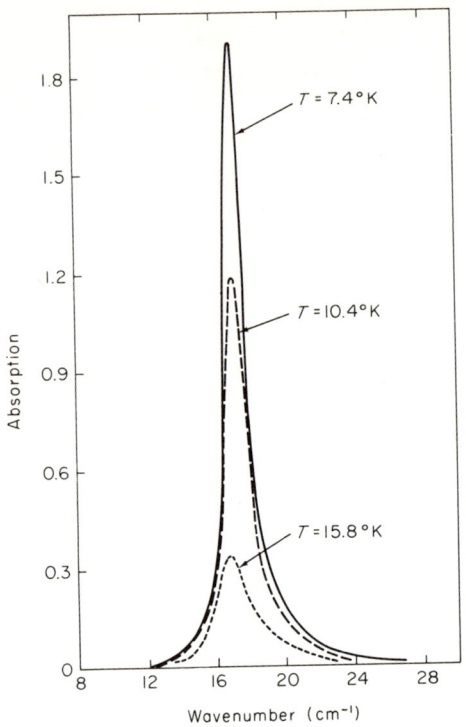

FIG. 7.34. Absorption spectrum of a resonant mode in KI:AgI at three different temperatures (after Sievers[53a]).

7.7. Ferroelectric Crystals

The phenomenon of ferroelectricity can be interpreted in terms of displacements of positive and negative ionic lattices of the crystal against very weak restoring forces. It may be regarded as an "instability" or "dielectric catastrophe" associated with a "soft" transverse optical mode of vibration in the far-infrared, arising from a situation where the forces on the displaced ions act in opposition and cancel. The ferroelectric material then has a permanent electric polarization, a ferroelectric domain structure whose growth of reorientation leads to hysteresis in the polarization versus applied field curve, a Curie–Weiss law temperature dependence, and other characteristics closely paralleling the better-known phenomena of ferromagnetism. In this section we describe the essentials of some of the physical principles underlying theoretical interpretations and experimental observations of the dielectric properties related to ferroelectricity, particularly as they relate to the far-infrared. The development of the far-infrared spectroscopic aspect of the ferroelectric state was stimulated by the theory given by Cochran ten years ago; on the timescale of the subject and the analogous field of ferromagnetism, this is a short history indeed.

Of the many known ferroelectric materials, barium titanate has been most extensively studied by a variety of techniques, including spectroscopy and neutron diffraction. As such it is an appropriate crystal for particular reference. Along with $SrTiO_3$, $KNbO_3$, and $PbTiO_3$, $BaTiO_3$ is a member of the group of ionic crystals with structure closely related to that of the mineral perovskite ($CaTiO_3$). The perovskite crystal structure of barium titanate is illustrated in Fig. 7.35. Ba^{2+} ions are located

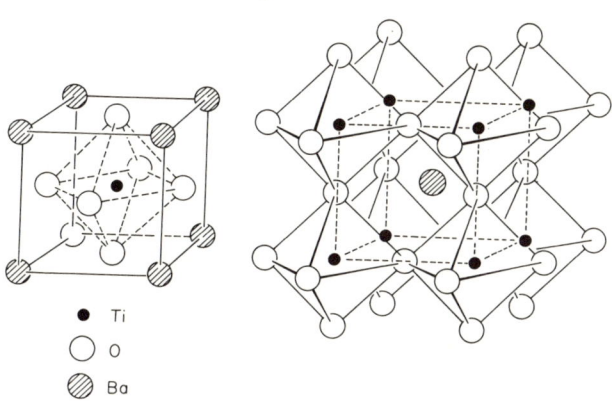

FIG. 7.35. The perovskite crystal structure.

at the corners of the cubic cell, O^{2-} ions at the face centers, and the Ti^{4+} ions at the center of the cube. The TiO_3 group can, as shown in Fig. 7.35, be considered to be arranged as a central Ti atom, octahedrally surrounded by six oxygen half-atoms. Above the Curie temperature $T_C = 393°K$, the crystal is cubic and its state is paraelectric rather than ferroelectric. Below the Curie point the crystal deforms slightly. Down to about 278°K it is tetragonal with the O^{2-} ions shifted by roughly 0.1 Å[52,55] in the 100 crystallographic direction (see Fig. 7.36) relative to the positive ions, and with a large permanent polarization P directed parallel to the cube edge. The macroscopic dielectric constant at low frequencies is several thousand. It is clear from experimental studies that in the tetragonal phase both

[55] G. Shirane, F. Jona and R. Pepinsky, *Proc. IRE* **43**, 1738–1793 (1955).

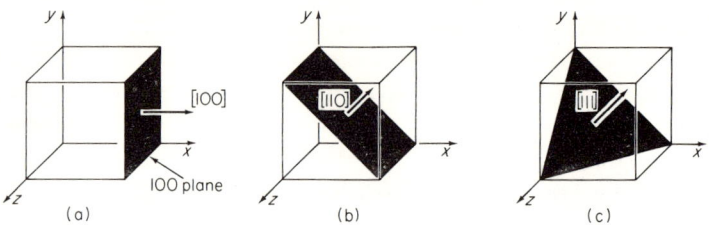

FIG. 7.36. Designation of planes in a crystal.

the Ba and Ti ions undergo displacements in a direction opposite to that of the oxygens.[52,55,56] Ferroelectric behavior persists as the temperature is decreased, but two further structural changes occur in the crystal and with them the direction of polarization reorientates. In the region of 278°K it becomes polarized in the direction of a face diagonal, and below about 203°K it is parallel to a body diagonal.[52]

Very large values of the static dielectric constant require that the electric field acting on a dipole be much larger than the applied field **E**. A Lorentz local field of value $\mathbf{E} + \mathbf{P}/3\varepsilon_0$ will play this essential role, and it will predict a dielectric constant

$$\varkappa = 1 + \frac{P}{\varepsilon_0 E} = 1 + \frac{N\alpha/\varepsilon_0}{1 - N\alpha/3\varepsilon_0}, \tag{7.121}$$

where N is the number of ions per unit volume and α their polarizability. In this model a "polarization catastrophe," or infinity in the dielectric constant, associated with the ferroelectric state occurs when

$$N\alpha = 3\varepsilon_0. \tag{7.122}$$

The growth of $\varkappa$ predicted here, when the condition Eq. (7.122) is approached, is the result of an inbuilt feedback process within the crystal. The polarization produced by a small initial charge displacement increases the local field, and this in turn acts to increase the displacement further. The instability thus started grows, but is ultimately limited at high fields by a breakdown of the linearity between the induced moment and the polarizing field. The runaway process is finally arrested leaving the crystal with a selfgenerated internal polarization locked in to its structure.

We note that the ferroelectric requires a finite polarizability, given by Eq. (7.122) if the Lorentz value of the local field is assumed to be

---

[56] G. Shirane, H. Danner and R. Pepinsky, *Phys. Rev.* **105**, 856–860 (1957).

$\mathbf{E} + \mathbf{P}/3\varepsilon_0$. But the derivation of the correction $P/3\varepsilon_0$ is based on an assumed spherical symmetry which does not prevail in the barium titanate crystal. Slater[57] has calculated an accurate Lorentz correction for the $BaTiO_3$ crystal and found that the oxygen ions which are in the same line as the Ti ion (that is, the line parallel to the electric field), exert very strong fields on each other, and the local field at the Ti ion is much greater than that obtained on the assumption of spherical symmetry. The field thus enhanced can act on the ions and produce large polarization and ferroelectricity, while the atomic polarizability need be only a small fraction of that given by Eq. (7.122).

Additionally, one must introduce an appropriate temperature dependence into the lattice motion to account for the dependence on temperature of the dielectric constant. This matter has been discussed by Devonshire[58] and others.[55,59] If restoring forces on the displaced ions are entirely linear functions of displacement, thermal expansion does not adequately account for observed behavior with temperature. But if, in addition to a linear term, appropriate anharmonic terms are included, the Curie–Weiss law

$$\varkappa = C/(T - T_C) \tag{7.123}$$

can be derived. For $BaTiO_3$ the Curie constant C is 170,000°K. Temperature dependence of the form Eq. (7.123) is the effect of anharmonic interactions involving terms in the energy with fourth and second powers of the displacement; restoring forces on displaced ions increase as the temperature is raised, and the polarizability undergoes a small linear decrease.[57] Devonshire's treatment of thermal effects expands the free energy in powers of the polarization and includes the contributions of elastic strains in the crystal. He obtains the Curie–Weiss law with the constant C related to the coefficient of the fourth power term.

Transverse Optical Phonon Approach

Cochran[60] has developed a theory related to that of Slater, but more general. Ferroelectric transitions occur in Cochran's theory as the result of an instability of the crystal associated with a low-frequency normal mode of vibration. Of central importance here is the Lyddane–Sachs–

---

[57] J. C. Slater, *Phys. Rev.* **78**, 748–761 (1950).

[58] A. F. Devonshire, *Phil. Mag.* [7], **40**, 1040 (1949); *Advan. Phys.* **3**, 85 (1954).

[59] E. Fatuzzo and W. J. Merz, "Ferroelectricity." North-Holland Publ., Amsterdam, 1967.

[60] W. Cochran, *Advan. Phys.* **9**, 387–423 (1960).

Teller relation for it predicts that a catastrophe will arise in the static dielectric constant $\varkappa_s$ when the frequency of a transverse optical mode of lattice vibration approaches zero. This, in turn, arises from a cancellation of forces acting on the ion. Two forces are distinguished. There is a short-range repulsive force between nearest neighbors of quantum-mechanical "overlap" origin, and a coulomb force of long range throughout the crystal produced by the movement of ions from their equilibrium positions, that is, by polarization. In a superposition of these two force fields, the ion dynamical equations show that equal and opposite effects can arise for one transverse optical branch; the local field created by polarization provides, as it were, feedback, for it tends to increase the amplitude of the displacement. For longitudinal optical vibrations the forces do not oppose but rather aid each other.

Cochran employs the shell model where an ion is represented as a core coupled to a surrounding shell of outer electrons. A model of a diatomic crystal then has both ionic and electronic polarizability which may be calculated for a diagonally cubic crystal (i.e., a crystal in which every atom has surroundings of tetrahedral symmetry). This crystal structure is a useful approximation and serves as a guide for more complicated ferroelectric crystals like $BaTiO_3$. A molecule in a diatomic crystal with diagonal cubic symmetry vibrating in a TO mode is acted upon by a local field $\mathbf{P}/3\varepsilon_0$ due to polarization. From the theory of dynamics[61] we know the equation of motion of a vibrating two-body molecule with central force is exactly the same as that of a single particle of mass μ, the reduced mass of the molecule, displaced by a distance x from the fixed center of force. The frequency ω_T of the transverse optical vibration is given by

$$\mu\ddot{x} + fx = qP/3\varepsilon_0. \tag{7.124}$$

The polarization P is the sum, $P_E + P_I$, of the ionic polarization P_I and electronic polarization

$$P_E = N\alpha_{el}\frac{P}{3\varepsilon_0} \equiv B_{el}P,$$

where α_{el} includes electron distortion of both the positive and negative ion, and $B_{el} = N\alpha_{el}/3\varepsilon_0$, by definition. The ionic polarization P_I is Nqx. We may thus substitute $P = Nqx/(1 - B_{el})$ into Eq. (7.124), and obtain a displacement varying as $\exp i\omega_T t$. We obtain the transverse optical

[61] R. K. Wangsness, "Introduction to Theoretical Physics." Wiley, New York, 1963.

phonon frequency

$$\omega_T^2 = \frac{f}{\mu}\left(1 - \frac{B_{\text{ion}}}{1 - B_{\text{el}}}\right), \tag{7.125}$$

with $B_{\text{ion}} = Nq^2/3\varepsilon_0 f$. The effect of crystal polarization is to effectively reduce the interlattice force constant, and to give cancellation and the limiting case of infinitely long wavelength TO phonons when $B_{\text{ion}} = 1 - B_{\text{el}}$.

Equation (7.124) also gives the natural frequency ω_L of longitudinal optical vibrations when the value of P applicable to this mode is used. In addition to the polarizing field $\mathbf{P}/3\varepsilon_0$ there is now a depolarizing field because with each reversal of direction of the longitudinally directed field within a lattice some wavelengths thick there is a bound charge compression or rarefaction. The constituitive relation $\mathbf{D} = \varepsilon_0\mathbf{E} + \mathbf{P}$ gives, in the absence of free charge, $\nabla \cdot \mathbf{E} = -(1/\varepsilon_0)\nabla \cdot \mathbf{P}$, and a depolarizing field $-\mathbf{P}/\varepsilon_0$ acting to restore the oscillating molecule towards its equilibrium position. The equation of motion of longitudinal optical vibration,

$$\mu\ddot{x} + fx = q[(P/3\varepsilon_0) - (P/\varepsilon_0)], \tag{7.126}$$

then shows harmonic oscillations with frequency ω_L given by

$$\omega_L^2 = \frac{f}{\mu}\left(1 + \frac{2B_{\text{ion}}}{1 + 2B_{\text{el}}}\right). \tag{7.127}$$

In this case, crystal polarization acts to oppose the vibrational motion and to raise the LO phonon frequency.

From Eqs. (7.125) and (7.127), we have

$$\frac{\omega_T^2}{\omega_L^2} = \frac{1 + 2B_{\text{el}}}{1 - B_{\text{el}}} \cdot \frac{1 - (B_{\text{el}} + B_{\text{ion}})}{1 + 2(B_{\text{el}} + B_{\text{ion}})}. \tag{7.128}$$

We can recognize the terms on the right-hand side of Eq. (7.128) by referring back to Eq. (7.102) of Section 7.5.2, and taking $f_j = 1$ for this case of N oscillators per unit volume. At optical frequencies where we denote the dielectric constant as $\varkappa_\infty$, the ions are too heavy to follow the oscillations; only one value of j is involved and this must give the term B_{el} on the right of Eq. (7.102), describing the electronic polarizability of the molecule. Straightforward transposition gives

$$\varkappa_\infty = (1 + 2B_{\text{el}})/(1 - B_{\text{el}}), \tag{7.129}$$

where the subscript ∞ designates the dielectric constant in the visible region of the spectrum below the natural frequency (in the ultraviolet and beyond) of the bound electrons. In the extreme of low frequencies, $\omega = 0$ in Eq. (7.102), the static dielectric constant $\varkappa_S$ is given by the term B_{el} plus the ionic term B_{ion}, and we obtain

$$(\varkappa_S - 1)/(\varkappa_S + 2) = B_{el} + B_{ion},$$

or

$$\varkappa_S = \frac{1 + 2(B_{el} + B_{ion})}{1 - (B_{el} + B_{ion})}. \tag{7.130}$$

From Eqs. (7.129) and (7.130), we then have

$$\omega_T^2/\omega_L^2 = \varkappa_\infty/\varkappa_S = \varepsilon_\infty/\varepsilon_S, \tag{7.131}$$

which is the Lyddane–Sachs–Teller (or LST) relation for a diatomic molecule. The LST relation has been generalized by Cochran to diagonally cubic crystals with n atoms per unit cell.

The significance of the LST relation is that through it the very large value of the static and low-frequency dielectric constant in a ferroelectric can be associated with a low frequency TO mode of vibration. On the basis of this association, dynamic analysis of the vibration leads Cochran to a dispersion equation similar in form to Eq. (7.97) of Section 7.5.2.[60,62,63] For a system with one mode of oscillation much lower in frequency than the others, Eq. (7.97), in the limit of low frequency $\omega \to 0$, shows the product of the square of this frequency and $\varkappa_S$ to be a constant, provided $\varkappa_S$ is large. In a ferroelectric crystal, $\varkappa_S$ is very large indeed, and its association with the lowest TO phonon frequency ω_F has enabled Cochran to reduce his dispersion equation to yield

$$\varepsilon_S \omega_F^2 = \text{constant}. \tag{7.132}$$

The subscript F is chosen to emphasize the ferroelectric associations of the phonon.

We know the temperature dependence of the low-frequency permittivity to be the Curie–Weiss law Eq. (7.123); hence, by Eq. (7.132), the ferroelectric phonon frequency must vary with temperature as

$$\omega_F^2 \propto T - T_C. \tag{7.133}$$

[62] B. D. Silverman, *Phys. Rev.* **125**, 1921–1930 (1962).

[63] W. G. Spitzer, R. C. Miller, D. A. Kleinman and L. E. Howarth, *Phys. Rev.* **126**, 1710–1721 (1962).

Comparison of Eqs. (7.131) and (7.133) now shows that as T approaches T_C, the TO phonon frequency becomes minimal and the static dielectric constant approaches a maximum. Of course, the dielectric constant even of a single crystal does not literally become infinite at the Curie point. Nonlinearities limit the charge displacement. There is damping associated with lattice movement, and the curve of $\varkappa$ versus temperature is rounded off somewhat but is still peaked in the near vicinity of T_C. Even so, the Curie–Weiss law holds to within a few degrees of the Curie point. Below the Curie point the crystal is in a state of permanent polarization with the dielectric constant falling with temperature. As Fig. 7.37 shows, the dielectric constant of $BaTiO_3$ approaches 10^4 at the Curie point, decreases with temperature, but remains as high as two or three thousand at room temperature and below.

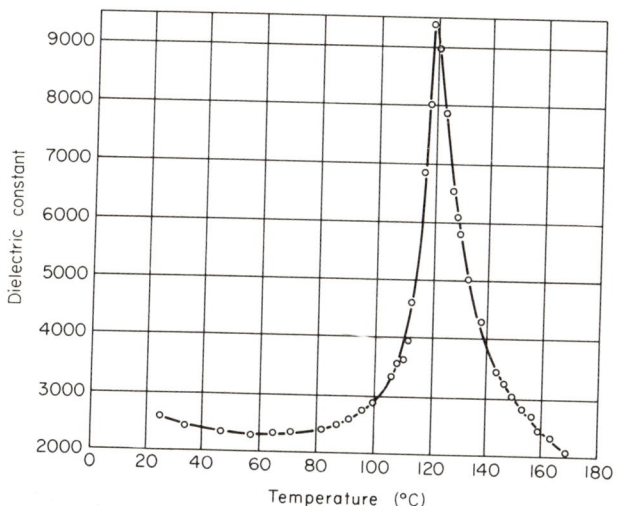

FIG. 7.37. Temperature dependence of the dielectric constant of barium titanate (after Von Hippel[64]).

The temperature dependence of ω_F is an important ingredient of the transverse optical phonon interpretation of the ferroelectric state, which has enabled experimenters to check a central consequence of the Cochran theory. By neutron spectrometry techniques, Cowley[65] has measured the frequency of the $k = 0$ lowest frequency TO mode at different tempera-

---

[64] A. Von Hippel, "Dielectrics and Waves." Wiley, New York, 1954.
[65] R. A. Cowley, *Phys. Rev. A* **134**, 981–997 (1964).

tures of a $SrTiO_3$ crystal. His straight line plot (Fig. 7.38) of frequency squared as a function of temperature supports Cochran's prediction, and extrapolation to intercept the temperature axis gives the estimated Curie temperature as $T_C = 32 \pm 5°K$. The movement of the ions in this normal mode was found to be predominantly a vibration of the titanium ion against the oxygen tetrahedron. Cowley has measured several branches of the frequency versus wavevector dispersion curves for this crystal and found them to be in quite reasonable agreement with theoretical curves

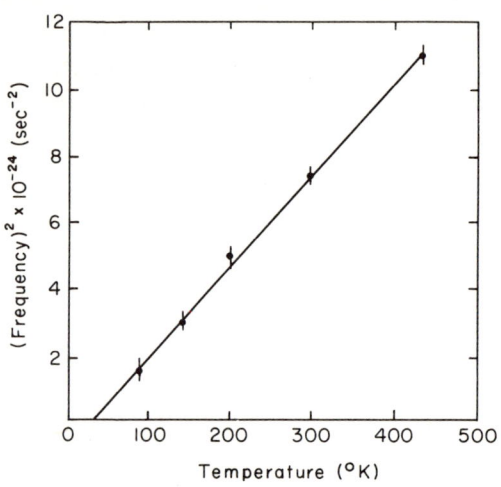

FIG. 7.38. Plot of the frequency squared against temperature for the $k = 0$ lowest frequency TO mode of $SrTiO_3$ (after Cowley[65]).

based on shell models with ions interacting with one another, both through short-range forces between neighbors and through long-range dipolar forces determined by the ionic displacement. The treatment of ferroelectricity as an instability of the crystal against one of the normal modes is shown by Cowley's results to be a valid approach to the phenomenon. Barker and Tinkham[66] have measured the reflection of far-infrared waves from $SrTiO_3$, and by a Kramers–Kronig dispersion analysis have obtained the real and imaginary parts of the dielectric constant shown in Fig. 7.39. The low-Q dispersion characteristic shows the resonant frequency to increase with temperature as predicted, and by roughly the predicted amount. The shift in the maximum of $\varkappa''$ from 40 to 100 cm^{-1} with temperature is reasonably consistent with the factor of

[66] A. S. Barker and M. Tinkham, *Phys. Rev.* **125**, 1527–1530 (1962).

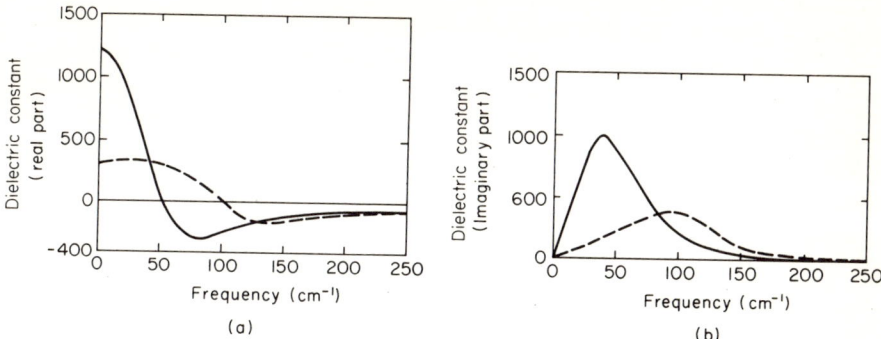

FIG. 7.39. The real part (a) and the imaginary part (b) of the dielectric constant of $SrTiO_3$ at temperatures of 93°K and 300°K. The results are calculated from reflectivity measurements. Solid curve $T = 93°K$; dashed curve $T = 300°K$ (after Barker and Tinkham[66]).

approximately two given by the Curie–Weiss shift, $(T - T_C)^{1/2}$. Spectroscopic techniques of observation as employed by Barker and Tinkham, display rather directly the dispersive behavior of the ferroelectric materials, and so too do the experimental plots published by Ballantyne[67] and by Spitzer et al.[63] Ballantyne measured the frequency and temperature response of $BaTiO_3$ over the wavelength range 2.5 μm to 12 mm, using six different spectrometers based on diffraction gratings, interferometry with Fourier transform analysis, and millimeter wavelength carcinotrons and klystrons. The measurements, when reevaluated by Barker,[68] supported the temperature dependence Eqs. (7.133) in the range 401–473°K.

Spitzer et al.[59,63] have shown experimentally that the dispersion spectra of $BaTiO_3$ and $SrTiO_3$ can be fairly well represented by the formulas

$$\varkappa' = \varkappa_\infty + \sum_{j=1}^{3} \frac{A_j \omega_j^2 (\omega_j^2 - \omega^2)}{(\omega_j^2 - \omega^2)^2 + \gamma_j^2 \omega^2}, \quad (7.134)$$

$$\varkappa'' = \sum_{j=1}^{3} \frac{A_j \omega_j^2 \gamma_j \omega}{(\omega_j^2 - \omega^2)^2 + \gamma_j^2 \omega^2}, \quad (7.135)$$

where γ_j is a damping constant and A_j gives a measure of the strength of the resonance. Three fundamental modes of vibration are assumed, following Last.[69] Of the $15N$ degrees of freedom in a crystal containing N

---

[67] J. M. Ballantyne, *Phys. Rev. A* **136**, 429–436 (1964).
[68] A. S. Barker, *Phys. Rev.* **145**, 391–399 (1966).
[69] J. T. Last, *Phys. Rev.* **105**, 1740–1750 (1957).

unit cells, each with 5 atoms, $3N$ are connected with translational motion, $3N$ with torsional motion, leaving $9N$ vibrational modes. N normal modes of the crystal correspond to a single normal mode of a unit cell. Thus there are 9 vibrations of the unit cell, and these can be classified as 3 vibrations of Ba against the TiO_3 group and 6 internal TiO_3 vibrations. The Ba–(TiO_3) vibration can be treated as diatomic, so to speak, by considering the TiO_3 group as a single atom situated at the Ti position. Last points out that, in the absence of mode coupling, the Ba–(TiO_3) vibrations are degenerate; in fact they are the triply degenerate pair illustrated in Fig. 7.40.

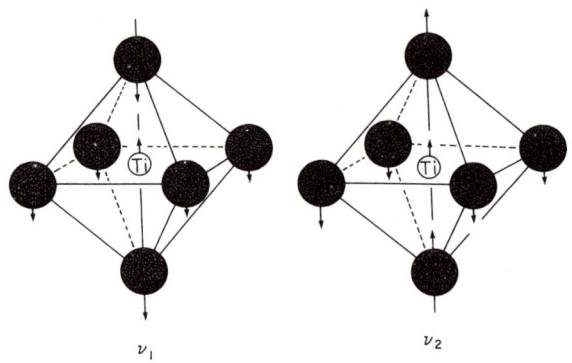

FIG. 7.40. Model illustrating infrared active normal modes of vibration of TiO_6 octahedron in $BaTiO_3$. ν_1 is a stretching mode and ν_2 a bending mode (after Last[69]).

Several groups[63,66,70] have observed and studied all three infrared active modes predicted on the basis of symmetry of the unit cell. Two of them are shown on each of the far-infrared dispersion curves of Figs. 7.41 and 7.42 as quite sharp resonances, while the lowest frequency vibration, thought by Spitzer et al. to be responsible for ferroelectricity,[63] is very heavily damped. This is particularly the case in $BaTiO_3$. The lowest frequency mode in $BaTiO_3$ is at 33.8 cm^{-1} (296 μm), and in $SrTiO_3$ it is at 87.7 cm^{-1} (114.3 μm) (see Spitzer et al.[63] for a formula for the resonant frequency of heavily damped modes). It has been ascribed to the Ba–(TiO_3) mode, but there is some disagreement between different workers which Fatuzzo and Merz[59] have discussed. Miller and Spitzer[71] have measured the reflectivity of single-crystal $KTaO_3$ at room temperature, where it is paraelectric. Observational data together with the be-

[70] C. H. Perry, B. N. Khanna and G. Rupprecht, *Phys. Rev. A* **135**, 408–412 (1964).
[71] R. C. Miller and W. G. Spitzer, *Phys. Rev.* **129**, 94–98 (1963).

7.7. FERROELECTRIC CRYSTALS

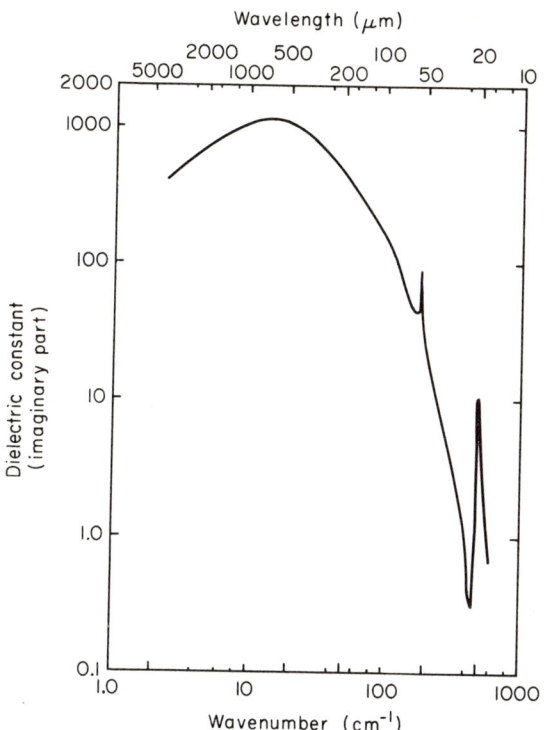

FIG. 7.41. Imaginary part of the dielectric constant of $BaTiO_3$ for the ordinary ray, as a function of frequency (after Spitzer et al.[63]).

havior predicted by classical dispersion theory are plotted in Fig. 7.43. Kramers–Kronig analysis of the data gives the imaginary part of the dielectric constant, with its three resonances, shown in Fig. 7.44.

Perry et al.,[70,72] have studied dielectric dispersion at room temperature and have discussed the question of mode assignment for the perovskite titanates $CaTiO_3$, $SrTiO_3$, $BaTiO_3$, and $PbTiO_3$, and the zirconates $CaZrO_3$, $SrZrO_3$, $BaZrO_3$, and $PbZrO_3$. Some of these crystals are ferroelectric and some are paraelectric, but $PbZrO_3$ is an *antiferroelectric* crystal below the transition temperature 506°K.[52,55] The dielectric constant of $PbZrO_3$ has a sharp peak at this temperature and shows low-frequency dielectric dispersion that results from a low-frequency infrared-active mode.[72] Notwithstanding, it shows no hysteresis and no net per-

---

[72] C. H. Perry, D. J. McCarthy and G. Rupprecht, *Phys. Rev. A* **138**, 1537–1538 (1965).

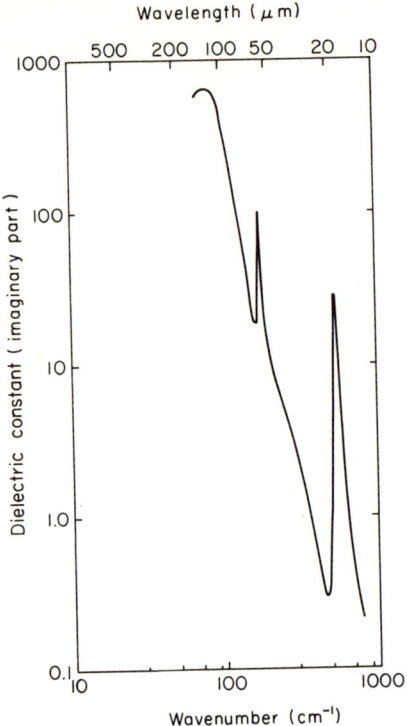

FIG. 7.42. Imaginary part of the dielectric constant of $SrTiO_3$ as a function of frequency (after Spitzer et al.[63]).

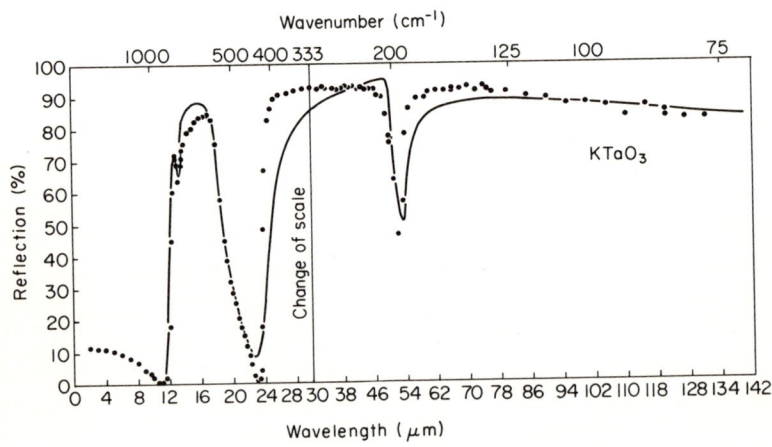

FIG. 7.43. Reflectivity of $KTaO_3$ as a function of wavelength. The solid curve was calculated from classical dispersion theory (after Miller and Spitzer[71]).

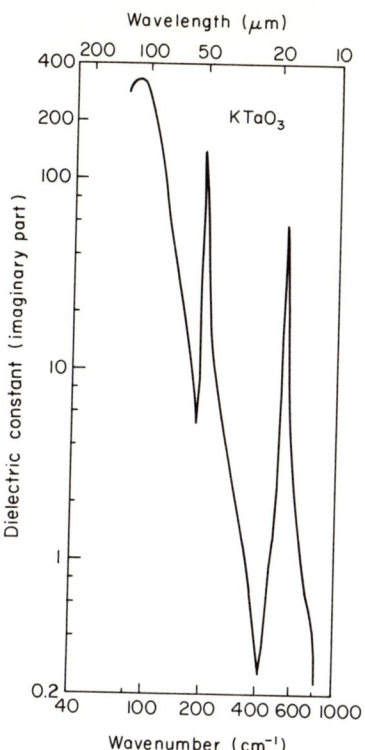

FIG. 7.44. Imaginary part of the dielectric constant of $KTaO_3$ as a function of frequency (after Miller and Spitzer[71]).

manent polarization. Its characteristics derive from the fact that the crystal is arranged in antiparallel arrays of dipoles which give zero resultant dipole moment over the crystal as a whole. Structural studies using single crystals have confirmed this antiferroelectric arrangement.[55]

7.8. Magnetic Resonances

7.8.1. Paramagnetic Resonance

Permanent magnetic moments and paramagnetic properties are exhibited by some atoms, molecules and free radicals (see Section 7.3 for O_2, NO, NO_2, OH, etc.), ions with partially filled electron shells (transition elements, rare earth and actinide ions), and other materials with unpaired electrons in their structure. In a static magnetic induction field **B** the dipoles of moment μ experience a torque, $\mu \times \mathbf{B}$, causing them to precess uniformly about **B**. The dipole has energy,

$$W = -\mu \cdot \mathbf{B}, \tag{7.136}$$

which can be exchanged with an electromagnetic wave via paramagnetic resonance interaction with its oscillating magnetic field.

The magnetic moment we speak of is causally related to the intrinsic electron spin and electron orbital motion and is proportional to the total angular momentum. This proportionality is written

$$\mu = g\beta \mathbf{J}, \tag{7.137}$$

where $\beta = q\hbar/2m$ is the Bohr magneton (equal to 9.27×10^{-24} A m² or 9.27×10^{-21} erg G⁻¹), and g is the spectroscopic splitting factor, or g-factor. Landés empirical expression gives the g-factor for a free atom in terms of J, S, and L, the total, spin, and orbital angular momentum quantum numbers,

$$g = 1 + \frac{J(J+1) + S(S+1) - L(L+1)}{2J(J+1)}. \tag{7.138}$$

The ratio of the magnetic moment to the angular momentum $[J(J+1)]^{1/2}\hbar$, or gyromagnetic ratio, is $\gamma = gq/2m$ and is twice as large for spin as for orbital momentum. That is, for $L = 0$, $g = 2$, and for $S = 0$, $g = 1$. Dirac[73] has removed the empirical nature of electron spin and its anomalous g-factor by showing that they emerge naturally from the relativistic quantum theory of the electron; indeed, spin is essential for the relativistic invariance of the quantum-mechanical equations of motion. The g-factor can be accurately calculated and has the precise value $g = 2.00232$.

According to quantum theory, the vector **J** has magnitude $[J(J+1)]^{1/2}$ and is subject to spatial quantization. In the absence of an applied field, the energy states are $(2J+1)$-fold degenerate, but the application of a magnetic field in the z-direction causes Zeeman splitting into $2J+1$ separate levels corresponding to integer values $M_J = J, J-1, \ldots, -J+1, -J$ along the z-direction. The energy levels can be written

$$W = g\beta M_J B. \tag{7.139}$$

When subject to the selection rule $\Delta M_J = \pm 1$, the frequency of tran-

---

[73] P. A. M. Dirac, "The Principles of Quantum Mechanics," 2nd ed. Oxford Univ. Press (Clarendon), London and New York, 1935.

sitions between Zeeman levels is seen to be

$$\nu = 1.401gB, \qquad (7.140)$$

where ν is in MHz and B in gauss. Fields up to 10 kG produce resonance in the microwave region.

Paramagnetic spectroscopy in solids deals with ions normally imbedded as a dilute concentration in diamagnetic host crystals. The paramagnetic spectrum of the ion is then influenced by the solid-state forces and interactions, and through this influence it can give detailed information about the solid itself. In the first place, the lattice environment will determine the damping of the precessional motion. Via spin–orbit coupling it can interact with lattice phonons to give what is called spin–lattice relaxation, but also the spins can affect one another directly in a process described in the equations of motion as the spin–spin relaxation time. This latter mechanism arises from the influence of the dipole field of one spin on the motion of another and it results in the decay of in-phase precessional spin motion caused by the application of a field to the equilibrium condition of random phases. Typically, the spin–lattice relaxation time $\tau_1 \approx 10^{-10}$ sec, and the spin–spin relaxation time $\tau_2 \approx 10^{-6}$ sec.

Classically, precession can be described by Newton's equation of angular motion, which in terms of the magnetization $\mathbf{M}$ $(= \Sigma \mathbf{\mu})$, or dipole moment per unit volume, is

$$\frac{d}{dt}\mathbf{M} = \gamma(\mathbf{M} \times \mathbf{B}). \qquad (7.141)$$

Here $\mathbf{B}$ is a superposition of a dc component B_z, and an oscillating component in the x-y plane, and losses are neglected. Spin–lattice and spin–spin relaxation times can be included in Eq. (7.141) phenomenologically to give the well-known Bloch equations[74] describing the precessional motion of $\mathbf{M}$. The solutions give magnetic dispersive characteristics about the paramagnetic resonance frequency with lineshapes that are functions of, and can yield, the relaxation times. In the limiting case of extremely small microwave fields, the lineshape, for τ_2 finite, is Lorentzian, and the width of the resonance is determined by τ_2 alone. But, in general, the shape is a mixture of Lorentzian and gaussian shapes.[75,76]

---

[74] See C. W. Haas and H. B. Callen, in "Magnetism" (G. T. Rado and H. Suhl, eds.), Vol. 1, Chapter 10. Academic Press, New York, 1963.

[75] G. T. Rado and H. Suhl, eds., "Magnetism," Vol. 1. Academic Press, New York, 1963.

[76] R. A. Levy, "Principles of Solid State Physics." Academic Press, New York, 1968.

In addition to the relaxation interaction, the paramagnetic ions in a solid are acted on by inhomogeneous electric fields produced by their diamagnetic neighbors. For the iron group elements (Ti^{3+}, V^{3+}, Cr^{3+}, Mn^{3+}, Fe^{2+}, Co^{2+}, Ni^{2+}, Cu^{2+}) the crystalline field, or "ligand" field[†] as it is commonly called, has a pronounced effect, but for the rare earth elements (Ce^{3+}, Pr^{3+}, Nd^{3+}, Pm^{3+}, Sm^{3+}, Eu^{3+}, Gd^{3+}, Tb^{3+}, Dy^{3+}, Ho^{3+}, Er^{3+}, Tm^{3+}, Yb^{3+}), the effect is much less. For the trivalent rare earth ions dilutely concentrated in a host crystal and at temperatures that are not too low, the ions behave essentially as though they were free, and their g-factor is given reasonably well by the free ion expression Eq. (7.138). The reason for this lies in the shell structure of the ions. Unfilled inner shells occur, as we go up the periodic table, when the orbital quantum numbers are so large that the energy required to put another electron into it is larger than that needed to start a new shell. In the case of the rare earths the uncompensated spins are associated with electrons in the incomplete $4f$ shell (1 for Ce, 2 for Pr, ..., 13 for Yb) within the ion. The closed outer electronic shells of the ions (two $5s$ and six $5p$ electrons, that is, $5s^2 p^6$ configuration) do not have a magnetic effect, and they screen the $4f$ electrons from external perturbations by neighboring molecules. In the case of the iron group ions, the $3d$ electrons are responsible for the paramagnetism, and these are influenced by crystalline fields to a much greater degree. This is partly due to the large $3d$ shell radius and partly to the absence of any outer electronic shell to screen the $3d$ shell.

The degeneracy of the free ion can be strongly changed by the fields introduced by the lattice. In the free ion, **L** and **S** couple to form a resultant **J**, which, in the absence of electric or magnetic fields, describes a single energy level that is $(2J + 1)$-fold degenerate. An electric field that is relatively weak and so does not break down the LS coupling, will cause J to precess about the field direction and split the states corresponding to the $2J + 1$ values of M_J along the crystal field direction. For the rare earths this situation of unbroken LS coupling prevails, because the $4f$ electrons are screened from the full strength of the disruptive forces. The crystalline field does, nevertheless, cause a level splitting of some 10^2 cm^{-1}; this is relatively small and of the order of room-temperature lattice phonon energies. Screening of the $4f$ shell then has the consequence that the spin–lattice interaction tends to be strong,

---

[†] The word "ligand" was invented by chemists for the cluster of ions or molecules about a paramagnetic ion.

7.8. MAGNETIC RESONANCES

except at low temperatures; the g-values are influenced by orbital angular momentum to depart widely from the free-spin value $g \approx 2$, and are usually anisotropic. For the magnetic field parallel to the crystal field axis, the g-value is denoted $g_{\|}$, and the value perpendicular to this $g_{\perp}$.

When the crystalline electrostatic field is very strong, LS coupling will be broken, and an ion will undergo a Stark splitting of the orbital levels into $2L + 1$ levels. This occurs in the case of the iron group where the $3d$ electrons feel the full strength of the interatomic forces, and Stark levels have a zero (magnetic)-field splitting of about 10^4 cm^{-1}.

Splitting of $3d$ shell degeneracy by a ligand field of cubic symmetry can be appreciated qualitatively by considering the charge cloud distributions of $3d$ hydrogen orbitals relative to the six negative atoms as illustrated in Fig. 7.45. We suppose each of the six diamagnetic neighbors to have a spherical charge distribution, and the paramagnetic ion to be represented by the hydrogen wavefunctions of Section 7.1. Representing the radial wavefunction by $R(r)$, we can write

$$\psi_1 = \left(\frac{15}{16\pi}\right)^{1/2} \sin^2\theta \sin 2\phi \, R(r) = \left(\frac{15}{4\pi}\right)^{1/2} xy \, \frac{R(r)}{r^2},$$

$$\psi_2 = \left(\frac{15}{4\pi}\right)^{1/2} \sin\theta \cos\theta \sin\phi \, R(r) = \left(\frac{15}{4\pi}\right)^{1/2} yz \, \frac{R(r)}{r^2},$$

$$\psi_3 = \left(\frac{15}{4\pi}\right)^{1/2} \sin\theta \cos\theta \cos\phi \, R(r) = \left(\frac{15}{4\pi}\right)^{1/2} xz \, \frac{R(r)}{r^2},$$

$$\psi_4 = \left(\frac{15}{16\pi}\right)^{1/2} \sin^2\theta \cos 2\phi \, R(r) = \left(\frac{15}{16\pi}\right)^{1/2} (x^2 - y^2) \, \frac{R(r)}{r^2},$$

$$\psi_5 = \left(\frac{5}{16\pi}\right)^{1/2} (3\cos^2\theta - 1) R(r) = \left(\frac{5}{16\pi}\right)^{1/2} (2z^2 - x^2 - y^2) \, \frac{R(r)}{r^2}.$$

In the absence of any perturbing electric or magnetic field, these five levels are degenerate. However, when the forces exerted by the six ligand atoms are considered, one sees that the various charge configurations are affected differently. For ψ_1, ψ_2, and ψ_3 the electron clouds avoid the charge cloud of the neighboring negative ions, while for ψ_4 and ψ_5 the electron clouds more nearly approach the ions and experience a greater repulsion. Thus we expect the orbitals to split into two groupings with ψ_1, ψ_2, and ψ_3 degenerate and at lower energy than the other two. Further splitting within each of these groups can be shown to arise when one removes the restriction of a spherically symmetric charge distribution.

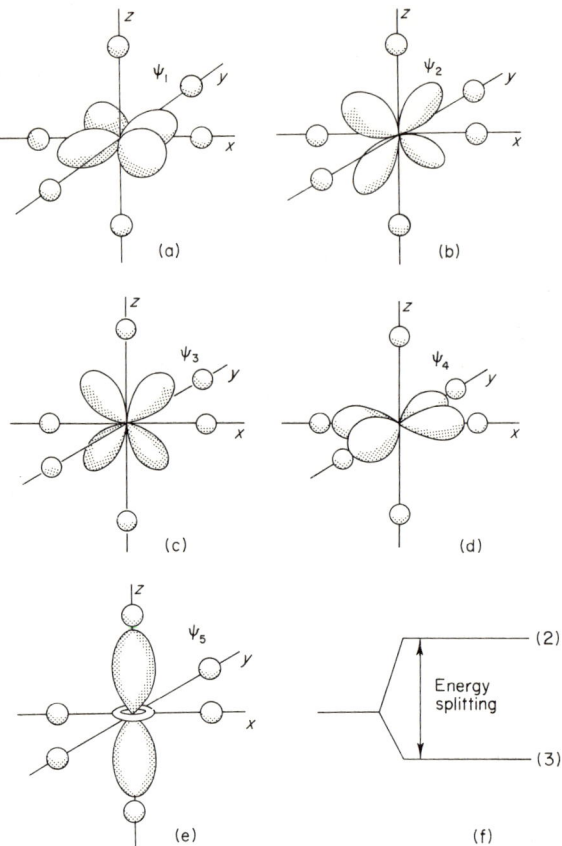

FIG. 7.45. Charge-cloud distribution of the 3d orbitals of a magnetic ion with six diamagnetic neighbors with spherical charge distribution. (a), (b), and (c) represent the wavefunctions ψ_1, ψ_2, and ψ_3, and (d) and (e) the wavefunctions ψ_4 and ψ_5. There is an energy splitting into two degenerate groups [illustrated in (f)], because the charge distributions are in closer proximity in (d) and (e) than in (a), (b), and (c). The numbers (2) and (3) in (f) indicate the degeneracy.

The effect of p orbitals associated with the ligand atoms is discussed in this context by Ingram.[77]

Frequently the iron group ions have a g-factor quite close to the free-spin value of 2. The reason for this is that the orbital angular momentum is *quenched* by the crystalline fields of the solid. Quenching is a phenomenon in noncentral fields associated with motion of the plane of the

---

[77] D. J. E. Ingram, "Spectroscopy at Radio and Microwave Frequencies," 2nd ed. Butterworth, London, 1967.

7.8. MAGNETIC RESONANCES

orbit which leads to the component of orbital momentum L_z having an average value of zero. In quantum-mechanical language, the Stark field mixes the states to generate wavefunctions which give L_z a zero average value. A simple example of the quenching of angular momentum for p-states in a field of rhombic symmetry is given by Morrish[78] and Kittel.[52] They show, for example, that the linear admixture ψ of the two p-states with $M_L = +1$ and $M_L = -1$ in a field of rhombic symmetry has the property $\int \psi^* L_z \psi \, d\tau = 0$. When quenching is complete, a g-factor close to that given by the Landé expression with $L = 0$ is to be expected. The angular momentum component of electrons in the $3d$ shell is substantially quenched by the crystalline field, but a small level of spin–orbit interaction generally remains to give g-values on either side of 2.[52]

Spin degeneracy still remains. However, it can be shown that, despite the fact that the electrostatic field cannot interact directly with the spin, it can operate indirectly through LS coupling to give zero-field splitting[79] of some of the $2S + 1$ spin states. Ultimately a certain amount of degeneracy remains which an electric field cannot reduce. This is the Kramers degeneracy. Kramers has proven the following theorem: If a system contains an odd number of electrons, the application of an electric field leaves at least two-fold degeneracy in the levels. A magnetic field can, of course, split this degeneracy.

Most experimentation with paramagnetic resonance has been carried out in the microwave band, where magnetic fields of only a few kilogauss need be applied. When the g-factor is around 2, the magnetic field sensitivity of paramagnetic resonance is generally not very different from that required for free electron cyclotron resonance, namely, 2.8 MHz G^{-1}. For some materials, however, it may be much larger than 2, at least for certain directions in the crystal. Tb^{3+} and Dy^{3+} are examples. Baker and Bleaney[80] have measured the g-value when the magnetic field is parallel to the crystalline field axis of ytterbium ethyl sulphate containing 0.1% terbium ethyl sulphate and have obtained the value $g_\| = 17.72$. Dysprosium ethyl sulphate has values of $g_\|$ typically of the order of 10, so that with a field of 100 kG a tuning range of 50 wavenumbers will be available.[81]

---

[78] A. H. Morrish, "The Physical Principles of Magnetism." Wiley, New York, 1965.
[79] R. S. Anderson, *Methods Exp. Phys.* **3**, 441–524 (1962).
[80] J. M. Baker and B. Bleaney, *Proc. Phys. Soc. London Sect. A* **68**, 257 (1955).
[81] N. Bloembergen, *in* "High Magnetic Fields" (H. Kolm, B. Lax, F. Bitter and R. Mills, eds.). MIT Press, Cambridge, Massachusetts, and Wiley, New York, 1962.

TABLE 7.7. Energy Levels and g-Factors for Dy^{3+} Ions in Dysprosium Ethyl Sulphate[a]

Crystalline field component	Energy splitting at 4.2°K (cm⁻¹)	$g_\parallel$	$g_\perp$
I		10.82	0
II	16.04	5.50	8.43
III	21.22	13.76	1.73
IV	60.40	19.74	0
V	69.59	16.58	1.61
VI	144.80	2.84	8.43
VII	198.24	3.08	0
VIII	238.82	1.18	10.46

[a] Calculations by Powell and Orbach.[82] Reprinted by permission of the Institute of Physics and the Physical Society.

Dysprosium has a $4f^9$ ground state configuration. According to Hund's rules the spin is the maximum allowed by the Pauli principle, and L is then the maximum allowed for this value of S. Thus $S = \frac{5}{2}$ and $L = 5$. As the shell is more than half filled, Hund's rules also give $J = L + S = \frac{15}{2}$. The ground state J manifold is thus denoted $^6H_{15/2}$. In the crystal field of ethyl sulphate this manifold splits into eight Kramers doublets. Their energy separations and g-values as calculated by Powell and Orbach[82] are given in Table 7.7. Gramberg has obtained some Stark separations and g-factors from experimental optical data, and has published results that agree well with the theoretical values. The values in Table 7.7 can be used to indicate the Zeeman splitting of Dy^{3+} in dysprosium ethyl sulphate. For a magnetic field applied parallel to the trigonal crystal axis, the energy levels of the four lowest Kramers doublets separate as shown in Fig. 7.46. With fields of hundreds of kilogauss, the Zeeman energy can become larger than the crystalline field splitting, and a number of microwave and far-infrared transitions become available. From such observations the properties of the crystal field and the relaxation processes between the excited ions and lattice vibrations can be evaluated.

Observations of direct transitions between the Stark components of the ground J manifold, and the Zeeman split Kramers doublets, have been made by Hill and Wheeler[83] for Dy^{3+}, Er^{3+}, and Sm^{3+} in ethyl

[82] M. J. D. Powell and R. Orbach, *Proc. Phys. Soc.* **78**, 753–758 (1961).
[83] J. C. Hill and R. G. Wheeler, *Phys. Rev.* **152**, 482–494 (1966).

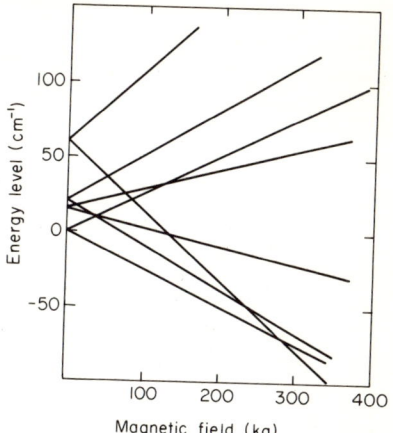

Fig. 7.46. The energy levels of the four lowest Kramers doublets of Dy^{3+} in dysprosium ethyl sulphate, for a magnetic field applied in the direction of the trigonal axis (after Bloembergen[81]).

sulphate. Using Fourier transform spectroscopic techniques, and fields up to 80 kG, they have measured and carefully identified the far-infrared phonon spectrum of the host crystal. In order to separate observed absorption lines into phonon and crystal field lines, Hill and Wheeler first obtained a spectrum for lanthanum ethyl sulphate (see Fig. 7.47). Dia-

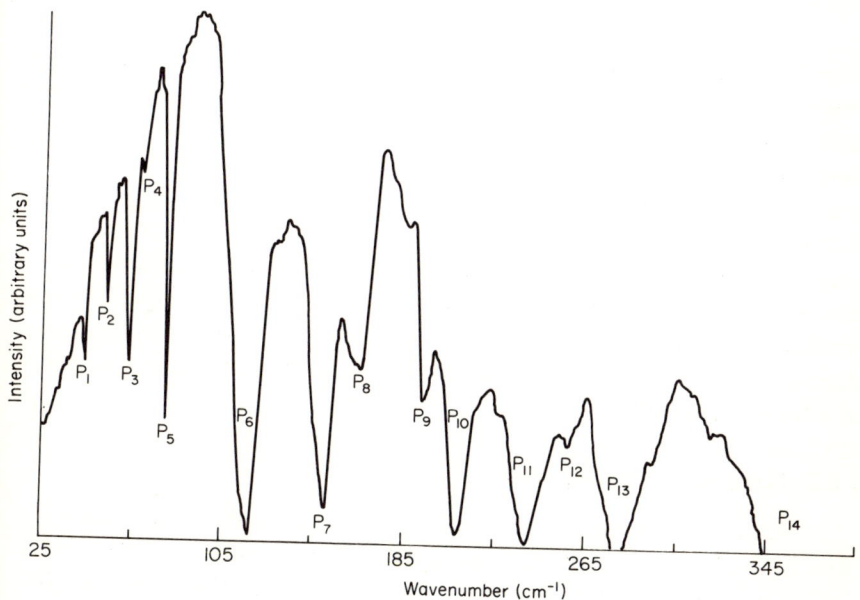

Fig. 7.47. Transmission spectrum of lanthanum ethyl sulphate showing 14 phonon modes, labelled P_1–P_{14} (after Hill and Wheeler[83]).

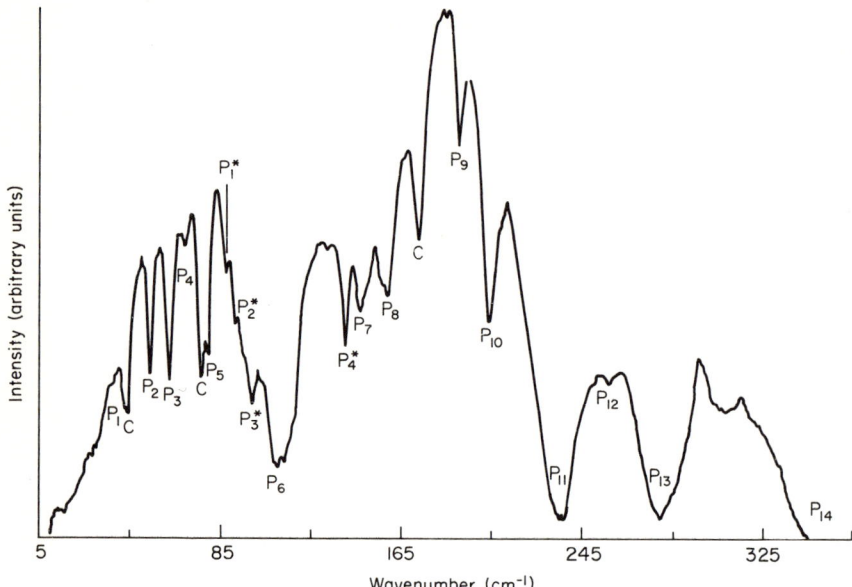

FIG. 7.48. Erbium ethyl sulphate spectrum. P_1–P_{14} are the 14 lanthanum ethyl sulphate lines shown in Fig. 7.47, P_1^*–P_4^* are additional phonon lines, and the lines labeled with a C are crystal field lines (after Hill and Wheeler[83]).

magnetic lanthanum has no 4f electrons and therefore no Stark levels. With paramagnetic ions introduced into the crystal, Stark level transitions are observed and distinguished from phonon transitions on the basis of their Zeeman splitting. For the case of erbium ions the composite spectrum is shown in Fig. 7.48.

Hadni[41] and others[84] have studied the far-infrared spectra of paramagnetic ions in various host crystals and have measured transitions at the wavelengths recorded in Table 7.8. By cooling to liquid helium temperatures, the competing effects of lattice vibrational modes are reduced, and the population distribution is such that most transitions involve the ground state.

7.8.2. Ferromagnetic Resonance

In the preceding section we considered magnetically dilute materials. The individual paramagnetic ions acted independently, and their interactions in the far-infrared were determined by the Stark energy (10^2–

---

[84] J. H. M. Thornley, *Phys. Rev.* **132**, 1492–1493 (1963).

TABLE 7.8. Some Transitions Observed for Dilute Paramagnetic Ion Concentrations in Crystals

Paramagnetic ion	Host crystal	Measured transition wavelengths (μm)
Pr^{3+}	Chloride	100; 325
Pr^{3+}	Floride	100; 160
Sm^{3+}	Chloride	150; 250
Sm^{3+}	Floride	95
Er^{3+}	Chloride	144
Cr^{3+}	Al_2O_3	100; 270
Ti^{3+}	Al_2O_3	263
Ce^{3+}	Zinc nitrate	204; 476

10^4 cm^{-1}) and the spin–orbit energy (10^2–10^3 cm^{-1}). Classical magnetic dipolar coupling between electron spins in dilute systems is quite small, as we can see from the dipole–dipole energy expression

$$\Delta W \simeq \mu B = \mu_0 \mu^2 / 4\pi r^3.$$

Here r is the dipole (ion) separation, $B = \mu_0 \mu / 4\pi r^3$ is the magnetic induction field due to one ion at the position of its neighbor, and $\mu_0 = 4\pi \times 10^{-7}$ hm^{-1}. For a concentration of, say, 10^{20} paramagnetic ions per cubic centimeter each with a dipole moment of 1 Bohr magneton,

$$B = 10^{-7} \times 10^{-23}/10^{-26}$$
$$= 10^{-4} \quad \text{weber m}^{-2}$$
$$= 1 \text{ G}.$$

The interaction energy between nearest neighbors is then $\Delta W \sim 10^{-4}$ cm, and the temperature at which $kT = \Delta W$ is 1.4×10^{-4} °K. If the ions have the undiluted concentration 10^{23} cm^{-3}, the energy of the classical dipole–dipole interaction is 10^{-1} cm^{-1}. This can be disrupted by thermal motions at a temperature of 0.4°K, and is inadequate as a candidate for the strong dipole alignment in ferromagnetic materials.

Exchange and Anisotropy Fields

As the concentration of magnetic ions is increased, the wavefunctions of the different ions overlap and an *exchange interaction* sets in. Cooperative phenomena rather than individual ion interactions then enter the picture and give rise to ferromagnetism with its various manifestations.

The exchange interaction is a strictly quantum-mechanical phenomena which we can illustrate in a simple way by considering the problem of a two-electron system. The system is described by the Schrödinger equation

$$\Delta_i^2 \psi + \Delta_j^2 \psi + (2m/\hbar^2)[W - V_i - V_j + (q^2/r_{ij})]\psi = 0, \quad (7.142)$$

where r_{ij} is the separation of the electrons, and q^2/r_{ij} the coulomb potential energy between them. As a first approximation, we suppose the electrons are so far from each other that we can neglect the mutual interaction potential energy q^2/r_{ij}. In other words, we regard the electrons as moving independently of each other, electron i in a potential V_i/q and electron j in a potential V_j/q. Equation (7.142) readily separates into two independent one-electron wave equations if we substitute

$$\psi = u_a(i)u_b(j). \quad (7.143)$$

Here $u_a(i) \equiv u_a(x_i, y_i, z_i)$, and $u_b(j) \equiv u_b(x_j, y_j, z_j)$, the subscripts a and b being used to distinguish particular kinds of solutions, and the symbols in parenthesis to distinguish between the two electrons. Associated with these one-electron wavefunctions are eigen energies W_a and W_b, and associated with the product wavefunction Eq. (7.143) is the eigen energy $W = W_a + W_b$.

In addition to Eq. (7.143), the product wavefunction

$$\psi = u_a(j)u_b(i) \quad (7.144)$$

is also a solution of Eq. (7.142), in the approximation where q^2/r_{ij} is neglected. The existence of two wavefunctions for a given energy means that we have two-fold degeneracy; this is called *exchange degeneracy*. Solutions (7.143) and (7.144) correspond to the states of the system with the two electrons interchanged. Any linear combination of them is also a solution of Schrödinger's equation, but we must satisfy the requirement that the system is physically unchanged when the electrons are interchanged, for they are indistinguishable. Suitable combinations are the sum and difference of Eqs. (7.143) and (7.144), which are symmetrical (ψ_S) and antisymmetrical (ψ_{AS}) with respect to interchange of i and j:

$$\psi_S = (1/\sqrt{2})[\psi_a(i)\psi_b(j) + \psi_a(j)\psi_b(i)], \quad (7.145)$$

$$\psi_{AS} = (1/\sqrt{2})[\psi_a(i)\psi_b(j) - \psi_a(j)\psi_b(i)]. \quad (7.146)$$

7.8. MAGNETIC RESONANCES

The physically meaningful quantities $\psi_S\psi_S^*$ and $\psi_{AS}\psi_{AS}^*$ are both unaffected by the interchange of electrons, despite the change of sign of ψ_{AS}.

Let us now take electron spin into account. We introduce the two spin wavefunctions α and β, corresponding to the two values of the z-component of spin angular momentum $m_{sz} = \pm\tfrac{1}{2}$. For example, $\alpha(i)$ is the wavefunction for electron i if it has the "spin up" value $m_{sz} = +\tfrac{1}{2}$. The total spin wavefunctions are then constructed of products of the two independent spin functions, and linear combinations of the products. Three symmetrical wavefunctions representing the triplet state ($m_{sz} = 1, 0, -1$), and one antisymmetrical wavefunction representing the singlet state $S = 0$, result:

$$(1/\sqrt{2})\alpha(i)\alpha(j), \quad (1/\sqrt{2})\beta(i)\beta(j),$$
$$(1/\sqrt{2})[\alpha(i)\beta(j) + \alpha(j)\beta(i)], \quad (1/\sqrt{2})[\alpha(i)\beta(j) - \alpha(j)\beta(i)]. \tag{7.147}$$

Total wavefunctions of the electrons are formed by the product of orbital and spin functions which satisfy the Pauli exclusion principle. On the basis of the principle it was shown by Dirac[73,85,86] that the total wavefunction must change sign when the two electrons are exchanged, that is, they are always antisymmetrical. They must then be the combinations

$$\psi_1 = \tfrac{1}{2}[u_a(i)u_b(j) + u_a(j)u_b(i)][\alpha(i)\beta(j) - \alpha(j)\beta(i)],$$
$$\psi_2 = \tfrac{1}{2}[u_a(i)u_b(j) - u_a(j)u_b(i)]\alpha(i)\alpha(j),$$
$$\psi_3 = \tfrac{1}{2}[u_a(i)u_b(j) - u_a(j)u_b(i)]\beta(i)\beta(j), \tag{7.148}$$
$$\psi_4 = \tfrac{1}{2}[u_a(i)u_b(j) - u_a(j)u_b(i)][\alpha(i)\beta(j) + \alpha(j)\beta(i)].$$

ψ_1 is the singlet state in which the two spins are antiparallel, and ψ_2, ψ_3, and ψ_4 represent the triply degenerate levels of the spin-parallel state which we can simply describe by a wavefunction $\psi_{2,3,4}$.

If the spins are parallel, the Pauli principle requires that the electrons be spatially separated, whereas for the antiparallel arrangement their orbital wavefunctions may overlap. The difference in the charge distributions means that the Coulomb interaction energies of the singlet and

---

[85] J. H. Van Vleck, *Phys. Bull. Inst. Phys. and Phys. Soc.*, London **19**, 167–175 (1968).

[86] D. H. Martin, "Magnetism in Solids." Iliffe, London, 1967.

triplet states are different. The difference is

$$W_{ex} = W_{singlet} - W_{triplet} = \int \psi_1^*(q^2/r_{ij})\psi_1 \, d\tau_i \, d\tau_j$$

$$- \int \psi_{2,3,4}^*(q^2/r_{ij})\psi_{2,3,4} \, d\tau_i \, d\tau_j$$

$$= 2 \int u_a^*(i) u_b^*(j)(q^2/r_{ij}) u_a(j) u_b(i) \, d\tau_i \, d\tau_j. \tag{7.149}$$

The energy difference is called the *exchange energy*. While it is entirely electrostatic, it arises via the Pauli principle from the relative orientation of the two spins, so we can write it in the form

$$W_{ex} = -2J_{ij} \mathbf{S}_i \cdot \mathbf{S}_j. \tag{7.150}$$

From Eqs. (7.149) and (7.150)

$$J_{ij} = \int u_a^*(i) u_b^*(j)(q^2/r_{ij}) u_a(j) u_b(i) \, d\tau_i \, d\tau_j. \tag{7.151}$$

J_{ij} is the *exchange integral* between the two electrons. We reiterate that it has a strictly quantum-mechanical origin associated with wavefunction overlap and the exclusion principle. Its value can be as large as 10^2–10^3 cm^{-1}, and it furnishes the strong coupling mechanism for ferromagnetism that is wanting in purely classical electromagnetic theory.

Heisenberg introduced the exchange mechanism in 1927, and he went on to derive the basic equations of magnetism on the basis of spin coupling. For ferromagnetism the spins are parallel, that is, $\mathbf{S}_i \cdot \mathbf{S}_j = S^2$, and J_{ij} must be positive for this to be the state with minimum exchange energy. Negative J_{ij} corresponds to antiferromagnetism, for here W_{ex} is a minimum when the spins are antiparallel or $\mathbf{S}_i \cdot \mathbf{S}_j = -S^2$. In ferrimagnets the spins are antiparallel but unequal, as the illustration in Fig. 7.49 implies. The "molecular" or "effective" field postulated by Weiss as proportional to the magnetization can be described as an exchange field in the Heisenberg theory.[52] That is, $\mathbf{B}_{ex}$ is related to $\mathbf{M}$ by[†]

$$\mathbf{B}_{ex} = \lambda \mathbf{M}, \tag{7.152}$$

[†] We use $\mathbf{B}$ rather than $\mathbf{H}$ with the understanding that quantities such as γ, λ, and the demagnetization coefficients in Eq. (7.174) differ by the factor μ_0, the permeability of free-space, from the values they have in formulations using $\mathbf{H}$.

7.8. MAGNETIC RESONANCES

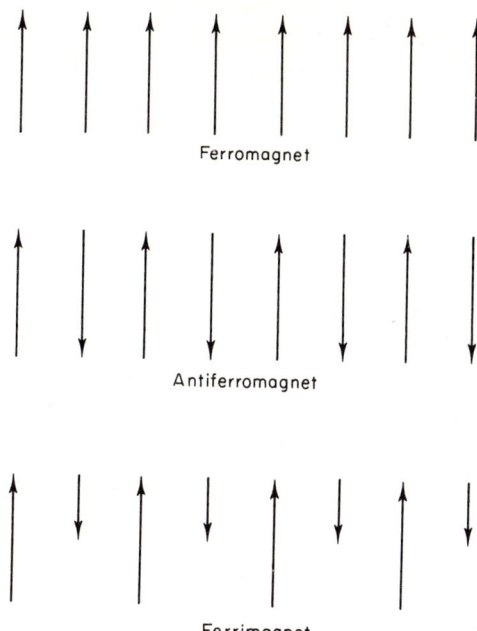

FIG. 7.49. Representation of spin alignments in ferromagnetic, antiferromagnetic, and ferrimagnetic materials.

and it is interpretable in terms of the exchange energy. In a ferromagnet the total exchange energy of spin i with its neighbors j, of which there are an effective number Z, can be written

$$-\boldsymbol{\mu} \cdot \mathbf{B}_{\text{ex}} = -2 \sum_{j}^{Z} J_{ij} \mathbf{S}_i \cdot \mathbf{S}_j$$
$$= -2J_{ij}ZS^2,$$

where $\boldsymbol{\mu}$ represents the spin dipole magnetic moment. That is,

$$B_{\text{ex}} = 2J_{ij}ZS^2/\mu = 2J_{ij}ZS/g\beta, \tag{7.153}$$

since

$$\mu = \gamma\hbar S = g\beta S. \tag{7.154}$$

The Weiss constant λ is

$$\lambda = 2J_{ij}ZS/g\beta\mu N = 2J_{ij}Z/g^2\beta^2 N, \tag{7.155}$$

where N is the density of spins.

The energy difference between the states with parallel and antiparallel spins is J_{ij} and, as we have seen, the interaction can be assigned to an exchange field. For $Z = 1$ and $S = \frac{1}{2}$, its value is $B_{\text{ex}} = J_{ij}/g\beta$; for $g = 2$ and β equal to 1 Bohr magneton, the ratio between the field and the transition frequency is 2.8 MHz G^{-1}. $J_{ij} \approx 10^2$ cm^{-1} thus corresponds to $B_{\text{ex}} \approx 10^6$ G. The transition frequency, 3×10^{12} Hz in this case, is in the far-infrared.

In a ferromagnetic crystal there is, in addition to the rather large exchange energy, a magnetocrystalline or *anisotropy energy*. Or, if we wish, we can speak of an anistropy field B_{an}, just as we can use the notion of an exchange field. As the name implies, this energy is associated with the magnetic anisotropy of crystals as demonstrated by the different magnetic susceptibilities in the various crystallographic directions. For example, in body-centered cubic iron the [100] directions, that is, the directions along the cube edges, are *easy axes* for magnetization, while

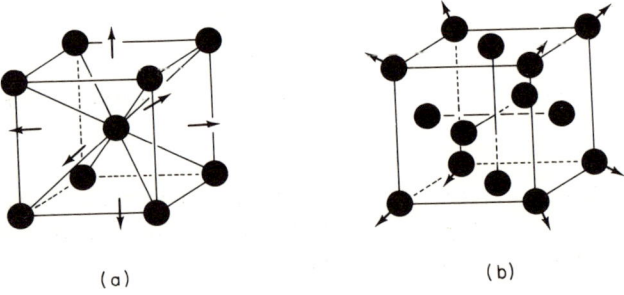

Fig. 7.50. Arrows indicate the directions of easy magnetization in (a) iron and (b) nickel.

the [111] directions are *hard axes*. For face-centered cubic nickel, the cube diagonals, the [111] directions, are easy axes. Figure 7.50 illustrates these directions. The energy of anisotropy can originate either in the crystal field or, more often, from anisotropic exchange.[85] This latter effect arises through the coupling of the spins to incompletely quenched orbital angular momentum. Orbital motion can interact electrostatically with the crystal structure through overlapping wavefunctions of electrons in neighboring ions. The spin–orbit wavefunction is asymmetric and the overlap is different for different directions in the crystal. Anisotropy energy tends to be of the order of 1 cm^{-1}, and the anisotropy field B_{an} of the order of 10^4 G.

Spin Waves

Study of the modes of precessional motion along a chain of spin vectors can do much to aid the understanding of magnetic resonance phenomena. The model of a magnetic solid as an array of coupled precessing spins is not unlike the model of a crystalline dielectric where the atoms of the solid can vibrate in normal modes of coupled motion, and the forms of the solutions in the two cases have much in common. As we saw in Section 7.6.1, the ionic dynamics can be analyzed in terms of lattice waves which follow from a simple linear chain model. In the case of magnetic crystals we can, in a similar way, use the one-dimensional chain with spins coupled by nearest neighbor exchange interactions.

In a ferromagnet at absolute zero temperature, all spins along the chain are lined up, and in a small field they all precess in phase about the field direction. This precession is the zero-point motion of the spins. When the chain is perturbed by an applied oscillatory field, or by thermal motion when $T > 0°K$, the system of coupled spins is excited, and the excitation takes the form of wavelike precessional motion with phase variation that is periodic in time and in displacement along the chain. These are called *spin waves*. Just as we can interpret lattice waves in terms of phonons, so too can we describe the spin motion quantum mechanically; in this description a quantum of spin motion is called a *magnon*.

We can treat the problem of spin wave excitation classically, starting from Newton's law relating to angular momentum, or quantum mechanically, starting from the commutation of the angular momentum with the Hamiltonian $[H, \mathbf{S}]$. To see that these approaches are equivalent we start with the commutation equation

$$-i\hbar \, d\mathbf{S}/dt = [H, \mathbf{S}] = H\mathbf{S} - \mathbf{S}H, \tag{7.156}$$

where

$$H = -\boldsymbol{\mu} \cdot \mathbf{B} = -\gamma \mathbf{S} \cdot \mathbf{B}.$$

Substitution gives

$$(\mathbf{S} \cdot \mathbf{B})\mathbf{S} - \mathbf{S}(\mathbf{S} \cdot \mathbf{B}) = (i\hbar/\gamma) \, d\mathbf{S}/dt.$$

Consider only the x-component. The left-hand side gives

$$[(S_z B_z)S_x - S_x(S_z B_z)] + [(S_y B_y)S_x - S_x(S_y B_y)]$$
$$+ [(S_x B_x)S_x - S_x(S_x B_x)] = i\hbar(S_y B_z - S_z B_y),$$

as the commutation relationships for S_x, S_y, and S_z are

$$[S_xS_x] = 0, \qquad [S_zS_x] = i\hbar S_y, \qquad \text{and} \qquad [S_xS_y] = i\hbar S_z.$$

We then have

$$\gamma(S_yB_z - S_zB_y) = dS_x/dt,$$

and similar equations for the y and z components. Thus Eq. (7.156) is in agreement with the classical equation

$$d\mathbf{S}/dt = \gamma \mathbf{S} \times \mathbf{B}. \tag{7.157}$$

Let us proceed from the classical equation for the spin chain and simplify the problem by assuming the coupling between spins to be nearest neighbor exchange fields only. We neglect the anisotropy field and refer the reader to Sievers[87,53a] for an analysis including this field. The energy of the nth spin is

$$W_n = -2J\mathbf{S}_n \cdot (\mathbf{S}_{n-1} + \mathbf{S}_{n+1}), \tag{7.158}$$

where J denotes the exchange integral between nearest neighbors. From Eq. (7.153), the exchange field acting on the nth spin is

$$\mathbf{B}_n = (2J/\gamma\hbar)(\mathbf{S}_{n-1} + \mathbf{S}_{n+1}). \tag{7.159}$$

Substitution into the equation of spin motion Eq. (7.157) then gives

$$\hbar \, d\mathbf{S}_n/dt = 2J\mathbf{S}_n \times (\mathbf{S}_{n-1} + \mathbf{S}_{n+1}). \tag{7.160}$$

This equation is nonlinear, but it may be linearized for small amplitudes by writing

$$\mathbf{S}_n = \mathbf{S}_z + \boldsymbol{\sigma}_n, \tag{7.161}$$

where $\mathbf{S}_z$ is a constant vector in the direction z of magnetization, and $\boldsymbol{\sigma}_n$ is a small vector in the x-y plane. It should be noted that the z direction has no particular relation to the direction of the linear chain. When Eq. (7.161) is substituted in Eq. (7.160), and only first-order terms in $\boldsymbol{\sigma}$ are retained, we obtain

$$\hbar \, d\boldsymbol{\sigma}_n/dt = 2J\mathbf{S}_z \times (\boldsymbol{\sigma}_{n-1} - 2\boldsymbol{\sigma}_n + \boldsymbol{\sigma}_{n+1}), \tag{7.162}$$

---

[87] A. J. Sievers, NATO Advanced Study on Elementary Excitations and Their Interactions, Cortina d'Ampezzo, Italy, July 11–23, 1966; Rep. No. 562. Materials Sci. Center, Cornell Univ., Ithaca, New York, 1966.

7.8. MAGNETIC RESONANCES

or, in vector component form,

$$\hbar(d\sigma_n/dt)_x = -2JS_z(\sigma_{n-1} - 2\sigma_n + \sigma_{n+1})_y, \quad (7.163)$$

$$\hbar(d\sigma_n/dt)_y = 2JS_z(\sigma_{n-1} - 2\sigma_n + \sigma_{n+1})_x. \quad (7.164)$$

Precessional motion of $\mathbf{S}_n$ implies circular motion in the x-y plane of σ_n, and this is constructed as $\sigma_n^+ = (\sigma_n)_x + i(\sigma_n)_y$, and $\sigma_n^- = (\sigma_n)_x - i(\sigma_n)_y$. Accordingly we multiply Eq. (7.163) by $i = \sqrt{-1}$ and subtract Eq. (7.164) to obtain

$$i\hbar \, d\sigma_n^+/dt = -2JS_z(\sigma_{n-1}^+ - 2\sigma_n^+ + \sigma_{n+1}^+). \quad (7.165)$$

Similarly, we can obtain

$$i\hbar \, d\sigma_n^-/dt = 2JS_z(\sigma_{n-1}^- - 2\sigma_n^- + \sigma_{n+1}^-). \quad (7.166)$$

Equation (7.165) is similar to the equation of a monatomic linear chain describing the modes of vibration of a crystal lattice. The consequences are, then, immediately clear from the results of Section 7.6.1 specialized to the case of equal atomic masses and one atom per unit cell. There exist traveling spin-wave solutions of Eq. (7.165) with σ_n^+ varying with distance along the chain and with time as $e^{i(\omega t - kna)}$, where a is the lattice spacing, and $k = 2\pi/\lambda$. When we substitute, we obtain

$$-\hbar\omega = -2JS_z(e^{-ika} - 2 + e^{ika})$$
$$= 4JS_z(1 - \cos ka). \quad (7.167)$$

Similarly, Eq. (7.166) gives the dispersion curve for ω negative. We note that S_z is limited to the values $-S, -S+1, \ldots, S-1, S$, which are the components of the spatially quantized vector of length $[S(S+1)]^{1/2}$. For small values of k, Eq. (7.167) reduces to

$$-\hbar\omega \simeq Ja^2k^2, \quad (7.168)$$

for the linear chain with spin $\tfrac{1}{2}$.

For a three-dimensional ferromagnetic lattice of cubic symmetry and with arbitrary S, the form of Eq. (7.168) still applies but with a multiplicative constant $ZS/3$, where Z is the number of nearest neighbors.[86]

The dispersion relation for a linear chain of spins with nearest neighbor coupling, and also a field B_0 applied in the $-z$ direction, can be obtained from a semiclassical treatment based on the equation

$$\hbar \, d\mathbf{S}_n/dt = \gamma\hbar\mathbf{S}_n \times \left[\mathbf{B}_0 + (2J/g\beta) \sum_j \mathbf{S}_j\right]. \tag{7.169}$$

This describes the precessional motion of spin S_n located between spins S_{n-1} and S_{n+1} represented by $\sum_j \mathbf{S}_j$ as illustrated in Fig. 7.51. Alter-

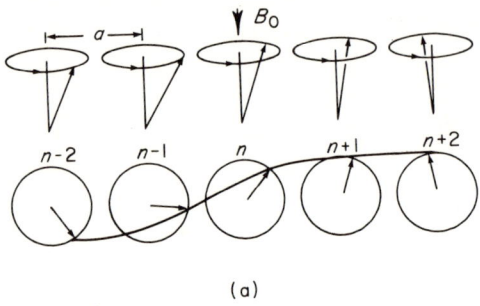

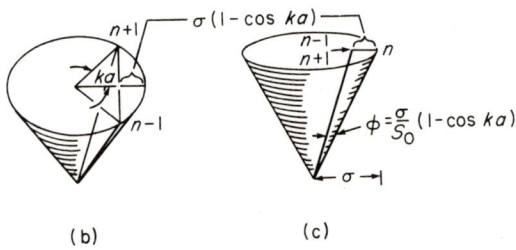

FIG. 7.51. Spin-wave precessional motion along a chain of spins with lattice spacing a. (a) shows two views of the spin precession; (b) and (c) show spin n and its nearest neighbors $n - 1$ and $n + 1$ collapsed onto the same precessional circle. k is the wave-vector, ϕ is the phase angle between spin n and spins $n - 1$ and $n + 1$ (after Levy[76]).

natively, one may use the quantum mechanical equation of motion $i\hbar\dot{\mathbf{S}}_n = [\mathbf{S}_n, H_n]$ with the Hamiltonian

$$H_n = -g\beta\mathbf{S}_n \cdot \left[\mathbf{B}_0 + (2J/g\beta) \sum_j \mathbf{S}_j\right]$$
$$= -g\beta\mathbf{S}_n \cdot \mathbf{B}_{\text{eff}}. \tag{7.170}$$

Here we have defined B_{eff} as the sum of the applied and exchange fields

$$\mathbf{B}_{\text{eff}} = \mathbf{B}_0 + \mathbf{B}_{\text{ex}} = \mathbf{B}_0 + (2J/g\beta) \sum_j \mathbf{S}_j. \tag{7.171}$$

7.8. MAGNETIC RESONANCES

The interested reader is referred to Herring and Kittel[88] for a reduction of the commutator equation.

To reduce Eq. (7.169), we use a geometric model devised by Keffer et al.[89] for the evaluation of $S_n \times \sum_j S_j$. We refer to Fig. 7.51. If we excite a spin wave of wavenumber k in our linear chain, the spins will precess and, at any instant of time, there will be a pattern of progressively increasing phase along the chain. At any instant of time spin n will make an angle ϕ with the plane determined by spins $n-1$ and $n+1$, which are equally spaced ahead and behind n by a phase angle ka. The amplitude of precession is measured by the radius σ of the circular projection of the motion, and this increases as the excitation is increased. From Fig. 7.51 it is clear that the sagitta is $\sigma(1 - \cos ka)$ and that $\phi = (\sigma/S_0)(1 - \cos ka)$, where $S_0 = [S(S+1)]^{1/2}$. Thus, we have

$$S_n \times \sum_j S_j = S_n \times (S_{n-1} + S_{n+1})$$
$$= -2S_0^2 \sin \phi, \text{ in magnitude}$$
$$\simeq -2S_0 \sigma (1 - \cos ka).$$

The equation of motion Eq. (7.169) can then be written

$$\hbar \, dS_n/dt = \gamma \hbar S_0 [B_0 \sin(S_n, B_0) - (2J/g\beta)(2S_0 \sin \phi)]$$
$$= \gamma \hbar [\sigma B_0 - (4JS_0\sigma/g\beta)(1 - \cos ka)]. \quad (7.172)$$

The right-hand side of Eq. (7.172) is the torque acting on spin n. The torque is $T = \boldsymbol{\omega} \times \hbar S_n$, and its magnitude

$$T = \hbar \sigma \omega,$$

can be equated to the right side of Eq. (7.172) to give

$$\hbar \omega = \hbar \gamma B_0 - 4JS_0(1 - \cos ka). \quad (7.173)$$

If $B_0 = 0$,

$$-\hbar \omega = 4JS_0(1 - \cos ka)$$
$$\simeq 2JS_0 k^2 a^2, \quad \text{for} \quad k \to 0,$$

as expected.

---

[88] C. Herring and C. Kittel, *Phys. Rev.* **81**, 869–880 (1951).
[89] F. Keffer, H. Kaplan and Y. Yafet, *Amer. J. Phys.* **21**, 250–257 (1953).

An oscillating magnetic field couples to a ferromagnet when it acts simultaneously and identically on all spins to drive them in the uniform mode of precessional motion. That is, it couples to the $k = 0$ spin wave, and in transferring energy it bends the spin vectors away from the z-direction. It follows from Eqs. (7.172) and (7.173) that the exchange field $\mathbf{B}_{\text{ex}}$ does not contribute to the frequency of the $k = 0$ ferromagnetic resonance mode. The reason is that $\mathbf{S}_n \times \sum_j \mathbf{S}_j = 0$, as we have shown; in other words, the net torque on spin n by its nearest neighbors $n - 1$ and $n + 1$ is zero when they are aligned parallel and precess precisely in phase.

If the exchange field were effective in a ferromagnet then, even in a zero external field, the resonance frequency would be in the vicinity of 10^{13} Hz. But because it is ineffective, ferromagnetic resonance tends to occur at microwave frequencies. Our discussion does, however, oversimplify the interaction. The effective field acting on the spins is not, in general, simply the applied field but rather the resultant of the applied field, the anisotropy field, and demagnetizing fields, the latter being determined by the geometry of the sample. For some materials and geometries $\mathbf{B}_{\text{eff}}$ is essentially the applied field $\mathbf{B}_0$ and, with the g-factor near 2, the resonance frequency $\omega = \gamma B_0 = gqB_0/2m$ is similar in value to that produced by magnetic fields applied to paramagnetic materials.

In general, one must include anisotropy and demagnetizing effects in the interpretation of ferromagnetic resonance. Kittel[52] has shown that for a field applied along one of the axes (z, say) of an ellipsoid, cut with a preferred crystallographic axis in the z-direction, the ferromagnetic resonance frequency is given by

$$\omega = \gamma\{[B_0 + (N_x - N_z)M_S + B_{\text{an}}][B_0 + (N_y - N_z)M_S + B_{\text{an}})\}^{1/2}, \tag{7.174}$$

where N_x, N_y, and N_z are demagnetizing coefficients. For the special case of $\mathbf{B}_0$ parallel to the easy axis of a spherical sample, $N_x = N_y = N_z$, and so[76]

$$\omega = \gamma[B_0 + B_{\text{an}}]. \tag{7.175}$$

The term γB_{an} has an angular dependence and can be distinguished experimentally from γB_0 on this basis. Thus microwave resonance observations can yield γ and hence the g-factor; they can also lead to a measure of the anisotropy field. Such measurements have, for example, given $g = 2.10$ for Fe, and $g = 2.21$ for Ni.[52]

7.8.3. Antiferromagnetic Resonance

The magnon dispersion relation of an antiferromagnet can be derived from a one-dimensional chain model with successive spins directed alternately up and down as we move along the chain. We take this to be the representation of an antiferromagnet composed of two interpenetrating sublattices A and B, the former with spins up and the latter with spins down. J is now negative. The antiferromagnetic form of Eq. (7.165) for a spin n on sublattice A is

$$i\hbar \, d\sigma_n^+/dt = -2JS_z(\sigma_{n-1}^+ + 2\sigma_n^+ + \sigma_{n+1}^+). \tag{7.176}$$

For a spin at position $n+1$ on the chain, belonging to sublattice B, we have

$$i\hbar \, d\sigma_{n+1}^+/dt = 2JS_z(\sigma_n^+ + 2\sigma_{n+1}^+ + \sigma_{n+2}^+). \tag{7.177}$$

We seek solutions of the form

$$\sigma_n^+ = u_a e^{i[\omega t - kna]}, \qquad \sigma_{n+1}^+ = u_b e^{i[\omega t - k(n+1)a]},$$

and obtain on substitution

$$(\omega - \omega_{\text{ex}})u_a - \omega_{\text{ex}} \cos ka \, u_b = 0, \tag{7.178}$$

and

$$\omega_{\text{ex}} \cos ka \, u_a + (\omega + \omega_{\text{ex}})u_b = 0, \tag{7.179}$$

where

$$\omega_{\text{ex}} = 4|J|S_z/\hbar.$$

Solutions exist if

$$\begin{vmatrix} \omega_{\text{ex}} - \omega & \omega_{\text{ex}} \cos ka \\ \omega_{\text{ex}} \cos ka & \omega_{\text{ex}} + \omega \end{vmatrix} = 0,$$

that is, for

$$\omega^2 = \omega_{\text{ex}}^2(1 - \cos^2 ka), \tag{7.180}$$

or

$$\omega = |\omega_{\text{ex}} \sin ka| \tag{7.181}$$

$$\simeq 4|J|S_z ka/\hbar, \quad \text{for} \quad k \to 0. \tag{7.182}$$

In an approximation that neglects the anisotropy field a spin-wave vector model similar to that used in Section 7.8.2 can be developed. In considering this model Keffer et al.[89] have found that stability of the exchange interaction requires that the lattice of "up spins" and the lattice of "down spins" be inclined to one another and precess in the same sense with different precessional amplitudes, as illustrated in Fig. 7.52. To every permissible value of k there are two antiferromagnetic precessional modes, as we must expect for a structure with two magnetic ions per unit magnetic cell. Referring to Fig. 7.52 (and ignoring $\mathbf{B}_0$), mode 1 corresponds to the clockwise precession seen from the top with the spin

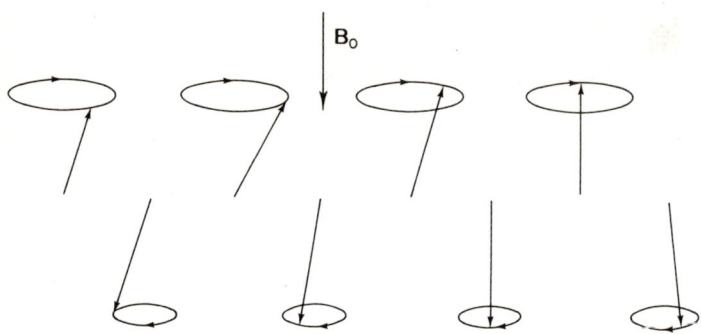

FIG. 7.52. Representation of the spin precessional motion on the two sublattices of an antiferromagnet.

up lattice having larger precessional radius. Mode 2 is the counterclockwise motion obtained by inverting Fig. 7.52. The precession is influenced strongly by the exchange interaction, for this produces a large torque between an ion and its neighbors when they are turned from their antiparallel alignment.

In the absence of an applied field the two modes are degenerate, but when a field is applied they are affected differently. That is, $\mathbf{B}_0$ adds to the precessional frequency of one mode and subtracts from the other. Keffer et al.[89] show that the resonance frequency of mode 1 is

$$\omega_1 = \gamma B_0 + (4J/\hbar)S_0 ak, \tag{7.183}$$

and that of mode 2 is

$$\omega_2 = \gamma B_0 - (4J/\hbar)S_0 ak. \tag{7.184}$$

The anisotropy field $\mathbf{B}_{\text{an}}$ is most significant in antiferromagnetic resonance and cannot be neglected. Its effect on the dynamics of electron spin can be found by introducing an anisotropy field $+\mathbf{B}_{\text{an}}$ in the z-direction acting on the A spins, and an oppositely directed field $-\mathbf{B}_{\text{an}}$ acting on the B spins. We take the easy axis of magnetization of the crystal as the

z-direction. The two sublattices of spins are constrained to be antiparallel by exchange fields

$$\mathbf{B}_{\text{exA}} = -\lambda \mathbf{M}_B, \quad \text{and} \quad \mathbf{B}_{\text{exB}} = -\lambda \mathbf{M}_A, \quad (7.185)$$

which each sublattice exerts on the other. Here the Weiss constant λ is positive, and the exchange fields are therefore antiparallel to, and proportional to, the sublattice magnetizations producing them. With no external field applied, sublattice A is acted on by a field $-\lambda \mathbf{M}_B + \mathbf{B}_{\text{an}}$, and sublattice B by $-\lambda \mathbf{M}_A - \mathbf{B}_{\text{an}}$, as shown in Fig. 7.53.

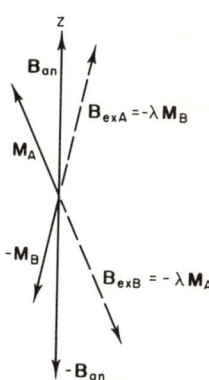

Fig. 7.53. Diagram showing the fields acting on the lattices A and B in an antiferromagnet. The anisotropy fields, $\pm \mathbf{B}_{\text{an}}$, act in the direction of easy magnetization (the z-direction), and each sublattice of spins acts on the other through an exchange field (after Kittel[54]).

We follow the treatment of Kittel[52] and set $(M_A)_z = M$, and $(M_B)_z = -M$. The linearized equations of precessional motion are

$$\frac{d}{dt}(M_A)_x = \gamma\{(M_A)_y[\lambda M + B_{\text{an}}] - M[-\lambda(M_B)_y]\},$$
$$\frac{d}{dt}(M_A)_y = \gamma\{M[-\lambda(M_B)_x] - (M_A)_x[\lambda M + B_{\text{an}}]\}, \quad (7.186)$$

$$\frac{d}{dt}(M_B)_x = \gamma\{(M_B)_y[-\lambda M - B_{\text{an}}] - (-M)[-\lambda(M_A)_y]\},$$
$$\frac{d}{dt}(M_B)_y = \gamma\{(-M)[-\lambda(M_A)_x] - (M_B)_x[-\lambda M - B_{\text{an}}]\}. \quad (7.187)$$

With the construction $M_A^+ = (M_A)_x + i(M_A)_y$, and $M_B^+ = (M_B)_x + i(M_B)_y$, and with $B_{\text{ex}} \equiv \lambda M$, we obtain the secular determinant

$$\begin{vmatrix} \gamma(B_{\text{an}} + B_{\text{ex}}) - \omega & \gamma B_{\text{ex}} \\ \gamma B_{\text{ex}} & \gamma(B_{\text{an}} + B_{\text{ex}}) + \omega \end{vmatrix} = 0. \quad (7.188)$$

The antiferromagnetic resonance frequency is then given as

$$\omega = \gamma[B_{an}(B_{an} + 2B_{ex})]^{1/2}. \tag{7.189}$$

The angular frequency ω of Eq. (7.189) is that of a uniform mode of precessional motion of the sublattices. It corresponds to the limiting case $k = 0$. The magnon spectrum of a two-sublattice antiferromagnet can be shown to be[86,89a]

$$\omega = \gamma\left[(B_{ex} + B_{an})^2 - B_{ex}^2\left(\frac{1}{Z}\sum_m \cos \mathbf{k}\cdot\mathbf{a}_m\right)^2\right]^{1/2} \pm \gamma B_0, \tag{7.190}$$

where the summation is over all the vector displacements $\mathbf{a}_m$ joining an ion to its neighbors. $\mathbf{B}_0$ is applied in the direction of easy magnetization. For a cubic crystal, and small k, Eq. (7.190) reduces to

$$\omega = \gamma\left[(B_{ex} + B_{an})^2 - B_{ex}^2\left(1 - \frac{2k^2a^2}{Z}\right)\right]^{1/2} \pm \gamma B_0, \tag{7.191}$$

and, for $B_0 = 0$ and $k = 0$, to Eq. (7.189).

In contrast to the situation in ferromagnets, the effect of the exchange field on the antiferromagnetic resonance persists even when $k \to 0$. We note also that the effect of the exchange interaction comes in with the anisotropy and disappears when $B_{an} \to 0$. The anisotropy of antiferromagnetic crystals then has a large effect for B_{ex} tends to be very much larger than B_{an}.

We have discussed antiferromagnets composed of two sublattices, for these systems have been investigated more extensively than systems with larger numbers of sublattices. As we have said the exchange field in an

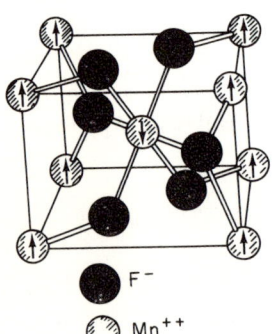

FIG. 7.54. Crystal structure of the antiferromagnet MnF_2, with arrows showing the directions of spin orientation of the manganese atoms.

[89a] J. Van Kranendonk and J. H Van Vleck, *Rev. Mod. Phys.* **30**, 1-23 (1958).

TABLE 7.9. Resonance Data on Antiferromagnetic Crystals at $T \simeq 0°K$ [a]

Material	Frequency	Wavelength	Néel temperature (°K)	Reference
FeF_2	52.7 cm^{-1}	190 μm	78.4	Ohlmann and Tinkham[90]
MnO	27.6 cm^{-1}	362 μm	120	Keffer et al.[89]
NiO	36.5 cm^{-1}	274 μm	523	Sievers and Tinkham[91]
NiF_2	31.1 cm^{-1}	322 μm	73.2	Richards[92]
α-$CoSO_4$	20.6, 25.4, and 35.8 cm^{-1}		12	Silvera et al.[93]
MnF_2	261.4 GHz	1.15 mm	67.7	Johnson and Nethercot[94,94a]
UO_2	17.8 cm^{-1}			Sievers[87]
CoF_2	28.5 and 36 cm^{-1}	351 and 278 μm	37.7	Richards[92]
$KNiF_3$	48.7 cm^{-1}	205 μm		Richards[92]
$CuCl_2 \cdot 2H_2O$	17.5 GHz	1.7 cm	4.3	Morrish[78] and Johnson and Nethercot[94]
Cr_2O_3	158 GHz	1.9 mm	307.5	Morrish[78]
$MnTiO_3$	150 GHz			
$MnCO_3$	123.2 GHz	2.44 mm	31.6	Richards[95]

[a] After Morrish.[78]

antiferromagnet is generally large and the resonance given by Eq. (7.189) tends to be at millimeter and submillimeter wavelengths, as the results recorded in Table 7.9 show. For example, the MnF_2 crystal of Fig. 7.54 is a two-sublattice antiferromagnet in which the anisotropy field is 8.8 kG and the exchange field is 556 kG[94,94a]; its resonance frequency is 8.7 cm^{-1}

[90] R. C. Ohlmann and M. Tinkham, *Phys. Rev.* **123**, 425–434 (1961).
[91] A. J. Sievers and M. Tinkham, *Phys. Rev.* **129**, 1566–1571 (1963).
[92] P. L. Richards, *J. Appl. Phys.* **34**, 1237–1238 (1963).
[93] I. F. Silvera, J. H. M. Thornley and M. Tinkham, *Phys. Rev. A* **136**, 695–710 (1964).
[94] F. M. Johnson and A. H. Nethercot, *Phys. Rev.* **114**, 705–716 (1959).
[94a] D. Bloor and D. H. Martin, *Proc. Phys. Soc. London* **78**, 774–776 (1961).
[95] P. L. Richards, *J. Appl. Phys.* **35**, 850 (1964).

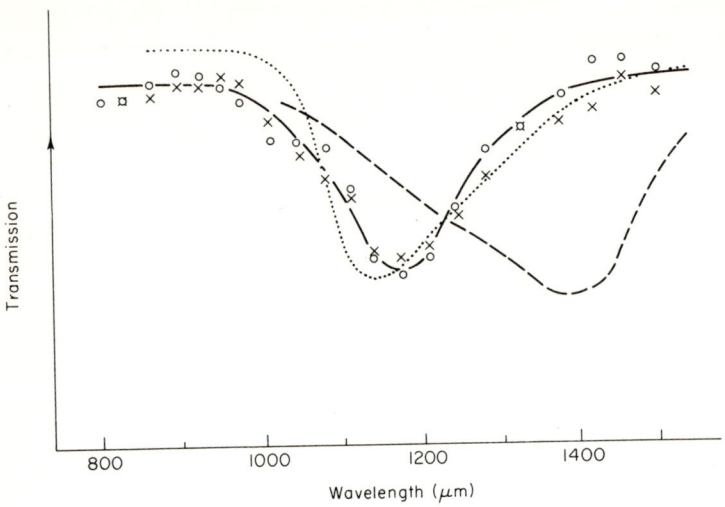

Fig. 7.55. Transmission of MnF$_2$ showing antiferromagnetic resonance at three temperatures. Dotted curve is 5.5°K, solid curve is 16°K, and dashed curve is 42.5°K. (After Bloor and Martin[94a]. Reproduced by permission of the Institute of Physics and the Physical Society.)

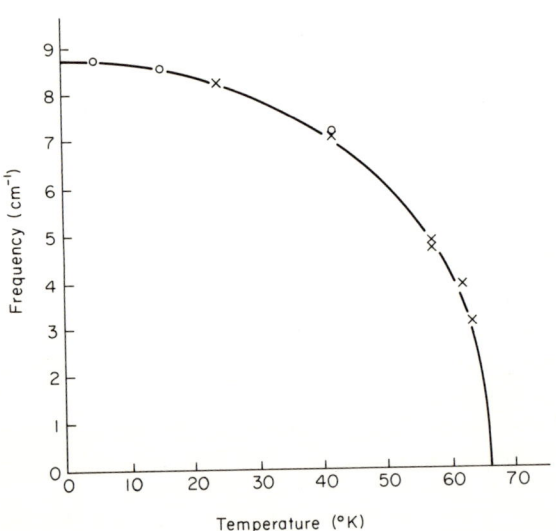

Fig. 7.56. Temperature dependence of the antiferromagnetic resonance frequency of MnF$_2$. Circles are infrared, and crosses are microwave measurements (after Bloor and Martin[94a]).

at absolute zero (see Figs. 7.55 and 7.56). FeF_2 is again a two-sublattice system; its anisotropy field has the exceptionally high value 200 kG, and the effective internal field is some 500 kG, which means a very short resonance wavelength, ~ 200 μm.[90] This antiferromagnetic resonance and also its temperature dependence are shown in Figs. 7.57 and 7.58. The anisotropy energy of this crystal can be written in the form

$$W_{an} = -(K/2)(\cos^2\theta_1 + \cos^2\theta_2),$$

where θ_1 and θ_2 are the angles between the two sublattice magnetizations and the easy (C) axis, and K is the anisotropy constant. Ohlmann and Tinkham[90] have inferred the value of K at absolute zero to be 1.1×10^{18} erg cm^{-3} (40 cm^{-1} per atom). The magnetic ordering of the arrays of

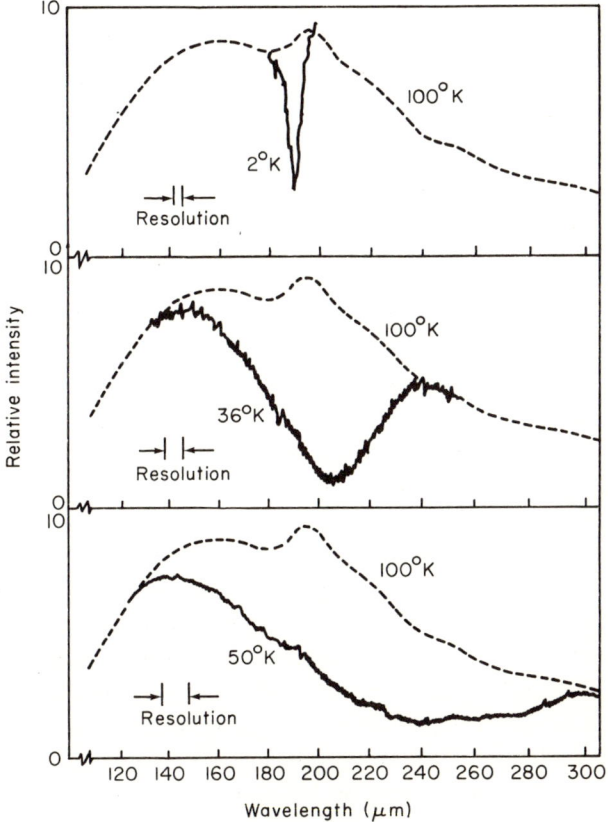

FIG. 7.57. Far-infrared spectrum of FeF_2 showing the antiferromagnetic resonance line and its temperature dependence (after Ohlmann and Tinkham[90]).

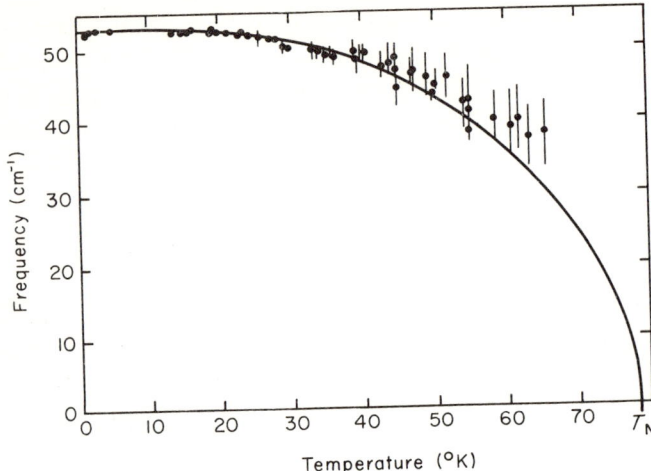

FIG. 7.58. Temperature dependence of the antiferromagnetic resonance line of FeF_2. The Néel temperature is 78.4°K (after Ohlmann and Tinkham[90]).

antiparallel spins is, of course, temperature dependent and finally disappears above the Néel temperature $T_N = 78.4$°K (see Fig. 7.57).

The double degeneracy of the two-sublattice modes noted earlier, can be split into two distinct frequencies by the application of a magnetic field. Such a splitting has been observed by Heller et al.[96] in Cr_2O_3 (see Fig. 7.59); as the external field is increased two modes with opposite senses of circular polarization separate linearly with B_0, with the mode of positive circular polarization decreasing in frequency. By using very high magnetic fields one can shift this antiferromagnetic mode into the range of standard millimeter wavelength experimentation. With very high pulsed magnetic fields, Foner[97] has used this approach to study Cr_2O_3 and MnF_2 at the standard frequencies 35 and 70 GHz.

The relaxation times in antiferromagnetic crystals are usually quite short, and the resonant line widths tend to be large, usually a few hundred to several thousand gauss. When the resonance frequency is not sufficiently greater than $1/\tau$, it may be necessary to apply a large external field in order to distinguish or resolve lines. Low temperatures and far-infrared radiation can be important to the experimentalist. In Table 7.10 the temperature dependence of the antiferromagnetic resonance frequency and line width of MnF_2 as measured by Johnson and Nethercot[94] are recorded.

[96] G. S. Heller, J. J. Stickler and J. B. Thaxter, *J. Appl. Phys.* **32**, 3075 (1961).
[97] S. Foner, *J. Phys. Radium* **20**, 336 (1959); *Phys. Rev.* **107**, 683 (1957).

7.8. MAGNETIC RESONANCES

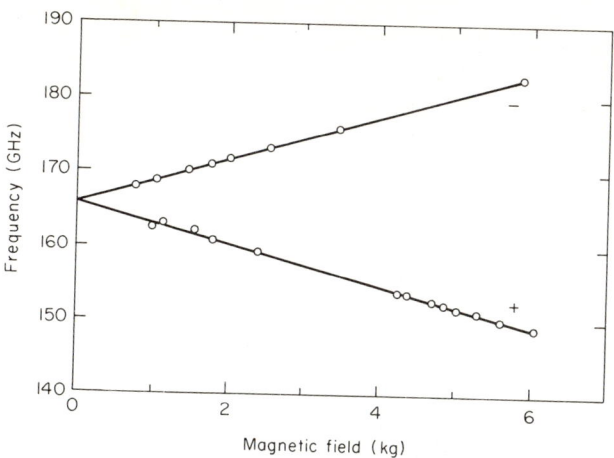

Fig. 7.59. Splitting of degenerate levels of the antiferromagnetic crystal Cr_2O_3 by a magnetic field applied along the easy (C) axis of the crystal. The temperature of the crystal is 77°K, and its Néel temperature 307.5°K (after Heller et al.[96]).

Table 7.10. Temperature, Frequency, and Linewidth for Observed Antiferromagnetic Resonance of MnF_2 with Zero Applied Magnetic Field[a]

Temperature (°K)	Frequency (GHz)	Approximate linewidth (GHz)
63.6	96	10.6
62.3	117	9.5
57.5	142	8.5
57.5	146	8.4
41.8	212.7	2.0
24.7	247.2	1.4

[a] After Johnson and Nethercot.[94]

7.8.4. Ferrimagnetic Resonance

A ferrimagnetic crystal is composed of oppositely directed spins which are either unequal in magnitude or different in number, and therefore exhibits a residual magnetization. The prototype of ferrimagnetic materials, the ferrites, have chemical form XFe_2O_3 (where X may be Mn, Ni, Fe, etc.), and spinel crystal structure.[52] They are commonly used to construct wave-handling devices (isolators, polarizers, etc.) at microwave frequencies. Certain perovskite-type oxides, garnets, and other structures are also ferrimagnetic,[98] and some of them have high internal fields leading to resonances in the far-infrared spectrum.

A simple ferrimagnet may have but two sublattices of opposing but unequal spins and may be represented diagramatically by vector arrays as in Fig. 7.60. Like the spin motion in an antiferromagnet, the precessional behavior of a ferrimagnet can be derived from the classical or quantum-mechanical equations of motion, the analysis being only slightly

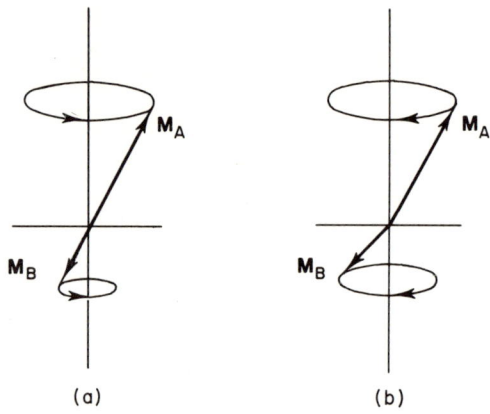

FIG. 7.60. Representation of the oppositely directed, but unequal, magnetization vectors of a two-sublattice ferrimagnet. In (a) M_A and M_B are collinear and act as a single vector $M_A - M_B$ uninfluenced by exchange. This resonance is a low-frequency mode precessing about an applied field in the same direction as a ferromagnet. In (b) the magnetization vectors are not collinear, and the exchange interaction exerts a torque to precess the spin in a high-frequency mode with the opposite sense to the low-frequency mode.

more general than that in Section 7.8.3.[78] As we must expect with two types of magnetic ions per unit cell, the dispersion curves consist of two branches. In the lower uniform precessional mode, the sublattice magnetizations are collinear and precess clockwise when viewed in the direction of the applied magnetic field, while in the higher frequency mode the sublattice magnetization vectors have unequal precessional angles and precess in the counterclockwise sense. This latter mode is called the *ferrimagnetic exchange resonance* mode, for the torque acting on the spins is largely of exchange origin; the low-frequency mode is called the "ferri-

[98] S. Chikazumi, "Physics of Magnetism." Wiley, New York, 1964.

magnetic resonance" or, sometimes, "ferrimagnetic resonance," because it involves a net magnetization $M_A - M_B$ with precessional motion quite like that in a ferromagnet.[78]

It can be shown[78] that the ferrimagnetic mode frequency can be written in the form

$$\omega = -\gamma_{\text{eff}}(B_0 + B_{\text{eff}}),$$

where B_{eff} and γ_{eff} are such that ω tends to be in the microwave region.[75,78] The ferrimagnetic exchange resonance frequency as calculated by Kaplan and Kittel can be written[75]

$$\omega \simeq \lambda(|M_A|\gamma_B - |M_B|\gamma_A), \tag{7.192}$$

where λ is the Weiss constant (or "exchange coupling parameter"), γ_A and γ_B are the gyromagnetic ratios of the magnetic ions on lattices A and B, and M_A and M_B are the magnetizations of the two lattices which are in opposite directions when the spin system is at rest. As it involves exchange interaction, the frequency given by Eq. (7.192) tends to be in the high-frequency region, 10^{11}–10^{12} Hz.

Most ferrimagnets are more complicated than the simple two sublattice system. There may be three[99] or more sublattices, and more than two magnetic ions per unit magnetic cell, with the result that many more than two branches may occur in the dispersion spectrum. In some of the modes of precessional motion the spins will be aligned, so that exchange interactions make no contributions to their energies, and the branches have low frequencies in the spin wave dispersion limit $k = 0$. In the motions associated with other branches, the spins are misaligned and exchange contributions result in high limiting frequencies at $k = 0$ which may be in the far-infrared or high-frequency microwave spectrum.

High-frequency modes may be excited by radiation with frequency equal to the intercept of the ω-axis at $k = 0$, provided the wavelength of the spin wave and the wavelength of the radiation are equal. If the phase velocities and hence wavelengths are different, there will be zero energy transfer. Quantum mechanically, the conditions for effective interaction are

$$(\mathbf{k})_{\text{spin wave}} = (\mathbf{k})_{\text{radiation}} \quad \text{and} \quad (\omega)_{\text{spin wave}} = (\omega)_{\text{radiation}}, \tag{7.193}$$

the former being the principle of conservation of momentum ($\hbar\mathbf{k}$) and

---

[99] M. Tinkham, *Phys. Rev.* **124**, 311–320 (1961).

the latter the principle of conservation of energy ($\hbar\omega$) for conversion of a photon to a magnon, in the absence of phonons or other quanta of energy.

Ferrimagnetic resonance in rare earth iron garnets (general formula $3R_2O_3 \cdot 5Fe_2O_3$, where R is a trivalent rare earth) have been extensively studied by means of far-infrared spectroscopy. Sievers and Tinkham[100] have observed the Kaplan–Kittel exchange resonance [Eq. (7.192)] in various garnets. This mode brings into play both the rare earth and the iron sublattice magnetizations and these, as we see from Eq. (7.192) determine the mode frequency. The spins on a sublattice precess in phase all with about the same amplitude so it is only the net sublattice moment that enters into the mode and accordingly the measurement is one of macroscopic magnetic properties.

In the garnets, the iron ions are very strongly coupled below the Curie point ($\sim 550°K$), so that the ordering is very complete at room temperature and below. The rare earth ions exchange–couple to the iron ions and tend to align themselves against the nett moment of the ferric sublattice. This ordering of the rare earth ions is significant only below about $30°K$, for the iron–rare earth coupling is relatively weak. Weaker still is the coupling between rare earth ions.

Transitions wherein an individual rare earth spin reverses in the exchange field produced by the ferric sublattice have also been observed in garnets. Such single ion transitions are split by the crystalline electric field and the iron exchange field, and through this splitting one can measure the fields acting on the ion and also its g-factor. Microscopic magnetic properties are thus accessible to single ion resonances. An example is the work of Sievers and Tinkham[100] who have observed the angular dependence and Zeeman shift of the exchange–split Kramers doublet of Yb^{3+} in YbIG and have obtained measures of the anisotropic g-value and the anisotropy of the exchange field. For the principal crystallographic directions, they have found effective fields of 0.62×10^5, 1.8×10^5, and 1.7×10^5 G, and g-values of 3.9, 3.2, and 4.1, respectively.

Exchange resonance frequencies in yitterbium, erbium, samarium, and holmium iron garnets at $2°K$ are recorded in Table 7.11. The sensitivity of the rare earth lattice magnetization to temperature means that the exchange resonance frequency will be temperature dependent, and may be identified on this basis. This feature contrasts with the behavior of

---

[100] A. J. Sievers and M. Tinkham, *Phys. Rev.* **129**, 1995–2004 (1963); also *Phys. Rev.* **124**, 321–325 (1961).

7.8. MAGNETIC RESONANCES

TABLE 7.11. Ferrimagnetic Exchange Resonance Frequencies of Rare Earth Iron Garnets at 2°K

Material	Frequency (cm^{-1})	Wavelength (μm)
YbIG	14.0	716
ErIG	10.0	1000
SmIG	33.5	298
HoIG	38.5	260

single ion excitations, for these, being determined essentially by the ferric lattice magnetization, are relatively insensitive to temperature.

The infrared transmission through a sample of pressed YbIG powder at $T = 2°K$ is shown in Fig. 7.61. In examining the far-infrared data on this ferrimagnet, and extrapolating to 0°K, Richards[92] has assigned the resonances at 23.2 and 27.0 cm^{-1} to single Yb ions, and a line at 26.3 cm^{-1} to a Yb–Yb combination mode. At 0°K the exchange resonance occurs at 13.65 cm^{-1}. Richards has also found at a resonance frequency of 3.00 cm^{-1} absorption due to the resultant spin of the ferric and rare earth ions precessing around the easy axis of the crystal. The resonant frequency of this mode gives a direct measure of the strength of the anisotropy favoring the easy direction of magnetization.

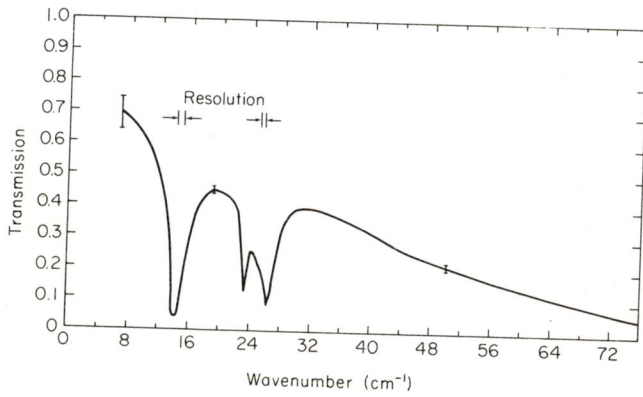

FIG. 7.61. Transmission spectrum of a YbIG pressed powder sample of thickness 1.47 mm at a temperature of 2°K (sample thickness, 1.47 mm; $T = 2°K$) (after Sievers and Tinkham[100]).

7.9. Raman Effect

Although most of the resonances discussed in this chapter, and in previous chapters, are generally studied by direct far-infrared spectroscopy, many of them can be observed through the modulating effect they produce when they scatter light. When a beam of monochromatic light is scattered in a gas, liquid, or solid the oscillators comprising the material may reflect some of the radiation without change of frequency, or they may undergo quantum transitions and either extract discrete amounts of energy from the beam or add energy to it. When the light is unchanged in frequency we have *Rayleigh scattering*, when it is changed by transitions involving the absorption or emission of quanta it is *Raman scattering*.

Raman radiation carries detailed information about the quantum states of the scatterer for it enables the measurement of fundamental and overtone frequencies, as well as frequencies that are combinations of normal mode frequencies. It has been applied particularly to the study of molecular vibrations, and in many ways has been a supplementary technique to those of direct infrared emission and absorption spectroscopy. However, whereas the latter method measures the three components of the vector polarization, we shall see that the Raman spectrum is determined by, and can measure, the nine components of the tensor polarizability. Moreover, it may happen, especially with symmetrical molecules, that some lines that are absent in the infrared spectrum are present in the Raman spectrum, and vice versa. This is, in fact, a general rule for molecules with a center of symmetry, e.g., CO_2.

Traditionally the technique of Raman spectroscopy has been based on the use of the incoherent emission spectrum of mercury, and notably the strong Hg e-line at 4358 Å. But since the advent of the laser the techniques have been refined, and lines from ruby, He–Ne, argon ion, and CO_2 lasers have been scattered from a variety of oscillators, including vibrations of molecules and crystal lattices, oscillations of plasmas, spin waves in magnetic materials, and electron cyclotron resonance in semiconductors.

Two Raman processes involving vibrational states are illustrated in Fig. 7.62. One is called *Stokes scattering*; it involves the absorption of a laser photon by the oscillator in its ground vibrational state and the creation of a photon with frequency ω_S which is lower than the laser frequency ω_l. Conservation of energy requires also that a vibrational quantum be created with frequency ω_v such that the Stokes frequency is $\omega_S = \omega_l - \omega_v$. The second process is *anti-Stokes scattering*. Here the

7.9. RAMAN EFFECT

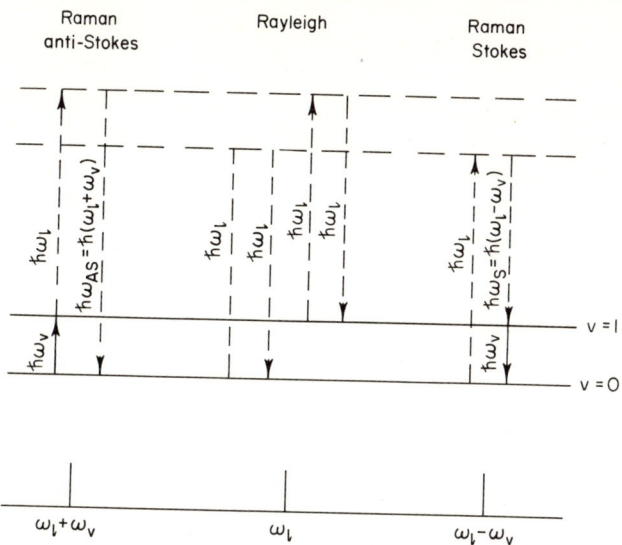

FIG. 7.62. Energy level diagram representing the quantum transitions of Raman and Rayleigh scattering. The solid arrows represent molecular vibrational transitions, and the broken arrows virtual transitions. The spectral lines shown at the bottom indicate an anti-Stokes line where the incident laser photon $\hbar\omega_1$ results in emission of a photon of energy $\hbar(\omega_1 + \omega_v)$, a Rayleigh line where the incident and scattered photons have equal energy, and a Stokes line with photon energy $\hbar(\omega_1 - \omega_v)$. ω_v is the vibrational frequency.

laser quantum $\hbar\omega_1$ raises the oscillator from its first vibrational state, while a photon $\hbar\omega_{AS}$ is emitted. A vibrational quantum $\hbar\omega_v$ must also be absorbed to provide the anti-Stokes photon energy $\hbar\omega_{AS} = \hbar(\omega_1 + \omega_v)$. Because the anti-Stokes process involves those oscillators that are excited above the ground state, the anti-Stokes lines are weaker than the Stokes lines by a factor $\exp -\hbar\omega_v/kT$. In general, a number of Raman lines appear and their displacements from the Rayleigh line give the energy differences in the molecule.

The origin of the Raman effect in the scattering of light can be seen by considering the change of polarizability associated with a normal mode of vibration of a molecule. In general, the potential energy of a molecule is not simply a quadratic function of the nuclear separation, but a function containing higher power terms. The variation of polarizability with the normal coordinate q_i can be written as a Taylor series expansion

$$\alpha = \alpha_0 + (\partial\alpha/\partial q_i)_0 q_i + (\partial^2\alpha/\partial q_i \partial q_j)_0 q_i q_j + \cdots, \quad (7.194)$$

where α_0 is the polarizability when the nuclei are in their equilibrium positions. To first approximation we can neglect terms higher than first order. From what we have said in Section 7.2.1 we can write $q_i = Q_i \times \cos \omega_v t$, and we can see immediately from Eq. (7.194) that variation of the polarizability at the vibrational frequency requires the derivative $(\partial \alpha / \partial q_i)_0$ to be different from zero. If we assume an electric field $E = 2E_0 \times \cos \omega_1 t$ incident on the vibrator from a laser, we can write the induced electric dipole moment as

$$\mu_i = \alpha E = 2\left[\alpha_0 + \left(\frac{\partial \alpha}{\partial q_i}\right)_0 Q_i \cos \omega_v t\right] E_0 \cos \omega_1 t$$

$$= 2\alpha_0 E_0 \cos \omega_1 t + E_0 \left(\frac{\partial \alpha}{\partial q_i}\right)_0 Q_i [\cos(\omega_1 + \omega_v)t + \cos(\omega_1 - \omega_v)t]. \quad (7.195)$$

The three terms on the right describe, respectively, the Rayleigh scattering and the Stokes and anti-Stokes lines of the Raman spectrum of the vibrator. These transitions obey the selection rule $\Delta v = 0, \pm 1$. If one includes higher terms of the Taylor series expansion Eq. (7.194), higher addition and combination frequencies appear in the induced dipole moment and (since the scattered intensity is proportional to the square of this moment) in the Raman scattered radiation. When $(\partial^2 \alpha / \partial q_i^2)_0 \neq 0$ the Raman spectrum selection rule includes $\Delta v = \pm 2$ transitions.

The symmetrical linear molecule CO_2 can illustrate the role of the derivative $\partial \alpha / \partial q_i$ in determining the Raman activity. The normal mode ν_1 involves the vibrational stretching of the C—O bonds symmetrically on either side of the oxygen (see Fig. 7.63). By symmetry, there is zero dipole moment and the mode is infrared inactive. However, as illustrated in curve 1 of Fig. 7.64, the polarizability is larger than the equilibrium value in one half period of vibration and smaller in the other half, so that $(\partial \alpha / \partial q_i)_0$ is nonzero and the mode is Raman active. On the other hand, the bond bending mode ν_2 and the unsymmetrical stretching mode ν_3 both give rise to changing dipole moments and are infrared active. The

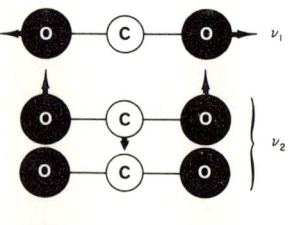
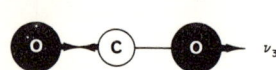

FIG. 7.63. The normal modes of vibration of the CO_2 molecule.

7.9. RAMAN EFFECT

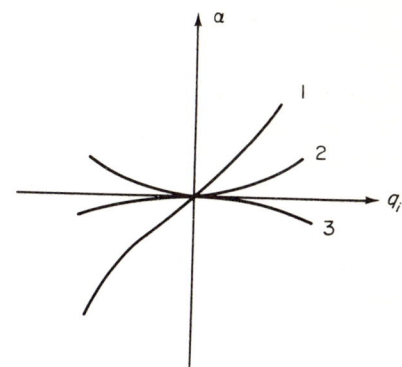

FIG. 7.64. Curves illustrating the variation of the polarizability α of a molecule such as CO_2 with the normal coordinate q_i. Curve 1 shows the behavior for the ν_1-mode, where α is larger on one side of the equilibrium position than on the other and $(\partial\alpha/\partial q_i)_0 \neq 0$. Curves 2 and 3 describe the ν_2-mode and ν_3-mode of CO_2 with $(\partial\alpha/\partial q_i)_0 = 0$.

polarizability as a function of the normal coordinates of the ν_2 and ν_3 modes are, however, both symmetrical about the equilibrium positions. Hence, for these two modes

$$(\partial\alpha/\partial q_i)_0 = 0,$$

and both are Raman inactive.

The property of Raman spectroscopy demonstrated for the ν_1-mode of CO_2, namely the occurrence of Raman lines in the absence of a permanent electric dipole moment, has special value for homopolar molecules such as N_2, O_2, and H_2. By symmetry, these molecules have zero electric dipole moment so that transitions cannot be excited directly by an oscillating electric field. They do, however, exhibit polarizabilities which change with the vibrational and rotational motion, and which modulate the scattered light with the Raman spectra of the allowed transitions.

In general, the polarizability of a molecule is anisotropic, and if we ignore rotation we may write the induced dipole moment in tensor form

$$\begin{bmatrix} \mu_x \\ \mu_y \\ \mu_z \end{bmatrix} = \begin{bmatrix} \alpha_{xx} & \alpha_{xy} & \alpha_{xz} \\ \alpha_{yx} & \alpha_{yy} & \alpha_{yz} \\ \alpha_{zx} & \alpha_{zy} & \alpha_{zz} \end{bmatrix} \begin{bmatrix} E_x \\ E_y \\ E_z \end{bmatrix}. \qquad (7.196)$$

The polarizability is a symmetric matrix and as such it can be diagonalized. The three diagonal elements α_1, α_2, and α_3 then determine a polarizability ellipsoid with orientation set by the symmetry axes of the molecule.

Except for a spherical molecule the three elements α_1, α_2, and α_3 of the diagonalized matrix are not all equal, and when the molecule rotates it modulates the scattered radiation and imprints on it the rotational Raman spectrum. For a linear molecule the allowed transitions are given by the selection rule $\Delta J = 0, \pm 2$. Raman lines shifted by twice the rotational frequency are to be expected classically because the rotation of the molecule through an angle of π clearly results in a polarizability indistinguishable from the original value. For the particular case of a spherical molecule, the polarizability does not change with rotation and only the Rayleigh line occurs. For the rotation of other molecules, and for combined molecular vibrational–rotational motion, transitions with $\Delta J = 0, \pm 1, \pm 2$, may be allowed.[16]

In the case of crystalline solids the unit cells are all oriented in the same manner and fixed with respect to the laboratory frame of reference,

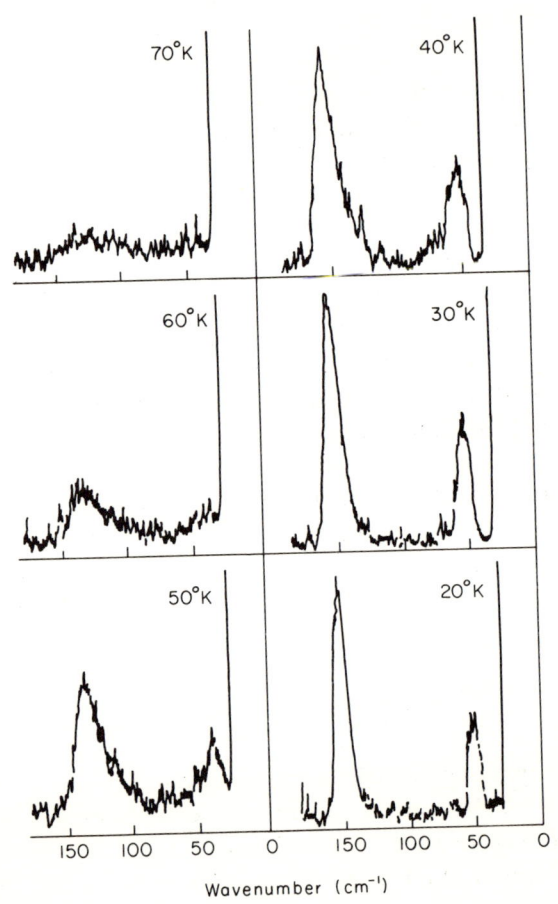

FIG. 7.65. Frequency shift of the Stokes scattered light from FeF_2 at various temperatures. Observation is via the component α_{zy} of the polarizability tensor, and the lines near 52 and 154 cm^{-1} are due to photons scattered by one and two magnons, respectively (after Fleury et al.[102]).

and under these circumstances Raman scattering can be a very effective probe. A tensor of the form Eq. (7.196) describes the polarizability. In this case one can determine all elements of the tensor by suitably arranging the plane of polarization of the laser radiation and selecting the various components of polarization of the Raman radiation.

Experimental Raman scattering with lasers has been used to investigate many of the resonances discussed earlier in this volume, and an extensive series of papers describing a range of the work has been published.[100] Argon ion laser radiation has been Raman scattered from spin wave magnons in MnF_2 and FeF_2, from single and multiple phonons in ZnO, CdS, CdS_xSe_{1-x}, and other materials, and from many other mechanisms. Figure 7.65 illustrates the type of spectrum observed for the case of the antiferromagnet FeF_2.[101,102] Figure 7.66 shows typical phonon spectra for $xx + xy$ and xz polarizations of 4880 Å argon ion laser radia-

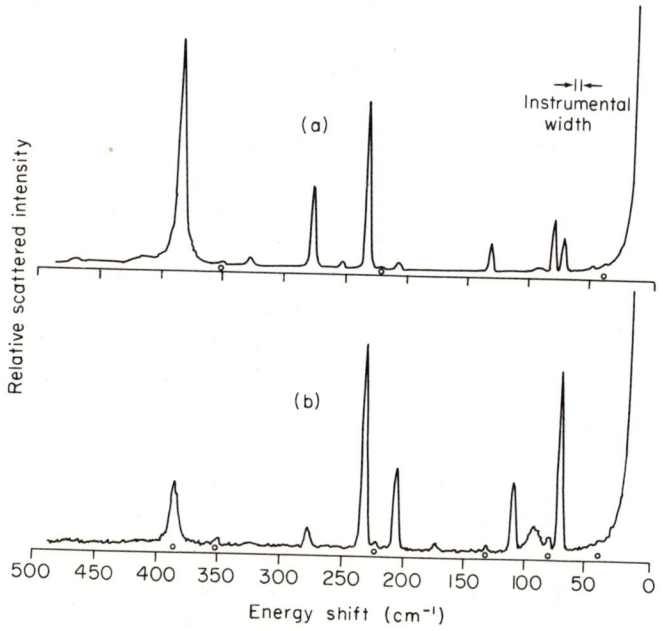

FIG. 7.66. Phonon spectrum of the crystal $CsMnF_3$ at 5°K observed by Raman scattering. (a) involves α_{xx} and α_{xy}, and (b) involves α_{xz} (after Chinn[103]).

[100] G. B. Wright, ed., "Light Scattering Spectra of Solids." Springer-Verlag, Berlin and New York, 1969.

[101] P. A. Fleury, S. P. S. Porto, L. E. Cheesman and H. J. Guggenheim, *Phys. Rev. Lett.* **17**, 84–87 (1966).

[103] S. R. Chinn, *Phys. Rev. B* **3**, 121–128 (1971).

tion scattered from $CsMnF_3$.[102] Ruby lasers are now in frequent use in the measurement of plasma oscillations in gaseous and solid-state plasmas, while the He–Ne laser has contributed to molecular vibrational Raman work.[100] High-power CO_2 laser beams have been scattered from electrons in cyclotron motion (Landau-level scattering) in semiconductors. As this laser has a frequency ($\lambda = 10.6$ μm) well below the argon ion, ruby, and He–Ne laser frequencies, one can probe the electron carriers without encountering complications from scattering by optical phonons in the semiconductor crystal, for the cross section for optical phonon scattering decreases as the fourth power of the laser frequency.[102]

Raman scattering of high power CO_2 laser radiation from Landau levels and electron spins in semiconductors has particular interest to infrared and far-infrared physics, for it enables the measurement of the physical parameters determining cyclotron resonance motion, and it has led to a new laser which is tunable in the infrared. The possibility of observing inelastic light scattering from mobile carriers in Landau levels of semiconductors was proposed by Wolff,[104] and the theory was extended by Yafet[105] to include spin-flip transition. In following up the proposal, Slusher et al. found that CO_2 laser radiation (frequency ω_1) scattered from n-type InSb in a magnetic field exhibited peaks at three distinct frequencies: a spin-flip mode at $\omega_1 - \beta g_{\text{eff}} B/\hbar$, and Landau lines at $\omega_1 - 2\omega_c$ and $\omega_1 - \omega_c$, $\omega_c = qB/m^*$ being the cyclotron frequency or Landau level spacing, and $\beta g_{\text{eff}} B/\hbar$ is a splitting of a Landau level. The Raman processes corresponding to $\Delta n = 2$ and $\Delta n = 1$ transitions between Landau levels scatter with a cross section $\sim 10^{-24}$ cm^{-2} sr^{-1} in a sample with an electron concentration of 5×10^{16} cm^{-3}, while the spin-flip Raman scattering between the Landau sublevels $s = \frac{1}{2}$ and $s = -\frac{1}{2}$ (that is, $\Delta s = 1$) has a cross section $\sim 10^{-23}$ cm^2 sr^{-1}, and is strong. Figure 7.67 shows the scattered light spectrum recorded by Slusher et al., and Fig. 7.68 plots the magnetic field dependences of the three transitions.

Raman observations of this type provide a versatile method of studying electronic parameters in semiconductors, such as the effective mass m^* and g-factor g_{eff}, the relaxation time τ, etc. Unlike direct absorption spectroscopy which gives similar physical information, with the Raman technique transitions widely different in frequency can be conveniently studied, since there is no requirement that the frequency of the incident radiation match that of the transition of interest.

[104] P. A. Wolff, *Phys. Rev. Lett.* **16**, 225–228 (1966).
[105] Y. Yafet, *Phys. Rev.* **152**, 858–862 (1966).

7.9. RAMAN EFFECT

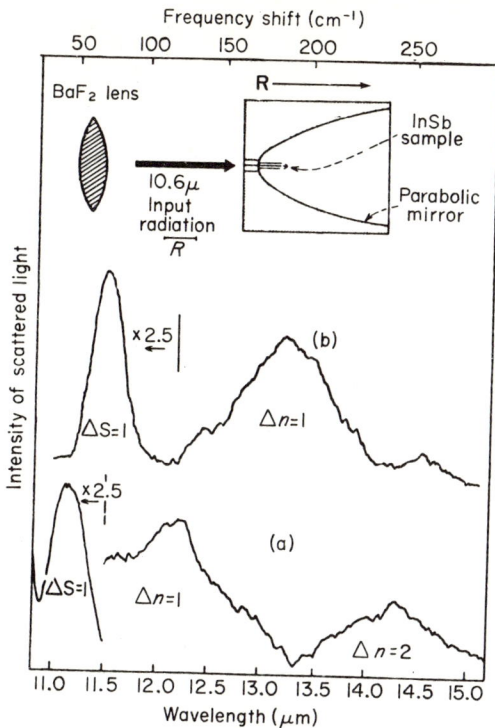

FIG. 7.67. Spectrum of Raman light scattered at right angles from n-type InSb with a carrier concentration of 5×10^{16} cm^{-3}, in a magnetic field. (a) is at 26.2 kG, and (b) at 36.7 kG. $\Delta n = 1, 2$ are Landau level transitions, and $\Delta S = 1$ is a spin-flip transition. The experimental arrangement is shown in the inset sketch (after Slusher et al.[106,106a])

In the scattering of a 10.6-μm laser beam from InAs Patel and Slusher[107] have observed modulation on the scattered radiation not only from single-particle Landau-level transitions, but also collective scattering effects from the carriers. Their observations show collective plasma modes caused by the electron gas of the semiconductor, and with the application of a magnetic field their observed frequency shift shows the upper hybrid plasma resonance to be the scatterer.

Generally, the scattered light is much weaker than the incident laser light because the reradiated photons at the Raman frequency are emitted

---

[106] R. E. Slusher, C. K. N. Patel and P. A. Fleury, *Phys. Rev. Lett.* **18**, 77–80 (1967).
[106a] C. K. N. Patel, in "Modern Optics" (Microwave Res. Inst. Symp. Ser.), Vol. XVII. Polytech. Press of the Polytech. Inst. Brooklyn, New York, 1967.
[107] C. K. N. Patel and R. E. Slusher, *Phys. Rev.* **167**, 413–415 (1968).

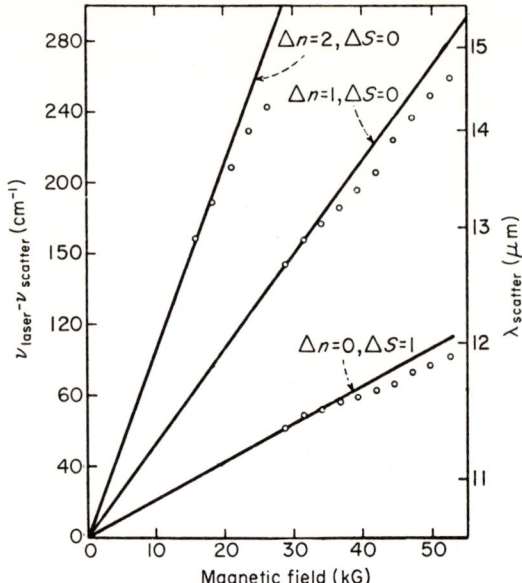

Fig. 7.68. Magnetic field dependences of the Raman scattered light for the spin-flip transition $\Delta S = 1$, and the Landau level transitions $\Delta n = 1$ and 2. The scatterer is n-type InSb with a carrier concentration of 5×10^{16} cm^{-3} (after Slusher et al.[106]).

randomly in all directions. There is no coherence of phase between the waves from the separate oscillators and the scattering efficiency is low, of the order of 10^{-6} scattered photons per incident photon. The spectral linewidth is that of the scattering oscillators.

With the use of very intense laser sources, such as the ruby laser, the densities of both laser and Raman photons can be at a level to cause stimulated transitions of the scattering oscillators. In this manner, intense coherent Raman emission can be generated with a photon conversion efficiency approaching unity.

Stimulated Raman scattering was investigated by Eckhardt et al. when they irradiated nitrobenzene with a ruby laser beam.[107a] It was distinguished from the spontaneous effect not only through its intensity but also by having a high degree of collimation, and a narrower linewidth. Soon after the discovery, stimulated Raman scattering was observed for many gases, liquids and solids. Table 7.12 gives a list of materials with shifts up to 1000 cm^{-1}.

[107a] G. Eckhardt, R. W. Hellwarth, F. J. McClung, S. E. Schwartz, D. Weiner and E. J. Woodbury, *Phys. Rev. Lett.* **9**, 455–457 (1962).

7.9. RAMAN EFFECT

TABLE 7.12. Substances Found to Exhibit Stimulated Raman Effect[a]

Liquids	Frequency shift (cm^{-1})	Solids	Frequency shift (cm^{-1})
Bromoform	222	Quartz	128
Tetrachloroethylene	447	Lithium niobate	152
Carbon tetrachloride	460	α Sulphur	216
Hexafluorobenzene	515	Lithium niobate	248
Bromoform	539	Quartz	466
Trichlorethylene	640	α Sulphur	470
Carbon disulfide	656	Lithium niobate	628
Chloroform	667	Calcium tungstate	911
Orthoxylene	730	Stilbene	997
α Dimethyl phenethylamine	836		
Dioxan	836		
Morpholine	841		
Thiophenol	916		
Nitro methane	927		
Deuterated benzene	944		
Cumene	990		
1:3 Dibromobenzene	990		
Benzene	992		
Pyridine	992		
Aniline	997		
Styrene	998		
m Toluidine	999		
Bromobenzene	1000		

[a] After Pantell and Puthoff.[108]

Stimulated Raman scattering has given a means of far-infrared wave generation by conversion from optical photons. Gelbwachs et al.[109] have generated stimulated Raman emission by scattering a Q-switched ruby laser beam from a transverse optical lattice mode of vibration in a LiNbO$_3$ crystal. The mode energy is partially mechanical and partially electromagnetic, and the quanta of such mixed modes are called *polaritons*. The mode is infrared active and the frequency of the scattered light can

---
[108] R. H. Pantell and H. E. Puthoff, "Fundamentals of Quantum Electronics." Wiley, New York, 1969.
[109] J. Gelbwachs, R. H. Pantell, H. E. Puthoff and J. M. Yarborough, *Appl. Phys. Lett.* **14**, 258–262 (1969).

be tuned by altering the angle of the scattered light relative to the direction of the incident beam. Yarborough et al.[110] have reported tunable optical emission from lithium niobate with an efficiency up to 70%. With a 1-MW ruby laser, they have obtained continuously tunable coherent emission from 50–200 μm at the polariton frequency, with pulsed power levels exceeding 10 W.

Spin-flip transitions in InSb have been studied in stimulated Raman scattering by Patel et al.[111,112] and by Allwood et al.[113] By irradiating the semiconductor with a CO_2 laser Patel et al., have generated magnetic-

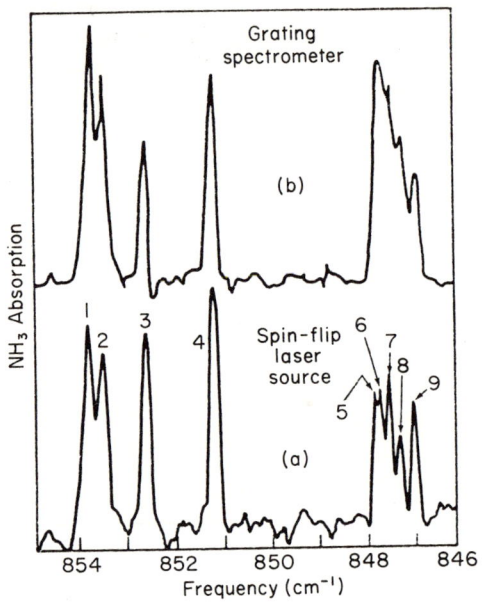

FIG. 7.69. (a) Absorption spectrum of NH_3 at a pressure of 10 torr taken with a spin-flip Raman-laser spectrometer. (b) The NH_3 spectrum taken with a high-resolution grating spectrometer. The Raman spectrometer is seen to resolve lines (e.g., 5 and 6) not resolvable by the grating instrument (after Patel et al.[112]).

[110] J. M. Yarborough, S. S. Sussman, H. E. Puthoff, R. H. Pantell and B. C. Johnson, *Appl. Phys. Lett.* **15**, 102–105 (1969).

[111] C. K. N. Patel and E. D. Shaw, *Phys. Rev. Lett.* **24**, 451 (1970).

[112] C. K. N. Patel, E. D. Shaw and R. J. Kerl, *Phys. Rev. Lett.* **25**, 8–11 (1970).

[113] R. L. Allwood, S. D. Devine, R. G. Mellish, S. D. Smith and R. A. Wood, *Proc. Phys. Soc. London (Solid State Phys.)* **3**, L186–L189 (1970); see also R. L. Allwood, R. B. Dennis, S. D. Smith, B. S. Wherrett and R. A. Wood, *Proc. Phys. Soc. (Solid State Phys.)*, **4**, L63–67 (1971).

field tuned emission extending from 10.9–13.0 μm with fields up to 100 kG. The extensive frequency interval covered here is facilitated by the large g-factor of electrons in InSb. The linewidth of the stimulated Raman radiation has been estimated as less than 0.03 cm^{-1}, and the linearity and resettability of tuning are better than one part in 3×10^4. With these characteristics the magneto-Raman laser gives superior resolution to that obtained with the best conventional grating spectrometers. This is clearly shown by comparison of the spectrum of NH_3 recorded by Patel with the Raman laser, and one measured with a high-resolution grating instrument. As the absorption spectra of Fig. 7.69 show, the transitions marked 5 and 6 are resolved by the Raman laser spectrometer but not by the grating spectrometer.

Allwood et al.[113] have recently reported double-Stokes spin-flip scattering of 10.6-μm Q-switched CO_2 laser radiation in InSb and have demonstrated that the potential tuning range extends from 9–16 μm for magnetic fields to 100 kG. At the double-Stokes frequency $\omega_1 - 2\beta g_{\text{eff}} B/\hbar$, they observed peak scattered power of 20 mW.

7.10. The Energy Gap in Superconductors

7.10.1. High Purity Superconductors

In this final section we give an outline of the role played by millimeter and submillimeter wave experimentation in establishing the existence of an energy gap between the superconducting state and the normal state of electrons in a superconductor. As we have seen in the chapter dealing with radiation detection, this is not the only application of far-infrared to superconductivity, for the Josephson effect has important application to this region; nor is it the only experimental method of showing the existence of an energy gap. We shall, however, restrict our discussion to far-infrared evidence for the energy gap. A review of evidence provided by a wider range of experimentation is given by Biondi et al.[114] Among the alternative methods, that based on electron tunneling across a superconductor–insulator–normal metal junction is particularly significant, for this, like the method of far-infrared spectroscopy, gives a most direct means of investigating the energy gap.[115]

---

[114] M. A. Biondi, A. T. Forrester, M. P. Garfunkel and C. B. Satterthwaite, *Rev. Mod. Phys.* **30**, 1109–1136 (1958).

[115] I. Giaever and K. Megerle, *Phys. Rev.* **122**, 1101–1111 (1961).

Prior to 1955 the measurements on superconductors at microwave frequencies up to 10^{10} Hz[114,116] had shown the absorption to approach zero as the temperature T approached zero. Yet at infrared and visible frequencies it was known that the superconducting state was little different from the normal state. Somewhere in the range 10^{10}–2×10^{13} Hz a change of behavior suggestive of a quantum jump across a forbidden energy gap appeared to occur, which far-infrared experimenters set out to locate and measure. Between 1956 and 1960 transmission and reflection experiments with millimeter and submillimeter radiation provided detailed measurements for photon energies spanning the gaps of several metals.

TABLE 7.13. Critical Temperature T_c of Some Superconducting Elements

Element	T_c (°K)
Titanium	0.39
Zirconium	0.55
Aluminium	1.17
Rhenium	1.7
Zinc	0.90
Vanadium	5.05
Tantalum	4.39
Niobium	9.2
Tin	3.73
Indium	3.40
Thallium	2.36
Lead	7.2
Mercury	4.15

On the theoretical side Bardeen, Cooper and Schrieffer (BCS) in 1957 published their theory of superconductivity predicting an energy gap $\sim 3.5 \, kT_c$ at absolute zero,[116a] where T_c is the critical temperature above which the metal becomes normal. Some values of T_c for pure elements are listed in Table 7.13. The BCS theory predicted also that the gap should become narrower as T increases towards T_c, and it showed the shape of the absorption edge for frequencies above the gap energy where the absorption should rapidly approach that of the normal state.

[116] A. B. Pippard, *Advan. Electron. Electron Phys.* **6**, 1–45 (1954).
[116a] J. Bardeen, L. N. Cooper and J. R. Schrieffer, *Phys. Rev.* **108**, 1175–1204 (1957).

7.10 THE ENERGY GAP IN SUPERCONDUCTORS

Microwave and far-infrared experiments may be based on the change of Q of a superconducting cavity, on the heating effect of the radiation (calorimetric method), or on the measurements of power absorption during transmission through a superconducting waveguide or thin film. Frequently the measurements are interpreted in terms of the surface impedance, $Z \equiv R + iX$—the ratio of electric and magnetic fields at the surface of the metal—to give R_s/R_n, the ratio of the surface resistance R_s in the superconducting state to that (R_n) in the normal state. The power absorption coefficient at reflection is

$$A = 1 - \left(\frac{1 - Z/Z_0}{1 + Z/Z_0}\right)^2,$$

where Z_0 is the impedance of free space, and $Z_0 \gg |Z|$. To a very close approximation $A = 4R/Z_0$, and the surface resistance is directly proportional to the fraction of the incident power that is absorbed.

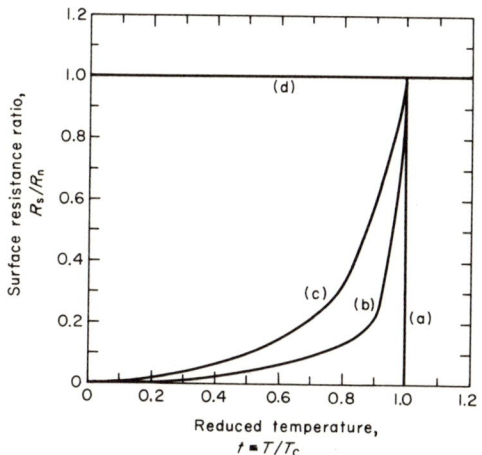

FIG. 7.70. Sketch illustrating the general behavior of the surface resistance ratio as a function of temperature T in the region of the critical temperature T_c. (a) is the dc curve, (b) and (c) are for microwave wavelengths, and (d) is for short infrared wavelengths (after Biondi et al.[114]).

For radiation at microwave and optical frequencies the surface resistance ratio R_s/R_n is expected to behave as in Fig. 7.70. While the dc curve (a) drops abruptly to zero at $T = T_c$, the short infrared or optical frequency curve (d) is unaffected by the superconducting phase transition. The effect expected with millimeter-wave photons is a low surface resis-

434   7. SPECTRA OF GASES, LIQUIDS, AND SOLIDS

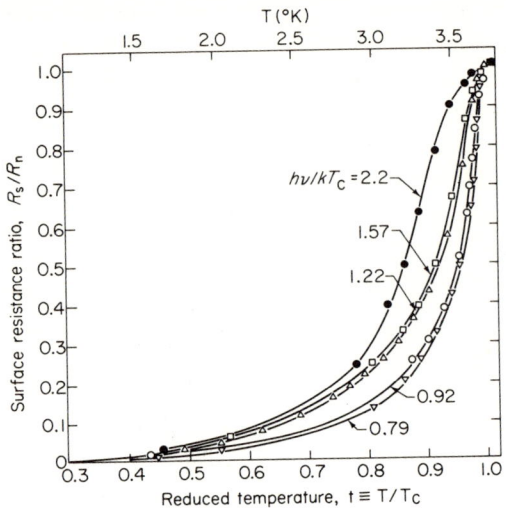

Fig. 7.71. Measurements of the surface resistance ratio R_s/R_n of superconducting tin as a function of temperature T in the region of the critical temperature T_c (after Biondi et al.[114]).

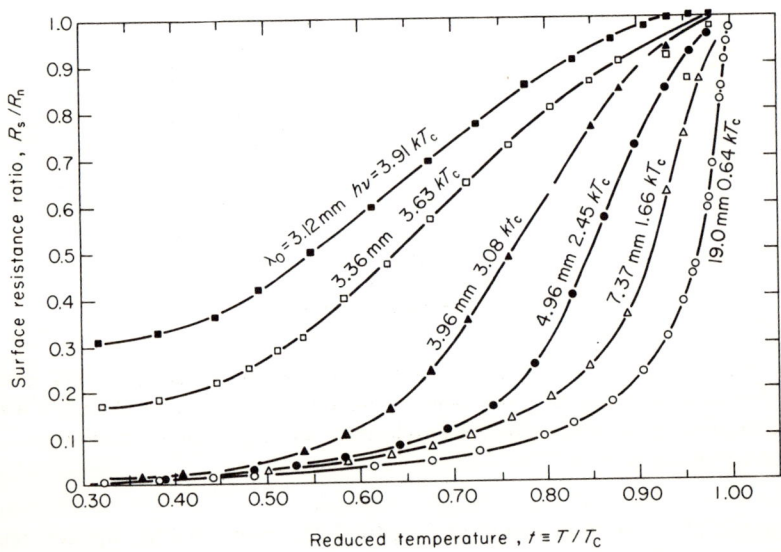

Fig. 7.72. Measured values of the surface resistance ratio of superconducting aluminium as a function of temperature T in the region of the critical temperature T_c. The wavelengths of the radiation and the corresponding photon energies in terms of kT_c are indicated on the curves (after Biondi and Garfunkel[117]).

7.10 THE ENERGY GAP IN SUPERCONDUCTORS

tance at low temperatures rising as the gap narrows and the photon energy starts to excite transitions across it. Plots in Fig. 7.71 of the surface resistance ratio as a function of temperature from $T = 0.3T_c$ to $T = T_c$ show the behavior observed at microwave frequencies. The data was obtained by Biondi et al.[116b] from measurements of the transmission of an extruded tin waveguide.

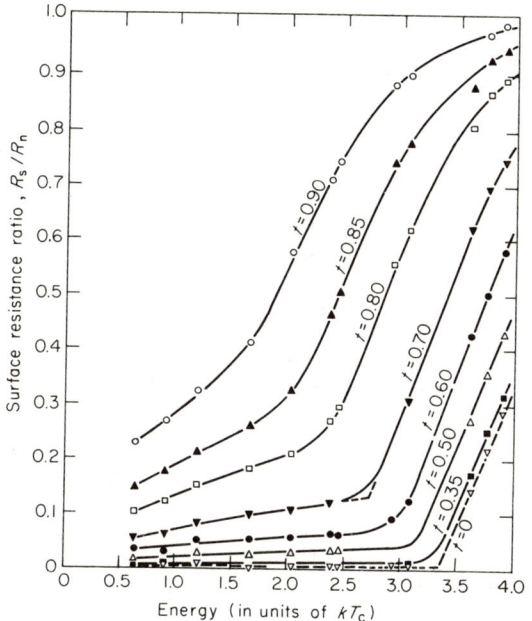

FIG. 7.73. Isotherms of surface resistance ratio of aluminium versus photon energy derived from absorption measurements at a number of millimeter wavelengths (after Biondi and Garfunkel[117]).

In the case of aluminium the energy gap is narrow ($T_c = 1.178°K$) and can be spanned by millimeter-wave photons. On this metal Biondi and Garfunkel[117] have measured the surface resistance with radiation wavelengths of 1.9 cm to 3.12 mm. Their results, plotted in Figs. 7.72 and 7.73, show the frequency and temperature dependence in the vicinity of the energy gap and, by extrapolation to $t = T/T_c = 0$, give an energy gap of $(3.25 \pm 0.1)kT_c$ at absolute zero. The width of the gap is shown

---

[116b] M. A. Biondi, A. T. Forrester and M. P. Garfunkel, *Phys. Rev.* **108**, 497–498 (1957).

[117] M. A. Biondi and M. P. Garfunkel, *Phys. Rev. Lett.* **2**, 143–145 (1959).

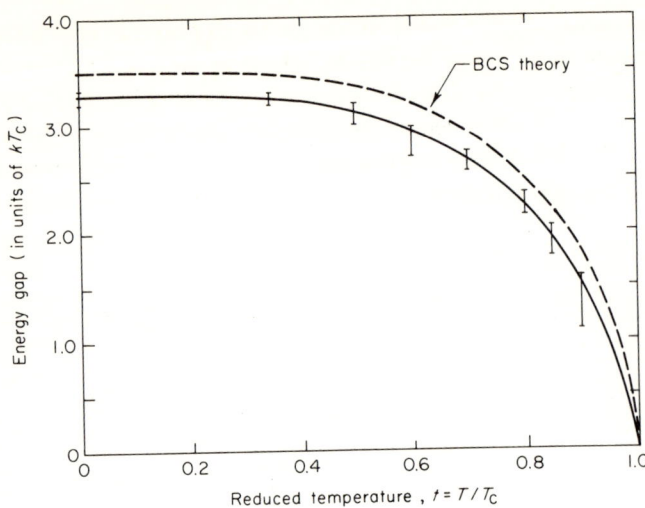

Fig. 7.74. Superconducting energy gap in aluminium as a function of temperature. The experimental results (solid curve) are derived from millimeter-wave surface resistance measurements, and the dashed curve is given by the theory of Bardeen, Cooper and Schrieffer. At absolute zero the measured gap is $(3.25 \pm 0.1)kT_c$ (after Biondi and Garfunkel[117]).

both experimentally and theoretically in Fig. 7.74 to decrease as T increases towards T_c.

Studies of the energy gap by direct far-infrared spectroscopic methods have been made by Tinkham and his associates[118–121] both by transmission through thin films and by surface reflection. The approach to the spectroscopy of superconductors through the radiation transmission properties of thin films lends itself to relatively straightforward interpretation. The films are thin compared to the superconducting penetration depth so that the field strength is constant throughout the film. In contrast, wave reflection measurements lead to a surface resistance which is essentially the resistance of a layer one skin-depth thick. Because of the anomalous skin effect, the skin depth is a complicated function of frequency and temperature which makes detailed interpretation difficult.

In working with films 20 Å thick Glover and Tinkham[119] have found that for $h\nu > 20\, kT_c$ the ratio of the transmission in the superconducting

[118] M. Tinkham, *Phys. Rev.* **104**, 845–846 (1956).
[119] R. E. Glover and M. Tinkham, *Phys. Rev.* **108**, 243–256 (1957).
[120] P. L. Richards and M. Tinkham, *Phys. Rev. Lett.* **1**, 318–320 (1958).
[121] D. M. Ginsberg and M. Tinkham, *Phys. Rev.* **118**, 990–1000 (1960).

7.10 THE ENERGY GAP IN SUPERCONDUCTORS

state to that in the normal state is of the order of unity, as is to be expected. As the photon energy decreased towards the energy gap the transmission rose substantially above that in the normal state (see Fig. 7.75). It is found that this behavior is predicted by the BCS theory. At about the gap frequency, and below it, the transmission decreases towards zero. The measurements of Glover and Tinkham yield a gap energy between 3 and 4 kT_c in lead and tin. This result and the shape of the absorption edge are in good agreement with the predictions of Bardeen, Cooper and Schieffer, and, in fact, have formed one of the crucial supports of the BCS theory.

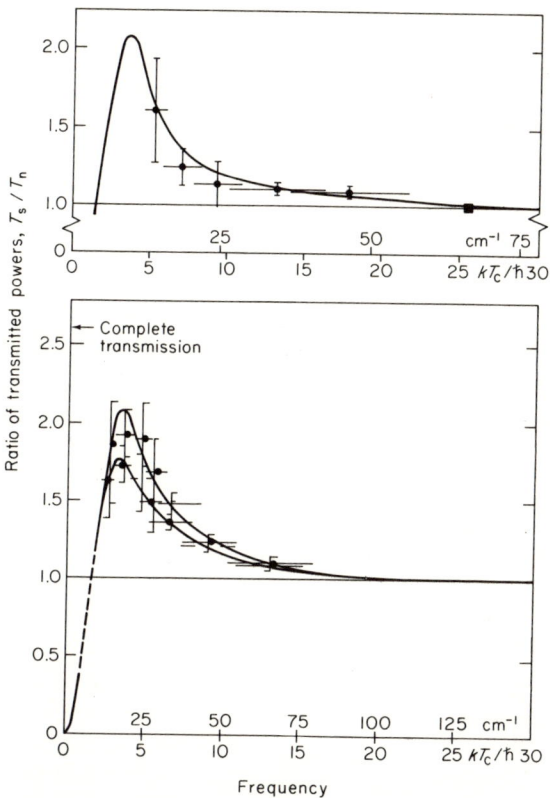

FIG. 7.75. Experimental ratios T_s/T_n of power transmitted through tin (top: $R_n = 176 \ \Omega/\Box$, $T/T_c = 0.58 \pm 0.10$) and lead (bottom: $R_n = 192 \ \Omega/\Box$, $T/T_c = 0.30 \pm 0.05$, $T/T_c = 0.67 \pm 0.03$) films in the superconducting and normal states, as a function of frequency. The curves show an initial rise which is consistent with BCS theory, and a fall towards the transmission of the normal state when the photon energy exceeds the energy gap which is between 3 and 4 kT_c (after Glover and Tinkham[119]).

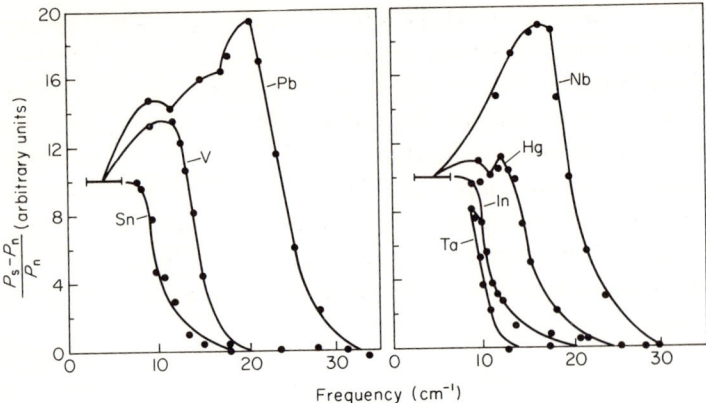

FIG. 7.76. Difference in the power reflected from the surfaces of seven metals in the superconducting state (P_s) and in the normal state (P_n) as a function of frequency. The curves have been normalized to a common value at the lowest frequency (T_c °K: lead 7.15, vanadium 5.1, tin 3.73, indium 3.39, mercury 4.15, tantalum 4.39, niobium 9) (after Richards and Tinkham[122]).

Richards and Tinkham[122] have measured the far-infrared reflection properties of a range of superconductors, and from them have derived the energy gap measurements listed in Table 7.14. Figure 7.76 shows measurements of $(P_s - P_n)/P_n$ from which this gap information is derived, where P_s and P_n are the reflected powers in the superconducting and normal states. The vertical scale of these curves is roughly propor-

TABLE 7.14. Values of the Energy Gap in Superconductors[a]

Superconductor	Energy gap
Indium	$(4.1 \pm 0.2) kT_c$
Tin	$(3.6 \pm 0.2) kT_c$
Mercury	$(4.6 \pm 0.2) kT_c$
Tantalum	$\leq 3.0 \, kT_c$
Vanadium	$(3.4 \pm 0.2) kT_c$
Lead	$(4.1 \pm 0.2) kT_c$
Niobium	$(2.8 \pm 0.3) kT_c$

[a] As measured by Richards and Tinkham.[122]

[122] P. L. Richards and M. Tinkham, *Phys. Rev.* **119**, 575–590 (1960).

tional to $R_n - R_s$, the difference between the surface resistances of the superconducting and normal states.[121] For high-frequency photons, the superconductor behaves like a normal metal, and $R_s = R_n$. Below the energy gap, R_s is small, and the frequency dependence is that of the surface resistance in the normal state. The losses increase roughly as the square root of the frequency as the photon energy approaches the gap width.

7.10.2. Superconductors Containing Magnetic Impurities

The measurements described so far have been concerned with high purity (~99.99%) metals, and the results have been in essential agreement with the BCS theory. Anderson[123] has generalized the BCS theory and extended it to superconductors containing nonmagnetic impurities, and experimental studies have confirmed that the presence of such impurities only slightly influences the behavior of the superconductor. This is not the case when magnetic impurities are present in the superconductor. Drastic influences arise which affect both the energy gap and the critical temperature.

Several experiments[124-126] have shown that the critical temperature and the energy gap both decrease with the increase in magnetic-impurity concentration. Indeed, the energy gap can close completely with impurity levels of only a few percent. Abrikosov and Gorkov[127] have predicted such behavior theoretically. Electron tunneling measurements have been a prime source of experimental information on the effects of impurities, but they have been open to the objection that the results might be artifacts caused by the tunnel junction itself. Far-infrared methods also display the properties of these materials quite directly, but they have the advantage that they measure the properties of the magnetically impure superconductors not at a junction but rather over an extended layer, and are therefore free from the possible objection to the tunneling experiments.

Experimental measurements of the ratio of the real parts of the conductivity of thin films of lead in superconducting and normal states are given in Fig. 7.77. Two types of impurity atoms are used. In Fig. 7.77(b) the rare earth gadolinium is introduced in increasing concentrations up

---

[123] P. W. Anderson, *Phys. Rev.* **112**, 1900–1916 (1958).
[124] N. Barth, *Z. Phys.* **148**, 646 (1952).
[125] M. A. Woolf and F. Reif, *Phys. Rev. A* **137**, 557 (1965).
[126] G. J. Dick and F. Reif, *Phys. Rev.* **181**, 774–783 (1969).
[127] A. A. Abrikosov and L. P. Gorkov, *Sov. Phys. JETP* **12**, 1243 (1961).

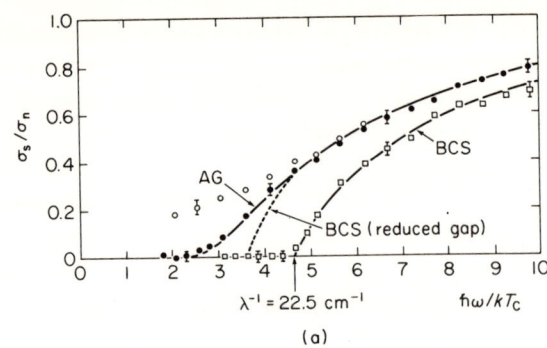

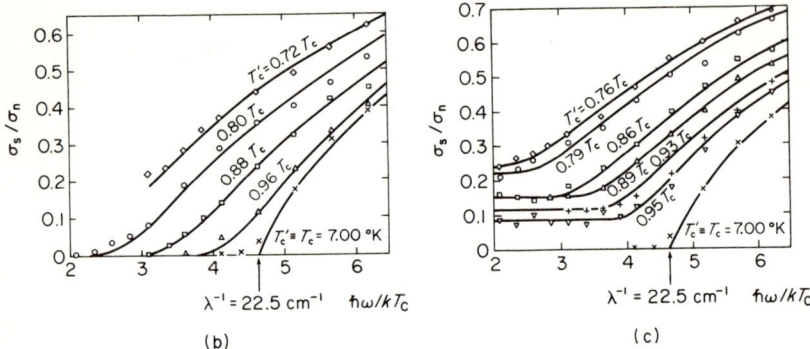

FIG. 7.77. (a) The ratio σ_s/σ_n of the real part of the conductivity of a lead film in its superconducting state (σ_s) to that in its normal state (σ_n), as a function of frequency. For pure lead, the experiments fit BCS theory with the energy gap at 22.5 cm^{-1}. For Gd impurities, Abrikosov–Gorkov (AG) theory fits the observations, both showing a reduction in gap width by the magnetic impurities. Mn impurities have a pronounced effect on the gap properties. As indicated, the impurity levels are such as to reduce the critical temperature T_c to 0.80 of the pure lead value (T_c) in the case of Gd, and to $0.79 T_c$ in the case of Mn. (☐ pure Pb, $T_c' = T_c = 7.00°K$; ● Pb + Gd impurities, $T_c' = 0.80 T_c$; ○ Pb + Mn impurities, $T_c' = 0.79 T_c$.) (b), (c) The frequency dependence of σ_s/σ_n for various concentrations of Gd and Mn impurity atoms, respectively. The solid curve are based on the theory of Abrikosov and Gorkov, modified in the case of Mn to give best fit (after Dick and Reif[126]).

to a level which reduces the critical temperature to $T_c' = 0.72 T_c$, that is, to 0.72 of the critical temperature of pure lead. The impurity is coupled to the conduction electrons of the superconductor via the exchange interaction. As the spins are due to the 4f electrons in Gd, and these are screened by surrounding electrons, the spin-dependent coupling to the conduction electrons is weak. This case is found to be in satisfactory

7.10. THE ENERGY GAP IN SUPERCONDUCTORS

agreement with the theory of Abrikosov and Gorkov, but to diverge from BCS theory. We see from Fig. 7.77(a) and (b) that the energy gap (which occurs at 22.5 cm^{-1} in the pure metal) decreases as the impurity level is raised.

In the case of manganese impurities the spin is associated with unscreened $3d$ electrons which couple strongly to the conduction electrons. As Fig. 7.77(a) and (c) show, the Mn impurity brings about a pronounced effect on the energy gap properties, and the observed behavior diverges from both BCS and Abrikosov–Gorkov predictions.

AUTHOR INDEX

Numbers in parentheses are footnote numbers. They are inserted to indicate that the reference to an author's work is cited with a footnote number and his name does not appear on that page.

A

Abrikosov, A. A., 439
Adams, E. M., 182
Adams, N. I., 17
Ade, P. A. R., 101
Alexander, F. B., 67
Alishouse, J. C., 15, 16
Alkemade, C. T. J., 221
Allen, C., 171, 184, 184(28)
Allis, W. P., 266, 278
Allwood, R. L., 430, 431
Ambegaokar, V., 75
Anderson, P. W., 439
Anderson, R. S., 391
Arams, F., 170, 171, 184
Aström, E., 273
Ayres, W. P., 27
Azbel', M. Ya., 229, 244

B

Baker, J. M., 391
Baldwin, D. E., 298
Ballantyne, J. M., 381
Baratoff, A., 75
Bardeen, J., 432
Barker, A. S., 380, 381, 382(66)
Barr, E. S., 5
Barth, N., 439
Bassov, N. G., 12
Beck, A. H. W., 27, 31, 41, 200
Becker, R. C., 27, 28(34)
Becklake, E. J., 165

Bedard, F. D., 333, 334(22, 30), 335, 336(22)
Bekefi, G., 266, 290, 292, 298(3), 299, 300
Bell, R. J., 101
Beloglasov, V. I., 45
Benjamin, B., 335
Berman, P. R., 29
Bernard, M. G. A., 66
Bernstein, I. B., 260
Bers, A., 266, 278(2)
Besson, J. M., 67, 68, 69
Beynon, J. D. E., 286
Biondi, M. A., 431, 432(114), 433, 434, 435, 436
Blanc, J., 353, 354(38)
Blatt, J. M., 201
Bleaney, B., 391
Bled, J., 101, 126
Bleekrode, R., 48
Blevins, G. S., 335
Bloembergen, N., 391, 393
Blomfield, D. L. M., 126
Bloor, D., 100, 110(14), 170, 226(23), 312, 315, 319, 411, 412
Blum, F. A., 72
Bohm, D., 279
Boivin, A., 112, 113(32)
Born, M., 112, 132(30), 351, 355
Bott, I. B., 77, 80(128), 81(131), 260
Boyle, W. S., 169, 249, 250(29)
Brailsford, A. D., 249, 250(29)
Brannen, E., 28, 44(40)
Brecher, C., 353, 354
Bresson, A., 126
Bridges, T. J., 51

Brillouin, L., 242, 358, 368(48)
Broadskiy, I. R., 14
Brown, F. C., 357, 364
Brown, J., 112
Brown, M. A. C. S., 179, 185(45), 186(45), 187, 188
Brown, S. C., 290, 292, 298, 299
Brueck, S. R. J., 72
Buchsbaum, S. J., 254, 266, 278(2)
Budenstein, P. P., 21
Burkhardt, E. G., 51
Burns, G., 66
Burnstein, E., 247, 248
Burrus, C. A., 21, 23(24), 150, 163, 335
Butler, J. F., 67, 68(111, 113), 69(113)
Butternecke, H. J., 126, 152
Button, K., 171, 184(28)

C

Calawa, A. R., 67, 68(111, 113), 69(113)
Callen, H. B., 387
Caplan, P. J., 91
Cardona, M., 357
Carruthers, P., 236, 237(14)
Carson, J. R., 223
Chamberlain, J. E., 48, 286
Chanal, D., 354, 355, 356, 364, 366(43)
Chandrasekar, S., 191
Chang, T. Y., 51, 52
Chang, W. S. C., 93
Chantry, G. W., 101, 102
Chasmar, R. P., 91, 168, 172(20), 221(20), 223(20), 225(20)
Cheesman, L. E., 424(102), 425, 426(102)
Cheung, A. C., 318
Chikazumi, S., 416
Chinn, S. R., 425
Chynoweth, A. G., 19
Cleeton, C. E., 11, 150, 317
Cochran, W., 375, 378(60)
Coleman, P. D., 27, 28(34, 37), 47, 48, 49, 50(80)
Collins, G. B., 11, 40(5)
Connes, J., 140, 144
Cooley, J. P., 17
Cooper, J., 172, 173, 174(34)

Cooper, L. N., 432
Cowan, M., 153, 335
Cowley, R. A., 379, 380
Crocker, A., 12, 48, 54
Culshaw, W., 121, 123(39), 125, 126
Cupp, J. D., 208
Cupp, R. E., 165
Czerny, M., 137, 311

D

Dail, H. W., 249, 250
Daneu, V., 22
Danner, H., 374
Davydov, A. S., 235
Dean, T. J., 100, 110(14), 312, 315(7), 319(7)
Decamps, E., 354, 355, 356
Delcroix, J. L., 269
Dellis, A. N., 283, 284, 287
De Lucia, F., 165, 310, 311, 312(3), 334(3, 31), 335
De Maria, A. J., 64
Denenstein, A., 206
Dennis, R. B., 430
Dennison, D. M., 320, 321(18a)
de Ronde, F. C., 126, 152
Derr, V. E., 125, 156
Devine, S. D., 430, 431(113)
Devonshire, A. F., 375
Dexter, R. N., 244, 245
Dianov, E. M., 94
Dick, G. J., 439, 440
Dicke, R. H., 167, 233, 235(6)
Dimitrenko, I. M., 75
Dimmock, J. O., 178
Dingle, R. B., 229
Dirac, P. A. M., 386, 397
Dixon, R. N., 323
Dorfman, Ya. G., 229
Dousmanis, G. C., 334(27), 335, 337
Dreicer, H., 301
Dressel, H. O., 41
Dresselhaus, G., 128, 246
Drude, P., 94
Dufour, C., 141, 145(58)
Duraffourg, G., 66

E

Earl, W. H. F., 283, 284(12), 287(12)
Eck, R. E., 75
Eckhardt, G., 428
Eckhardt, W., 150, 163, 164(11)
Einstein, A., 46, 210
Eisberg, R. M., 235, 332
Esaki, L., 174
Evans, H. M., 101
Evenson, K. M., 208

F

Fan, H. Y., 243, 256, 257
Faries, D. W., 29
Fatuzzo, E., 375, 381(59), 382(59)
Faust, W. L., 47, 48
Favero, P. G., 334(24), 335, 336
Fellgett, P., 6, 141
Fermi, E., 107, 309
Feshbach, H., 357
Feynman, R. P., 201
Findlay, F. D., 48
Firestone, F. A., 167, 172(18)
Fleming, J. W., 101
Fleury, P. A., 424, 425, 426(102), 427, 428
Foner, S., 414
Forrester, A. T., 431, 432(114), 433(114), 434(114), 435
Fowler, R. H., 212
Fox, A. G., 51, 52(83)
Frayne, P. G., 53
Frenkel, L., 53
Fröhlich, H., 357
Froelich, H. R., 28, 44(40)
Froome, K. D., 23, 25
Fujita, S., 104, 135(20)
Fuller, R. W., 236, 237

G

Gallagher, J. J., 333, 334(22, 30), 335, 336(22)
Galt, J. K., 249, 250(29), 251
Garfunkel, M. P., 431, 432(114), 434(114), 435, 436
Garrett, G. C. B., 48
Gebbie, H. A., 48, 92, 101, 184, 247, 248(27), 286
Gehring, K. A., 29
Gelbwachs, J., 429
Gentle, K., 314
Genzel, L., 105, 106, 112, 149, 150, 163, 164(11), 370
George, A., 286
Giaever, I., 431
Ginsberg, D. M., 436
Ginsburg, N. G., 320, 321(18a)
Ginsburg, V. L., 269
Ginzton, E. L., 113, 127(36), 150
Glagolewa-Arkadiewa, A., 11
Glauber, R. J., 236
Glenn, W. H., Jr., 64
Glover, R. E., 436, 437
Golay, M. J. E., 167
Goodwin, D. W., 179, 185(44)
Gordon, J. P., 12, 45, 318
Gordy, W., 7, 12, 21, 150, 151, 153(68), 163, 165, 310, 311, 312(3), 317, 333, 334(3, 8, 24, 31, 32, 32a), 335(8), 336(24), 337(8), 338(23)
Gorelik, L. L., 171
Gorkov, L. P., 439
Goubau, G., 113
Gould, R. W., 314
Grimes, C. C., 206, 207, 208, 209, 210
Grishaev, I. A., 45
Gross, E. P., 279
Grove, A. S., 180
Guggenheim, H. J., 424(102), 425, 426(102)

H

Haas, C. W., 387
Hadni, A., 112, 123(33), 354, 355, 356, 364, 366, 394
Halford, R. S., 353, 354(38)
Hansen, W. W., 198
Harding, G. N., 293, 295
Harding, W. R., 92
Harman, T. C., 67
Harman, W. W., 38, 43(52)
Harris, S. M., 236, 237(13)

Harrison, R. I., 172
Harvey, A. F., 91, 113(6)
Hassler, J. C., 48, 49, 50(80)
Heald, M. A., 266, 269, 278, 279(4), 290(4)
Hebel, L. C., 254
Heitler, W., 89
Heller, G. S., 155, 414, 415
Hellwarth, R. W., 428
Helminger, P., 334(31), 335
Henisch, H. F., 19
Henley, E. M., 3, 236
Henry, P., 356, 364, 366(43)
Herman, F., 243
Herring, C., 405
Hermann, G. F., 314
Hertzberg, G., 317, 320(16), 420(16)
Hill, A. E., 63
Hill, J. C., 392, 393, 394
Hill, R. M., 26, 314, 334(29), 335
Hilsenrath, S., 255
Hilsum, C., 92
Hinckelmann, O. F., 122
Hinckley, E. D., 67
Hindin, H. J., 122
Hirshfield, J. L., 77, 80(127), 191, 260, 298
Hochberg, A. K., 54
Hocker, L. O., 22, 53
Houghton, J. T., 90, 91(3), 94(3), 217, 364, 368(50)
Howarth, L. E., 378, 381(63), 382(63), 383(63), 384(63)
Hsu, T. W., 263, 264
Huang, K., 355
Hubner, G., 48, 49, 50(80)

I

Ichiki, S. K., 314
Ingram, D. J. E., 390
Inoue, K., 29
Irisova, N. A., 94

J

Jache, A., 335
Jacquinot, P., 141, 145, 149(60)
Jain, A. L., 174
Javan, A., 22, 53

Jeffers, W. Q., 47
Jenkins, F. A., 133
Johnson, B. C., 72(117), 73, 430
Johnson, C. M., 333, 334(22, 30), 336, 336(22)
Johnson, F. M., 411, 415
Jona, F., 373, 374(55), 375(55), 383(55)
Jones, F. E., 91, 168, 172(20), 221(50), 223(20), 225(20)
Jones, G., 12
Jones, G. O., 100, 110(14), 312, 315(7), 319(7)
Jones, H., 111, 112(28)
Jones, R. C., 158, 213(2), 223(2)
Jones, R. H., 179, 185(44)
Joos, G., 212
Josephson, B. D., 75, 204

K

Kamper, R. A., 76
Kaner, E. A., 229
Kapitza, S. P., 44
Kaplan, D. E., 314
Kaplan, H., 405, 407(89), 411(89)
Karp, A., 37
Kaufman, I., 28
Keffer, F., 405, 407, 411
Kelley, P. L., 67
Kennard, E. H., 160
Kerl, R. J., 70, 71(115), 430
Keyes, R. J., 67
Keyes, R. W., 182
Khanna, B. N., 382, 383(70)
Kimmitt, M. F., 48, 99(19a), 103, 135(19a) 137, 139, 172, 179, 185(45), 186(45), 187, 188, 293, 294
Kinch, M. A., 180, 226(46)
King, G. W., 317, 320(15), 333(15)
King, W. C., 21
Kip, A. F., 128, 246
Kittel, C., 128, 158, 159(3), 243, 246, 369, 373(52), 374(52), 383(52), 391(52), 398(52), 405, 406, 409, 415(52)
Klein, J. A., 154
Kleinman, D. A., 378, 381(63), 382(63), 383(63), 384(63)
Kneubühl, F. K., 50(79, 80), 53, 61

Koenig, S. H., 172
Kogelnik, H., 53
Kolosov, V. I., 45
Kon, S., 54, 61
Krikorian, E., 353, 354(38)
Kroon, D. J., 122, 126(40), 127, 128
Kuipers, G. A., 312, 313
Kuypers, W., 28, 29, 32(42, 43)

L

Lambert, J. - P., 356, 364, 366(43)
Landau, L. D., 233
Landecker, K., 42, 44(59)
Langenberg, D. N., 75, 201, 206, 250, 251
Langley, S. P., 157
Langmuir, R. V., 115
Larsen, T., 105
Last, J. T., 381
Lax, B., 67, 240, 243, 244(18), 245(17, 22), 246(17), 250(17), 255, 260(17)
Lax, M., 178
Lebedew, P. N., 11
Lecomte, J., 6, 7(5b)
Lee, E., 172
Leighton, R. B., 201
Lengyel, B. A., 45, 46(65)
Levy, R. A., 387, 404, 406(76)
Lewis, W. B., 212
Li, T., 51, 52(83), 53
Lichtenberg, A. J., 298, 299(24)
Lide, D. R., Jr., 53
Lifshitz, E. M., 233, 244
Livingston, R., 335
Lobikov, E. A., 171
Lodge, O., 11
Loubser, J. H. N., 154
Louisell, W., 236
Low, F. J., 171
Lowenstein (also Loewenstein), E. V., 6, 140, 141(55)
Ludlow, J. H., 172, 173
Lyddane, R. H., 363, 368(49)

M

McCarthy, D. J., 383
McClung, F. J., 428
McCoy, C. T., 162
McDonald, D. G., 208
MacDonald, D. K., 222
McDowell, C. A., 53
Macfarlane., G. G., 222
McFarlane, R. A., 47, 48
McFee, J. H., 47
McKnight, J. S., 333
McKnight, R. V., 103, 104(19), 338(23)
McLean, T. P., 211
Maiman, T. H., 12
Maki, A. G., 53
Malein, A., 283, 284(12), 287(12)
Manabe, A., 96, 98(13a), 108(13a)
Manley, J. M., 18
Marcuse, D., 50, 310, 333(2)
Margenau, H., 269
Mariner, P. F., 117, 119(37a), 120(37a)
Martin, D. H., 100, 110(14), 135, 170, 184, 226(23), 312, 315(7), 319(7), 397, 410(86), 411, 412
Mathias, L. E. S., 12, 48, 54
Mavroides, J. G., 240, 243, 245(17), 246(17), 250(17), 260(17)
Mawer, P. A., 100, 110(14), 312, 315(7), 319(7)
Megerle, K., 431
Mellish, R. G., 430, 431(113)
Melngailis, I., 67, 69, 178
Meredith, R., 163, 164, 216
Merritt, F. R., 250, 251(32a)
Merz, W. J., 375, 381(59), 382(59)
Messenger, G. C., 162
Michelson, A. A., 140
Miesch, R. A., 165
Miller, R. C., 378, 381(63), 382(63), 383(63), 384(63,71), 385
Mirri, A. M., 334(24), 335, 336(24)
Mitchell, W. H., 173
Mitra, S. S., 353, 354(39), 355, 356, 364, 365
Mitsuishi, A., 96, 98, 102, 104, 107, 108, 109(25), 135(20)
Mizushima, M., 334(29), 335
Möller, K. D., 103, 104(19)
Montgomery, D. G., 26
Mooradian, A., 72
Moore, T. W., 250, 251

Morlot, G., 356, 364, 366(43)
Morrish, A. H., 391, 411, 416, 417(78)
Morse, P. M., 357
Moser, J. F., 50(79, 80), 61
Mott, N. F., 111, 112(28)
Motz, H., 42, 44(60), 45
Murai, A., 53
Myakota, V. I., 45

N

Nathan, M. I., 66
Nethercot, A. H., 154, 411, 415
Nicols, E. F., 6, 11
Nielsen, A. H., 312, 313(9), 320
Nieto, M. M., 236, 237(14)
Nunnink, H. J. C. A., 28, 32(41), 34

O

Ohl, R. S., 21
Ohlmann, R. C., 123, 411, 413, 414
Olsen, J. N., 286
Orbach, R., 392
Oster, L., 292

P

Packard, R. F., 165
Page, C. H., 18
Page, L., 17
Palik, E. D., 247, 258
Panofsky, W. K. H., 287, 296(16)
Pantell, R. H., 72(117), 73, 74, 429, 430
Papoular, R., 126
Parker, C. D., 178
Parker, W. H., 206
Parks, W. F., 101
Patel, C. K. N., 29, 47, 48, 70, 71, 427, 428, 430
Paul, W., 67, 68(113), 69(113)
Pauling, L., 185, 323
Payne, C. D., 165
Pearson, E. F., 334 (32, 32a), 335
Pepinsky, R., 373, 374(55), 375(55), 383(55)
Perry, C. H., 100, 110(14), 312, 315(7), 319(7), 382, 383(70)

Petritz, R. L., 221
Peyton, B., 184
Phelan, R. J., 67, 68(111), 69
Phillips, M., 287, 296(16)
Phillips, R. G., 171
Phillips, R. M., 41
Picus, G. S., 247, 248(27)
Pierce, J. R., 11, 34(7)
Pippard, A. B., 432
Plantinga, G. H., 41
Poehler, T. O., 54, 285
Porto, S. P. S., 424(102), 425, 426(102)
Potak, M. H. N., 17
Powell, M. J. D., 392
Prewer, B. E., 165
Prior, A. C., 293, 294(19)
Prokhorov, A. M., 12
Pryce, A. W., 92
Puthoff, H. E., 72(117), 73, 429, 430
Putley, E. H., 163, 172, 173, 174(29), 175, 177, 179(41), 183, 184, 211, 216, 221(29), 225

Q

Quist, T. M., 67

R

Rado, G. T., 387, 417(75)
Ramachandra Rao, D., 53
Ramo, S., 120, 220
Ramsey, W. Y., 15, 16
Randall, H. M., 6, 7(6), 167, 172(18), 318, 320, 321
Rank, D. M., 318
Ratcliffe, J. A., 266
Rediker, R. H., 65, 66, 67(104), 68(111, 113), 69(113)
Reif, F., 439, 440
Renk, K. F., 105, 106
Richards, P. L., 29, 123, 124, 128, 129, 137(43), 138, 140, 144, 146, 147, 148, 171, 206, 207(77), 208(77), 209(77), 210, 411, 419, 436, 438
Richtmyer, R. D., 198
Risley, A. S., 208
Roberts, R. W., 27
Roberts, V., 92, 293, 294(19), 295

AUTHOR INDEX

Robinson, L. C., 20, 24, 31, 35, 36(49), 37, 40, 55, 58, 59, 60(98), 61, 78, 188, 191, 192, 193(61), 196(57), 197(57), 199, 201, 259, 282
Robinson, T. S., 358
Robson, P. N., 263, 264
Rodgers, K. F., 169
Rohrbaugh, J. H., 17
Rollin, B. V., 179, 180(43), 226(46)
Rowe, H. E., 18
Rubens, H., 141
Rubin, L., 171, 184(28)
Rupprecht, G., 382, 383(70)

S

Sachs, R. G., 363, 368(49)
Salomonovic, A. E., 335, 338(26)
Sanchez, A., 22
Sanders, T. M., 335, 337(25)
Sands, M., 201
Sard, E., 184
Satterthwaite, C. B., 431, 432(114), 434(114)
Sauter, F., 260
Scalapino, D. J., 75, 201, 206
Schawlow, A. L., 151, 312, 317(10), 320(10), 333(10), 335, 336(10), 338(10), 344, 351(10)
Schiff, L. I., 4
Schmidt, P. H., 250, 251(32a)
Schneider, J., 77, 80, 260
Schottky, W., 222
Schrieffer, J. R., 201, 432
Schwartz, S. E., 428
Schwering, F., 113
Schwinger, J., 297
Seitz, F., 203
Sells, V., 28, 44(40)
Sesnic, S., 298, 299
Shapiro, S., 30, 206, 207(77), 208(77), 209(77), 210(77)
Sharpless, W. M., 161, 164
Shaw, E. D., 70, 71(115), 430
Shaw, N., 173, 184
Shen, Y. R., 29
Shimazu, M., 172
Shirane, G., 373, 374(55), 375(55), 383(55)
Shockley, W., 229, 243(3)
Sievers, A. J., 369, 370, 371, 372, 402(53a), 411, 418, 419, 425(100), 426(100)
Silver, A. H., 75, 76, 206
Silvera, I. F., 411
Silverman, B. D., 378
Sladek, R. J., 182, 183
Slaggie, E. L., 236, 237(13)
Slater, J. C., 35, 375
Slough, W., 48
Slusher, R. E., 427, 428
Small, J. G., 22
Smerd, S. F., 297(25), 300
Smith, D. F., 312, 313(9)
Smith, G. E., 128, 129, 254
Smith, R. A., 91, 168, 171, 172(20), 221(20), 223(20), 225, 227
Smith, S. D., 90, 91(3), 94(3), 217, 233, 364, 368(50), 430, 431(113)
Smith, W. V., 151, 153(68), 317
Smythe, W. R., 17
Sokoloff, D., 22
Solomon, S. S., 223
Sperling, G., 163, 164(11)
Spitzer, L., 269
Spitzer, W. G., 256, 257, 378, 381, 382(63), 383(63), 384(63, 71), 385
Stanevich, A. E., 14, 91, 103
Steel, W. H., 82, 86(1), 145(1)
Steffen, H., 50(79, 80), 53, 61
Steffen, J., 50(79,80)
Sterling, S. A., 208
Stetser, D. A., 64
Stewart, T. W. W., 28
Stickler, J. J., 155, 414, 415(96)
Stillman, G. E., 178
Stone, N. W. B., 48, 184
Stone, S. M., 41
Strauch, R. G., 165
Strauss, A. J., 67, 68(111)
Strimer, P., 356, 364, 366(43)
Strong, J., 134, 140, 148, 149, 172, 311
Strutt, M. J. O., 158
Stubbs, H. E., 171
Suhl, H., 387, 417(75)
Sullivan, T., 53
Sussman, S. S., 72(117), 73, 430
Sutherland, G. B. B. M., 172, 311, 317(6)

Suzaki, Y., 172
Svistunov, V. M., 75
Swan, C. B., 26
Szekeres, G., 192

T

Tait, G. D., 61
Takami, K., 172
Takatsuji, M., 172
Tannenwald, P. E., 178
Taub, J. J., 122
Taylor, B. N., 75, 201, 206
Tear, J. D., 6, 11
Teitler, S., 247, 258
Teller, E., 363, 368(49)
Templeton, I. M., 222
ter Haar, D., 233, 236
Tetenbaum, S. J., 26
Thaxter, J. B., 414, 415(96)
Theriault, J. P., 171
Thiessing, H. H., 91
Thirring, W., 3, 236
Thon, W., 45
Thornley, J. H. M., 394, 411
Thornton, D. D., 318
Tidman, D. A., 26
Tien, P. K., 47
Timofeev, V. N., 94
Tinkham, M., 9, 123, 137, 138, 380, 381, 382(66), 411, 413, 414, 418, 419, 425(100), 426(100), 436, 437, 438
Torrey, H. C., 163
Townes, C. H., 151, 154, 312, 317(10), 318, 320(10), 333(10), 334 (27, 28), 335, 336(10), 337(25), 338(10), 344, 352(10)
Trambarulo, R. F., 151, 153(68), 317
Tremblay, R., 112, 113(32)
Trivelpiece, A. W., 298, 299(24)
Trubnikov, B. A., 298
Tuma, O., 298, 299(24)
Turner, A. F., 137
Turner, R., 54, 285
Twiss, R. Q., 11, 77, 260, 300
Tyte, D. C., 62, 63

U

Ulrich, R. 105

V

Valkenburg, E. P., 125, 156
Vanasse, G. A., 140, 148, 149
van Iperen, B. B., 28, 32(41, 42), 34
Van Kranendonk, J., 410
Van Nieuwland, J. M., 122, 126(40), 127, 128
Van Tran, N., 29
Van Vleck, J. H., 91, 343, 397, 410
Van Vliet, K. M., 221
Vartanian, P. H., 27
von Happ, H., 163, 164
Von Hippel, A., 318, 351, 379

W

Wachtel, J. M., 77, 80(127), 191, 260
Wagoner, G., 246
Wallis, R. F., 247, 258
Wang, M., 170, 171, 184(28)
Wangsness, R. K., 110, 181, 215(47), 376
Ward, S., 283, 284(12), 287
Warner, F. L., 163, 164, 216
Warren, J. W., 53
Weber, L. R., 320, 321(18a)
Weber, R., 163, 164(11)
Webster, D. L., 27, 28
Wegrowe, J. G., 126
Weibel, G. E., 41
Weiner, D., 428
Weiss, A. A., 297(25), 300
Weisskopf, V. F., 343
Welch, W. J., 318
Wendling, H., 354, 355, 356
Werthamer, N. R., 206
Wharton, C. B., 266, 269, 278, 279(4), 290(4)
Wheeler, R. G., 392, 393, 394
Wherrett, B. S., 430, 431(113)
Whiffen, D. H., 312, 348
Whinnery, J., 120, 220
Whitbourn, L. B., 55, 58, 59, 60(98), 61, 201
White, H. E., 133
White, J. U., 110
Whitehurst, R. N., 45
Whitmer, C. A., 163
Wild, J. P., 297, 300

Wilkinson, G. R., 135
Williams, N. H., 11, 150, 317
Williams, R. A., 93
Wills, M. S., 48, 54
Wilson, E. B., 185, 323
Wingerson, R. C., 78, 191
Witteman, W. J., 48
Wittke, J. P., 45, 46(64), 233, 235(6)
Wolf, E., 112, 132(30), 351
Wolfe, C. M., 178
Wolff, P. A., 260, 426
Wood, R. A., 430, 431(113)
Wood, R. W., 134, 141
Woodbury, E. J., 428
Woolf, M. A., 439
Wright, D. A., 90
Wright, G. B., 255, 425
Wright, M. L., 122
Wright, N., 318

Y

Yafet, Y., 182, 405, 407(89), 411(89), 426
Yager, W. A., 249, 250
Yajima, T., 29
Yakinova, B. V., 45
Yamada, Y., 102, 104, 107, 108(25), 109(25), 135(20)
Yamanaka, M., 54, 61(97)
Yamamoto, J., 54, 61(97)
Yanson, I. K., 75
Yarborough, J. M., 72, 73, 429, 430
Yariv, A., 45, 367
Yaroslavsky, N. G., 14(13), 91, 103
Yata, K., 96, 98(13a), 108(13a)
Yoshinaga, H., 54, 96, 98(13a), 102, 104(18), 107, 108(13a, 25), 109(25), 135(20)

Z

Zeiger, H. J., 155, 244, 245(22), 255, 318
Zernicke, F., 29
Ziman, J. M., 250
Zimmerman, J. E., 76
Zucker, J., 172
Zwerdling, S., 171

SUBJECT INDEX

A

Absorption
  air, 91–93
  coefficient, 94, 342, 344, 345
  free electrons in cyclotron motion, 189, 190, 230–238, 259–263
  superconductors, 433–441
  water, 354–355
  water vapor, 91–93, 321
AgCl transmission, 95
Ammonia
  laser, 49–50
  molecule, 318–319
  spectrum, 318–319
Amplifiers, post detector, 225–226
Anisotropy energy, 400, 411, 413
Anisotropy field, 400, 411, 413
Antiferroelectric crystal ($PbZrO_3$), 383
Antiferromagnetic resonance, 407–415
  frequency, 410, 411
Antireflecting films, 96–98
Anti-Stokes scattering, 70, 420, 421
Apodization, 142–144
Arc discharge in mercury, 13–15
Arc harmonic generator, 23–26
Asymmetrical top, 304, 320–321
Atmospheric transmission, 91–93
Atmospheric windows, 91–93
Attenuation
  atmospheric, 91–93
  light pipes, 124
  oversize waveguides, 123
  plasma, 280
  standard waveguides, 118, 120, 121, 123
  superconducting films, 433–441
Azbel'–Kaner cyclotron resonance, 249–251

B

Backward wave oscillator, 36–38
Band mode, 368
Band-pass filters, 105–109
Bands
  parallel, 327
  perpendicular, 327
Barrier layer, 19, 20, 160
  capacitance, 164, 165
$BaTiO_3$, 373–375, 379, 381–383
Beam splitters, 146, 147, 149
Blackbody radiation, 13, 16, 288, 289
  from plasmas, 288, 289, 291–295
Black polyethylene, 99, 100, 101
Blazed grating, 134, 135, 138, 139
Bohr magneton, 386
Bohr radius of impurity atoms in semiconductors, 177
Bolometer detectors, 157, 168–172
Boltzmann distribution, 46
Boltzmann equation, 266
Boltzmann's constant, 2
Bremsstrahlung emission from plasmas, 287–295
Brewster angle polarizers, 104
Brillouin zone, 360
Bromine cyanide laser emission, 50

C

Carbon bolometer, 169, 170
Carbon dioxide
  laser, 61–64
  Raman scattering, 422, 423
  vibrational modes, 323, 324
Carcinotron, see Backward wave oscillator
Cavities, 124–126, 156
Cerenkov radiation, 28, 300

SUBJECT INDEX

Circular polarizer, 127, 128
CO_2, see Carbon dioxide
CoF_2 antiferromagnetic resonance, 411
Coherent detection, 227, 228
Coherent emission, 10, 18–81
Coherent state, 236, 237
Commutation equation, 3, 401
Commutator bracket, 4
Complex conductivity of plasma, 268, 271
Complex dielectric constant, 351–358
Complex effective dielectric constant of plasma, 271
Complex refractive indices of plasma, 272–279
Conductivity of plasma in rotating coordinates, 271
Conversion efficiency of diodes, 21–23
Coriolis coupling, 326, 327
Critical temperature for superconductivity, 76, 77, 432
Cr_2O_3 antiferromagnetic resonance, 411
Crystal detectors, see Diode detectors
Crystal field splitting, 388–394
Crystal lattice vibrations, 358–372
CsI transmission, 95
$CuCl_2 \cdot 2H_2O$ antiferromagnetic resonance, 411
Curie–Weiss law, 375
Current noise, 221, 222
Cutoff
  plasma, 275, 280
  waveguide, 116–119, 121
Cyclotron radiation, 259–264, 295–299
  harmonic, 297–299
  cadmium, 250, 251
  copper, 250, 251
  detection, 188, 201
  free electrons, 77–81, 259–263
  InSb, 128, 129, 187, 188
  maser, see Electron cyclotron maser
  semiconductors, 249–251
  spectrometer–detector, 188–201
Czerny–Turner monochromator, 84, 137

D

D^*, 224, 225
Depletion layer, see Barrier layer
Detectivity D, 224, 225
Detectivity "D star," see D^*
Detectors
  bolometer, 157, 168–172
  crystal diode, see Diode detectors
  cyclotron resonance, see Cyclotron resonance detection
  doped germanium, 171
  gallium arsenide, 178
  Golay cell, 167–168, 178
  InSb, 179, 188
  Josephson junction, 201–210
  photoconductive, 174–188
  pyroelectric, 172–174
Deuterium dioxide laser, 48, 49
Diamond, 96
Diatomic chain, 359–361
Dielectric constant, see also Complex dielectric constant
  dense media, 352
  dilute media, 352
  plasma, 271
  polymer, 97
Diffraction gratings, see Grating
Dimethylamine laser emission, 48–50, see also HCN laser
Diode detectors, 158–166
Diode harmonic generator, 18–23
Diode lasers, 64–69
Dipole moment
  electric, 310, 313, 314
  matrix element, 304, 341
Dirac notation, 304
Dispersion
  classical theory, 351–357
  diatomic chain, 359–363
  relations for dielectrics, 94, 350–355
  relations for temperate plasma, 273–278
  relations for warm plasma, 278–279
Distortional polarization, 347
D_2O laser, see Deuterium dioxide laser
Dominant mode in waveguide, see Principal mode in waveguide
Doppler linewidth, 345
Dysprosium ethyl sulphate, 391–393

E

Easy axes of magnetization, 400
Echelette grating, 134, 135
  filters, 138–140
Effective dielectric constant of plasma, 271
Effective mass, 241–243
Effective pathlength of a cavity, 156
Einstein coefficient of absorption, 341
Einstein transition probabilities, 46, 341
Electron bunching, 27, 28, 30–33, 35, 39, 42–44
Electron cyclotron
  maser, 77–81
  resonance, see Cyclotron resonance
  spectrometer–detector, see Cyclotron resonance spectrometer–detector
Electron tubes, 27–29, 30–45
Electronic angular momentum of molecules, 328–338
Electronic polarization, 347, 348
Ellipsoid of inertia, 303
Energy gap, see Superconductor energy gap
Energy transmittance of a monochromator, 136
Equation of radiation transfer, 290
Equivalent circuit of point-contact diode, 162
Equivalent noise resistance, 225, 226
Erbium ethyl sulphate, 394
ErIG ferrimagnetic exchange resonance, 419
Étendue, 82, 83, 136, 145
  advantage, 145
Exchange degeneracy, 396
Exchange energy, 398
Exchange field, 398–400
Exchange integral, 398
Exchange interaction, 395–400
Extraordinary wave in plasma, 275, 276
Extrinsic photoconductivity, see Impurity photoconductivity

F

Fabry–Perot
  filter, 105, 107
  resonator, 50, 51, 125, 126
  spectrometer, 156

Faraday rotation
  plasma, 286, 287
  semiconductors, 130, 257–259
FeF_2
  antiferromagnetic resonance, 411–414
  Raman scattering, 424, 425
Fellgett advantage, 141
Fermi–Dirac distribution, 64, 65, 243
Fermi level, 65, 159
Ferrimagnetic exchange, 416–419
Ferrimagnetic resonance, 416–419
Ferrite harmonic generator, 27
Ferroelectric crystals, 372–385
Ferromagnetic resonance, 394–407
Filters
  absorption and reflection, 102, 110
  band-pass, 105, 109
  echelette grating, 138, 140
  interference, 105, 107
  low-pass, 102–104, 109, 110
  metal grating, 105, 107
  replica grating, 103, 104, 139
  waveguide, 121
Finesse, 86, 87
Flicker noise, 224
F-number, 84, 137
Fourier transform spectroscopy, see Multiplex spectrometry
Free carrier
  absorption, 181, 182
  photoconductivity, 179–188
  reflectivity minimum, 255, 257
Froome harmonic generator, see Arc harmonic generator

G

Gallium arsenide diodes, 20, 165, 166
Gas lasers, 45–64
Gas mantle, 15, 16
Germanium
  bolometer, 171
  diodes, 163–165
  photoconductors, 178
  windows, 95, 96, 97
g-factor (or g-value), 386, 388, 389, 391, 392
Glass transmission, 94

Globar, 15, 16
Golay cell, 167–168
Grating
  equation, 130
  filter, 105–107, 139
  monochromator, 130–140
  polarizer, 105, 130
  polyethylene, 103, 104, 139
Grid
  filter, see Grating filter
  polarizer, see Grating polarizer
g-value, see g-factor

H

Hagen–Rubens relation, 112
Hard axes of magnetization, 400
Harmonic conversion efficiency, 18
  arc harmonic generator, 25, 26
  diodes, 21–23
  plasma, 26
Harmonic generation
  diodes, 18–23
  electron beams, 27–29, 44
  ferrites, 27
  Josephson junction, 30
  linear accelerators, 27, 28, 44
  microtron, 28, 44
  nonlinear reactors, 18–23
  nonlinear resistors, 18–23
  plasma arcs, 23–26
  plasmas, 26–30
Harmonic mixing, 21–23, 165
HCl, 89, 310–313
HCN
  laser, 48–61, 328
  maser, 50
  molecule, 310, 324, 325, 328
Heisenberg representation, 3
Helium laser emission, 48, 49
HF rotation spectrum, 312
HI spectrum, 312, 314, 337
High-pass waveguide filter, 121
HoIG ferrimagnetic exchange resonance, 419
H_2O
  laser, 48, 49
  spectrum, 91–93, 320, 321

Hot-body radiators, 13–16
Hund's cases (a) and (b), 332, 333
Hydrogen chloride, see HCl
Hydrogen cyanide, see HCN
Hydroxyl lines, 334, 335

I

ICN laser, 48, 50
Impurities in superconductors, 439–441
Impurity induced absorption in alkali halides, 370–371
Impurity induced lattice absorption, 368–372
Impurity photoconductivity, 175–179
Incoherent detection, 227
Incoherent sources, 10, 11, 13–18
Indium antimonide
  circular polarizer, 128, 129
  detectors, see Detectors, InSb
  junction laser, 67–69
  Landau level Raman scattering, 426–428
  spin-flip Raman scattering, 426–428
Indium arsenide
  junction laser, 67, 68
  Raman scattering, 427
InSb, see Indium antimonide
Instrumental function of interferometer, 143
Interference filters, 105–107
Interferogram, 140–143, 282–284
Interferometers for plasma measurements, 281–286
Interferometric spectrometers, 146–149, 156
Inversion doublet, 318, 319
Iodide cyanide laser, see ICN laser
Ionic polarization, 347, 348
Ionization energy of impurity atoms in semiconductors, 177
Iron fluoride, see FeF_2
Irtran, 95, 96
Isolator, 129, 130

J

Johnson noise, 219–221
Josephson effect, 74–77, 201–210

Josephson junction detector, 201–210
Josephson junction mixing, 30
Josephson junction generator, 74–77
Junction lasers, 67–69

K

KBr transmission, 95
Kirchhoff's radiation law, 290
Klystron, 28–34
KNiF$_3$ antiferromagnetic resonance, 411
Kramers degeneracy, 391–393, 418
Kramers–Kronig relations, 357–358
KRS transmission, 95

L

LA branch, see Longitudinal acoustic branch
Lambda doubling, 335, 336
Lamellar grating interferometer, 147–149
Landau level, 80, 185, 186, 188, 234, 236, 260, 261, 426–428
    Raman scattering, 426–428
Laser resonators, 50–53, 56, 64, 68, 71
Laser mixing
    diodes, 21, 22
    nonlinear crystals, 29
Lasers, see also Masers
    CO$_2$, 29, 52, 61–64
    gas, 45–64
    HCN, 48–61
    p-n junction, 64–69
    Raman, 70–74
Lattice vibrations, 358–372
Light gathering power, see Étendue
Light pipes, 123, 124
LiNbO$_3$ Raman laser, 72, 73, 429, 430
Linear accelerator, 27, 28, 44, 45
Linear chain of atoms, see Diatomic chain
Linear molecules, 303, 306–315
Line shapes, 232–235, 239, 240, 343, 344
Linewidth, 240, 343, 345
    of crystal modes, 370, 371
Lithium niobate, see LiNbO$_3$ Raman laser
LO branch, see Longitudinal optical branch
Local mode absorption, 368–370

Longitudinal acoustic branch, 361
Longitudinal optical branch, 361
Longitudinal plasma waves, 278, 279, 300, 301
Low-pass filters, see Filters
LST relation, see Lyddane–Sachs–Teller relation
l-type doubling, 325, 328
Lyddane–Sachs–Teller relation, 375, 376, 378

M

Magnetic resonances, 385–419
Magneto-Kerr effect, 259
Magnetooptical effects
    plasmas, 286, 287
    semiconductors, 257–259
Magnetoplasma effects, 249–259, 273–287
Magneto-Raman laser spectrometer, 430
Magnetron, 38–41
    harmonics, 29, 154
Magnon, 401
Manganese fluoride, 410–412, 414, 415
Masers, see also Lasers
    electron cyclotron, 77–81
    HCN, see HCN maser
Mass radiator, 11, 16–18
Matrix element
    electric dipole transitions, 304, 341
    harmonic oscillator transitions, 235
Maxwell–Boltzmann distribution, 46
Maxwell's equations, 2, 114, 267
Melinex, see Mylar (polyethylene terephthalate)
Mercury arc discharge lamp, see Arc discharge in mercury
Metallic reflection, 110–112
Metal–metal diode, 22
Metal–semiconductor diodes, 18–23, 158–166
Methyl alcohol laser emission, 49, 50
Methyl fluoride laser emission, 50
Michelson interferometer, 146–148
Microwave plasma interferometer, 281–283
Microwave spectrometers, 150–155
Mixing
    diodes, 21, 22, 163–165

SUBJECT INDEX

InSb, 29, 184
Josephson junction, 30
$MnCO_3$ antiferromagnetic resonance, 411
MnF_2 antiferromagnetic resonance, 411
MnO antiferromagnetic resonance, 411
$MnTiO_3$ antiferromagnetic resonance, 411
Modes
  acoustical, 361
  dominant in waveguides, 113, 117, 118
  optical, 361
  transverse, 362
Molecular lasers, see Gas lasers
Molecular rotation spectra, 306–321
Molecular vibrations, 321, 328
Monochromator, 130–140
Multiplex gain (or Fellgett advantage), 141
Multiplex spectrometry, 140–149
Mylar (polyethylene terephthalate), 97, 146
  beam splitter, 146–148

N

NaCl, 88, 90–95, 359, 370, 371
Narrow-band photoconductive detector, 185–188
Néel temperatures of antiferromagnetic crystals, 411
Negative absorption by electrons, 77–81, 259–263
Neon laser emission, 48, 49
NEP (noise equivalent power), 158
  bolometers, 169–171
  diode superheterodyne detectors, 164
  diode video detectors, 163
  germanium bolometer, 171
  Golay cell, 167
  InSb detector, 187
  Josephson junction, 208
  pyroelectric detector, 173
  superconducting bolometer, 170
Nernst glower, 15, 16
NH_3
  laser emission, 49, 50
  spectrum, 318, 319
NiF_2 antiferromagnetic resonance, 411
NiO antiferromagnetic resonance, 411

Nitrous oxide spectrum, 334, 337
Noise
  current, 211, 221, 222
  equivalent power, see NEP
  flicker, 211, 224
  Johnson, 211, 219–221
  photon; see Noise radiation
  radiation, 211, 217
  recombination, 211, 221
  shot, 211, 222–224
  temperature, 211, 217–219
Noise figure of superheterodyne receiver, 164, 165
N_2O laser, 49, 51, 52
Normal vibrations, 322, 325
Notation
  HCN laser states, 328
  molecular states with electronic angular momentum, 330–339

O

OCS
  lines, 313, 338
  molecule, 310, 313, 314
OH lines, 334, 335, 337
Optical constants, 96, 101, 102, 104, 365, 366
Optical extent, see Étendue
Ordinary wave in plasma, 255, 256, 273–275, 279
Orientational polarization, 347–351
O_2 spectrum, 336
Overlapping orders, 134, 135
Oversize waveguide, 122, 123

P

Paramagnetic ion transitions, 392–395
Paramagnetic resonance, 385–394
Parallel bands, 327
P branch of vibration-rotation spectrum, 327, 328
PbSe junction laser, 67–69
Periodic structure, 35, 37
Perovskite crystal structure, 373
Perpendicular bands, 327, 328
Perturbation calculation of transition probability, 339–342

SUBJECT INDEX

Phase-sensitive, or synchronous, detection, 226, 227
Phase shift in plasmas, 280–286
Phonons, 367, 368
Photoconductive detectors, 174–188
Photoconductivity
  impurity (or extrinsic), 175–179
  free-carrier, 179–188
Photon noise, *see* Noise
Planck's constant, 2
Planck's radiation law, 288
Plasma
  conductivity, 268–271
  frequency, 110
  interferometry, 280–286
Plasma-arc harmonic generator, *see* Arc harmonic generator
p-n junction laser, 64–69
Point-contact diodes, *see* Diode detectors
Polarizability ellipsoid, 423
Polarizers
  Brewster angle, 104
  waveguide, 104, 105
  wire grid, 105–107
Polyethylene, 96, 97, 99–101
  grating filters, 103, 104, 139
Polymer films, 97, 98, 100
Post-detector amplification, 225–228
Power reflectivity, 86, 87, 95, 105–107, 112, 253–257
Pressure broadening, 345
Principal mode in waveguide, 113, 117, 118
Pyroelectric detectors, *see* Detectors

Q

Q branch of vibration–rotation spectrum, 327, 328
Q of a resonator, 85–87
Quantization of lattice vibrations, 367
Quartz, 94, 96, 99
Quasi-optical components, 112, 113

R

Radiation, *see* Blackbody radiation
Radiation noise, *see* Noise, radiation

Raman active vibrators, 422, 423
Raman effect, 420–431
Raman laser, 70–74
  spectrometer, 430
Rayleigh criterion, 132, 133
Rayleigh–Jeans law, 13, 288
Rayleigh scattering, 420, 421
R branch of vibration–rotation spectrum, 327, 328
Recombination noise, *see* Noise, recombination
Reflection filters, *see* Filters
Reflection from crystals, 363–365
Reflection from superconductors, 436–438
Reflection minimum, 255–257
Reflectivity, 95
  metals, 112
  reststrahlen crystals, 107–109, 363–365
Refractive index
  crystal quartz, 96
  diamond, 96
  germanium, 96
  irtran, 96
  polyethylene, 96
  sapphire, 96
  silicon, 96
  TPX, 96
Relativistic electrons, 42–45
Relaxation times in liquids, 348
Replica grating, *see* Grating
Resolution
  cyclotron resonance spectrometer, 193, 194
  Fabry–Perot spectrometer, 156
  grating spectrometer, 133, 139
  microwave spectrometer, 153
  multiplex spectrometer, 144
Resolving power, *see* Resolution
Resonance
  antiferromagnetic, *see* Antiferromagnetic resonance
  cyclotron, *see* Cyclotron resonance
Resonant cavities, 124–126
Resonant mode absorption, 368, 371, 372
Response time
  bolometer, 169–171
  diode detector, 165, 166
  electron cyclotron detector, 193, 194

SUBJECT INDEX

free-carrier photoconductive detector, 187
germanium detector, 171
Josephson junction, 208
pyroelectric detector, 173, 174
superconducting bolometer, 170
Reststrahlen
   absorption, 94, 95
   crystals, 94, 95, 99, 100, 103, 107, 109, 364, 365
   in polyethylene, 102, 103
   filters, 103, 107–109
   reflection, 107–109, 363, 364
Rock salt, see NaCl
Rotational levels of HCl, 89, 311, 312
Rotational spectra of molecules, 306–321

S

Sapphire, 96, 99
Schottky barrier diode, 165, 166
Schrödinger equation, 3, 233, 323, 329, 339, 367
Selection rules, 89, 304, 309, 316, 323, 327, 328, 423, 424
Semiconductor lasers, see Lasers, p-n junction
Shot noise, see Noise
Silicon, 95, 96, 98, 99, 100
   diodes, 20, 21, 163–165
Skin-depth in plasma, 289
SmIG ferrimagnetic exchange resonance, 419
Sodium chloride, see NaCl
Sources
   coherent, 18–81
   incoherent, 13–18
Space charge waves in plasma, 278, 279
Spatial harmonics, 35, 36, 38
Spatial quantization, 308
Specific detectivity D^*, see D^*
Spectral line shapes, see Line shapes
Spectral response of Josephson junction, 209
Spectrometers, 130–140
Spectroscopic splitting factor, see g-factor
Spherical top, 304, 319–320
Spin–flip frequency, 71, 426

Spin–flip Raman laser, 70–73, 430, 431
Spin–flip Raman scattering, 70, 426–428, 430–431
Spin–lattice relaxation, 387
Spin–spin relaxation, 387
Spin waves, 401–406
Splitting by ligand field, 388–394
$SrTiO_3$, 373, 380–384
Standard waveguide, 113–121
Stimulated Raman laser, 70–74, 428–431
Stimulated Raman scattering, 70, 428–431
Stokes scattering, 70–74
Superconducting bolometer, 170, 171
Superconducting junction detector, see Josephson junction detector
Superconducting junction radiation, see Josephson junction generator
Superconductor energy gap, 431–441
Superheterodyne detection, 163–165, 184
Surface resistance of superconductors, 433–435, 439, 440
Symmetric top, 303, 315–319
Synchrotron radiation, 295–299

T

Teflon, 97, 98
Temperature noise, see Noise
Tensor polarizability, 423
Thermal detectors, 166–174
Thermal population distribution, 345, 346
Thermal time constant, 166
TO modes, see Transverse optical modes
TPX, 101, 102
Transmission
   atmospheric, 91–93
   diamond, 96
   germanium, 96
   glass, 94
   irtran, 95, 96
   polyethylene, 96, 99, 101
   polymer films, 98, 100
   quartz, 96, 99
   sapphire, 96, 99
   silicon, 96, 98, 99, 100
   superconducting films, 437
   TPX, 96, 102
Transverse electric modes in waveguides, 114, 116–121

Transverse magnetic modes in waveguides, 114, 117–119
Traveling wave oscillator, 34–38
Transverse acoustic modes, 362
Transverse optical modes, 362, 363, 378–380
Tunable cyclotron resonance detector, 185–188, 188–201
Tunable photoconductive detector, 185–188
Tuning sensitivity of Josephson junction, 74

U

Undulator, 44, 45
UO_2 antiferromagnetic resonance, 411
Upper hybrid frequency, 276

V

Vibrational absorption
  local modes, 368–370
  resonant modes, 368, 371, 372
  reststrahlen, 90, 94, 95, 100, 103, 108, 109, 363–365
Vibration–rotation bands, 327–328
Vibration-rotation states of HCN laser transitions, 53, 328
Vibrations
  CO_2, 323, 324
  HCN, 325
  H_2O, 326
  linear triatomic molecules, 324, 325

Video detection, 162, 163, 165, 184
Vinyl chloride laser emission, 50

W

Water molecule spectrum, see H_2O spectrum
Water vapor laser, see H_2O laser
Wave equation, 114
Wavefunction of rigid rotor, 307
Waveguide, see Standard waveguide, Oversize waveguide
  components, 126–128
  filters, 121
Wavenumber, 1
Windows
  atmospheric, 91–93
  dielectric, 94–102
Wire grid polarizer, see Grating polarizer

X

Xenon laser emission, 49

Y

YbIG ferrimagnetic exchange resonance, 418, 419

Z

Zeeman splitting, 386, 392, 418